T0207381

Chaos and Time-Series Analysis

Chaos and Time-Series Analysis

Julien Clinton Sprott

Department of Physics
University of Wisconsin–Madison

This book has been printed digitally and produced in a standard specification in order to ensure its continuing availability

OXFORD
UNIVERSITY PRESS

Great Clarendon Street, Oxford OX2 6DP
United Kingdom

Oxford University Press is a department of the University of Oxford. It furthers the University's objective of excellence in research, scholarship, and education by publishing worldwide. Oxford is a registered trade mark of Oxford University Press in the UK and in certain other countries

First published 2003
Reprinted 2013

British Library Cataloguing in Publication Data
Data available

Library of Congress Cataloging in Publication Data
Data available

ISBN 978-0-19-850840-3

Preface

This book grew out of a survey course by the same title that I developed and teach at the University of Wisconsin–Madison. The course consists of fifteen 100-minute weekly lectures and a weekly programming project. These lectures form the fifteen chapters of the book, although I have included considerably more material in the book than I am able to cover in the lectures. The students are about half undergraduates and half graduates and other researchers, representing a wide variety of fields in science and engineering. More details about the course can be found on the World Wide Web at http://sprott.physics.wisc.edu/phys505/.

The book is an introduction to the exciting developments in chaos and related topics in nonlinear dynamics, including the detection and quantification of chaos in experimental data, fractals, and complex systems. I have tried to mention, however briefly, most of the important concepts in nonlinear dynamics. Most of the basic ideas are encountered several times with increasing sophistication. This is the way most people learn, and it emphasizes the interconnectedness of the various topics. Emphasis is on the physical concepts and useful results rather than mathematical proofs and derivations. The book is aimed at the student or researcher who wants to learn how to use the ideas in a practical setting, rather than the mathematically inclined reader who wants a deep theoretical understanding.

While many books on chaos are purely qualitative and many others are highly mathematical, I have tried to minimize the mathematics while still giving the essential equations in their simplest possible form. I assume only an elementary knowledge of calculus. Complex numbers, differential equations, matrices, and vector calculus are used in places, but those tools are described as required. The level should thus be suitable for graduate and advanced undergraduate students in all fields of science and engineering as well as professional scientists in most disciplines.

I feel that chaos is best learned by a hands-on approach, and because of its nature, this means writing simple computer programs. Thus, in addition to the usual algebraic exercises, I have included at the end of each chapter the computer project that my students turn in each week. The projects are open-ended and have an optional part meant to challenge the more ambitious students. I have found that the required part usually takes about one to four hours, depending on the student's computational skill.

Programming advice

I do not recommend any particular computer platform or programming language, and I do not provide much formal help with programming. I feel programming is a skill, like mathematics, that all science and engineering students should acquire somewhere during their training. Students who have never programmed can usually learn to do so while taking the course. In fact, chaos offers an enjoyable way to develop and hone these skills.

I recommend that students acquire a personal computer and a modern compiler in a language of their choice. Just as the best language for speaking is the one most familiar to you, the best computer language is the one you are most comfortable using. If you are skilled in a language such as BASIC, C, Java, Pascal, or FORTRAN, get a modern interactive compiler for that language and use it on your PC. Any language will suffice, and modern compilers in the various languages are so good that there is little reason to prefer one over another. If you have never done any serious programming, you might start by learning BASIC. It is easy to learn and more than adequate for the projects in this book. My personal favorite is PowerBASIC (http://www.powerbasic.com/) because it is easy to learn, powerful, and as fast as any C compiler I have encountered. I do most of my programming in DOS, but Windows versions of the PowerBASIC compiler are available.

Another possibility is one of the math packages such as Mathematica, Maple, Matlab, MathCAD, Derive, or Theorist, or even a modern spreadsheet such as Excel, Quattro Pro, or Lotus 1-2-3. This option would be most sensible if you are already highly skilled in its use. You should be able to complete most if not all of the projects in this way. In the long run, you will probably find a conventional programming language more versatile and useful, however.

In any case, I would advise you to develop your programs as modular subroutines and to document them so that they can be reused. There will be occasions while working through the book where you will need something you did several chapters before. Especially in Chapters 9–13, dealing with time-series analysis, you will develop routines that may be of use in analyzing data from your own research.

Web resources

A Web page for the book at http://sprott.physics.wisc.edu/chaostsa/ contains supplementary materials, computer programs, color versions of some of the figures, animations, errata, answers to the exercises, and links to Web resources. I will keep this updated as links change and as I become aware of new resources that may be of interest. You will also find information there on how to contact me in case you find errors in the book, want to comment on it, or make suggestions for future editions.

Acknowledgments

I am indebted to many people in the preparation of this work. First and foremost is George Rowlands of the University of Warwick who introduced me to the subject and continues to be a valued mentor and colleague. Cliff Pickover of IBM–Watson has been a source of inspiration and advice. I have had productive collaborations with Wajdi Ahmad of the University of Sharjah in the United Arab Emirates (electronics), Debbie Aks of the University of Wisconsin–Whitewater (psychology), Janine Bolliger of the Swiss Federal Research Institute (landscape ecology), Robin Chapman of the University of Wisconsin–Madison (communicative disorders), Dorina Creanga of Al. I. Cuza University in Romania (biophysics), Dee Dechert of the University of Houston (economics), Hans Gottlieb of Griffith University in Australia (physics), Wendell Horton of the University of Texas (physics), Ken Kiers of Taylor University (physics), Stefan Linz of the University of Münster in Germany (physics), Karl Lonngren of the University of Iowa (electrical and computer engineering), and Keith Warren of Ohio State University (social work).

I am also indebted to many students and former students who learned along with me and stimulated my thinking, most notably David Newman, Christopher Watts, Kevin Mirus, Adam Fleming, Brian Meloon, David Albers, Oguz Yetkin, Lucas Finco, Nicos Savva, and Del Marshall. I am grateful to my colleagues Stewart Prager, Paul Terry, Cary Forest, and countless others in the University of Wisconsin–Madison plasma physics group for maintaining a stimulating and flexible working environment.

Sönke Adlung, Anja Tschörtner, Marsha Filion, Richard Lawrence, and Lydia Davis of Oxford University Press provided help, guidance and encouragement. Geoff Brooker of Oxford University answered questions about LaTeX, in which this book was written. David Albers, Witold Kinsner, Del Marshall, George Rowlands, and Lenny Smith read the manuscript carefully and provided many useful comments. John and Celia Hall copy-edited the final manuscript. Nicos Savva looked carefully at the exercises and provided most of the solutions available on the Web. Finally, I want to acknowledge the late Donald Kerst (1911–1993), who was my graduate thesis advisor and mentor, and who launched my academic career.

Julien Clinton Sprott
Madison, Wisconsin
September 2002

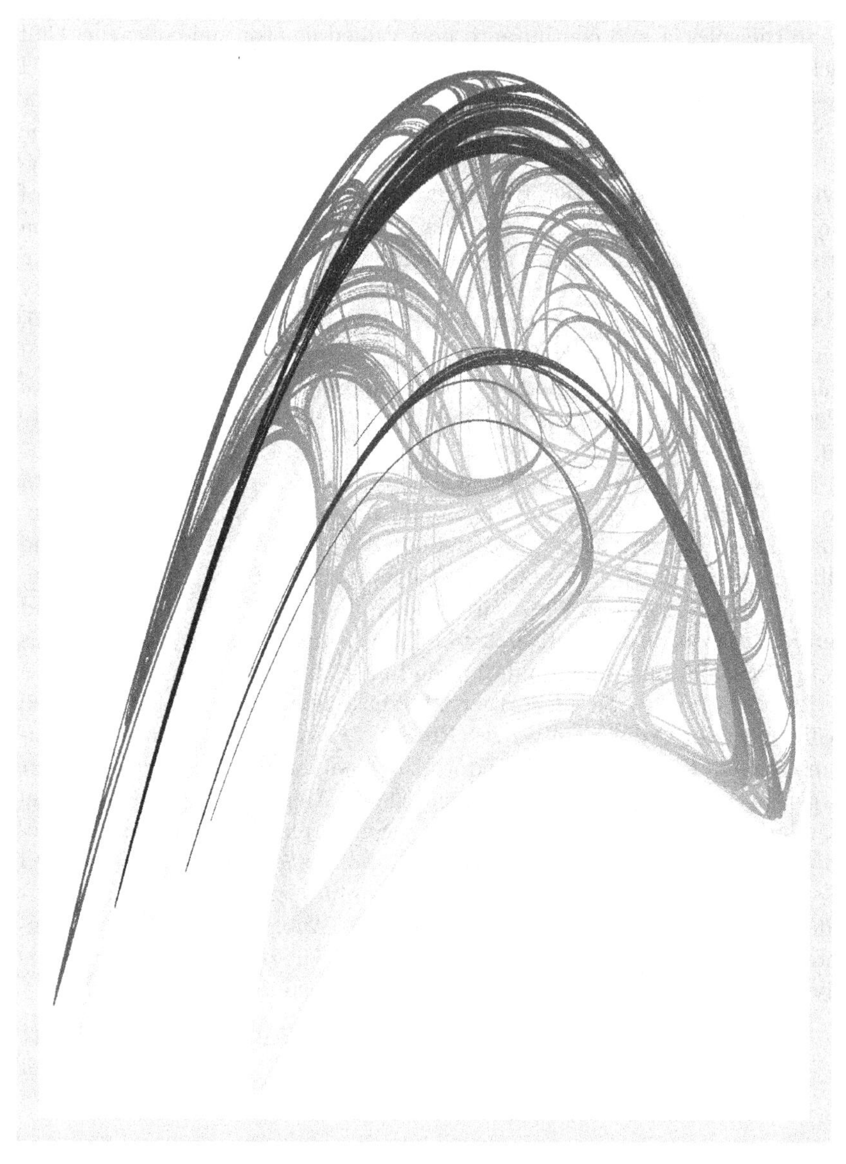

Strange attractor for the three-dimensional map $X_{n+1} = X_n^2 - 0.2X_n - 0.9X_{n-1} + 0.6X_{n-2}$ (see §6.10.2).

Contents

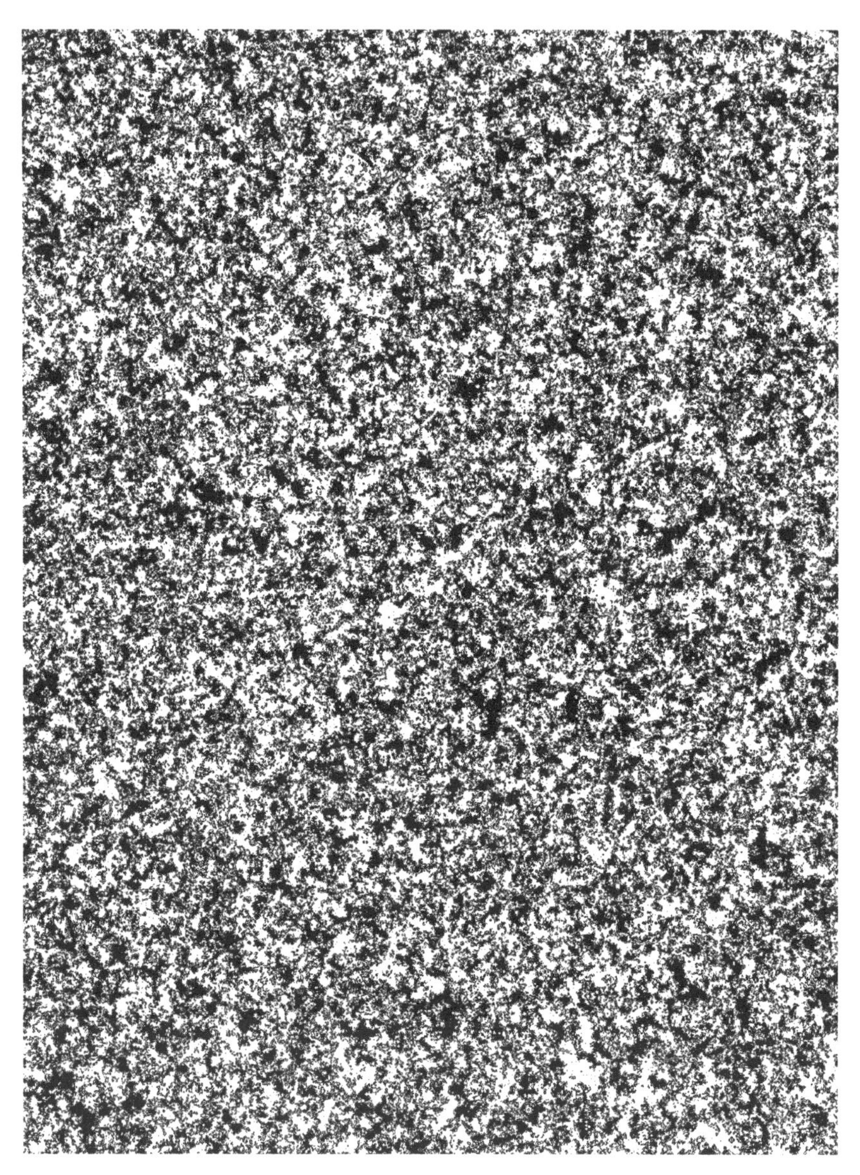

Complex two-dimensional pattern from the ten-thousandth generation of a deterministic cellular automaton in which a dead cell remains dead if exactly six of its eight nearest neighbors are alive and otherwise gives birth, and a live cell remains alive if one, two, or six of its eight nearest neighbors are alive and otherwise dies (see §15.1.4).

1 Introduction

The scientist does not study nature because it is useful; he studies it because he delights in it, and he delights in it because it is beautiful. If nature were not beautiful, it would not be worth knowing, and if nature were not worth knowing, life would not be worth living — Henri Poincaré.

Fundamental to science is the presumption that experiments are predictable and repeatable. Thus it surprised most scientists when simple deterministic systems were found that were neither predictable nor repeatable. Instead, they exhibited *chaos*,[1] in which the tiniest change in the initial conditions produces a very different outcome, even when the governing equations are known exactly.

Aspects of chaos have been known for hundreds of years. Isaac Newton[2] was said to get headaches contemplating the three-body gravitational problem such as the Sun, Moon, and Earth. In 1887, King Oscar II of Sweden (and Norway until he renounced the Norwegian throne on June 7, 1905) offered a prize of 2500 crowns to anyone who could solve the n-body problem and hence demonstrate stability of the Solar System. The prize was awarded on January 21, 1889 to Jules Henri Poincaré, not for solving the problem, but for submitting a paper (Poincaré 1890) of over 200 pages showing that even the three-body problem was impossible to solve.[3] He also concluded that minute differences in the initial conditions could result

[1]The word 'chaos' dates from about 800 BC and derives from the Greek $\chi\alpha o\varsigma$, meaning a complete absence of order, which, as we shall see, is something of a misnomer.

[2]Sir Isaac Newton (1642–1727) was an English physicist and mathematician, who in addition to developing classical mechanics and gravitation, invented calculus (along with Leibniz) and showed that white light contains the colors of the rainbow, but he was entirely without humor and retired from research, spending the last third of his life as a government official following a nervous breakdown in 1693 (Westfall 1980).

[3]Jules Henri Poincaré (1854–1912), a French mathematician, physicist, and philosopher, whose career spanned the heyday of classical physics, worked in all areas of mathematics and developed the *qualitative* (geometric) method of analysis now called 'topology' to understand the stability of dynamical systems. His paper contained a serious error that was discovered after the prize was awarded and was hidden away, allowing him to submit a correct paper in 1890 for publication in the prestigious journal *Acta Mathematica*, leading to instant fame (Peterson 1993, Barrow-Green 1997). While the stability of the Solar System remains mathematically unproven, its long existence is compelling, although simulations using a specially designed parallel computer called the 'Digital Orrery' (Sussman and Wisdom 1992) indicate that chaos should manifest itself after a few million years, not necessarily ejecting any of the existing planets, however.

in very different solutions after a long time. In a 1903 essay in 'Science and Method,' Poincaré (1914) wrote

If we knew exactly the laws of nature and the situation of the universe at the initial moment, we could predict exactly the situation of that same universe at a succeeding moment. But even if it were the case that the natural laws had no longer any secret for us, we could still only know the initial situation approximately. If that enabled us to predict the succeeding situation with the same approximation, that is all we require, and we should say that the phenomenon had been predicted, that it is governed by laws. But it is not always so; it may happen that small differences in the initial conditions produce very great ones in the final phenomena. A small error in the former will produce an enormous error in the latter. Prediction becomes impossible, and we have the fortuitous phenomenon.

Poincaré thus anticipated much of modern chaos, but his discoveries lay dormant for over half a century awaiting the advent of computers that allowed scientists of solve problems whose solutions are chaotic and to visualize those solutions. This chapter introduces the phenomenon of chaos and describes some simple demonstrations in which it can be easily observed.

1.1 Examples of dynamical systems

Chaos is one subject area in the field of *nonlinear dynamics*,[4] which is part of the broader field of *dynamical systems*. A dynamical system is one that evolves in time. Dynamical systems can be *stochastic*, in which case they evolve according to some random process such the toss of a coin, or *deterministic*, in which case the future is uniquely determined by the past according to some rule or mathematical formula. Henceforth, we will assume deterministic dynamics unless otherwise specified. The chaos we describe is often called *deterministic chaos*, and it can only occur when the governing equations are nonlinear.

1.1.1 Astronomical systems

An example is the motion of astronomical bodies governed by the *universal law of gravitation* ($F = Gm_1m_2/r^2$) and *Newton's second law of motion* ($F = ma$). The inverse square dependence of the force on the separation of the masses is the *nonlinearity* that allows chaos, and yet it does not guarantee it. In fact, astronomical motions on human time-scales are some of the most predictable processes in nature. Planetary motion was quantitatively described by Johannes Kepler[5] four hundred years ago through the

[4]The Polish mathematician Stanislaw Marcin Ulam (1909–1984) liked to say that 'the study of nonlinear dynamics in nature is like the study of nonelephant mammals in zoology' (Ulam 1983), which is to say that most processes in nature are nonlinear.

[5]Johannes Kepler (1571–1630) was a deeply religious and mystic German astronomer whose calculations of planetary motion dispelled the widely accepted Ptolemaic geocentric model and brought on the wrath of the Church.

Fig. 1.1 Trajectory of a planet orbiting a binary star.

analysis of astronomical observations by Tycho Brahe.[6] It is possible not only to predict thousands of years in advance when an eclipse of the Sun will occur, but where on the Earth to stand to get the best view of it. Whereas a planet or comet orbits the Sun in a highly predictable ellipse, a planet orbiting a pair of stars exhibits chaos, as in Fig. 1.1, even when the stars are fixed and the motion is restricted to a plane. Such a planet would be an unlikely habitat for extraterrestrial life.

1.1.2 The Solar System

Chaotic motions occur within the Solar System. Hyperion, one of the smaller moons of Saturn, is irregularly shaped, as in Fig. 1.2, and tumbles chaotically as it orbits Saturn (Wisdom *et al.* 1984). The asteroid belt between Mars and Jupiter has gaps, discovered by the American mathematician and astronomer Daniel Kirkwood (1888), in which asteroids with orbital periods a simple fraction (such as 1:2, 1:3, 3:5, and 2:7) of Jupiter's period were ejected into orbits that intercept Mars and the Earth because of chaotic motion (Wisdom 1987). This theory explains why special meteorites called *chondrites*, known to have come from the vicinity of the 1:3 resonance region, are found on Earth. It is comforting to know that relatively few asteroids remain in those gaps. Furthermore, objects in highly elliptical orbits around the Sun, such as Halley's Comet, are expected to be chaotic because of their different interactions with the massive planets such as Jupiter on successive orbits (Chirikov and Vecheslavov 1989). It is likely that the Solar System once contained other planets that were ejected by close encounters with one another or with the existing planets (Laskar

[6]Tycho Brahe (1546–1601) was a Danish astronomer who hired Kepler as an assistant to perform mathematical calculations and thus prove his geocentric model, but he had a contentious relationship with Kepler and died eighteen months later. Kepler absconded Tycho's data after his death and spent seventeen years analyzing it before announcing his laws of planetary motion, although he called them 'relationships.'

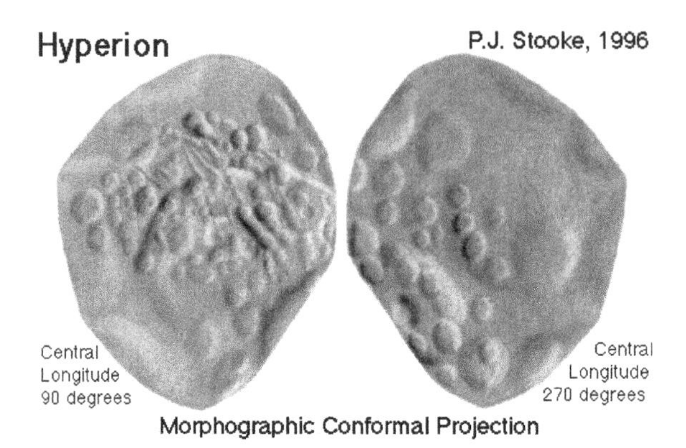

Fig. 1.2 Shaded relief map of Hyperion, a satellite of Saturn (from Stooke 1996).

1996). The Solar System is apparently 'full' of planets, containing the most possible while retaining stability on time-scales of a few billion[7] years.

1.1.3 Fluids

Another example is the motion of a *fluid* (liquid, gas, or plasma) consisting of very many interacting molecules or infinitely many infinitesimal fluid elements moving in response to their neighbors. The mixing that occurs when you stir the cream in your coffee is a chaotic process. Fluids can flow in a *laminar* (regular) or *turbulent* (chaotic) manner, as in Fig. 1.3.[8] Two sticks placed in a smoothly flowing river will remain side by side for a long time, but they will quickly separate if the river is turbulent. Chaotic atmospheric motion prevents long-term weather prediction[9] and illustrates the sensitive dependence on initial conditions known as the *butterfly effect* in which a butterfly flapping its wings in Brazil can cause tornadoes in Texas (Lorenz 1993). Other examples of dynamical systems that can exhibit chaos include the motion of electrons in atoms, the spread of forest fires, the flocking of animals, the behavior of crowds, the propagation of earthquakes, and the movement of automobile and air traffic.

1.1.4 Nonphysical systems

Dynamical systems do not require a literal spatial motion. The variables can be more abstract. The economy is a dynamical system in which money flows from buyer to seller. The body contains dynamical systems, such as the heart, lungs, and limbs. Information flows in the brain, inside computers, across the Internet, and within social networks. Ecological systems are dynamic, as is the growth of cancer, the spread of epidemics, the spread

[7] Following standard American usage, 1 billion = 10^9, 1 trillion = 10^{12}, and so forth.

[8] Uncropped color versions of this and other figures from the book are available at http://sprott.physics.wisc.edu/chaostsa/.

[9] Actually, *predicting* the weather is easy, but predicting it *correctly* is difficult.

Fig. 1.3 Turbulent flow of water (Rheinfall, Switzerland).

of computer viruses, the distribution of food, the evolution of language, the movement of electricity through the power grid and within electrical circuits, the succession of notes in a musical composition, the reaction of chemicals, the buildup of armaments in competing nations, and the love expressed in romantic relationships. Much of the language used to describe dynamical systems was developed in the context of Newtonian mechanics, but the concepts are much more general, and hence the widespread interest in the subject.

1.2 Driven pendulum

Some of the most convincing examples of chaos are various driven pendulums. A pendulum can be mounted on a cart that oscillates periodically back and forth, driven by a variable-speed motor. The oscillating cart could also carry troughs of various shapes in which balls move chaotically, a vertical elastic ribbon, or a small aquarium in which a drop of food coloring added to the water mixes only when the fluid is turbulent. Various chaotic toys (Moon 1992), as in Fig. 1.4, contain an outer ring driven by a periodic pulse from an electromagnet in the base, which in turn drives an inner ring that exhibits chaos.

A less convincing but simpler demonstration is a double pendulum, with one pendulum suspended from the bottom of the other. If the bearings have sufficiently low friction, the upper pendulum oscillates nearly periodically and drives the lower pendulum transiently into chaos until the motion damps out. In each case, chaos appears when the pendulum goes nearly or fully 'over the top.' Otherwise the restoring force is too nearly linear. The friction can be counteracted by using a motor-driven cart to move the pivot point of the upper pendulum back and forth periodically, as in Fig.

Fig. 1.4 Chaotic toy.

1.5. The required motion is smaller than for the single pendulum because the upper pendulum amplifies the motion of the cart.

Each of these examples can be understood by assuming the motion results from a restoring force F proportional to $-\sin x$, a friction term proportional to $-dx/dt$, and a driving force $A \sin \Omega t$, which when substituted into Newton's second law ($F = md^2x/dt^2$) leads to the equation

$$\frac{d^2x}{dt^2} + b\frac{dx}{dt} + \sin x = A \sin \Omega t \tag{1.1}$$

where x is the angle the pendulum makes with the vertical (see §3.7 for a more thorough discussion). This equation can be solved by computer to show that there are combinations of the parameters (such as $A = 0.6$, $b = 0.05$, and $\Omega = 0.7$) at which chaos occurs.

The chaos is best exhibited in a *Poincaré section* (see §6.6.2), as in Fig. 1.6, in which the instantaneous angle x and angular velocity dx/dt are plotted at a constant phase of the drive, such as where Ωt is an integral multiple of 2π. This plot is a *fractal* (see Chapter 11), an object with structure on all scales and a dimension that is not an integer. It is a cross-section of something called a *strange attractor* (see Chapter 6) with a dimension of about 2.7 (see Chapter 12). The complete attractor lies within a *torus* (a doughnut-shaped[10] region of space). It can be animated, displaying a

[10]These are American doughnuts with a hole in the center, not the British variety filled with jam.

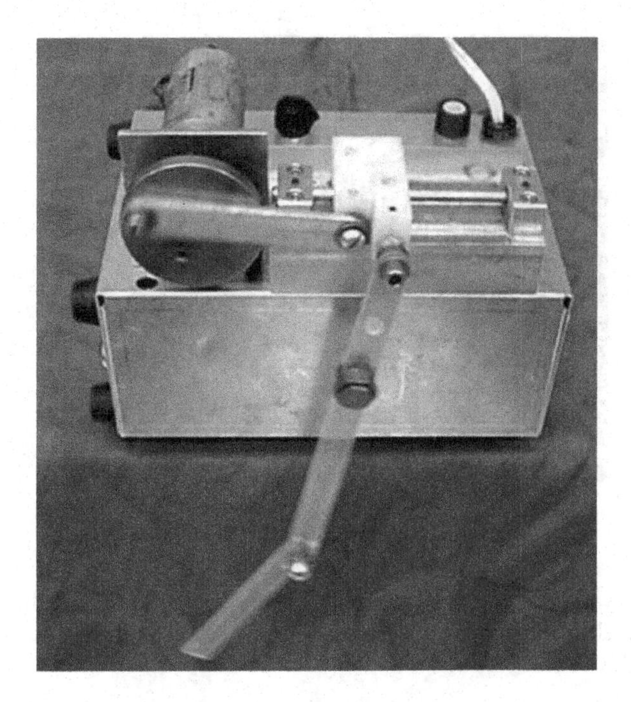

Fig. 1.5 A driven double pendulum.

Fig. 1.6 Poincaré section of a damped driven pendulum.

succession of plots at various phases of the drive, corresponding to different slices of the doughnut. The animation, which resembles a taffy-pulling machine, as in Fig. 1.7, illustrates the stretching that causes sensitive dependence on initial conditions and the folding that produces the fractal microstructure. The same effect can be demonstrated with a lump of silly putty.

Ueda (1979) studied a variant of this system with the $\sin x$ restoring force replaced by x^3 (see §4.8.1). This variant can be simulated with operational amplifiers whose output is connected to a loudspeaker so that the transition from periodic to chaotic behavior can be heard (Moon 1992).

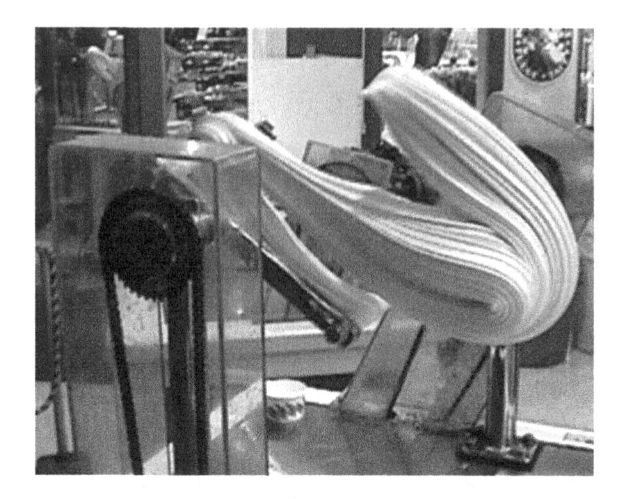

Fig. 1.7 A taffy-pulling machine.

The Ueda system is a special case of *Duffing's oscillator* (Duffing 1918)[11] that has both a linear and cubic restoring force, usually of opposite signs (see §4.8.1).

1.3 Ball on an oscillating floor

A simple chaos demonstration consists of a petri dish glued to the cone of a loudspeaker facing upward, as in Fig. 1.8. A ping-pong ball is placed on the dish, and the speaker is connected to an oscillator with a frequency of about 8 Hz and variable amplitude. The periodic bouncing at low amplitude changes to chaotic as the amplitude is increased. The transition can be heard as well as seen. The amplitude is one of several possible *control parameters*, a small change in which can turn the chaos on and off. The existence of such control parameters and their effect on the behavior is one of the signatures of chaos (see Chapter 7).

This system is relatively easy to simulate on a computer. Ignoring air resistance, the time required for the ball to return to the dish is proportional to its initial velocity, $\Delta t = 2v/g$, where g is the *acceleration due to gravity* (9.8 m/s^2). If the collision of the ball with the dish is perfectly elastic, then its speed is conserved during the collision in the frame of reference of the dish. Thus the change in velocity at each bounce is $\Delta v = 2\Omega A \sin \Omega t$, where A is the oscillation amplitude of the speaker. The dynamics are described by a pair of difference equations

$$v_{n+1} = v_n + 2\Omega A \sin q_n \tag{1.2}$$
$$q_{n+1} = q_n + 2\Omega v_{n+1}/g \tag{1.3}$$

[11]Georg Duffing (1861–1944) was a German engineer who studied nonlinear oscillations in machines in an effort to reduce noise and prolong their life.

Fig. 1.8 Ping-pong ball bouncing in a dish mounted on a loudspeaker.

in which the subscript n denotes the bounce number, a discrete measure of time, and q_n is Ωt at the time of the nth bounce. With a minor change of variables ($X = q, Y = 2\Omega v/g$), this system is the *Chirikov map* (Chirikov 1979), a two-dimensional chaotic system discussed in §8.8.5.

1.4 Dripping faucet

A simple demonstration of chaos available to everyone is the dripping faucet ('tap' in the UK), studied extensively by Shaw (1984), a version of which is in Fig. 1.9. By letting the drops fall on an aluminum pie plate, you can easily hear the dripping sound. A microphone at the plate connected to an amplifier enhances the demonstration. As you increase the drip rate, the transition from periodic to chaotic behavior is evident. Contrast the chaotic dripping sound with the regular ticks of a metronome. With patience, you can explore this transition and often hear the *period doubling* in which two rapid drips are followed by a longer delay before the next two drips. It should sound something like 'dripdrop–dripdrop–dripdrop....' Further period doublings, prior to the onset of chaos, exist but are hard to detect without special instrumentation.

The mechanism involves an oscillation of the water left behind at the faucet opening when a drop is released (Ambravaneswaran *et al.* 2000). At high drip rates, the oscillation produced by the release of one drop has not yet damped out when the next drop releases. The release time of the subsequent drop depends on the phase of oscillation of the previous drop in a manner similar to the ball on an oscillating floor. This demonstration can be instrumented with an electronic timer to measure the interval between

Fig. 1.9 A dripping faucet exhibits a chaotic drip rate.

Fig. 1.10 Laminar and turbulent flow from a faucet.

sound pulses or between pulses from a photogate through which the drops fall. A plot of the time between drips versus the time between the previous drips is approximately parabolic, and it can be modeled by the *logistic equation* discussed in Chapter 2.

At much higher flow rates you can observe the transition from *laminar* flow to *turbulent* flow, as in Fig. 1.10, and sometimes a periodic oscillation near the transition. Thus with a single faucet, you can observe chaos and the transition to chaos in both a *discrete* dynamical system (the dripping faucet) and in a *continuous* dynamical system (the water stream). The distinction between discrete and continuous dynamics will be a theme throughout the book, and we will return to the important and incompletely understood subject of turbulence in Chapter 15.

1.5 Chaotic electrical circuits

Chaos has also been observed in various electrical circuits (van Wyk and Steeb 1997), and these examples are especially suitable not only for demonstrations, but for collecting lengthy and accurate chaotic data records.

1.5.1 Inductor–diode circuit

One of the simplest chaotic electrical circuits, studied extensively by Testa *et al.* (1982), contains only an inductor and diode, as in Fig. 1.11. With a sinusoidal voltage source $V_0 \sin \Omega t$, the current I and the voltage V across the diode exhibit a transition from periodic to chaotic behavior with a succession of period doublings when the input voltage is increased. Early versions of the circuit used a varicap diode in which the small junction capacitance changes with reverse voltage. It turns out that any diode will work, and the best choice is a power diode with a relatively large junction capacitance, especially when forward biased.

With a sufficiently large inductor (several hundred millihenries), the optimum source frequency is in the audio range (5–10 kHz), in which case the output can be connected to an amplifier and speaker, allowing you to hear the successive period doublings and finally the sound of chaos. You can omit the audio oscillator by placing the circuit in the feedback loop of the audio amplifier and using the amplifier gain as the control parameter. In constructing the circuit, it is important to use an inductor with a high Q (low resistance), a voltage source with a peak voltage of at least 10 V and a low output resistance, and a detector for the voltage V with a high input impedance.

You can analyze this circuit using the equations

$$L\frac{dI}{dt} = V_0 \sin \Omega t - V \tag{1.4}$$

$$\{C_0 + I_0 \alpha T(1 - \alpha V)\mathrm{e}^{-\alpha V}\}\frac{dV}{dt} = I - I_0(1 - \mathrm{e}^{-\alpha V}) \tag{1.5}$$

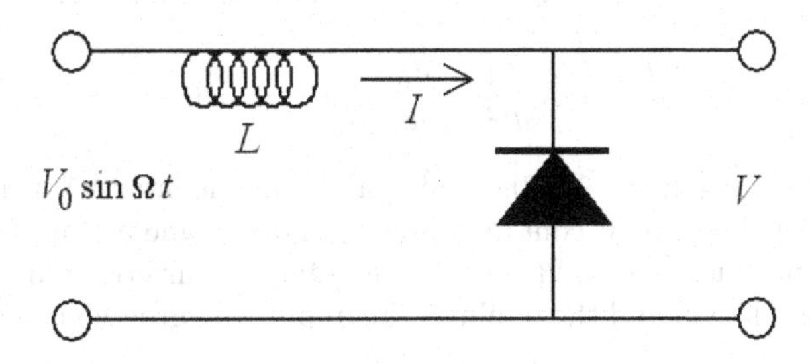

Fig. 1.11 Simple chaotic circuit driven by a sinusoidal voltage.

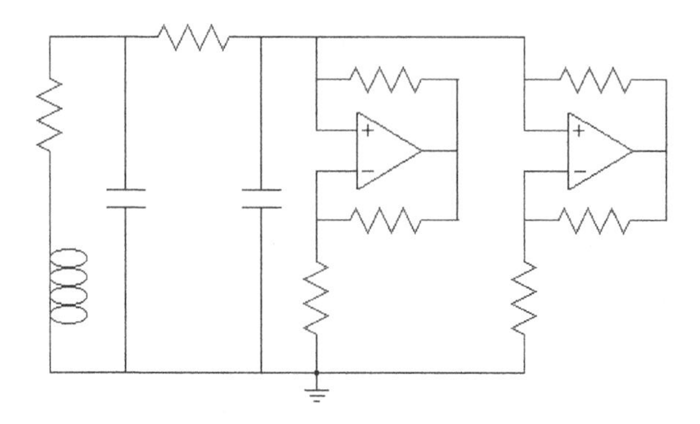

Fig. 1.12 Chua's circuit.

Typical parameters that lead to chaos are $L = 0.25$ H, $V_0 = 10$ V, $\Omega = 2\pi \times 10^4$ Hz, $C_0 = 1 \times 10^{-11}$ F, $I_0 = 1 \times 10^{-12}$ A, $\alpha = 38$ V^{-1}, and $T = 1 \times 10^{-5}$ s. The system is numerically *stiff* and difficult to solve because the exponential dependence on V causes relatively abrupt changes in the current I, requiring a variable time step (see §3.9.4), but the solution closely mimics the circuit behavior.

1.5.2 Chua's circuit

Many other chaotic electrical circuits have been constructed. One popular example is *Chua's circuit* (Matsumoto *et al.* 1984, Madan 1993), which contains an inductor and two capacitors coupled to a pair of operational amplifiers that provide a nonlinear *negative resistance*, as in Fig. 1.12. The operational amplifier can drive a loudspeaker in the audio range, making it an effective, self-contained, portable demonstration. Discussion of Chua's circuit with suggested component values and interactive simulations can be found on the World Wide Web.[12]

1.5.3 Jerk circuits

A particularly simple class of chaotic circuit is based on third-order differential equations (*jerk equations*), as described in §4.8.5 with piecewise linear nonlinearities (Sprott 2000a,b). For example, the system

$$\frac{d^3x}{dt^3} + A\frac{d^2x}{dt^2} + \frac{dx}{dt} - |x| + 1 = 0 \tag{1.6}$$

has a chaotic solution for $A = 0.6$ and can be implemented with the circuit in Fig. 1.13(a). The circuit contains three integrators and a pair of diodes (a *rectifier*) to produce the nonlinear $|x|$ term. Only the inverting inputs of the amplifiers are shown, and the noninverting inputs are grounded. If the fixed

[12]Links to descriptions of Chua's circuit and other Web resources can be found at http://sprott.physics.wisc.edu/chaostsa/.

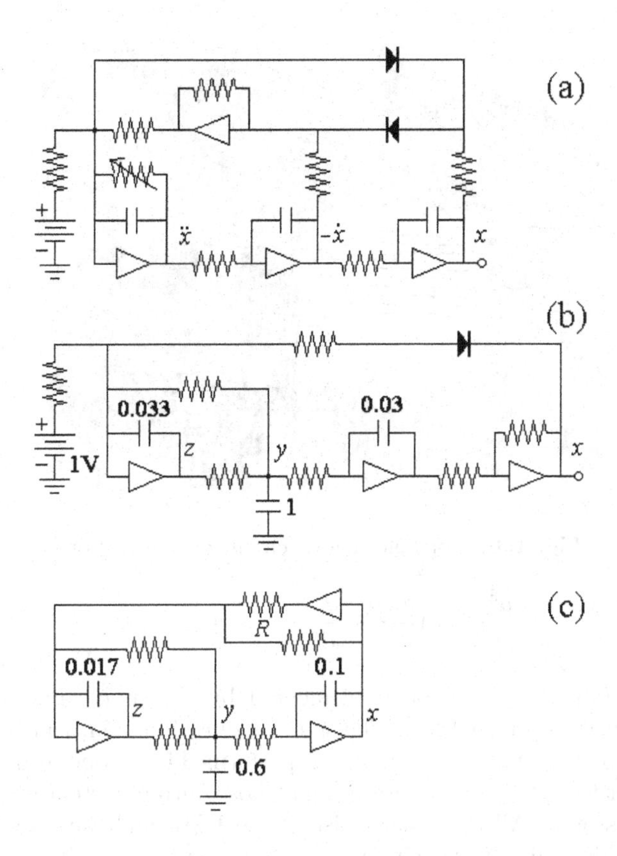

Fig. 1.13 Chaotic jerk circuits (see text for component values).

resistors are all 1 Ω and the capacitors are all 1 F, then the circuit works in real time and produces chaotic oscillations with a dominant frequency of $1/2\pi \simeq 0.16$ Hz when the variable resistor is adjusted to $1/A \simeq 1.67$ Ω. A more practical implementation uses 1-kΩ resistors and 0.1-μF capacitors to raise the frequency by a factor of 10^4 into the audio range (∼1600 Hz) so that the bifurcations and chaos can be heard. The battery and its series resistor only affect the oscillation amplitude, and the required current of 1 mA can be obtained from the power supply for the operational amplifiers. You can make this current as large as you wish provided the operational amplifiers do not saturate.

The circuit in Fig. 1.13(b) solves the same equation with slightly fewer components by performing one of the integrations passively and using a single diode, at the expense of being more difficult to analyze. All the resistors are 1 kΩ, and the capacitors are labeled in μF.

Equation (1.6) is only one of a family of piecewise linear chaotic jerk equations, another one of which is

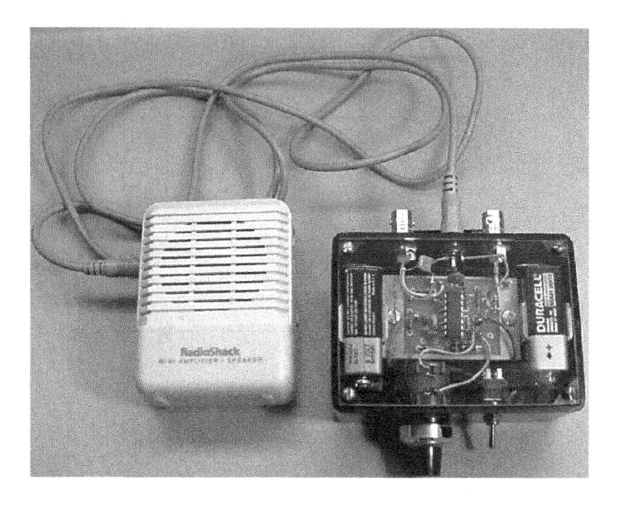

Fig. 1.14 Portable chaotic circuit demonstration.

$$\frac{d^3x}{dt^3} + A\frac{d^2x}{dt^2} + \frac{dx}{dt} - x + \operatorname{sgn} x = 0 \tag{1.7}$$

with $A \simeq 0.5$, which can be implemented by the simple but somewhat delicate circuit in Fig. 1.13(c). The sgn (*signum*) function, which gives ± 1 depending on the sign of it argument, is performed by an operational amplifier without feedback that saturates and whose output current is controlled by the resistor R. All the other resistors are 1 kΩ, and the capacitors are labeled in μF. This chaotic circuit may be the simplest of its class. Other similar circuits have been described by Elwakil and Kennedy (2001).

These inexpensive circuits can be easily constructed and make effective portable lecture demonstrations, as in Fig. 1.14. You can scale them to any desired frequency. In the audio range, you can hear the output, display it on an oscilloscope, and rapidly collect large amounts of experimental data. You can build them with high precision components and make detailed quantitative comparisons with theoretical predictions (Sprott 2000a,b).

Figure 1.15 shows on the left oscilloscope traces from the circuits in Fig. 1.13(b) and (c) respectively, and on the right the corresponding numerical calculation from eqns (1.6) and (1.7) on the same scales. The quantity x is plotted horizontally, and dx/dt is plotted vertically.

1.5.4 Relaxation oscillators

Other circuits use devices like neon bulbs with *negative resistance* or *hysteresis*. A single such device in combination with a resistor and capacitor will produce periodic *relaxation oscillations*, so-called because a slow rise in voltage is followed by a sudden relaxation to a lower voltage. Two or more such coupled oscillators can display complicated dynamics, including chaos, depending on the resistor values. The simple, two-bulb chaotic flasher in

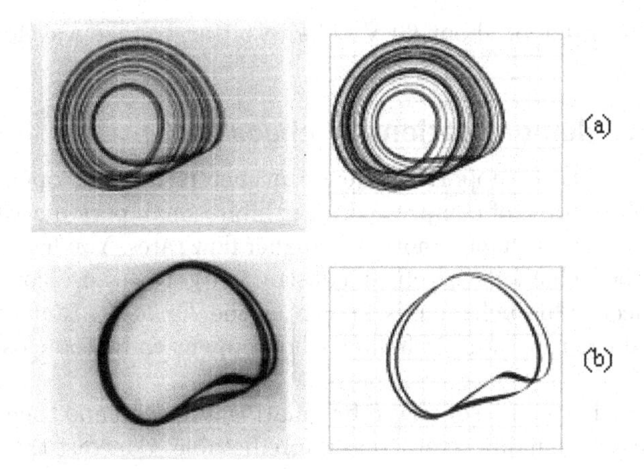

Fig. 1.15 Oscilloscope traces from the chaotic circuits in Fig. 1.13(b) and (c) respectively, and the corresponding numerical predictions.

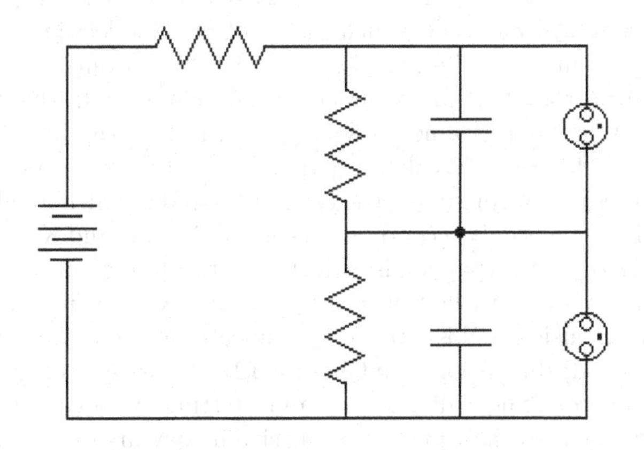

Fig. 1.16 Chaotic flasher using two neon bulbs.

Fig. 1.16 was described by Landauer[13] in an internal IBM memorandum entitled *Poor Man's Chaos* in 1977, and a variant using tunnel diodes was discussed by Gollub *et al.* (1978). You can extend the principle to more coupled oscillators, and each bulb can represent a piano key to produce chaotic music. You can analyze circuits with neon bulbs by assuming the bulb draws no current until the voltage across it exceeds a threshold of about 80 V, whereupon it acquires a very low resistance until the voltage

[13]Rolf Landauer (1927–1999) was a German-born theoretical physicist and computer scientist at IBM who showed that performing calculations does not require energy (Landauer 1996), but he was very critical of the prospects for optical and quantum computers as well as the unification of complex phenomena.

drops below a value of about 60 V. The resulting voltage waveforms have a sawtooth shape.

1.6 Other demonstrations of chaos

A small rubber hose supported a few centimeters from its open end and connected to a source of compressed air exhibits small periodic oscillations at low flow rates and chaotic motion at higher flow rates. You have probably seen the same behavior with an unconstrained garden hose connected to a water faucet. This behavior is known as the *firehose instability* and is potentially dangerous to a firefighter who happens to be too close to such a hose.

A similar effect can be observed by inflating a balloon and then releasing it. Its trajectory is highly erratic and unpredictable. Contrast this behavior to the extremely predictable motion described in most elementary physics texts.

An even simpler demonstration involves dropping a piece of paper. It flutters in a chaotic manner as it falls, and it is essentially impossible to predict where it will land. Two such pieces of paper released in the same way will end up in very different places on the floor. Contrast this to the nonchaotic trajectory of a ball dropped in the same way. In the two cases, the laws of physics are the same (gravity plus air resistance), but the motion is qualitatively different. Crush the paper into a ball and show that the chaos is suppressed. Surprisingly, the conditions under which a falling piece of paper is chaotic has only recently been solved (Tanabe and Kaneko 1994).

The crushed wad of paper also illustrates the fractal nature of many objects. Before it was crushed, it was essentially two-dimensional, but afterwards its dimension is closer to three. You can easily quantify the idea of fractal (fractional) dimension (see Chapter 12) by tightly crushing squares of paper of different linear dimensions and plotting the logarithm of their mass (proportional to the square of their size if they are of uniform thickness) versus the logarithm of the diameter of the corresponding crushed ball. The slope of the curve is the *mass dimension* and typically has a value of 2.3 to 2.4. Try to determine how the fractal dimension changes with the thickness of the paper.

You can place a small magnet inside a tennis ball suspended from above with a string to make a pendulum that oscillates periodically. When several other magnets are slid under it, the motion becomes chaotic. Depending on how the magnets are oriented, the pendulum will be attracted to some of the magnets and repelled by others. This demonstration illustrates the concepts of *attractors* and *repellors* as well as *basins of attraction* and *sensitive dependence on initial conditions.* It is an example of *transient chaos* since the motion eventually stops with the pendulum over one of the magnets, but it is hard to predict which. Another example of transient

Fig. 1.17 Smoke rising from a cigarette exhibits laminar and turbulent flow.

chaos is a pinball machine (or *Galton board*[14]) in which the ball always ends up in the trough at the bottom after executing a chaotic trajectory. These systems are relatively easy to simulate on a computer.

You can illustrate spatiotemporal chaos by the turbulent flow of water from a faucet, as in Fig. 1.10, the smoke rising from a cigarette, as in Fig. 1.17, the mixing of a dye being stirred in a tank of water, the patterns produced by a video camera aimed at the monitor to which it is connected and rotated slightly (Crutchfield 1988), or the patterns in chemical systems such as the *Belousov–Zhabotinsky* (BZ) *reaction* (Rinzel *et al.* 1984).[15]

These examples are mostly from physical systems so that you can easily assemble and study them, but the dynamics they exhibit are common to many other systems in the physical, biological, and social sciences. It helps to imagine a moving object as we study chaotic systems, but the concepts and results are much more general. Some of the most promising applications of chaos are to complicated systems where the underlying laws are not well

[14]The Galton board or *quincunx* (Galton 1894) is a vertical board with round pegs arranged in a triangular array in which balls dropped from the top apex of the triangle undergo collisions and form a Gaussian (normal) distribution at the bottom of the board. It was invented by the famous English scientist and explorer, Sir Francis Galton(1822–1911), cousin of Charles Darwin. Galton also developed fingerprinting, observed the cyclonic flow of air in weather patterns, and founded the eugenics movement (Gillham 2001).

[15]The oscillating BZ reaction, developed by the Soviet biophysicist, Boris Pavlovich Belousov (1893–1970) in 1959, and further studied by Anatol Markovich Zhabotinsky (1964), involves the oxidation of an organic species (malonic acid) from an acidified bromate solution in the presence of a metal ion catalyst such as cesium or iron.

understood. The hope is that at least some of the observed erratic behavior might be a consequence of deterministic chaos produced by simple nonlinear rules.

1.7 Exercises

Exercise 1.1 Derive a set of four first-order ordinary differential equations whose solution would give the three-body trajectory in Fig. 1.1.

Exercise 1.2 Write a set of equations similar to eqn (1.1) to describe two pendulums with frequencies differing by a factor a two and weakly coupled to one another.

Exercise 1.3 Derive a set of equations to describe the double pendulum in Fig. 1.5. Ignore friction and motion of the cart.

Exercise 1.4 Find the conditions under which Δt and Δv in eqns (1.2) and (1.3) are zero, and explain the meaning of your result.

Exercise 1.5 Find the conditions under which Δt is constant and $\Delta v = 0$ in eqns (1.2) and (1.3), corresponding to periodic motion.

Exercise 1.6 Choose realistic values for the parameters in eqns (1.2) and (1.3), and calculate the first few iterates v_n and q_n. Determine whether the result is periodic.

Exercise 1.7 Suppose the ball in §1.3 moves horizontally without friction, colliding elastically with two vertical walls a distance L apart with one wall fixed and the other oscillating with a small amplitude A ($\ll L$). Derive equations similar to eqns (1.2) and (1.3) for the time t_{n+1} and velocity v_{n+1} for each collision with the moving wall in terms of t_n and v_n.

Exercise 1.8 Investigate a real dripping faucet, and see if you can observe the period doublings that occur when the drip rate is such that chaos is about to onset.

Exercise 1.9 Label the components in Chua's circuit in Fig. 1.12, and derive a set of three first-order ordinary differential equations that describe its operation.

Exercise 1.10 Show that the circuit in Fig. 1.13(a) solves eqn (1.6) if the variable resistor has a value $R = 1/A$.

Exercise 1.11 Show that the circuit in Fig. 1.13(b) solves eqn (1.6) for $A \simeq 0.6$.

Exercise 1.12 Show that the circuit in Fig. 1.13(c) solves eqn (1.7) for $A = 0.5$.

Exercise 1.13 By defining two additional variables, y and z, write eqns (1.6) and (1.7) as systems of three first-order differential equations.

Exercise 1.14 Label the components of the chaotic flasher in Fig. 1.16, and derive a set equations that describe its operation.

1.8 Computer project: The logistic equation

This project gets you started using the computer to model chaotic processes and to explore some of the more obvious properties of the logistic equation, described in more detail in the next chapter. The logistic equation is perhaps the simplest example of a chaotic system. It models a process that exhibits initial exponential growth with a nonlinearity that ultimately stops the growth. Most of the common features of chaos are manifest in this simple example. The variable X is advanced successively in discrete time steps denoted by X_n for $n = 0, 1, 2, \ldots$. Start with an initial condition of $X_0 = 0.1$ and iterate the logistic equation

$$X_{n+1} = AX_n(1 - X_n) \tag{1.8}$$

many times.

1. Show that the solution approaches a fixed value for $0 < A < 3$, a periodic orbit for $3 < A < 3.56994\ldots$, and a chaotic orbit for $A = 4$.
2. What happens to the trajectory for $A < 0$ and for $A > 4$?
3. What happens in each case above if the initial condition is changed to $X_0 = 0.101$?
4. What happens if the initial condition is $X_0 < 0$ or $X_0 > 1$?
5. (optional) Can you find some initial conditions X_0 for which the solution is bounded (does not go off to infinity) but not chaotic for $A = 4$? (Hint: Calculate algebraically a value X_0 at which one of the successive iterates is also X_0, giving a cycle that repeats. There are infinitely many such values of X_0 in the range $0 < X_0 < 1$, but the probability that you will find one by randomly guessing is zero; they constitute a *set of measure zero.*)

2

One-dimensional maps

Chaotic systems such as those described in the previous chapter have a number of characteristics:

1. They are *aperiodic* (they never repeat).
2. They exhibit *sensitive dependence on initial conditions* (and hence they are unpredictable in the long term).
3. They are governed by one or more *control parameters*, a small change in which can cause the chaos to appear or disappear.
4. Their governing equations are *nonlinear*.

Whenever you encounter a phenomenon such as chaos that occurs in many different contexts, a useful scientific approach is to find and study the simplest system that exhibits the phenomenon. Perhaps the simplest chaotic mathematical system is the *logistic map*. It involves only a single variable and a single control parameter. Exact solutions can be found using only algebra, and it can be visualized graphically. It exhibits many aspects of more complicated chaotic systems, and hence it serves as a prototype for them. It has been used to model phenomena in fields as diverse as ecology, oncology, and finance.

One of the first persons to appreciate the importance and applicability of chaos in the logistic map was the theoretical physicist and mathematical biologist, Lord Robert May (1976), although the phenomenon had been earlier observed in studies of insects (Moran 1950) and fish populations (Ricker 1954). The interest was spurred by the proliferation of computers in the latter half of the previous century. This chapter describes the logistic map in some detail and mentions other one-dimensional maps. It develops much of the terminology used throughout the book.

2.1 Exponential growth in discrete time

Suppose that you deposited \$100 in a bank that gives 10% interest, compounded annually, and that you make no further deposits or withdrawals. You would have \$110 next year, \$121 the following the year, and so forth. More generally, if the amount in your account in year n is X_n, the amount the following year is

$$X_{n+1} = AX_n \tag{2.1}$$

where $A = 1.1$ in this example.

This is an example of a *deterministic dynamical system*, in that the next value of X is uniquely determined by the present value. It is a *linear system* because a graph of X_{n+1} versus X_n is a straight line. It is a *recursive relation*, since it is applied *recursively* (repeatedly) for successive values of n. It is an example of an *iterated map*. The verb 'to iterate' means to feed the output of the equation back to the input at the next time step, and the noun 'iterate' (pronounced with the accent on the first syllable) is the resulting value. 'Iteration' means repeating the process over and over in a kind of 'mathematical feedback.' The term 'map' (both a noun and a verb) derives from the process of transferring each point on the Earth to a corresponding point on a printed map. In the present example, we are mapping a point along the X axis into another point along the same axis, but we will soon be considering more complicated maps in two or more dimensions.

Another way of thinking about the process is to imagine that all possible X values are plotted along a straight line at a distance from some reference point ($X = 0$) proportional to X. Then at each time step, the whole line is stretched by 10% like a rubber band. For $A < 1$, the line shrinks rather than stretches, as would occur if you were paying interest rather than earning it. We say that all initial conditions X_0 (pronounced 'X naught') *attract* to the point $X = 0$ for $0 < A < 1$ and attract to infinity for $A > 1$. Systems whose solution attracts to infinity are called *unbounded.*

The value of X after n years is given by

$$X_n = A^n X_0 \tag{2.2}$$

This is an example of *exponential growth* (or *decay* if $0 < A < 1$) because the time step n appears as the exponent of A. Linear iterated equations usually have such exponential solutions. Hence the term 'linear' refers to the form of the equations, not the solution. The quantity A is the control parameter (or 'knob') that governs the nature of the dynamics. We assume that the control parameters remain fixed while the variables are changing, although in real experiments, the best we can usually expect is that they change much more slowly than the dynamical variables. The bank may change its interest rate from time to time. The value $A = 1$ separates two regions in which the behavior differs and is called a *bifurcation point*[1] (see Chapter 7). It is a point in an abstract one-dimensional *parameter space* (a line containing all possible A values), not to be confused with the one-dimensional space in which the variable X lives.

It is instructive also to consider the case with negative A. In that case, if X is originally positive, it is negative at the next time step, then positive, then negative, and so forth. The solution oscillates about $X = 0$. For $-1 < A < 0$ the oscillation amplitude decreases with time (*damps*), while

[1]The term 'bifurcate' comes from Latin words meaning 'two branches.'

for $A < -1$ it grows in time. Hence $X = 0$ is an attractor for all X_0 when $|A| < 1$. In the geometrical analogy, you can think of the line as shrinking (or stretching), while simultaneously flipping about the point $X = 0$ with each iteration.

2.2 The logistic equation

You know that exponential growth cannot continue for ever, whether it is money in the bank or people on the planet. Hence the linear dynamical system of eqn (2.1) with $|A| > 1$ is not a realistic model of any natural process. Typically, some nonlinearity stops or even reverses the growth. The nonlinearity is negligible at small values of X but dominates as X increases. It is usually a good strategy to analyze the linear system before introducing nonlinearities. Even when the linear system has obvious shortcomings, it is instructive and often indispensable to understand its properties.

Imagine a species of bug that lays its eggs in a given season each year and then dies. (The death avoids the complication of overlapping generations.) The next year, the eggs hatch, and the process repeats. We expect eqn (2.1) to describe this dynamical system when there is sufficient food for the bug population to grow. However, as the number of bugs increases, the food eventually becomes insufficient, and some bugs die before they can lay their eggs.

We must modify eqn (2.1) to include a term that reduces the growth as X increases. The simplest such mathematical form is the *logistic equation*

$$X_{n+1} = AX_n(1 - X_n) \tag{2.3}$$

We must now think of X, not as the number of bugs, but as the size of the population, scaled so that the maximum value (the *carrying capacity*) is $X = 1$. For $X \ll 1$, eqn (2.3) reduces to eqn (2.1), but for $X = 1$, the population is completely extinct the following year (no bugs survive). A similar equation has been used by Stutzer (1980) to model macroeconomics.

The nonlinearity is *quadratic*,[2] since multiplication gives $X_{n+1} = AX_n - AX_n^2$, which involves the square of X. The quadratic term gives nonlinear negative feedback and limits the growth. A graph of eqn (2.3), called the *logistic function* or *logistic curve*, is a *parabola*, as in Fig. 2.1. The figure shows the case $A = 4$, which has the special property that it maps the unit interval ($0 < X < 1$) onto itself twice (once for $0 < X < 0.5$ and again for $0.5 < X < 1$) in the reverse direction. In the rubber-band analogy, this case corresponds to stretching the rubber band to twice its length with each iteration, albeit nonuniformly, and then folding it back onto itself. The map is one-dimensional since there is a single variable X and the resulting curve is a line, even when it is plotted in a two-dimensional (or higher) space.

[2]Quadratic is derived from the Latin *quadratus*, meaning 'square.'

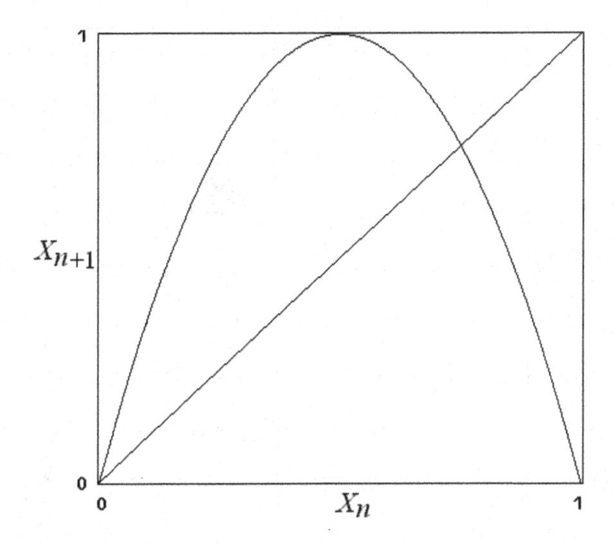

Fig. 2.1 The logistic map with $A = 4$.

An alternate form of the logistic equation (Grebogi *et al.* 1983) can be derived by defining a new variable $Y \equiv A(X - 0.5)$ and a new parameter $B \equiv A^2/4 - A/2$, giving

$$Y_{n+1} = B - Y_n^2 \tag{2.4}$$

which is also a parabola but centered on $Y_n = 0$. For $B = 2$, it maps the interval $-2 < Y < 2$ back onto itself twice in a manner similar to eqn (2.3) with $A = 4$. This map is called the *quadratic map* and is *conjugate*[3] to the logistic map. The terms 'quadratic map' and 'logistic map' are sometimes used interchangeably to describe both cases.

Also plotted in Fig. 2.1 is the 45° straight line $X_{n+1} = X_n$, whose intersections with the parabola give the values of X that do not change in time. For a quadratic map, there are two such intersections, corresponding to the solutions of the quadratic equation that result from setting $X_{n+1} = X_n = X^*$, although the solutions may coincide. The solutions $X^* = 0$ and $X^* = 1 - 1/A$ are called *fixed points* of the map.

It is interesting to examine how X approaches a fixed point starting from some initial condition $X_0 \neq X^*$. Figure 2.2 shows a simple graphical way to do this. Start from the initial value X_0 on the horizontal axis. Draw a vertical line to the parabola to determine X_1. Then from that point draw a horizontal line to the line $X_{n+1} = X_n$. Then draw a vertical line to the parabola to determine X_2. Repeat as necessary to determine the evolution of X. Such a diagram is called a *cobweb diagram* for obvious reasons. For the logistic map with $1 < A < 3$, all starting points in the range $0 < X_0 < 1$

[3]Technically, the maps $f(x)$ and $g(x)$ are conjugate if there exists a continuous, one-to-one, onto function $\phi(x)$ such that $\phi(f(x)) = g(\phi(x))$.

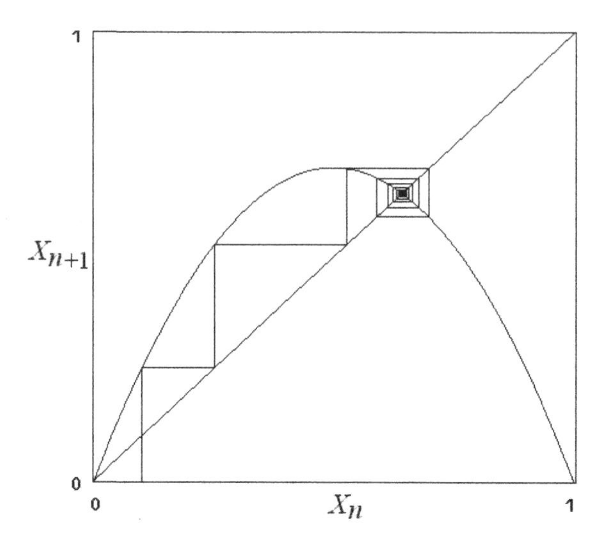

Fig. 2.2 Cobweb diagram for the logistic map with $A = 2.8$ and $X_0 = 0.1$.

approach the fixed point $X^* = 1 - 1/A$, although the solution may oscillate about it before settling to the final value. The behavior is reminiscent of a simple pendulum that oscillates about the vertical before friction brings it to rest at its final position.

2.3 Bifurcations in the logistic map

Recall that eqn (2.1) had values of $A = \pm 1$ where the behavior abruptly changed from bounded to unbounded. The logistic equation behaves similarly, except there are many more bifurcations and the behavior in the various regions is more diverse and interesting. It is useful to study the bifurcations of the logistic map in some detail because the general features are common to many chaotic systems. We will consider only positive values of A and X.

2.3.1 $0 < A < 1$ case

Since A is the slope of the parabola at $X = 0$, the parabola with $A < 1$ can only intersect the 45° line at one non-negative value, and there is a single fixed-point solution at $X^* = 0$. All initial conditions in the range $0 < X_0 < 1$ attract to the point $X^* = 0$. We say that these points lie within the *basin of attraction*[4] of $X^* = 0$ and that the fixed point $X^* = 0$ is *stable*. All points within the basin of attraction are *asymptotically fixed points*, because they approach the fixed point ever more closely after many iterations. The nonlinearity has little effect after the first few iterations.

[4]The term 'basin of attraction' comes from a wash basin or watershed and will be of more significance in higher-dimensional systems.

Values of X_0 outside the basin of attraction are unbounded (they escape to infinity).

2.3.2 $1 < A < 3$ case

Just as with eqn (2.1), for $A = 1$ a bifurcation occurs, and the fixed point at $X^* = 0$ becomes unstable. The attractor becomes instead a *repellor*. If X happens to be exactly zero, it would remain so, but if it is even slightly positive, it will grow initially at an exponential rate. The situation is like a pencil standing on its pointed end. Even the slightest push will make it topple. However, unlike eqn (2.1) in which the solutions are unbounded, the logistic map at $A = 1$ develops a new fixed point at $X^* = 1 - 1/A = 0$ that moves away from zero for $A > 1$.

If A is not too large, then that point is an attractor because all initial values in the range $0 < X_0 < 1$ are pulled toward it and eventually settle onto it. We say that the final state is a *period-1 cycle*, or simply a *1-cycle*, because each iterate is the same as the one before it. If the logistic equation were modeling the population of bugs, then it would predict an initial exponential growth for this range of A, but a final steady state in which the number of bugs is unchanging from year to year.

2.3.3 $3 < A < 3.44948\ldots$ case

For $A = 3$, the fixed point at $X^* = 1 - 1/A$ still exists but changes from stable to unstable, becoming a repellor. This bifurcation occurs when the slope of the parabola at the fixed point (called the *Floquet multiplier*) equals -1. For $A > 3$ we have exponential growth away from the point, rather than exponential decay toward it. Since the slope is negative, the solution oscillates on either side of the fixed point while moving away, just as it did for eqn (2.1) with $A < -1$. Hence the bifurcation at $A = 3$ is called a *flip*.

However, the growth does not continue for ever. Instead, it approaches a condition in which every other iterate is the same ($X_n = X_{n+2} = X_{n+4} = \ldots$) as in the cobweb diagram in Fig. 2.3. With a bit of algebra, this condition can be reduced to a *quartic* (fourth-degree polynomial) equation

$$A^3X^4 - 2A^3X^3 + A^2(A+1)X^2 - (A^2-1)X = 0 \qquad (2.5)$$

As with any quartic equation, there are four roots. One is obviously $X^* = 0$, corresponding to the unstable fixed point previously discussed. The other unstable fixed point, $X^* = 1 - 1/A$, is less obvious but is also a solution. Factoring out these terms reduces eqn (2.5) to a quadratic equation

$$A^2X^2 - A(A+1)X + A + 1 = 0 \qquad (2.6)$$

whose roots are

$$X^*_\pm = \frac{A + 1 \pm \sqrt{(A-3)(A+1)}}{2A} \qquad (2.7)$$

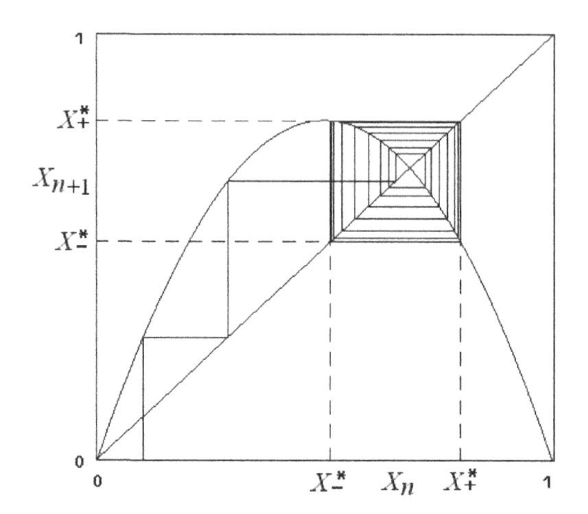

Fig. 2.3 Cobweb diagram for the logistic map with $A = 3.2$ and $X_0 = 0.1$.

For $-1 < A < 3$, the quantity under the square root is negative, and a real solution does not exist. For $A > 3$ there are two real roots between which X oscillates on successive iterations in the steady state. This is an example of a *2-cycle*. It is a *cyclic* or *periodic attractor* since almost any initial condition in the unit interval approaches it. In such a case the bugs would be plentiful one year, scarce the next, and then plentiful again. Note that for $A = 3$, eqn (2.7) has a single root at $X^* = 2/3$, which is the same as $X^* = 1 - 1/A$, meaning that the 2-cycle bifurcates *continuously*. Continuous bifurcations are also called *subtle*.

2.3.4 $3.44948\ldots < A < 3.56994\ldots$ case

The 2-cycle exists for all $A > 3$, but it becomes unstable when A reaches a value where the slope of the second-iterate map evaluated at $X = X^*$ as given by eqn (2.7) equals -1. The calculation leads to a quadratic equation

$$A^2 - 2A - 5 = 0 \tag{2.8}$$

whose positive root is $A = 1 + \sqrt{6} = 3.449490\ldots$. At this bifurcation, the 2-cycle becomes unstable, and a stable 4-cycle is born. The logistic map can have at most one stable periodic orbit for each value of A (Singer 1978). This property is not shared by all one-dimensional *unimodal* (single-humped) maps, however.

The process continues with successive *period doublings* (a *cascade*) with a new period appearing just as the previous one becomes unstable. The onset of these doublings is increasingly difficult to calculate, both analytically and numerically, but the next few values are

$A_3 = 3.544090\ldots$ (8-cycle)
$A_4 = 3.564407\ldots$ (16-cycle)

$A_5 = 3.568759\ldots$ (32-cycle)
$A_6 = 3.569692\ldots$ (64-cycle)
$A_7 = 3.569891\ldots$ (128-cycle)
$A_8 = 3.569934\ldots$ (256-cycle)
$A_9 = 3.569943\ldots$ (512-cycle)
$A_{10} = 3.569945\ldots$ (1024-cycle)
$\ldots$
$A_\infty = 3.5699456718\ldots$ (accumulation point)

The period doublings become successively closer, eventually accumulating at the point $A_\infty = 3.5699456718\ldots$, known as the *accumulation point*. At this point, the period becomes infinite (it never repeats), and the orbit visits infinitely many X values but fills a negligible portion of the unit interval ($0 < X < 1$). The iterates form a *Cantor set* (see §11.1), a fractal object with a noninteger dimension of about 0.538.

The spacing between successive bifurcations approaches a constant

$$\delta = \lim_{n\to\infty} \frac{A_n - A_{n-1}}{A_{n+1} - A_n} = 4.669201\ldots \tag{2.9}$$

known as the *Feigenbaum number* (Feigenbaum 1978, 1979, 1980), although it was earlier found by Grossmann and Thomae (1977). The bifurcations are given approximately by $A_k \simeq A_\infty - 1.542\delta^{-k}$. A remarkable property of the constant δ is its *universality*, having the same value for all unimodal maps with a quadratic maximum. It is a new mathematical constant, as basic to period doubling as π is to circles. It has been calculated to over a thousand digits,[5] and it appears not to be a combination of any other known constants. The Feigenbaum number arises as an *eigenvalue* of an operator (see §4.2), and hence it is sometimes whimsically called a 'Feigenvalue' (Briggs *et al.* 1991). Period doubling with apparently the same constant has been observed in many experiments including turbulent fluids (Libchaber 1982), electronic circuits such as Fig. 1.11 (Linsay 1981, Testa *et al* 1982), lasers (Harrison and Biswas 1986), chemical reactions (Simoyi *et al.* 1982), dripping faucets (Shaw 1984), musical instruments (Maganza *et al.* 1986), and even biological systems that are not obviously described by one-dimensional maps (Cvitanović 1984) such as heartbeats (Guevara *et al.* 1981).

2.3.5 $3.56994\ldots < A < 4$ case

When A is increased beyond the accumulation point, chaos onsets. The period is infinitely long, and finite regions of the unit interval are visited

[5] $\delta = 4.669201609102990671853203820466201617258185577475768632745651343 0\ldots$. Note that several books have given a value starting with $4.66920166\ldots$ in which the last repeated 6 is incorrect. Note also that there is no practical use for such precision, since 39 digits suffices to measure the size of the Universe to within the diameter of a hydrogen atom.

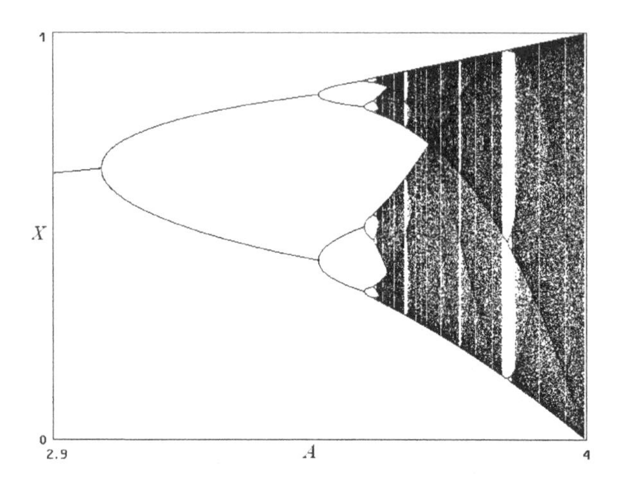

Fig. 2.4 Bifurcation diagram for the logistic map.

by the orbit. However, there are infinitely many windows (ranges of A) of periodicity. All periods are represented, but the width of the window decreases as the period increases. Each periodic window appears abruptly as A increases and contains its own period doubling route back into chaos with the same Feigenbaum number. For example, the prominent period-3 window onsets at $A = 1 + \sqrt{8} = 3.82842712\ldots$, although the proof is difficult (Saha and Strogatz 1994).

The period-3 orbit is special, because Sarkovskii (1964) and later Li and Yorke (1975) independently proved that a one-dimensional continuous map with a period-3 orbit for a particular parameter value has orbits of every period (including infinity) for that parameter and hence will be chaotic if all those orbits are unstable. The converse is not true; chaotic systems need not have a period-3 orbit. Each period higher than 3 occurs more than once. For example, there are two period-4 windows, three period-5 windows, five period-6 windows, and so forth (see §5.1.3).

The entire behavior can be summarized in a *bifurcation diagram*[6] as in Fig. 2.4. Such a diagram plots all possible values of X in the final state (after initial transients have died away) as a function of the control parameter A. The bifurcation points, sometimes incorrectly called 'pitchfork bifurcations' (see §7.3), are evident. This diagram is sometimes called a 'fig-tree diagram' because of its shape (viewed sideways) and because Feigenbaum means 'fig tree' in German.

There is another universal 'Feigenvalue' associated with the rate at which successive branches of the tree shrink in size with a value of $\alpha = 2.5029078750958\ldots$. Since there are multiple pairs of branches, and the

[6]Some authors (e.g., Strogatz 1994), call such a plot an 'orbit diagram' and reserve the term 'bifurcation diagram' for a plot that shows both the attractor and the unstable orbits, the latter usually with a dotted line, examples of which are in Chapter 7.

branches are not parallel, you must consider only the branches that intersect $X = 0.5$ and calculate their width exactly at the corresponding value of A (the superstable orbits described in §5.1.3). A simplified theory (Feigenbaum 1979), accurate to about 5%, gives $2\delta \simeq \alpha^2 + \alpha + 1$.

2.3.6 $A = 4$ case

As mentioned earlier, the case $A = 4$ is special because it maps the unit interval back onto itself. A map with such a property is called an *endomorphism*. You can imagine the X axis between zero and one being a rubber band, that with each iteration is stretched so that its midpoint ($X_n = 0.5$) reaches $X_{n+1} = 1$ and then the far end ($X_n = 1$) is folded back to $X_{n+1} = 0$. The stretching and folding are responsible for the chaos. Two nearby initial conditions separate on average because of the stretching, whereas the folding keeps them bounded. The stretching is not uniform, however, since it is large (a factor of four per iteration) at $X = 0$ and $X = 1$ but infinitely negative at $X = 0.5$. On average the stretching is a factor of two as suggested by the fact that it maps back onto itself twice. Each iterate X_n has two *preimages* X_{n-1} given by

$$X_{n-1} = 0.5 \pm \sqrt{0.25 - X_n/A} \tag{2.10}$$

that typically do not coincide. Consequently, one bit of information (a factor of 2) is lost with each iteration since there is no way of knowing from which preimage each value came (we say the map is *noninvertible*, in this case being *two-to-one*), in contrast to an *invertible map* (*one-to-one*) in which there is a unique preimage. This exponential loss of information is equivalent to the exponential growth of errors in the initial condition that is the hallmark of chaos. Noninvertibility is necessary for chaos in one-dimensional maps but not for maps in higher dimensions.

The chaos can be exhibited in a cobweb diagram, as in Fig. 2.5, or directly in a plot of successive iterates, as in Fig. 2.6. Although the behavior appears random, closer examination reveals the exponential growth at small values of X_n, small values of X_n following a large one, and a growing oscillation about the unstable fixed point at $X_n = 1 - 1/A = 0.75$. It is surprising that a simple quadratic equation can exhibit such complex behavior. If the logistic equation with $A = 4$ modeled the growth of bugs, then their population would exhibit erratic yearly fluctuations.

The logistic map with $A = 4$, sometimes called the *Ulam map* (Beck and Schlögl 1995) and written in the equivalent form $X_{n+1} = 1 - 2X_n^2$, was studied by Ulam[7] well before the modern chaos era. He and von Neumann proposed it as a computer random number generator (Ulam and von Neumann 1947, Phatak and Rao 1995).

[7]Stanislaw Marcin Ulam (1909–1984) was a Polish mathematician who emigrated to the United States in 1943 and was instrumental in developing the hydrogen bomb and the *Monte-Carlo method* for evaluating complicated integrals.

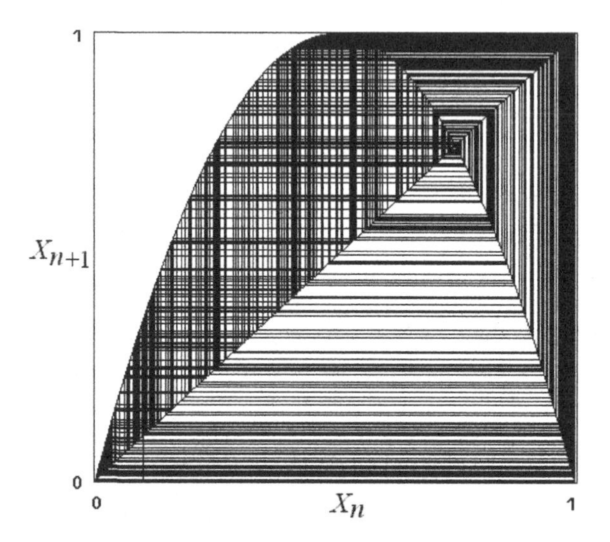

Fig. 2.5 Cobweb diagram for the logistic map with $A = 4$.

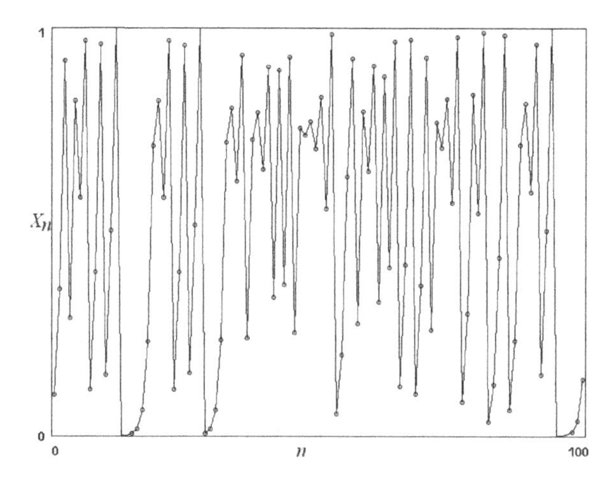

Fig. 2.6 Plot of the first hundred iterates of the logistic map with $A = 4$.

2.3.7 $A > 4$ case

For $A > 4$, the peak of the parabola exceeds one. Thus most initial conditions have iterates that eventually reach $X_n > 1$. When that happens, the next iterate is negative, and the repellor at $X = 0$ then pushes the orbit rapidly to minus infinity. Thus most orbits are unbounded for $A > 4$. However, the unstable fixed point at $X^* = 1 - 1/A$ persists for $A > 4$, as do all the unstable periodic orbits such as the 2-cycle given by eqn (2.7). The period-2 orbit cycles between a value slightly greater than zero and a value even slightly less than one. There is also a Cantor set (see §11.1) of *measure zero*[8] for which chaotic orbits exist for all $A > 4$. For example, with $A = 4.5$, every iteration of the map causes the middle third of the X values to be 'thrown out' of the interval 0 to 1, so the effect is that only the points within a Cantor set will stay in this interval; all others will escape. If A only slightly exceeds four and X is not on one of these orbits, it may take many iterations for the orbit to escape. In such a case, the orbit exhibits *transient chaos*.

We can estimate the number of iterations required for the orbit to escape. Suppose $A = 4 + \epsilon$, where ϵ is a small quantity ($\epsilon \ll 1$). The range of X_n for which $X_{n+1} > 1$ can be calculated from $AX(1 - X) = 1$. The resulting quadratic equation has roots $X \simeq 0.5 \pm 0.25\sqrt{\epsilon}$. If all X values were equally probable, then the probability that an iterate will fall in the region where $X_{n+1} > 1$ is $0.5\sqrt{\epsilon}$, and so the number of iterations before the orbit escapes is typically $2/\sqrt{\epsilon}$. A more accurate calculation requires knowing the density of points on the unit interval as discussed in §2.4.4 and gives a value of $\pi/\sqrt{\epsilon}$ (see also §7.6.1).

2.4 Other properties of the logistic map with $A = 4$

The logistic map with $A = 4$ is special in many ways. It is fully chaotic in the sense that almost every point on the unit interval is eventually visited for any initial condition, a property known as *topological transitivity*. This fact and the simplicity of the equation allows us to calculate several special properties of the logistic map with $A = 4$.

2.4.1 Unstable periodic orbits

Although almost every initial condition produces a chaotic orbit for $A = 4$, there are infinitely many that do not. For example, the point $X = 1 - 1/A = 0.75$ is a fixed point. Hence the iterates of that initial condition never change. Furthermore, there are unstable periodic orbits of every period. The period-2 orbit from eqn (2.7) with $A = 4$ has values $X = (5 \pm \sqrt{5})/8$, and so forth. Figure 2.7 shows all the periodic orbits up to period-10. There are 2^p values for which $X_{n+p} = X_n$ for each period p. An initial condition with any of these values will be forever periodic, and there are infinitely many of them, albeit mostly with very high periods. In fact, every point on the unit interval is arbitrarily close to one of these orbits. We say the interval is *dense* in periodic orbits, all unstable.

This property is similar to the existence of *rational numbers* (those that can be expressed as a ratio of two integers or, equivalently, those whose decimal expression eventually repeats) on the unit interval. There

[8]An object is said to have 'measure zero' or 'zero measure' if it occupies a zero fraction of the space it inhabits. For example, a line, even one of finite length, contains infinitely many points, but with no width it has zero area when drawn on a sheet of paper. The measure of a set with noninteger dimension is also called the *Lebesgue measure* after the French mathematician Henri Léon Lebesgue (1875–1941) who feared that generalizations would lead to the death of mathematics.

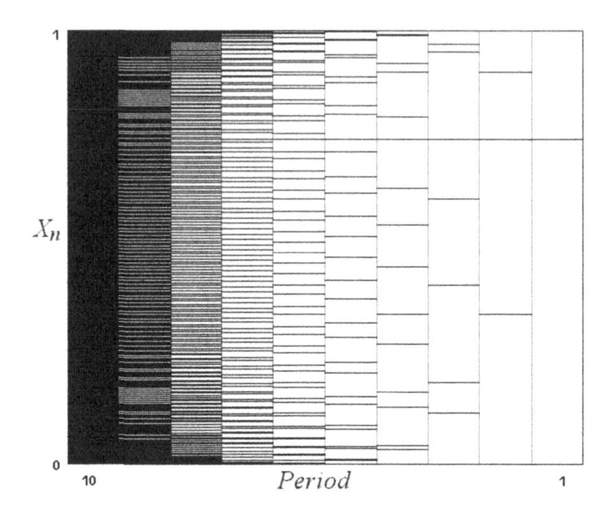

Fig. 2.7 Unstable periodic orbits for the logistic equation with $A = 4$.

are infinitely many of those, and every point is arbitrarily close to one, but almost every point is *irrational*. The rationals are a set of measure zero, and so are the dense unstable periodic orbits. Curiously, between any two rationals is an irrational, and between any two irrationals is a rational, but they certainly do not simply alternate. It is often argued that rational numbers are unlikely to occur in nature, but there are counterarguments based on the fact that many quantities are discrete, such as the mass of atoms, electric charge, angular momentum, and the population of animals.

2.4.2 Eventually fixed points

In addition to the infinity of initial conditions that lie on unstable periodic orbits, each periodic orbit has an infinite set of preimages, each of which is not chaotic. For example, the initial condition $X_0 = 0.5$ maps into $X_1 = 1$, and then into $X_2 = 0$, after which the value never changes. We say that $X = 0.5$ is an *eventually fixed point*, since the sequence of X values eventually stops changing. The initial condition $X_0 = 0.25$ maps into $X_1 = 0.75$, which is a fixed point, and so it is also an eventually fixed point. From eqn (2.10), it has two preimages at $X_{-1} = 0.5 \pm 0.25\sqrt{3}$, each of which has two preimages, and so forth. Even though there are only two fixed points for the logistic equation with $A = 4$, each has infinitely many preimages, each of which is an eventually fixed point, but they are a set of measure zero. Some of the eventually fixed points for $X^* = 0$ are shown in Fig. 2.8, and those for $X^* = 0.75$ are shown in Fig. 2.9.

The existence of eventually fixed points poses a potential numerical difficulty, particularly for the fixed point at $X = 0.5$ which is at the peak of the parabola. It is especially sensitive to round-off errors. Suppose the computer rounds values in the range $0.5 \pm \epsilon$ to 0.5. Then any value so

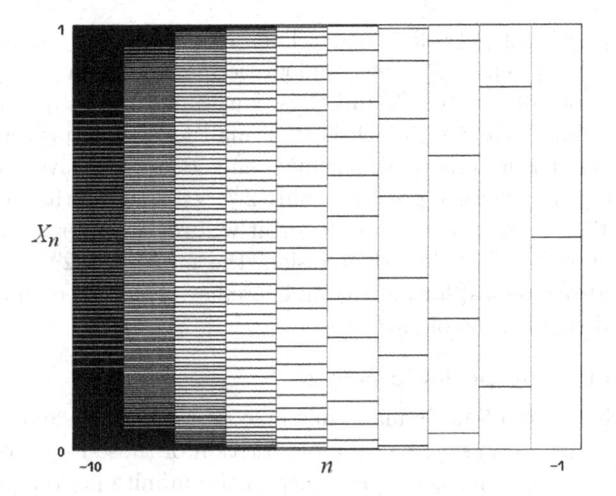

Fig. 2.8 Eventually fixed points for the fixed point $X^* = 0$ of the logistic map with $A = 4$.

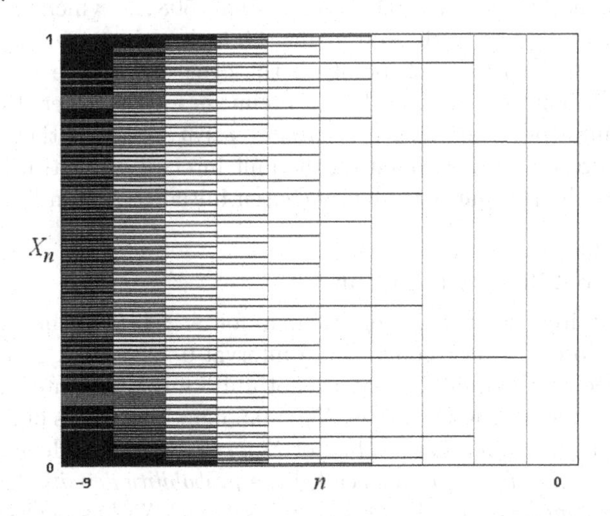

Fig. 2.9 Eventually fixed points for the fixed point $X^* = 0.75$ of the logistic map with $A = 4$.

rounded would go to zero two iterations later and would remain there for ever. The likelihood that an iterate falls within this range is on the order of $\sqrt{\epsilon}$. Thus with single precision (32-bit) arithmetic for which $\epsilon \sim 5 \times 10^{-8}$, we expect an arbitrary initial condition to reach the fixed point at zero after an average of about $1/\sqrt{\epsilon} \sim 4000$ iterations. Numerical experiments suggest a number closer to 2000.

Although you can overcome this particular difficulty by restricting A to values less than $4 - \epsilon$, there is still a round-off problem with other eventually fixed and periodic points. Thus it is advisable to use at least double precision when computing the logistic map. With double precision, ϵ is on the order of 5×10^{-16}, and thus you would expect difficulty after about 40 million iterations, which is still small for some purposes. In any case, a computer is a finite-state machine, and so any iterative process will eventually repeat. In single precision, only $2^{32} \sim 4 \times 10^9$ distinct values can be represented, guaranteeing that the orbit will repeat after no more than that many iterations. In double precision, the number is $2^{64} \sim 2 \times 10^{19}$. These limitations persist for more complicated systems and require caution in numerical simulations of chaos.

2.4.3 Eventually periodic points

Although there are infinitely many unstable periodic orbits and eventually fixed points, they represent a negligible fraction of those initial conditions for which chaos does not occur. For each of the infinite periodic solutions, there are infinitely many preimages whose iterates eventually become periodic. For example, the 2-cycle at $X = (5 \pm \sqrt{5})/8$ from eqn (2.7) with $A = 4$ has two preimages at $X_{-1} = 0.095491\ldots, 0.654508\ldots$, which in turn have two preimages, and so forth. You might think that the 2-cycle would have four preimages, two for each point on the cycle, but there are only two because each point on the 2-cycle is a preimage for the other. Despite the doubly infinite collection of such *eventually periodic points*, they still are a set of measure zero. Every point on the unit interval is arbitrarily close to one, but the chance that a randomly chosen initial condition lies on one is zero.

2.4.4 Probability distribution

With enough iterations of the logistic map for $A = 4$, the orbit approaches arbitrarily close to every point in the interval $0 < X < 1$. However, the points visited on the unit interval are not uniformly distributed. In particular, many values in the vicinity of $X_n = 0.5$ map into values of X_{n+1} close to one, which in turn map into values of X_{n+2} close to zero. Hence the *probability distribution function* (also called the *probability density distribution*, the *invariant measure*, or the *natural measure*) $P(X)$ has peaks at $X = 0$ and $X = 1$. $P(X)$ is the probability that a point X is within dX of X and is normalized so that the area under the $P(X)$ curve is $\int_0^1 P(X)dX = 1$. The point has unit probability of being somewhere in the interval $0 < X < 1$.

The probability distribution function can be calculated by applying a transformation to the binary shift map (see §2.5.4), for which the probability distribution is uniform. The result is

$$P(X) = \frac{1}{\pi\sqrt{X(1-X)}} \tag{2.11}$$

whose graph is in Fig. 2.10. The *ergodic hypothesis* (Ruelle 1976) asserts that this probability distribution is the same for many iterations of a single

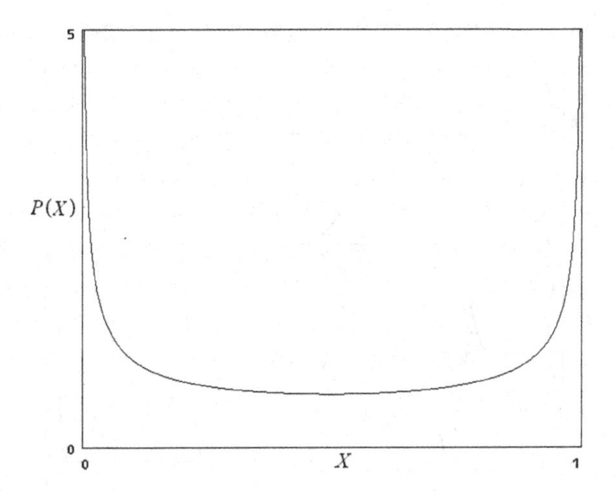

Fig. 2.10 Probability distribution function of the logistic map with $A = 4$.

orbit (time average) and for a high-order iteration of many orbits with a range of random initial conditions (ensemble average). Note that at the peak of the parabola ($A = 0.5$), the measure is $P(0.5) = 2/\pi$, which allows a more accurate calculation of the average number of iterations for the orbit to reach zero due to computer round-off errors. The highly nonuniform probability is a result of the nonuniform stretching. The singularities in $P(X)$ at $X = 0$ and $X = 1$ cause numerical difficulties when calculating quantities such as the Lyapunov exponent and correlation dimension (see Chapters 5, 12, and 13).

For an arbitrary value of A, no simple formula exists for the probability distribution function. However, $P(X)$ satisfies a variant of the *Frobenius–Perron*[9] *equation* (Ott 1993)

$$P(X) = \frac{P(X_A)}{|f'(X_A)|} + \frac{P(X_B)}{|f'(X_B)|} \tag{2.12}$$

where X_A and X_B are the preimages of X given by eqn (2.10) and f' is the derivative df/dX of the map function ($f' = A(1-2X)$ for the logistic map). Equation (2.12) simply asserts that the density of points around a given X value is inversely proportional to the local stretching of the space around each of its preimages. No general method exists for solving this equation analytically. By direct substitution, it is easy to show that $P(x)$ given by eqn (2.11) satisfies it. It can be approximated numerically by iterating a one-dimensional array of distinct $P(X)$ values subject to the boundary conditions that $P(X) = 0$ for $X > A/4$ and for $X < A^2(1-A/4)/4$. These boundary points are the first and second iterates of the value of X for

[9]Ferdinand Georg Frobenius (1849–1917) and Oskar Perron (1880–1975) were prolific German mathematicians of different generations.

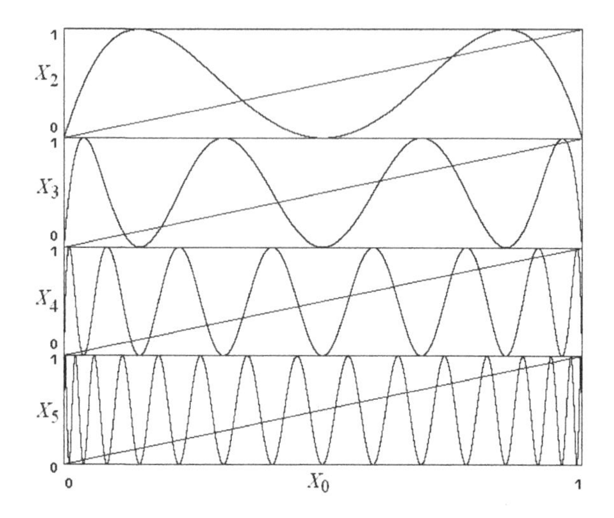

Fig. 2.11 First few iterates of the logistic map with $A = 4$.

which the map reaches its peak ($X = 0.5$), beyond which the orbits escape to infinity. It is simpler and more convenient to iterate the map many times from an arbitrary initial condition and plot a histogram of the resulting X values, which is equivalent according to the ergodic hypothesis.

2.4.5 Nonrecursive representation

Normally, for a chaotic system, it is not possible to derive an algebraic expression for the value of an arbitrary iterate of an arbitrary initial condition. However, because the logistic map with $A = 4$ is conjugate to the tent map, which is described by a simple mathematical operation (see §2.5.2), it is possible to derive such an equation (Schuster 1995)

$$X_n = \tfrac{1}{2}\{1 - \cos[2^n \cos^{-1}(1 - 2X_0)]\} \tag{2.13}$$

This equation is of limited use for large n because of the difficulty in calculating the cosine of a function whose argument grows as 2^n.

It is instructive to graph the first few iterates of X_n, as shown in Fig. 2.11. Note that the number of oscillations is 2^{n-1} and the number of intersections with the line $X_n = X_0$ is 2^n. Hence it is apparent why there are so many periodic orbits as the period increases, and why they are all unstable. The Floquet multiplier exceeds one in absolute value at every intersection. It is also evident why there is sensitive dependence on initial conditions. Two nearby values of X_0 lead to very different values of X_n for large n.

2.5 Other one-dimensional maps

The logistic map is *one-dimensional* in the sense that it has only one scalar variable X. There are infinitely many such one-dimensional maps, even for the interval $0 < X < 1$. We now describe a few of the more common ones.

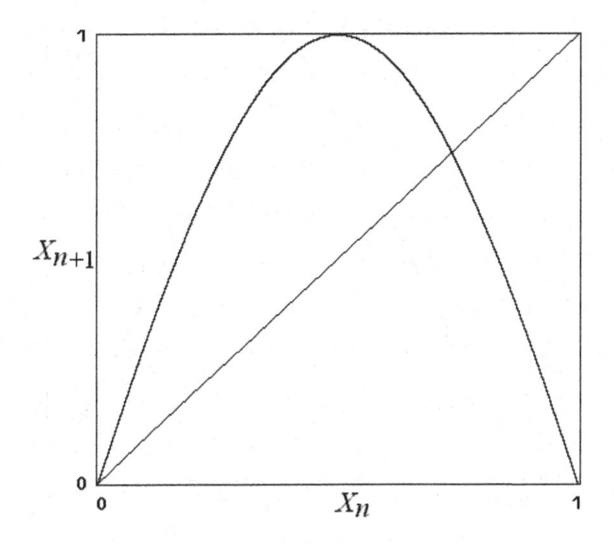

Fig. 2.12 Sine map with $A = 1$.

2.5.1 Sine map

The *sine map*

$$X_{n+1} = A \sin \pi X_n \tag{2.14}$$

with $A = 1$ maps the unit interval back onto itself in a manner very similar to the logistic map, as in Fig. 2.12. The sine map is quadratic near $X_n = 0.5$, just like the logistic map, and so it has an almost identical probability distribution and period-doubling route to chaos. The periodic windows occur in the same order, and with the same relative sizes. It has the same Feigenbaum number as the logistic map since the maximum is quadratic. Despite the similarities, there are quantitative differences. The Lyapunov exponent (see Chapter 5) is about half a percent smaller, the period-doubling bifurcations occur earlier, and the periodic windows are wider than in the logistic map. If you explore its properties with a calculator, be sure to calculate the argument of the sine function in radians, not degrees.

2.5.2 Tent map

A map that is simpler than the logistic map, and conjugate to it, is the *tent map*

$$X_{n+1} = A \min(X_n, 1 - X_n) \tag{2.15}$$

so named because of its tent-shape, as in Fig. 2.13. It is said to be *piecewise linear*, since a graph of X_{n+1} versus X_n consists of two straight lines that meet at $X_n = 0.5$. With $A = 2$ it maps the unit interval back onto itself twice, just as does the logistic map with $A = 4$. The stretching is a uniform factor of A for all X. However, it has a very different bifurcation sequence.

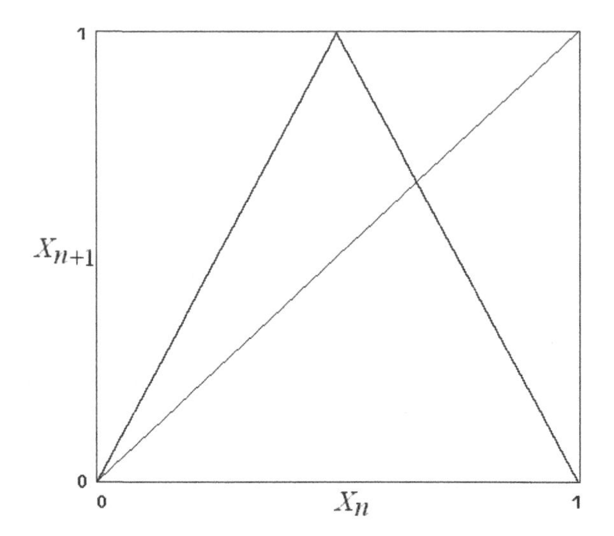

Fig. 2.13 Tent map with $A = 2$.

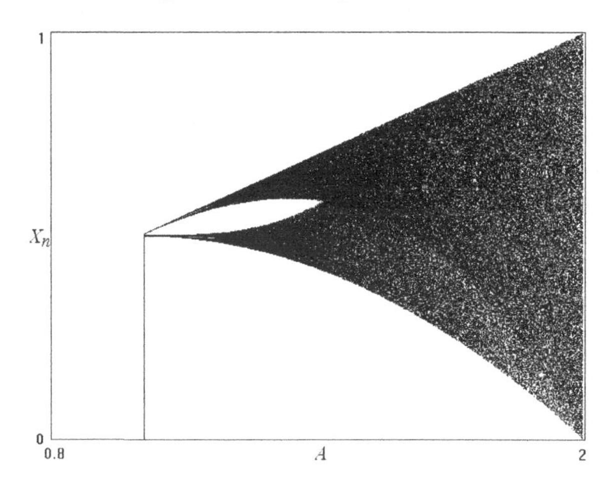

Fig. 2.14 Bifurcation diagram for the tent map.

It has a single stable fixed point at $X = 0$ for $A < 1$. At $A = 1$ the fixed point becomes unstable and gives birth to a 2-cycle that is also unstable. For higher values of A, other unstable cycles are born, producing chaos for nearly all values of X_0 with $1 < A < 2$. There are no periodic windows. For $A = 2$, unstable cycles of all periods occur. The bifurcation diagram for the tent map is in Fig. 2.14. It has a uniform (constant) probability distribution ($P(X) = 1$) for $A = 2$, and nearly all orbits are unbounded for $A > 2$.

A special numerical difficulty arises for the tent map with $A = 2$. Since

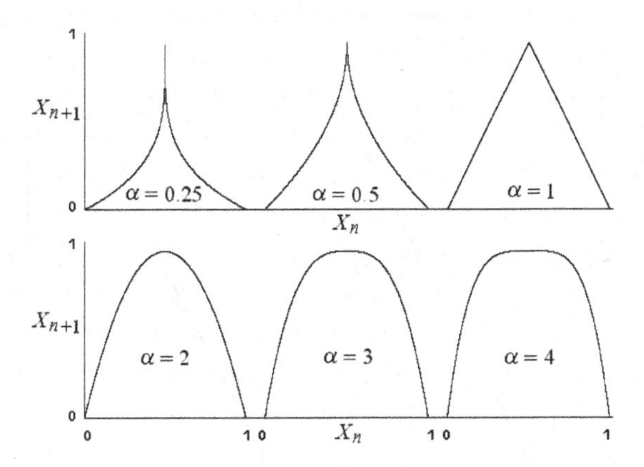

Fig. 2.15 General symmetric maps with $A = 1$ for various α.

each iterate is either exactly twice the previous iterate or one minus twice the previous iterate, all initial conditions that are specified by a finite number of binary digits[10] (bits) are eventually fixed points, and the iterates settle to zero after a number of iterations equal to the number of bits in the initial condition. You can overcome this difficulty by using a value of A that is slightly less than 2.0 ($A = 1.9999999$ in single precision). The fact that the map involves bit shifts is a simple proof that it is chaotic for most initial conditions since they contain an infinite number of arbitrary bits (Devaney 1989).

2.5.3 General symmetric map

The logistic map and the tent map are special cases of a more general symmetric map given by

$$X_{n+1} = A(1 - |2X_n - 1|^\alpha) \tag{2.16}$$

The case $\alpha = 1$ is the tent map, and $\alpha = 2$ is the logistic map, albeit with A in the range $0 < A < 1$. The parameter α is a measure of the smoothness[11] of the map, and typical examples are shown in Fig. 2.15. For each case, the maximum occurs at $X_n = 0.5$. An *extremum* (maximum or minimum) of a map is called a *singular point*. For $\alpha < 1$, the derivative f' is discontinuous at the singular point. For $\alpha \geq 1$, the derivative is continuous, and the function is smooth. For $\alpha > 2$, the second derivative f'' is zero, and we say the critical point is *degenerate*. Values of α less than 0.5 do not have chaotic solutions because the fixed point at $X = 0$ is stable for $0 < A < 1$.

[10] Binary numbers were reportedly invented by the German mathematician Gottfried Leibniz (1646–1716), co-inventor with Newton of calculus, while waiting to see the Pope to discuss reunification of Christianity.

[11] A function f is C^k-smooth if f and all its derivatives up to order k are continuous.

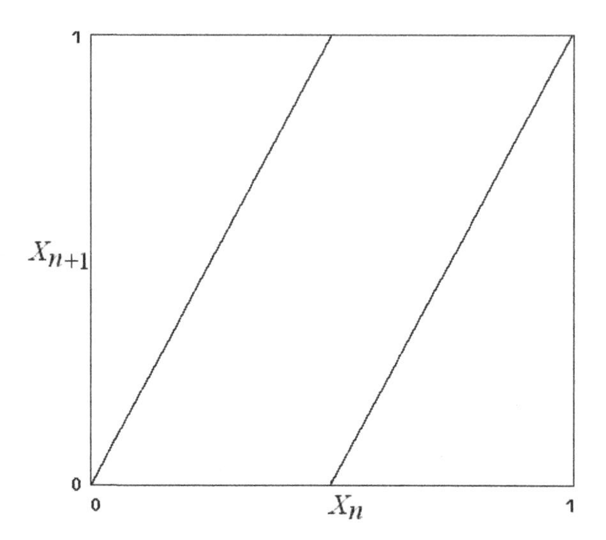

Fig. 2.16 Binary shift map.

The case with $\alpha = 0.5$ is called the *cusp map* and has a particularly simple probability distribution function given by $P(X) = (1 - X)/2$. For values of α greater than about three, it is difficult to study the map numerically because a wide range of values near $X = 0.5$ are rounded by the computer to exactly $X = 0.5$ and thus are eventually fixed points. You can counter this problem by avoiding integer values of α and limiting A to slightly less than one.

2.5.4 Binary shift map

Closely related to the tent map is the *binary shift map* (also called the *Bernoulli shift map*)[12]

$$X_{n+1} = 2X_n \pmod 1 \tag{2.17}$$

shown in Fig. 2.16. The 'mod 1' (*modulo*) function takes the remainder after dividing the quantity on the left ($2X_n$) by the quantity on the right (1). Hence the shift map consists of doubling X_n and then removing any digit to the left of the decimal point. In binary notation, multiplication by two is equivalent to shifting all the bits one place to the left, and hence the name. Note that in some computer languages the mod function works properly only for integers (1.1 mod 1 gives 0, not 0.1).

In the rubber-band analogy, the map is stretched, cut, and reattached with each iteration. Alternately, consider the interval $0 < X < 1$ as the circumference of a circle, where $2\pi X$ is the angle in radians. When the angle

[12]The Bernoulli shift map was named after the seventeenth-century Swiss brother mathematicians Jakob (1654–1705) and Johann (1667–1748) Bernoulli.

reaches 2π, it is relabeled as zero, so that no cutting and reattachment is needed.

The binary shift map is chaotic for almost all initial conditions because the stretching by a factor of two causes sensitive dependence on initial conditions. However, there are infinitely many initial conditions that do not give chaos. In fact, any rational value of X_0 is either an eventually fixed point or an eventually periodic point. Rational values that can be represented by a finite number of bits (such as $1/2^n$ for any finite integer n) iterate to zero after n iterations as all the nonzero bits are shifted left one-by-one and discarded until only zeros remain. Rational values that can be represented by a repeating pattern of bits (such as $1/3^n$ for any finite integer n) iterate to a periodic orbit. An irrational initial value such as $\sqrt{2}-1$ has infinitely many bits in its representation, and hence its iterates never repeat. Not only are almost all numbers irrational, but the bits (or digits) of almost all irrational numbers are random.

A digital computer always approximates irrational numbers with a finite number of bits, and so the result is only transiently chaotic for a number of iterations approximately equal to the number of bits in the mantissa of its floating-point representation (23 in single precision and 52 in double precision). In practice, you can circumvent this problem by replacing the factor of 2 with 1.9999999 in single precision, but even then the orbit will eventually repeat.

The binary shift map illustrates a profound connection between chaos and randomness. If you are given a random sequence of zeros and ones, you cannot tell whether they came from a sequence of coin flips or from the leading bit of a sequence generated by the shift map with an irrational initial condition. This process is an example of *symbolic dynamics* (Metropolis *et al.* 1973). Note that the probability distribution for the shift map, like the tent map with $A = 2$, is constant with $P(X) = 1$.

2.6 Computer random-number generators

You might think the shift map could be used to generate a uniform sequence of *pseudo-random* numbers in the range $0 < X < 1$, but it has a fatal flaw as a computer random-number generator. Each iterate is related in a simple way to its predecessor, and hence successive iterates can hardly be considered random. One way around this problem is to iterate many times to get the next value. If the values are stored as 64-bit numbers, then 64 iterations brings in a completely different string of bits since each iteration destroys one bit of information. Figure 2.17 shows that even six iterations of the binary shift map gives values that seem nearly random at the resolution of the graph.

However, this method also has drawbacks. Many mathematical operations are required for each random number generated, and the calculation has to be done using relatively slow floating-point operations to avoid running out of bits. Both of these problems are overcome by using a *linear*

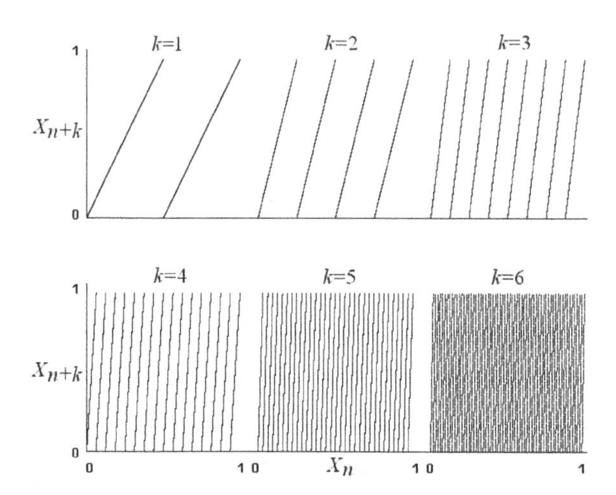

Fig. 2.17 First six iterates of the binary shift map.

congruential generator

$$X_{n+1} = AX_n + B \pmod{C} \tag{2.18}$$

with integer values of A, B, C, and X. This simple map will produce at most C values of X in the range $0 \leq X < C$ and will thus eventually repeat. If you want values in the range $0 \leq r < 1$, you need only take $r = X_n/C$.

The tricky part of applying this method is finding optimal values of A, B, and C that use all C of the values so that the period is as long as possible before repeating (which also guarantees that the probability distribution is uniform after a sufficient number of iterations), making C as large as possible without causing an overflow, and satisfying certain spectral tests for randomness. Knuth (1997) lists a number of 'good' choices such as $A = 7141, B = 54773$, and $C = 259200$, which does not overflow with 32-bit arithmetic.

With these values, the numbers repeat after about a quarter million iterations. Even worse is that if you generate sequences of k such numbers and plot them in a k-dimensional space (a hypercube with sides of unit length), they will lie on hyperplanes with dimension $k - 1$ in this space, and there are at most $C^{1/k}$ such planes. Thus if you are trying to sample randomly a three-dimensional space with triplets of numbers produced this way, your space will contain only $259200^{1/3} \simeq 64$ planes, which may be inadequate for your purpose.

One way to correct these deficiencies is to shuffle the numbers (Press *et al.* 1992). Suppose you generate an ordered list of 100 iterates of eqn (2.18). Then when you want a random number, choose one from the list randomly using eqn (2.18) and replace it with a new random number. This double randomization effectively removes the sequential correlation and makes the period extremely long. However, the number of distinct values is still limited to C.

If you use two random-number generators with different periodicities (values of C), one to provide the high-order bits and the other the low-order bits, then there are C_1C_2 distinct values, and the sequence is sufficiently random for most purposes. Alternately, add (or subtract) the output of two random-number generators with different periodicities (L'Ecuyer 1988). When combined with shuffling, the resulting numbers are so indistinguishable from randomness that Press *et al.* (1992) offer $1000 to the first person who can convince them otherwise.

You might wonder whether there is any true randomness in nature. This question is surprisingly profound, and the simple answer is that no one knows for sure. Physicists believe that atomic processes governed by the laws of quantum mechanics are inherently probabilistic and random.[13] Random number generators based on such processes have been built but often perform poorly compared with deterministic pseudo-random number generators (Hayes 2001a). One scheme samples the thermal voltage fluctuations across a resistor or other electronic device at even time intervals and returns a zero if the voltage is negative and a one if it is positive. Another uses the pulses from radioactive decay to sample randomly the polarity of a high-frequency square wave. Some Intel Pentium processors have hardware random-number generators built in. You can never prove that a string of numbers is random, but only that it is not random by finding some pattern in the numbers, and with so many possible patterns, testing them all is a hopeless undertaking.

2.7 Exercises

Exercise 2.1 Show that the quadratic map in eqn (2.4) reduces to the logistic map in eqn (2.3) for the appropriate transformation of variables.

Exercise 2.2 With $A = 2.8$ and $X_0 = 0.1$, calculate the first ten iterates of the logistic map, and show that they oscillate about and converge on the solution $X^* = 1 - 1/A$.

Exercise 2.3 By considering a nearby orbit, show that a fixed point is unstable if the local slope of the parabola (see Figs 2.1–2.3) exceeds one in absolute value. How does the behavior differ for positive and negative slopes?

Exercise 2.4 Derive the quartic eqn (2.5) and show that for $A > 3$ it has roots given by $X = 0$, $X = 1 - 1/A$, and the values given in eqn (2.7).

[13] Albert Einstein (1879–1955) never fully embraced the nondeterministic quantum theory and wrote to the Danish physicist and Nobel laureate (1922) Niels Bohr (1885–1962) in the late 1920s that 'God doesn't play dice [with the Universe]' to which Bohr retorted 'Quit telling God what to do!' Bohr also said that 'if you're not confused by quantum mechanics then you haven't understood it.'

Exercise 2.5 Derive eqn (2.8) for the onset of a stable period-4 solution of the logistic equation, and show that its solution is $A = 3.44948\ldots$.

Exercise 2.6 (difficult) Show that the onset of the period-3 window in the logistic map occurs at $A = 1 + \sqrt{8}$.

Exercise 2.7 Show that the probability distribution function for the logistic map with $A = 4$ given by eqn (2.11) is consistent with the Frobenius–Perron equation (2.12).

Exercise 2.8 Show that applying the *logit transform*

$$f(X) = \ln\left(\frac{X}{1 - X}\right) \tag{2.19}$$

to the iterates of the logistic map with $A = 4$ gives a probability distribution function

$$P(X) = \frac{1}{\pi(\mathrm{e}^{X/2} + \mathrm{e}^{-X/2})} \tag{2.20}$$

which is very close to a *Gaussian distribution*

$$P(X) = \frac{1}{\sqrt{2\pi}}\mathrm{e}^{-X^2/2} \tag{2.21}$$

Exercise 2.9 Show that the logistic map with $A = 4.5$ has a Cantor set of initial conditions that remain bounded upon repeated iteration.

Exercise 2.10 Show that the shape of the sine map is approximately parabolic near its peak.

Exercise 2.11 (difficult) Calculate the value of A at the onset of the period-2 region for the sine map in eqn (2.14).

Exercise 2.12 Use the Frobenius–Perron equation (2.12) to show that the probability distribution function for the tent map in eqn (2.15) is uniform (constant) for $A = 2$.

Exercise 2.13 Show that eqn (2.16) is smooth at the singular point for $\alpha \geq 2$ and not smooth for $\alpha < 2$.

Exercise 2.14 Calculate the fixed points X^* for the general symmetric map in eqn (2.16) as a function of A and α. Consider only the case with positive X^*, A, and α.

Exercise 2.15 Show that the general symmetric map in eqn (2.16) cannot have chaotic solutions for $\alpha < 0.5$.

Exercise 2.16 Calculate the value of A at the onset of the period-2 region for the general symmetric map in eqn (2.16) as a function of α.

Exercise 2.17 Show that the logistic map with $A = 4$ is equivalent to the binary shift map by substituting $X = (1 - \cos \pi\theta)/2$ into eqn (2.3) and finding a relation between θ_{n+1} and θ_n.

Exercise 2.18 Calculate the first bifurcation of the map $X_{n+1} = A(1 - 2|0.5 - \sqrt{X_n}|)^2$ suggested by Grossmann and Thomae (1977).

Exercise 2.19 Calculate the form of the probability distribution function $P(X)$ near $X = 0$ for the Grossmann–Thomae map in the previous exercise.

Exercise 2.20 Calculate a dozen iterates of the map $X_{n+1} = 106X_n + 1283 \pmod{6075}$ with $X_0 = 0$, and observe that the iterates are seemingly random integers in the range $0 \le X < 6074$.

Exercise 2.21 Show that the *process equation*, $X_{n+1} = X_n + A \sin X_n$ (Kauffman and Sabelli 1998), has an infinite number of fixed points, and calculate their location and stability.

2.8 Computer project: Bifurcation diagrams

This project is a continuation of the one in the previous chapter in which you will further explore the properties of the logistic equation. You will observe the common period-doubling route to chaos and will display your results graphically. The ability to produce and print graphs will be essential in all future projects.

1. For the logistic equation, produce a bifurcation diagram like Fig. 2.4. You can start with (almost) any initial condition in the range $0 < X_0 < 1$, but be sure to discard a sufficient number of the initial iterates before plotting the remaining ones.
2. Expand this plot in the range $3.0 < A < 3.57$, where successive period doublings occur, and use your plot to estimate the Feigenbaum number in eqn (2.9).
3. Expand your plot in the vicinity of the period-3 window at about $A = 3.84$, and show that it also exhibits a sequence of period doublings prior to the onset of chaos.
4. Stable orbits of all periods can be found somewhere in the range $3.0 \le A < 4.0$. Find a value of A at which the orbit has period 5.
5. (optional) See how many of the higher odd-integer stable periodic orbits (7, 9, 11, ...) you can find. (Hint: The periodic windows become successively more narrow as the period increases, but there are rules that you might discover and use to locate them.)

3 Nonchaotic multi-dimensional flows

The one-dimensional maps in the previous chapter illustrate many features of chaotic dynamical systems. However, they do not adequately model most natural processes. In this chapter we will relax two of their most limiting characteristics. We will consider models in which time advances continuously rather than discretely and models with more than one variable (dimensions higher than one). Most of the examples will be two-dimensional, since they are most easily visualized, but the extension to higher dimensions is straightforward. However, two-dimensional continuous-time systems cannot exhibit chaos, and so we will defer further discussion of chaos to the following chapters. In general, continuous-time systems are dynamically simpler than their discrete-time counterparts for a given dimension. Since continuous-time systems are described by differential equations, calculus rather than algebra is needed to describe them. They are the language in which the physical laws that govern the Universe are written. If you are unfamiliar with differential equations, you might consult one of the standard texts such as Boyce and DiPrima (1992).

3.1 Exponential growth in continuous time

Imagine a colony of ants that are continually reproducing and dying. Let $x(t)$ be the number of living ants at time t. We will assume the number of ants is sufficiently large that both x and t are continuous variables (they can have any value within some range). Note that we have switched notation from the previous chapter. We will use lower-case letters for continuous dynamical variables and upper-case letters for discrete variables. The time derivative dx/dt is the number of births minus the number of deaths per unit time. In the absence of external influences, we expect both of these rates to be proportional to the population of ants, leading to an equation

$$\frac{dx}{dt} = ax \tag{3.1}$$

where a is the control parameter (births minus deaths per ant per unit time or *fecundity*), analogous to A in the previous chapter. Equation (3.1) is sometimes called the *Malthus equation* after the British clergyman and professor[1] who wrote an essay (Malthus 1798) in which he argued that a population grows exponentially and thus will eventually outstrip its food supply, which grows only linearly. This essay supposedly inspired Charles Darwin[2] to develop his theory of evolution (Darwin 1859).

Equation (3.1) is a *first-order* (since the highest derivative is the first), *explicit* (since dx/dt has no factors multiplying it), *linear* (since dx/dt is proportional to x), *ordinary* (since there are no partial derivatives), *differential* (since it involves the derivative or differential) equation. In solving a differential equation, we seek a function $x(t)$ that satisfies the equation for all t. This task can be difficult and sometimes impossible. However, eqn (3.1) is easily solved by multiplying both sides of the equation by dt/x and integrating to get $\ln(x/x_0) = at$ or $x = x_0 e^{at}$, where x_0 is the initial condition (the value of x at $t = 0$). The point $x = 0$ is an *equilibrium point* (also called a *critical point* or *stationary point*) for the equation since $dx/dt = 0$ and $x = x_0$ for all t. It is analogous to the fixed point of an iterated map. In fact, many authors use the terms 'equilibrium point' and 'fixed point' interchangeably.

Equation (3.1) predicts exponential growth for $a > 0$ or exponential decay for $a < 0$. In one case the ant population increases without limit, and in the other, the population dies. The value $a = 0$ is a bifurcation point at which the equilibrium at $x = 0$ changes from stable (attracting) to unstable (repelling). The quantity a is the growth (or decay) rate and has units of inverse time. Unlike the discrete-time exponential growth in eqn (2.2), the value of x changes continuously rather than in jumps. We will refer to its motion along the x axis as a *trajectory*, analogous to the *orbit* of a discrete map. Many authors do not make such a distinction. A continuous-time differential equation such as eqn (3.1) represents a *flow* or *vector field*.[3] A flow can be considered as a special map in which successive iterates are very close in space and time, and a map can be considered as a flow sampled at discrete time intervals. Table 3.1 summarizes the differences between maps and flows.

Note that two nearby initial conditions separate at an exponential rate for $a > 0$. In fact, a is the rate of separation. We do not consider the dynamics to be chaotic, however, because the motion is unbounded. You can imagine all initial conditions as spread out along the x axis. As time evolves, the axis is stretched at a constant rate like a rubber band, and

[1]Thomas Robert Malthus (1766–1834) was a controversial but respected priest and scientist and England's first academic economist (Winch 1987).

[2]Charles Robert Darwin (1809–1882) was an English naturalist whose theories of evolution influenced biology similarly to the way Kepler influenced astronomy, Newton influenced physics, and Poincaré influenced mathematics.

[3]A vector is a set of quantities that go together. For example, the weather at a given location could be represented by a time-dependent vector whose components are temperature, humidity, wind speed and direction, barometric pressure, and so forth. It is a *field* in the sense that every location has its own such vector.

Table 3.1 Comparison of maps and flows.

Maps	Flows
Discrete time	*Continuous* time
Variables change *abruptly*	Variables change *smoothly*
Described by *algebraic* equations	Described by *differential* equations
Complicated 1-D dynamics	*Simple* 1-D dynamics
$X_{n+1} = f(X_n)$	$dx/dt = f(x)$
Capital letters	Lower case letters
Example: $X_{n+1} = AX_n$	Example: $dx/dt = ax$
Solution: $X_{n+1} = A^n X_0$	Solution: $x = x_0 e^{at}$
Growth for $A > 1$	Growth for $a > 0$
Decay for $A < 1$	Decay for $a < 0$
Solution is called an *orbit*	Solution is called a *trajectory*
$n \to t \Rightarrow A \to e^a$	$t \to n \Rightarrow a \to \ln(A)$

all points get progressively farther apart, with the more distant points receding more rapidly from the origin at $x = 0$. In addition to stretching, chaos requires a folding or bending to keep the trajectory bounded, just as with the one-dimensional maps in the previous chapter.

3.2 Logistic differential equation

Just as with maps, flows cannot grow exponentially for ever. In real situations, some nonlinearity intervenes and slows the growth. The simplest such nonlinearity is the same one that led to the logistic map, giving the *logistic differential equation*

$$\frac{dx}{dt} = ax(1 - x) \tag{3.2}$$

Figure 3.1 shows a plot of the *logistic function*, which is the righthand side of eqn (3.2). This equation was derived by Verhulst[4] (1845) to model population growth and is called the *Verhulst equation*. It is a rare case (and nearly the last in this book) where you can solve a nonlinear differential equation analytically, in this case using separation of variables, with the result

$$x = [1 + (1/x_0 - 1)e^{-at}]^{-1} \tag{3.3}$$

[4]Pierre François Verhulst (1804–1849) was a Belgian mathematician who worked on the theory of numbers and intended to publish the complete works of Euler before he became interested in social statistics, which consumed most of his career. For a good review of the history of the Verhulst equation, see Kingsland (1985).

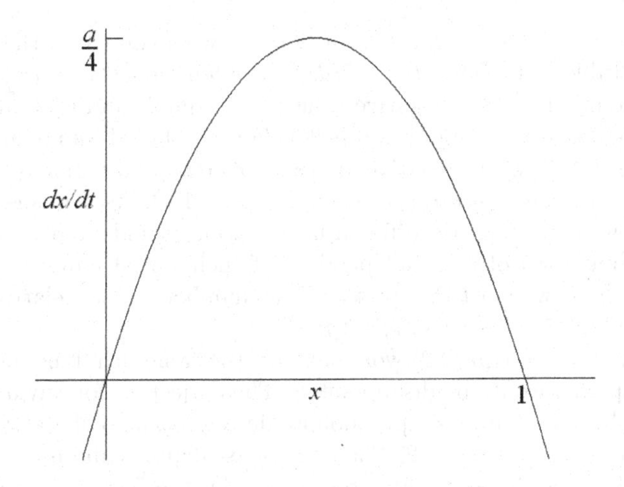

Fig. 3.1 The logistic differential equation function.

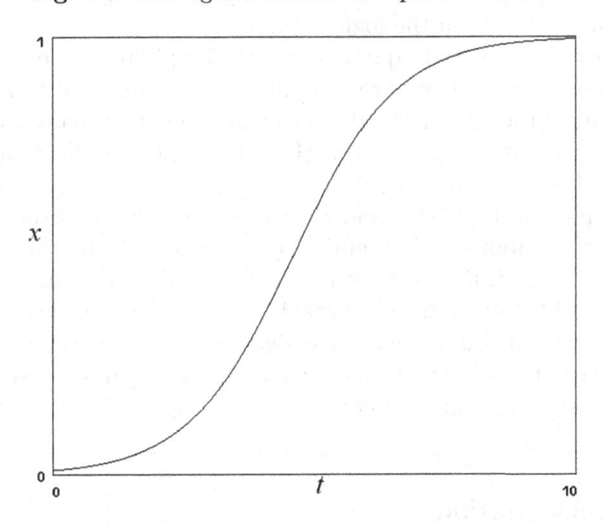

Fig. 3.2 Solution of the logistic differential equation with $x_0 = 0.01$ and $a = 1$.

The solution starts at $x = x_0$ when $t = 0$ and approaches $x = 1$ for large t as in Fig. 3.2. Such S-shaped curves are called *sigmoid curves* from the Greek sigma ($\sum$) or *logistic curves.*[5]

Even without knowing the solution, we can deduce many of its properties by studying the equilibrium points. Setting $dx/dt = 0$ in eqn (3.2) predicts equilibria at $x^* = 0$ and $x^* = 1$. The equilibrium at $x^* = 0$ is unstable,

[5]The name 'logistic' ('logistique' in French) was coined by Verhulst (1845), but without explanation. It comes from the Greek 'logistikos' (computational) and probably derives from the term 'logarithm' with which it was used synonymously in the 1700s. Since computation is needed to predict the supplies an army requires, 'logistics' has come to be used also for the movement and supply of troops.

and the one at $x^* = 1$ is stable for $a > 0$. As with eqn (3.1), the equilibrium is unstable if the slope $f' = df/dx$ of the function $f(x) = ax(1-x)$ is positive and stable if f' is negative. The stable equilibrium is an attractor, and the unstable equilibrium is a repellor. For $a > 0$ the basin of attraction of $x^* = 1$ is $x > 0$, while negative values of x attract to minus infinity. At $a = 0$, a *transcritical bifurcation* occurs (see §7.1.2). As a model of population growth, the logistic differential equation would predict a unique eventual population of ants independent of their initial number. As with the logistic *map*, we must interpret x as the number of ants relative to this maximum number (the *carrying capacity*).

If you had hoped eqn (3.2) would exhibit the same rich dynamics as the logistic map, you would be disappointed. The trajectory for any a is never more complicated than a simple monotonic approach to the attractor as given by eqn (3.3) and Fig. 3.2. There are no oscillations and no chaos. The reason is that the dynamics are constrained by the fact that $x(t)$ has to change continuously. The trajectory cannot jump abruptly from one value to another as could X_n in the logistic map.

The logistic differential equation is an example of a one-dimensional nonlinear flow with two equilibria. Nonlinear flows can have many equilibria. For example the system $dx/dt = \sin x$ has infinitely many equilibrium points at $x^* = n\pi$ with n an integer. Half of those (with odd n) are stable, and half (with even n) are unstable.

It is also possible for both f and f' to be zero at the equilibrium point. In such a case, the stability is determined by the sign of the first nonvanishing higher derivative of f. If that derivative is even (such as d^2f/dx^2), the point is always unstable, attracting on one side but repelling on the other. If the derivative is odd, it follows the same sign rules as f'. In both cases, the growth or decay is very slow in the vicinity of the equilibrium since $f' = 0$ there. Such cases are sometimes called *nonlinearly stable* (or unstable).

3.3 Circular motion

Now imagine a more complicated population dynamic consisting of two species, a predator and a prey, such as foxes and rabbits (Eigen and Schuster 1979). Suppose there is an equilibrium in which the fox and rabbit populations are constant in time. Now introduce an excess of rabbits $y = y_0$ at $t = 0$. The foxes then have more food, and their number begins to grow. As their excess x increases, they begin to kill the rabbits so that y decreases. We can represent this dynamic by a pair of linear differential equations

$$\frac{dx}{dt} = y \tag{3.4}$$

$$\frac{dy}{dt} = -x \tag{3.5}$$

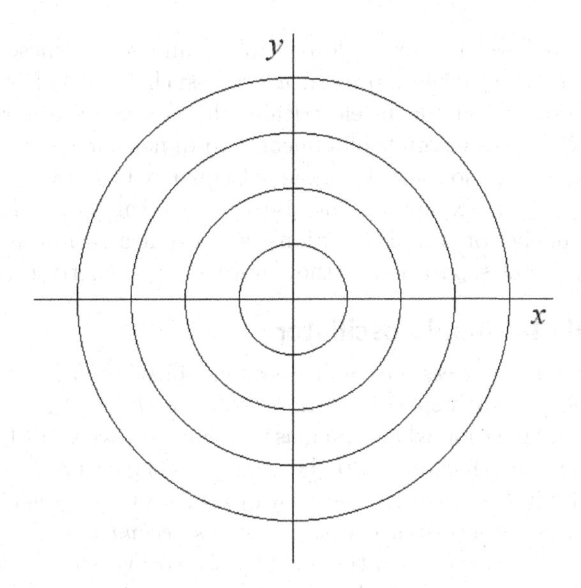

Fig. 3.3 Linear two-dimensional flow about a center.

In general, there would be a multiplicative constant on the righthand side of each equation, but you can scale the variables x, y, and t to make these constants unity.

Substituting y from eqn (3.4) into eqn (3.5) gives a single, *second-order*, linear, ordinary differential equation (ODE)

$$\frac{d^2x}{dt^2} + x = 0 \tag{3.6}$$

whose solution is some linear combination of $\sin t$ and $\cos t$. If x_0 is zero, only the sine term is present. The solution then is $x = y_0 \sin t$ and $y = y_0 \cos t$. This is the equation for a circle of radius y_0, as you can verify from $r = \sqrt{x^2 + y^2} = y_0\sqrt{\sin^2 t + \cos^2 t} = y_0$. The populations of foxes and rabbits undergo sinusoidal oscillations 90° out of phase with one another.

Note that there is no attraction to or repulsion from the equilibrium at $x^* = y^* = 0$. It is a *neutrally stable* (or *indifferent*) point and is called a *center* for obvious reasons. All trajectories circulate clockwise about it with radii determined by their initial conditions as in Fig. 3.3. Such a diagram is called a *state-space plot* since every point in the xy plane represents a possible state of the system (population of foxes and rabbits). This system is periodic since the trajectory repeats after a time $T = 2\pi = 6.283185307\ldots$, called the *period*, whose reciprocal is the *frequency*, $f = 1/T = 1/2\pi = 0.159154943\ldots$. If t has units of years, then the period is in years, and the frequency is in cycles per year. Any closed trajectory in a plane must enclose an equilibrium point.

Although we used the example of rabbits and foxes, these equations could model anything that moves in a circle such as a ball on a string, a planet orbiting a star, or an electrically charged particle gyrating in a magnetic field. It is an example of a linear, two-dimensional flow, and every point in the xy plane moves with the same frequency in a circular direction, like a swirling fluid, explaining the use of the term 'flow'. There is no sensitive dependence on initial conditions since two nearby initial conditions maintain a constant separation as they move along their trajectories.

3.4 Simple harmonic oscillator

Now consider a frictionless mechanical system consisting of a mass m suspended by a spring with *spring constant* (stiffness) k, as in Fig. 3.4. There is an equilibrium position where the mass remains at rest. Call this position $x = 0$, and imagine that at $t = 0$ the spring is relaxed ($x_0 = 0$) and the mass is moving in the $+x$-direction (downward, say) with velocity $v = v_0$. The mass moves in accordance with *Newton's second law*, $F = mdv/dt$, where F is the force exerted on the mass by the spring, which according to *Hooke's law* is $F = -kx$. This dynamical system is described by a pair of linear, first-order, ordinary differential equations

$$\frac{dx}{dt} = v \tag{3.7}$$

$$\frac{dv}{dt} = -\omega^2 x \tag{3.8}$$

where $\omega^2 = k/m$. These equations resemble eqns (3.4) and (3.5). Indeed, their solution is $x = (v_0/\omega)\sin\omega t$ and $v = v_0\cos\omega t$, and the trajectory in the vx plane is an ellipse with a ratio of v-axis to x-axis length of ω surrounding the center at $x^* = v^* = 0$. The motion is periodic with

Fig. 3.4 A simple harmonic oscillator.

period $T = 2\pi/\omega$ and frequency $f = \omega/2\pi$. This is an example of a *simple harmonic oscillator*. It could describe a wide class of systems that oscillate sinusoidally, such as a gently swinging pendulum or an electrical circuit containing a capacitor and inductor.

Even though the motion is one-dimensional (along the x axis), the dynamical system is two-dimensional since two first-order differential equations are required to describe it. The reason is that Newton's second law involves the acceleration, which is the second derivative of the displacement. In general, a mechanical system with n *degrees of freedom* has $2n$ variables, a position and velocity corresponding to each degree of freedom. This $2n$-dimensional state space is usually called *phase space* (Hirsch and Smale 1974), although the terms are unfortunately often used interchangeably. More typically, *momentum* (mass $\times$ velocity) is used instead of velocity as the second phase-space variable, since position and momentum play a special role in Hamiltonian dynamics (see Chapter 8) and quantum machanics. In particular, the *Heisenberg uncertainty principle*[6] states that no measurement can simultaneously determine the position x and momentum p such that the product of the uncertainties $\Delta x \Delta p$ exceeds a value on the order of *Planck's constant*[7] $h \simeq 6.63 \times 10^{-34}$ kg-m^2/s. Although the variables are more abstract, the trajectory of a harmonic oscillator in phase space, picturesquely called the *phase portrait* (or *phase-space portrait*), is a set of simple nested closed loops, each corresponding to a different initial condition, just as for the circular motion in the previous section. Only a representative set of trajectories are shown, since the portrait would otherwise be totally black.

This phase portrait can be understood in terms of energy conservation. The total energy consists of kinetic energy $(mv^2/2)$ and potential energy $(kx^2/2)$, whose sum is constant in the absence of friction. In this example, the initial position x_0 is zero, and the total energy must equal the initial kinetic energy $(mv_0^2/2)$. In general, the trajectory must satisfy

$$v^2 + \omega^2 x^2 = c^2 \tag{3.9}$$

where c^2 is a constant equal to $v_0^2 + \omega^2 x_0^2$. Equation (3.9) describes an ellipse (or a circle if time is rescaled so that $\omega = 1$).

3.5 Driven harmonic oscillator

Now consider a case where the frictionless harmonic oscillator in the previous section is subject to an additional periodic *driving force* $F_0 \sin \Omega t$. The equation of motion is then

[6]The uncertainty principle was discovered in 1927 by Werner Karl Heisenberg (1901–1976), a German theoretical physicist and Nobel laureate (1932) whose leadership role in the unsuccessful German nuclear weapons project during the Second World War is subject to historical debate.

[7]Planck's constant was named after Max Karl Ernst Ludwig Planck (1858–1947), a German physicist and Nobel laureate (1918) whose eldest son was killed in the First World War and whose other son was executed for plotting to assassinate Hitler.

$$\frac{d^2x}{dt^2} + \omega^2 x = A \sin \Omega t \tag{3.10}$$

where $A = F_0/m$ is an acceleration proportional to the maximum force F_0. This is a second-order, linear, *nonautonomous* (since t appears explicitly on the righthand side), ODE. Such an equation could be used to model an ecological or financial system with an external, seasonal influence. Think of the population of mosquitoes or the sale of air conditioners.

The solution consists of two parts, a *homogeneous* or *free* solution (obtained by setting the righthand side to zero), plus a *particular solution* whose frequency is Ω. The homogeneous solution was shown in the previous section to be $x = (v_0/\omega)\sin \omega t$. For the particular solution, assume $x = x_1 \sin \Omega t + x_2 \cos \Omega t$ and substitute into eqn (3.10) to get $x_1 = A/(\omega^2 - \Omega^2)$ and $x_2 = 0$. The complete solution is then

$$x = (v_0/\omega)\sin \omega t + \frac{A}{\omega^2 - \Omega^2}\sin \Omega t \tag{3.11}$$

Note several features of the solution. Firstly, there is a *resonance* for $\Omega = \pm\omega$ where the oscillation amplitude is infinite. Secondly, there is no static equilibrium unless $A = 0$. Thirdly, the solution has two frequencies (ω and Ω), and they need not be related. Hence the solution is not simply periodic. If ω/Ω is irrational, we call the frequencies *incommensurate* (or *incommensurable*) and the solution *quasiperiodic*. However, this is not an example of chaos because nearby initial conditions do not separate exponentially fast. Hence, absence of periodicity is a necessary but not a sufficient condition for chaos.

It is instructive to examine the solution of eqn (3.10) in phase space. For this purpose, rewrite eqn (3.10) in terms of x and v

$$\frac{dx}{dt} = v \tag{3.12}$$

$$\frac{dv}{dt} = -\omega^2 x + A \sin \Omega t \tag{3.13}$$

We can always convert a nonautonomous system into an autonomous one by defining a new variable, in this case $\phi \equiv \Omega t$. Then the two-dimensional system becomes three-dimensional

$$\frac{dx}{dt} = v \tag{3.14}$$

$$\frac{dv}{dt} = -\omega^2 x + A \sin \phi \tag{3.15}$$

$$\frac{d\phi}{dt} = \Omega \tag{3.16}$$

where x, v, and ϕ are the new state-space variables. This trick has the virtue that each point in the new three-dimensional space has a unique

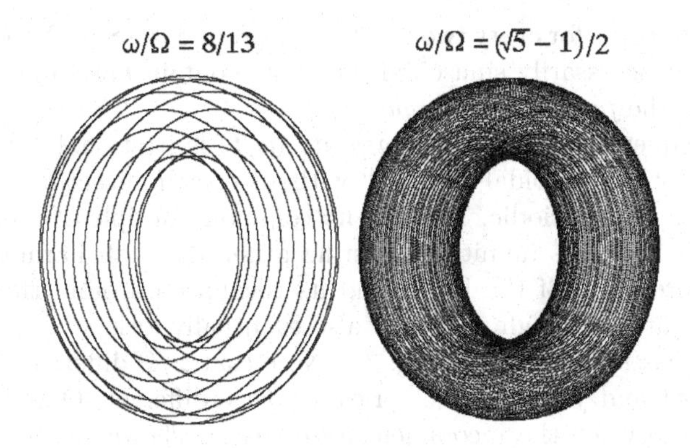

Fig. 3.5 Invariant torus for commensurate and incommensurate frequencies.

direction of flow associated with it for all times. Thus the dynamics can be represented in a purely geometric way. The penalty is that the *two-dimensional linear* system becomes a *three-dimensional nonlinear* system because of the $\sin\phi$ term.

Note that ϕ appears only as the argument of $\sin\phi$, and so it is periodic with period 2π. Thus the third dimension wraps onto itself, as in Fig. 3.5. The trajectory lies on a *torus* (the surface of a doughnut or inner tube). A cross-section of the torus (a slice through the doughnut the short way) shows the vx plane at a particular value of ϕ. A different cross-section (the way you would slice a bagel[8]) shows the $x\phi$ plane at a particular value of v. There are now two periods for the motion: $T_1 = 2\pi/\omega$ is the time required to transit the torus the short way, and $T_2 = 2\pi/\Omega$ is the time required to transit the torus the long way. The ratio ω/Ω is the number of times the trajectory transits the short way around for each time the long way and is called the *winding number* or *rotation number*. The angle through which it moves the short way for each transit the long way is called the *rotational transform*, which is thus 2π radians times the winding number.

If the frequencies are *commensurate* (ω/Ω is rational), then the trajectory eventually closes on itself, and the motion is simply periodic with a period equal to the smallest common multiple of the two periods. This condition is variously called *frequency locking*, *phase locking*, *mode locking*, *frequency pulling*, *synchronization*, *entrainment*, or *enslavement*. The Moon always shows the same face to the Earth because tidal forces[9] caused its rotation to lock to its revolution around the Earth, although it wiggles (*librates*) a bit. Eventually the Earth will show the same face to the Moon for the same reason. Mercury is locked into a 3:2 resonance with the Sun,

[8]Note that neither slice need be circular; they only must be closed loops.

[9]Tidal forces are also causing the Moon to lose orbital speed, making it recede from the Earth at a rate of about 4 cm per century, with the added gravitational energy coming from the rotational energy of the Earth.

rotating three times for every two revolutions around the Sun. Note that the motion is not necessarily sinusoidal, but may contain *harmonics* (integer multiples) of the *fundamental frequency*.

If the frequencies are *incommensurate* (ω/Ω is irrational), the trajectory fills the whole toroidal surface[10] without ever intersecting itself, and the motion is quasiperiodic. There is no common multiple of the frequencies, and the period is infinite, consisting generally of all harmonics of ω and Ω. The rotation of the Earth and its revolution around the Sun are incommensurate, although they may also eventually lock.

The two cases are shown in Fig. 3.5. Note that ω/Ω differs by less than half a percent and yet the trajectories are very different. Only a portion of the trajectory for the incommensurate case is shown in the figure to prevent filling in the entire torus for ease of visibility. This torus is not an attractor, since different initial conditions produce different nested tori; it is called an *invariant torus*. A set is *invariant* if a trajectory that starts on it remains on it for all time. All attractors are invariant sets, but not the converse.

Note that the curve on the left of Fig. 3.5 is highly *knotted* (called a *torus knot*). You cannot straighten it out and make it lie in a flat plane without intersections unless you cut and reconnect it. The number of times the curve links itself is called the *helicity*.

The autonomous representation in eqns (3.14)–(3.16) is not unique. The $\sin \Omega t$ function could arise from an auxiliary harmonic oscillator

$$\frac{d^2x}{dt^2} + \omega^2 x = y \tag{3.17}$$

$$\frac{d^2y}{dt^2} + \Omega^2 y = 0 \tag{3.18}$$

The first equation is coupled to the second, but the second is not influenced by the first. The two oscillators are sometimes called *master–slave*. The parameter A is now superfluous since it and the drive phase can be controlled by the initial values of y and dy/dt. This trick amounts to including a model of the driving force as part of the dynamical system rather than treating it as an external influence. Although the equations are simpler since they are linear, the simplicity comes at the expense of two extra dimensions, y and dy/dt, giving a four-dimensional system.

Perhaps the algebraically simplest autonomous system with an invariant torus is (Nosé 1991, Hoover 1995)

[10]Mathematicians always take a torus to be a two-dimensional object, whereas physicists often use the term to mean the three-dimensional space enclosed by it. You would never get fat eating a mathematician's doughnuts! Furthermore, a torus need not resemble a doughnut, but must only be *topologically equivalent*. A coffee mug is a torus as long as the handle is not broken, and the human body is a torus except for the fact that the hole through its center contains food in various stages of digestion.

$$\frac{dx}{dt} = y \tag{3.19}$$

$$\frac{dy}{dt} = -x + yz \tag{3.20}$$

$$\frac{dz}{dt} = 1 - y^2 \tag{3.21}$$

Most initial conditions near the origin give trajectories that lie on a torus, although some are chaotic. Initial conditions far from the origin give unbounded trajectories.

3.6 Damped harmonic oscillator

The driven (or *forced*) harmonic oscillator in the previous section is not physically realistic because it ignores friction, which is present in all physical and most other oscillators. Even small friction will drastically alter the long-term solution. There are many ways to include friction (Pippard 1989), but they all have a similar effect, and so we assume a simple phenomenological[11] linear form where the friction force is proportional to the velocity, but in the opposite direction, giving

$$\frac{d^2x}{dt^2} + b\frac{dx}{dt} + \omega^2 x = A\sin\Omega t \tag{3.22}$$

This form of friction is appropriate for the viscous damping of a small object moving slowly through a fluid.

As before, this equation has a homogeneous solution and a particular solution. We begin by considering the homogeneous solution[12] obtained by setting the righthand side to zero

$$\frac{d^2x}{dt^2} + b\frac{dx}{dt} + \omega^2 x = 0 \tag{3.23}$$

Such a *damped harmonic oscillator* is important in its own right, since it is probably more common in nature than driven oscillators. The shock absorbers on your car are damped harmonic oscillators; otherwise, the car would continue bouncing forever after hitting a bump.

Equation (3.23) resembles eqn (3.1), whose solution is an exponential function of time. Thus we assume its solution is also of the form $x = x_0 e^{at}$ and substitute into eqn (3.23) to get the *auxiliary equation*

$$a^2 + ba + \omega^2 = 0 \tag{3.24}$$

with solution

[11]The term 'phenomenological' is used in physics to describe an equation chosen to illustrate a phenomenon without regard to its accuracy or derivation.

[12]The first published solution of this equation was by Leonhard Euler (1707–1783) in 1743, but the solution was apparently known to the brothers Daniel (1700–1782) and Johann (1710–1790) Bernoulli in 1739.

$$a = -\frac{b}{2} \pm \frac{1}{2}\sqrt{b^2 - 4\omega^2} \tag{3.25}$$

There are two solutions corresponding to the plus and minus signs, and the full solution is a linear combination of them (the principle of *superposition*) with coefficients determined by the initial conditions. An nth-order linear ordinary differential equation has n initial conditions and n solutions, although they might coincide. Note that the square root is a real number only if $b \geq 2\omega$, in which case the system is *overdamped.*

3.6.1 Overdamped case

For the overdamped case with $b \gg 2\omega$ and $x_0 = 0$, the solution is

$$x \simeq \frac{v_0}{b}(\mathrm{e}^{-\omega^2 t/b} - \mathrm{e}^{-bt}) \tag{3.26}$$

as you can verify by substituting into eqn (3.23). The variable x starts at zero, rises rapidly in a time $\sim (2/b)\ln(b/\omega)$ to a maximum of $\sim v_0/b$, and then decays slowly back to zero with a time constant of $\sim b/\omega^2$.

3.6.2 Critically damped case

The case $b = 2\omega$ is *critically damped* and has a solution

$$x = v_0 t \mathrm{e}^{-bt/2} \tag{3.27}$$

as you can verify by substituting into eqn (3.23). Its form is similar to the overdamped case except that $x(t)$ approaches zero as fast as possible. If $v_0/x_0 \geq -b/2$, then the value of x does not overshoot zero, but with $v_0/x_0 < -b/2$, it will overshoot only once. The variable x reaches a maximum of $2v_0/eb$ in a time $2/b$.

3.6.3 Underdamped case

For the *underdamped* case with $b \ll 2\omega$, the quantity under the square root in eqn (3.25) is negative, and a is a *complex number* (a number with a *real* and *imaginary* part).[13] If you are familiar with complex numbers, you should be able to use the formulas in §B.4 to show that the solution with $x_0 = 0$ is

$$x \simeq \frac{v_0}{\omega}\mathrm{e}^{-bt/2}\sin\omega t \tag{3.28}$$

Otherwise, substitute eqn (3.28) into eqn (3.23) and verify that it is a solution. The solution undergoes a sinusoidal oscillation much like the undamped oscillator in §3.4, but with an amplitude that decreases exponentially with a time constant of $2/b$.

[13]The mathematical term 'complex' is used in the sense of having multiple components, rather than in the sense of 'complicated.' Complex numbers were invented when mathematicians realized that equations such as $x^2 + 1 = 0$ have no solutions in the real numbers.

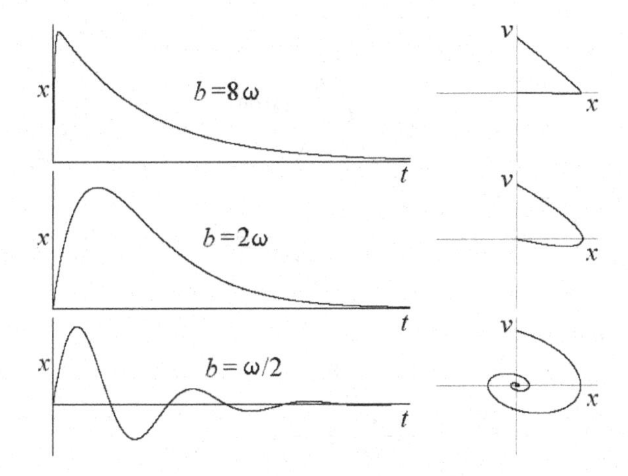

Fig. 3.6 Overdamped, critically damped, and underdamped behavior of a harmonic oscillator.

Note that the energy, which is proportional to the maximum of x^2, decays twice as fast as x, and hence its time constant is $1/b$. The *Q-factor* ('quality') of the oscillator is the number of radians of oscillations before the energy decays to $1/e$ of its initial value and is given by $Q = \omega/b$. In a mechanical system, the energy lost to friction is converted into heat. Systems that convert mechanical energy into heat are called *dissipative*. Systems that conserve mechanical energy are called *conservative*.

Figure 3.6 summarizes the behaviors of the damped harmonic oscillator. Note that in all cases, the origin $x = v = 0$ is an attractor, and all initial conditions in the vx plane are in its basin of attraction. Whereas we used to say 'the mass stops moving,' now in the pedantic jargon of dynamical systems theory we can instead say that 'all trajectories in the unbounded basin of attraction are attracted to the stable equilibrium point at the origin in phase space.' However, for $b < 0$ (*anti-damping*), the attractor becomes a repellor, and all orbits go to infinity. We will revisit this issue in the next chapter.

3.7 Driven damped harmonic oscillator

Now consider the *driven* damped harmonic oscillator of eqn (3.22). Since the homogeneous (or *transient*) solution decays to zero, we can ignore it if we are interested only in the long-term ($t \gg 1/b$) behavior of the system. Thus we calculate the particular (or *steady-state*) solution, which must be a sinusoidal function of the drive frequency Ω of the form $x = x_m \sin(\Omega t - \phi)$. Substitution into eqn (3.22) gives values for x_m and ϕ

$$x = \frac{A}{\sqrt{(\omega^2 - \Omega^2)^2 + b^2\Omega^2}} \sin(\Omega t - \phi) \tag{3.29}$$

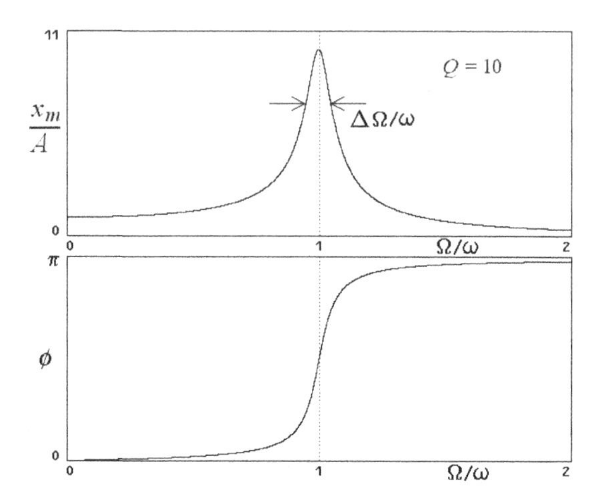

Fig. 3.7 Response function of a driven damped harmonic oscillator with $Q = 10$.

where

$$\phi = \tan^{-1} \frac{b\Omega}{\omega^2 - \Omega^2} \tag{3.30}$$

Note that the solution has resonances at $\Omega \simeq \pm\omega$ where the response function has its maximum amplitude of $x_m = A/b\Omega$ and phase shift of $\phi = \pi/2$ (90°). The actual maximum of x_m occurs at $\Omega = \sqrt{\omega^2 - b^2/2}$, but for the usual case of $b < \omega$ the difference is slight. The response x_m/A and the phase ϕ are plotted versus Ω in Fig. 3.7 for $Q = \omega/b = 10$. The Q is related to the width of the response function by $Q = \omega/\Delta\Omega$, where the width is determined at the points where the response is $1/\sqrt{2} = 0.7071067\ldots$ of its maximum, called *half-power points* since the power is proportional to x^2. In the limit of $Q \gg 1$, the power near the resonance at $\Omega \simeq \omega$ is

$$P(\Omega) \propto \frac{1}{(1 - \Omega/\omega)^2 + 1/4Q^2} \tag{3.31}$$

which is called the *Lorentzian*[14] line shape.

The time-dependent velocity is found by differentiating eqn (3.29) to obtain $v = -\Omega x_m \cos \Omega t$. Hence, $v^2/\Omega^2 + x^2 = x_m^2$ is a constant, and the trajectory in the vx plane is an ellipse with a size and shape determined by A, b, and Ω. This ellipse is a unique periodic solution to which all initial conditions attract. It is a *limit cycle*[15] and is a *dynamical equilibrium*, in contrast to the *static equilibrium* previously described. Imagine the torus in

[14]Not to be confused with Edward Lorenz (see §4.8.2), the Lorentzian line shape was named after the Dutch mathematical physicist Hendrik Antoon Lorentz (1853–1928) who won the Nobel Prize (1902) for his work in atomic physics and predicted the relativistic length contraction for which Albert Einstein later gave a more detailed explanation.

[15]The term 'limit cycle' was introduced by Poincaré, and is sometimes called a *Poincaré limit cycle*.

Fig. 3.5 shrinking in the small direction during the initial transient phase, becoming skinny like a bicycle tube and eventually shrinking to a single closed loop. For a torus with circular cross-sections, the ratio of the large radius to the small radius is called the *aspect ratio*, which in this example approaches infinity.

Limit cycles are a new and important type of attractor (or repellor if unstable). An example is the pendulum of a grandfather clock that settles into an oscillation of a given amplitude no matter how it is started. Hearts also exhibit limit-cycle behavior, and a defibrillator is used to kick a chaotic heart into the basin of attraction of its limit cycle. A limit cycle is a more realistic model of periodicity in ecological and economic systems than is the circular model in §3.3, which depends on initial conditions.

The stability of a limit cycle is studied by taking a slice through it so as to form a collection of points where the trajectory pierces the plane (a *Poincaré section*), and treating these points like static equilibria. A Poincaré section reduces a flow to a map and is the geometric equivalent of sampling the flow at discrete (but not necessarily equal) time steps. The time interval between successive visits of the trajectory to the same point in the Poincaré section is the *period* of the motion. The choice of Poincaré section is somewhat arbitrary. It need not be perpendicular to the trajectory, but it must not be tangent to it. Such an intersection is called *transverse*, and the resulting section is also called a *transverse section*.

3.8 Van der Pol equation

Limit cycles can also occur in two-dimensional *nonlinear* autonomous systems. Suppose the damping in eqn (3.23) is not constant but depends on x in such a way that it is negative for small amplitude but positive for large amplitude. Then the equilibrium at $x = v = 0$ would be unstable and would repel initial conditions in its vicinity, but it would become an attractor when the trajectory is more distant. The trajectory could then approach a limit cycle. Perhaps the simplest such system is

$$\frac{d^2x}{dt^2} + b(x^2 - 1)\frac{dx}{dt} + x = 0 \tag{3.32}$$

where the time t is scaled so that the natural frequency is $\omega = 1$. This system is clearly nonlinear (or *anharmonic*) since it contains a term $x^2 dx/dt$. Equation (3.32) is called the *van der Pol equation*, devised by van der Pol[16] (1926) to model oscillations in vacuum-tube ('electronic valves' in the UK) circuits. It has been used to model heartbeats (van der Pol and van der Mark 1928), sunspot cycles (Mininni *et al.* 2000), and pulsating

[16]Balthazar van der Pol (1889–1959) was a Dutch electrical engineer who initiated the systematic study of dynamical electrical circuits in the 1920s and 1930s.

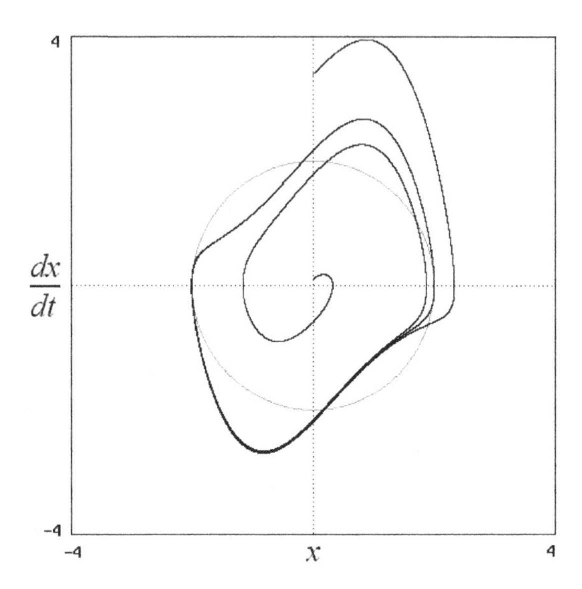

Fig. 3.8 Limit cycle for the van der Pol equation with $b = 1$.

stars called *Cepheids*.[17] It is an example of a *relaxation oscillator* (Grasman 1986). Other examples are *slip-stick* systems such as the motion of tectonic plates that lead to earthquakes and the vibration of a bowed violin string. Oscillators that alternately gain energy from their environment and return it are called *autocatalytic oscillators*.

The solution of nonlinear equations is usually difficult, but for $0 < b \ll 1$, eqn (3.32) has a nearly circular phase-space trajectory, as shown earlier. An initial condition near the origin slowly spirals outward until the average[18] of x^2 along the trajectory $\langle x^2 \rangle$ is near unity. Since the motion is nearly sinusoidal in time and circular in phase space, the limit cycle for $b \to 0$ is a circle of radius two

$$x^2 + v^2 \simeq 4 \tag{3.33}$$

For larger b, the trajectory weaves inside and outside the circle twice per cycle as energy is added and removed. Figure 3.8 shows the phase portrait for $b = 1$ superimposed on a circle of radius two with two different initial

[17]Cepheids (or Cepheid variables) are large yellow stars near the end of their luminous life with regular periods of 1 to 70 days. The name comes from the one discovered by the deaf British astronomer John Goodricke (1764–1786) in 1784, Delta Cephei (with a period of about 5.366341 days), in the constellation of Cepheus, but there had been earlier observations of variable brightness stars dating back at least to the 1596 discovery of Mira Ceti by the German astronomer, astrologer, and Lutheran pastor, David Fabricius (1564–1617). An earlier Cepheid, Eta Aquilae, was discovered in the same year by Goodricke's friend Edward Pigott.

[18]Angle brackets $\langle \cdots \rangle$ will be used in this text to denote the average of the quantity contained within, but the context will generally determine how to calculate it.

conditions. It will be left as a challenge for you to show that the lowest order correction for small but finite b is

$$x^2 + v^2 - \tfrac{1}{2}bxv^3 \simeq 4 \tag{3.34}$$

Points on the trajectories that approach the limit cycle are part of the *wandering set*, whereas the limit cycle itself is a *nonwandering set.*

A variant of the van der Pol equation with $b(x^2 - 1)v$ replaced by $b(x^2 + v^2 - 1)v$ has the nice feature that its limit cycle is exactly a unit-radius circle for any b. Sometimes, making an equation more complicated simplifies its analysis.

3.9 Numerical solution of differential equations

Although there are formal methods for solving systems of *linear* ordinary differential equations, the same is not true for *nonlinear* differential equations. In fact, for chaotic systems, analytical solutions almost never exist. Therefore, you must resort to approximations or numerical solutions. There are analog computers especially designed to solve differential equations, but they are expensive, difficult to program, and not very accurate. Digital computers, on the other hand, are inexpensive and accurate, but they necessarily represent the variables with finite precision and advance in discrete rather than continuous time steps. Thus they can iterate maps with a precision limited only by the number of bits used to represent the numbers, but they can only approximate flows. Hence the numerical solution of differential equations always involves finding a map that approximates the flow.

As an example, consider a two-dimensional flow such as the simple harmonic oscillator in §3.4 or the van der Pol equation in §3.8. The first step is to write the system in the form

$$\frac{dx}{dt} = f(x, y) \tag{3.35}$$

$$\frac{dy}{dt} = g(x, y) \tag{3.36}$$

with initial values x_0 and y_0 at $t = 0$. This is an example of an *initial-value problem.* The extension to higher dimensions is evident. Any system of ordinary differential equations can be reduced to a system of coupled first-order equations by a suitable definition of variables. If the equations are *nonautonomous* (f or g explicitly involve time), you can recast them into the above form using the method in §3.5.

The task is to find a map

$$X_{n+1} = F(X_n, Y_n) \tag{3.37}$$

$$Y_{n+1} = G(X_n, Y_n) \tag{3.38}$$

such that $x(t) \simeq X_n$ and $y(t) \simeq Y_n$ for $t = nh$, where h is some (presumably small) time step. Equations (3.37) and (3.38) are a special map whose successive iterates are very close together. The solution of the map approaches the solution of the flow for h sufficiently small. However, there are lower limits on h because of computation time and round-off errors. It is better to find a mapping that improves the fit with moderate h. Many books describe how to do this optimally. See, for example, Gear (1971), or use one of the many commercially available programs. We will describe four methods of increasing complexity and accuracy.

3.9.1 Euler method

The conceptually simplest numerical method for solving a differential equation follows from the definition of a derivative, $dx/dt = [x(t+h) - x(t)]/h$, in the limit $h \to 0$ and leads directly to the equations for the *Euler method*:

$$X_{n+1} = X_n + hf(X_n, Y_n) \tag{3.39}$$

$$Y_{n+1} = Y_n + hg(X_n, Y_n) \tag{3.40}$$

To see why this method is not generally recommended, apply it to the circular motion of eqns (3.4) and (3.5) with $X_0 = 1$ and $Y_0 = 0$. The solution after one iteration is $X_1 = 1$ and $Y_1 = -h$. Already, the radius of what should be a unit circle has grown from $R = 1$ to $R = \sqrt{1+h^2}$, which for small h is $R \simeq 1 + h^2/2$. The Euler method is *first-order* because the solution with each iteration is accurate to order h and the error is of order h^2. On successive iterations, the radius continues to grow by this same factor, causing the orbit to spiral outward. After a time t, the radius has grown to $R(t) = (1+h^2)^{t/2h} \simeq 1 + th/2$. After one cycle ($t = 2\pi$), the error in the radius is πh. The error after a fixed time (rather than a fixed number of iterations) in a first-order method is proportional to the first power of h.

For dissipative systems, the Euler method sometimes suffices since the error is equivalent to a reduction in the real damping. If the reduction is small compared with the damping, the solution is approximately correct. For example, the solution of eqn (3.23) for $\omega = 1$ and $b \ll 1$ spirals inward by a factor $e^{-\pi b} \simeq 1 - \pi b$ after one cycle. Hence, you can get an approximate solution provided $h \ll b$. Better yet, use an effective b equal to $b + h$, and the errors will nearly cancel, although the orbit is slightly distorted. For the same reason, the Euler method often suffices for systems whose solution is a limit cycle, although the limit cycle is somewhat distorted for large h.

At large h, the solutions can exhibit chaos, even when the original equations cannot. For example, the van der Pol equation with $b = 0.5$ solved by the Euler method with $h = 0.5$ gives the plot in Fig. 3.9. You know this solution is incorrect because the trajectory appears to intersect itself, which is impossible for a two-dimensional autonomous flow since each point in the plane has a unique flow direction. For slightly larger values of h, the orbit becomes unbounded, as is typical of the Euler method. Such *numerical instabilities* were known well before the study of chaos became popular. You

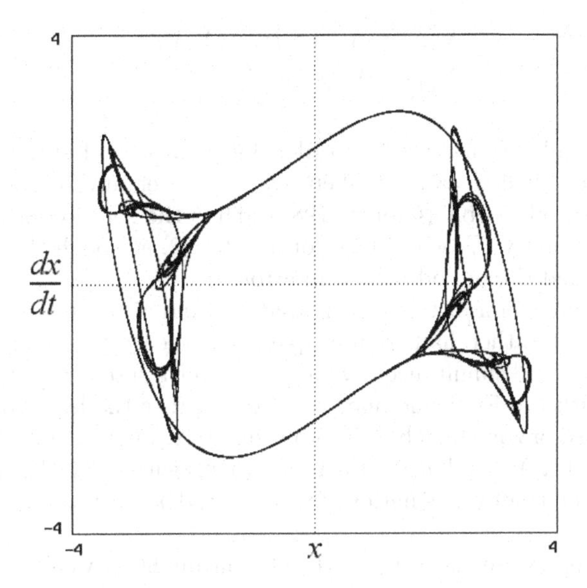

Fig. 3.9 The van der Pol equation with $b = 0.5$ solved by the Euler method with $h = 0.5$ exhibits numerical chaos.

should always verify that the numerical solution of a differential equation is independent of h and that the solution is unique as h approaches zero.

You might think the Euler method will always work if you just make h small enough. However, this is not a good strategy because small h requires more iterations to cover the same time interval and round-off errors become important. If the fractional round-off error in each step is ϵ and the errors are independent with mean zero, then the total error after time t is $\sim \epsilon\sqrt{t/h}$, whereas the iteration error (also called the *truncation error*) for a first-order method is $\sim th$. Thus the error is large at both large and small h, and there is a minimum where the two are comparable at $h \sim \sqrt[3]{\epsilon^2/t}$. With this h, you need $\sqrt[3]{t^4/\epsilon^2}$ iterations, which may try your patience and still not give a good result.

A problem with the Euler method is that the solution is advanced using only the initial value of the variables. To improve its accuracy, replace $f(X_n, Y_n)$ in eqn (3.39) with $f[(X_{n+1}+X_n)/2, (Y_{n+1}+Y_n)/2]$ and $g(X_n, Y_n)$ in eqn (3.40) with $g[(X_{n+1}+X_n)/2, (Y_{n+1}+Y_n)/2]$, and solve the resulting equations for X_{n+1} and Y_{n+1} in terms of X_n and Y_n. The resulting solution is more accurate, but the method requires an algebraic calculation whenever f and g are changed. Such a method is called *explicit*.

3.9.2 Leap-frog method

A common mistake in implementing the Euler method in eqns (3.39) and (3.40) is to inadvertently use the new value of X when iterating Y

$$X_{n+1} = X_n + hf(X_n, Y_n) \tag{3.41}$$
$$Y_{n+1} = Y_n + hg(X_{n+1}, Y_n) \tag{3.42}$$

The result is only accurate to first order, but it has the desirable property that the error exactly cancels when averaged over any half cycle for a symmetric periodic orbit (Cromer 1981). Thus the solution of eqns (3.4) and (3.5) will always close on itself for any $h < 2$, although the circle will be distorted and the period will be in error.

One interpretation of this case is that X_n and Y_n do not occur at the same time, but rather half a step apart, and hence the name *leap-frog method*. If you need simultaneous values of X and Y, use X_{n+1} and $(Y_{n+1} + Y_n)/2$. As with the Euler method, you can improve the leap-frog method at the expense of some algebra by replacing $f(X_n, Y_n)$ in eqn (3.41) with $f[(X_{n+1}+X_n)/2, Y_n]$ and $g(X_n, Y_n)$ in eqn (3.42) with $g[X_n, (Y_{n+1}+Y_n)/2]$, and solving the resulting equations for X_{n+1} and Y_{n+1} in terms of X_n and Y_n.

If you plan to use the Euler method, you might as well use the leap-frog method instead. It is even easier to implement, and it is usually more stable. It rarely gives a worse solution, and it can be dramatically better for some problems, especially where the solution is periodic.

3.9.3 Second-order Runge–Kutta method

The idea behind the *Runge–Kutta* or *midpoint* method (Runge 1895, Kutta 1901) is to use the Euler method to estimate f and g half a step ahead and then use these values to take the full step. Alternately, it can be considered as an Euler step followed by a leap-frog step

$$k_{1x} = hf(X_n, Y_n)$$
$$k_{1y} = hg(X_n, Y_n)$$
$$k_{2x} = hf\left(X_n + \frac{k_{1x}}{2}, Y_n + \frac{k_{1y}}{2}\right)$$
$$k_{2y} = hg\left(X_n + \frac{k_{1x}}{2}, Y_n + \frac{k_{1y}}{2}\right)$$
$$X_{n+1} = X_n + k_{2x} \tag{3.43}$$
$$Y_{n+1} = Y_n + k_{2y} \tag{3.44}$$

The Runge–Kutta method above is second order and returns simultaneous values of X and Y. It requires two evaluations each of f and g with each time step.

3.9.4 Fourth-order Runge–Kutta method

The Runge–Kutta method can be extended to higher order. The fourth-order representation is

$$
\begin{aligned}
k_{1x} &= hf(X_n, Y_n) \\
k_{1y} &= hg(X_n, Y_n) \\
k_{2x} &= hf\Big(X_n + \frac{k_{1x}}{2}, Y_n + \frac{k_{1y}}{2}\Big) \\
k_{2y} &= hg\Big(X_n + \frac{k_{1x}}{2}, Y_n + \frac{k_{1y}}{2}\Big) \\
k_{3x} &= hf\Big(X_n + \frac{k_{2x}}{2}, Y_n + \frac{k_{2y}}{2}\Big) \\
k_{3y} &= hg\Big(X_n + \frac{k_{2x}}{2}, Y_n + \frac{k_{2y}}{2}\Big) \\
k_{4x} &= hf(X_n + k_{3x}, Y_n + k_{3y}) \\
k_{4y} &= hg(X_n + k_{3x}, Y_n + k_{3y})
\end{aligned}
$$

$$X_{n+1} = X_n + \frac{k_{1x}}{6} + \frac{k_{2x}}{3} + \frac{k_{3x}}{3} + \frac{k_{4x}}{6} \tag{3.45}$$

$$Y_{n+1} = Y_n + \frac{k_{1y}}{6} + \frac{k_{2y}}{3} + \frac{k_{3y}}{3} + \frac{k_{4y}}{6} \tag{3.46}$$

This method requires four evaluations of each function at every time step. Higher-order formulas are rarely used because the gain is minimal for most problems. In fact, the accuracy sometimes degrades with increasing order beyond a certain point. The Runge–Kutta method does not guarantee stable solutions, as Fig. 3.10 illustrates. With slightly larger values of h, the solutions are unbounded in this example. Such behavior is typical of most numerical methods with large h. There is a chaotic region just before the orbits become unbounded. Compare this behavior with the logistic equation for $A \sim 4$. There are methods better than this for certain types of problems, but the fourth-order Runge–Kutta is relatively stable, easy to implement, and when extended to higher dimensions will suffice for all the examples in this book.

With all these methods, there is no necessity to keep h constant along the trajectory. Sometimes the flow is relatively smooth except in local regions, giving rise to two rather different time-scales, a fast time-scale where the flow changes abruptly and a slow time-scale for the trajectory to return to the region. An example of such a *stiff system* is the inductor–diode circuit in §1.5.1. In such a case, you can make large gains by using an *adaptive step size*. One way to do this is to check the accuracy of every iteration by repeating it in two steps with half the h. If the error is too large (and you get to decide what that means), then you continue halving the step size and recalculating the *present* step until your criterion is satisfied. If the error is smaller than your criterion, then you double h on the *next* time step.

With these numerical methods, we are ready to tackle arbitrary nonlinear flows in any dimension, including those that exhibit chaos.

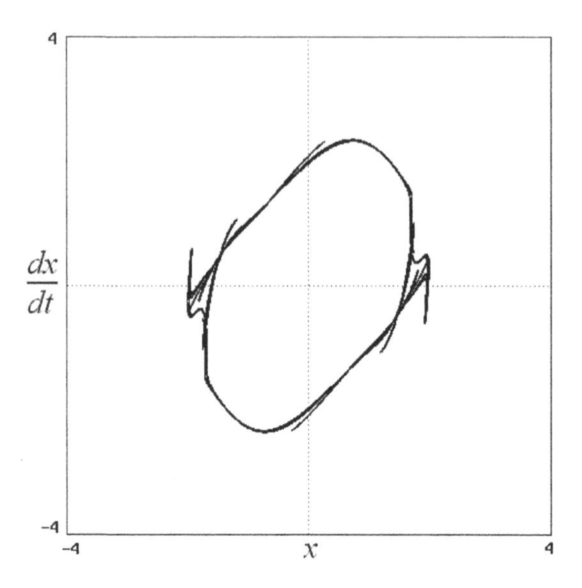

Fig. 3.10 The van der Pol equation with $b = 0.9$ solved by the fourth-order Runge–Kutta method with $h = 0.9$ exhibits numerical chaos.

3.10 Exercises

Exercise 3.1 Show that two nearby initial conditions for eqn (3.1) separate at an exponential rate, and show that the rate is given by the constant a.

Exercise 3.2 Show that eqn (3.3) is a solution of the Verhulst equation (3.2).

Exercise 3.3 Find the solution $x(t)$ for the equation $dx/dt = \sin x$ with $0 < x_0 < \pi$.

Exercise 3.4 Rewrite eqns (3.4) and (3.5) with multiplicative constants a and b, respectively, on their righthand side, and show that the oscillation about the center occurs at a frequency $f = \sqrt{ab}/2\pi$ and that the trajectory is an ellipse in xy-space with a ratio of axis lengths equal to $\sqrt{a/b}$.

Exercise 3.5 Verify that eqn (3.11) is a solution of the driven oscillator in eqn (3.10) for $x_0 = 0$, and calculate the corresponding $v(t)$.

Exercise 3.6 Verify that eqn (3.26) is a solution of the overdamped harmonic oscillator in eqn (3.23) for $b \gg 2\omega$ and $x_0 = 0$, and calculate the corresponding $v(t)$.

Exercise 3.7 Verify that eqn (3.27) is a solution of the critically damped harmonic oscillator in eqn (3.23) for $b = 2\omega$ and $x_0 = 0$, and calculate the corresponding $v(t)$.

Exercise 3.8 Verify that eqn (3.28) is a solution of the underdamped harmonic oscillator in eqn (3.23) for $b \ll 2\omega$ and $x_0 = 0$, and calculate the corresponding $v(t)$.

Exercise 3.9 Verify that eqn (3.29) is a solution of the driven damped harmonic oscillator in eqn (3.22), and calculate the corresponding $v(t)$.

Exercise 3.10 Calculate the value of Ω at which the response x_m is maximum for the driven damped harmonic oscillator in eqn (3.29).

Exercise 3.11 Show that for $Q \gg 1$, the width of the response function in Fig. 3.7 is given by $\Delta\Omega = \omega/Q$.

Exercise 3.12 Show that the driven damped harmonic oscillator in eqn (3.29) has a Lorentzian line shape as indicated in eqn (3.31) for $Q \gg 1$ and $\Omega \simeq \omega$.

Exercise 3.13 Show that eqn (3.33) is a limit-cycle solution of the van der Pol equation (3.32) in the limit $b \to 0$.

Exercise 3.14 Replace the damping term $b(x^2 - 1)$ with $b(x^2 + v^2 - 1)$ in the van der Pol equation, and show that the limit cycle is a circle in phase space with radius one.

Exercise 3.15 (difficult) Show that eqn (3.34) is a limit-cycle solution of the van der Pol equation (3.32) for $0 < b \ll 1$.

Exercise 3.16 Show that the period of the van der Pol equation (3.32) with $b \gg 1$ is $(3 - 2\ln 2)b$.

Exercise 3.17 Show that the system suggested by Lorenz (1993)

$$\frac{dx}{dt} = x - y - x^3 \tag{3.47}$$

$$\frac{dy}{dt} = x - x^2 y \tag{3.48}$$

has an equilibrium at the origin and a limit cycle with a radius of approximately 1.

Exercise 3.18 Show that in the limit of large step size, the Euler solution of the logistic differential equation gives the logistic map, and find a relation between h, a, and A.

Exercise 3.19 Show that the Euler method applied to the damped harmonic oscillator in eqn (3.23) with $b = h = 1$ and $\omega = 1$ gives a period-6 orbit.

Exercise 3.20 Sum the iteration and round-off errors for the Euler method, and find the exact value of h at which this sum is a minimum for fixed t and ϵ.

Exercise 3.21 Calculate the fractional error in the radius of a circular orbit after one time step of size h for the Euler method modified so that f and g are evaluated at the mean of the previous and current values of X and Y.

Exercise 3.22 Write out the equations you would solve for the damped harmonic oscillator using the Euler method modified so that f and g are evaluated at the mean of the previous and current values of X and Y.

Exercise 3.23 Show that the leap-frog method applied to the circular motion in eqns (3.4) and (3.5) has a fractional error of order h^2 after a single time step.

Exercise 3.24 Show that the leap-frog method applied to the circular motion in eqns (3.4) and (3.5) with $h = 1$ gives a period-6 orbit. Compare this period with the exact period.

Exercise 3.25 Show that the second-order Runge–Kutta method has an error of order h^3.

Exercise 3.26 Show that the fourth-order Runge–Kutta method has an error of order h^5.

Exercise 3.27 Calculate the value of h required for each of the four numerical methods described such that the fractional truncation error in each time step for a circular orbit is 10^{-8}, and state how many iterations are required with this value to follow the orbit for 10 cycles.

3.11 Computer project: Van der Pol equation

In this project, you will learn to solve a system of differential equations numerically. You will demonstrate the accuracy of your method by applying it to a simple harmonic oscillator and then to the van der Pol equation whose solution is a limit cycle. The numerical method will be needed for several future projects.

The van der Pol oscillator is given by

$$\frac{dx}{dt} = y \tag{3.49}$$

$$\frac{dy}{dt} = b(1 - x^2)y - x \tag{3.50}$$

1. Start with $b = 0$, in which case the van der Pol oscillator reduces to a simple harmonic oscillator. With an initial condition of $x_0 = 1$ and $y_0 = 0$, the trajectory should then be a circle in the xy plane of radius 1. Use the fourth-order, Runge–Kutta or another suitable method to solve this system from $t = 0$ to at least $t = 100$, and verify the reliability of your result by showing that the trajectory does not spiral either inward or outward. You might do this either graphically or by calculating the value of $x^2 + y^2$ at each time step.

2. With $b = 1$, show that various initial values of x_0 and y_0 are attracted to a limit-cycle curve as in Fig. 3.8.
3. What is the effect of changing the value of b (both positive and negative values)?
4. (optional) Add a sinusoidal drive term $A \sin \Omega t$ to the righthand side of eqn (3.50), and explore the behavior of the system for various values of A and Ω. You should be able to observe conditions under which the oscillation frequency ω locks onto the frequency of the drive (*entrainment*) and other conditions where the limiting solution is not simply periodic, but rather is either quasiperiodic or chaotic (see §4.6 and §6.3.2).

4
Dynamical systems theory

We have seen examples of dynamical systems for iterated maps and continuous flows. Maps are simpler to analyze numerically and have a rich variety of dynamical behaviors, even in one dimension. By contrast, except in the simplest cases, flows require approximate numerical methods, and in one dimension the solutions cannot do anything more complicated than grow or decay to an equilibrium point. Even in two dimensions, the most complicated behavior is a growth or decay to a periodic limit cycle.

In this chapter we will develop a more general theory of dynamical systems and extend the ideas to three dimensions where the flows can exhibit chaos. Although it is often difficult to calculate the trajectory, much can be gleaned from identifying the equilibrium points and examining the flow in their vicinity. Since the flow is usually smooth near these equilibria, we can make a linear approximation and use the ideas developed in the previous chapter. From this knowledge, we can construct a good qualitative picture of how the flow must behave throughout the entire state space. The material in this chapter is slightly more formal than usual and makes some use of complex numbers and matrix algebra.

4.1 Two-dimensional equilibria

We have already encountered the concept of a *static equilibrium*, which is a point (x^*, y^*) where a flow described by a system such as eqns (3.35) and (3.36) has $f(x^*, y^*) = 0$ and $g(x^*, y^*) = 0$. A linear system will have exactly one such point, whereas a nonlinear system may have none, many, or even infinitely many such points. We also know that an equilibrium can be either *stable* or *unstable* depending on whether nearby initial conditions are attracted to it or repelled from it. We saw in the context of the damped harmonic oscillator in §3.6 that the state-space trajectory sometimes spirals around the point (underdamped behavior), in which case the equilibrium is called a *spiral point* or *focus*. Other times the trajectory approaches it along a nearly straight line (overdamped behavior), in which case the equilibrium is called a *radial point* or *node*.

It is also possible for some trajectories to be attracted to the equilibrium point and others to be repelled from it, in which case the equilibrium is called a *saddle point*, which is a type of *hyperbolic point*. The term 'saddle'

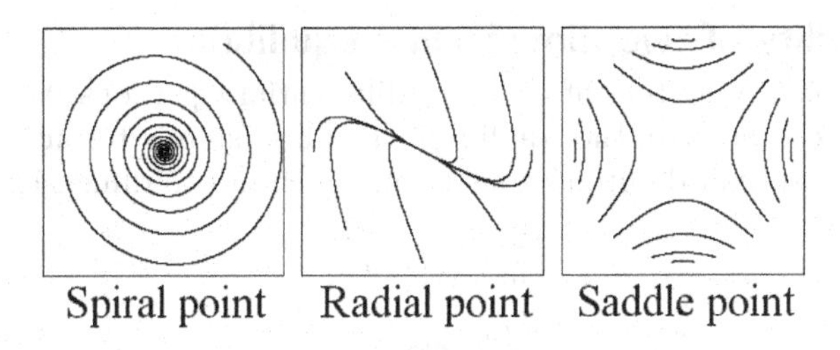

Fig. 4.1 Types of two-dimensional static equilibria.

comes from the saddle of a horse, which is shaped so that a marble would roll toward the center of the saddle from two directions but away in the other two. Potato chips sometimes have this shape. The term 'hyperbolic' comes from the fact that the flow lines are *hyperbolas*. Foci and nodes are called *antisaddles*, and they are also hyperbolic points except when they are neutrally stable. Figure 4.1 shows the phase portraits for these three types of equilibria. In each case the equilibrium point is at the center of the plot.

Each of the examples in Fig. 4.1 comes from a linear system of equations. The first two come from the system $d^2x/dt^2 + bdx/dt + x = 0$ with $|b|$ less than or greater than 2, respectively. The case $b = 0$ neither spirals in nor out but is a *center*, as in Fig. 3.3. The sign of b determines whether the point is an *attractor* ($b > 0$) or a *repellor* ($b < 0$). Note that running time backwards (changing the sign of t) will typically convert an attractor into a repellor and vice versa. The saddle point in Fig. 4.1 comes from the linear system $d^2x/dt^2 - x = 0$ and attracts from some directions but ultimately repels all trajectories. We will return to this interesting and important case in §4.4.

Each of these behaviors can also occur in *nonlinear* systems. Nonlinear systems can also have *limit cycles*, which are periodic attractors or repellors such as the van der Pol example in §3.8, whose stability can be analyzed by calculating the stability of the equilibrium points that occur in a Poincaré section through the trajectory. With two-dimensional continuous flows, these cases exhaust the possibilities. In particular, chaos cannot occur because the *no-intersection theorem* (Hilborn 2000) forbids trajectories from crossing one another (Rosen 1970), and they can only approach a point or limit cycle or go to infinity. After the transient has settled, bounded solutions can either remain fixed or oscillate periodically, but nothing else. This result is called the *Poincaré–Bendixson*[1] *theorem* (Hirsch and Smale 1974) and is a cornerstone of dynamical systems theory.

[1]Ivar Bendixson (1861–1935) was a Swedish mathematician, skilled teacher and writer, and a left-wing political activist who used his mathematical skills to reform the voting system in Sweden.

4.2 Stability of two-dimensional equilibria

Given an arbitrary system of ordinary differential equations, how do you identify and characterize the equilibria? The first step is to put the equations into the standard autonomous form, which in two dimensions is

$$\frac{dx}{dt} = f(x, y) \tag{4.1}$$

$$\frac{dy}{dt} = g(x, y) \tag{4.2}$$

and calculate x^* and y^* from the algebraic equations $f(x^*, y^*) = 0$ and $g(x^*, y^*) = 0$. This calculation may not be easy, since the equations are nonlinear, but it only requires algebra.

Now we construct a new set of linear equations that describe the flow in the vicinity of each equilibrium point and use the methods of the previous chapter to deduce their properties. In one dimension, this is relatively easy, and leads directly to $dx/dt = \lambda(x - x^*)$, where $\lambda = df/dx$ is evaluated at $x = x^*$ (see §3.1). Hence, the sign of df/dx determines whether the equilibrium is stable ($\lambda < 0$) or unstable ($\lambda > 0$), and the slope of $f(x)$ at the equilibrium determines the exponential growth (or decay) rate, $x(t) = x^* + (x_0 - x^*)e^{\lambda t}$.

In two dimensions, the situation is more complicated because there are two variables (x and y), two functions (f and g), and two solutions, which we guess are exponential functions of time. We begin by evaluating the *Jacobian*[2] *matrix* (see §B.8)

$$J = \begin{pmatrix} \frac{\partial f}{\partial x} & \frac{\partial f}{\partial y} \\ \frac{\partial g}{\partial x} & \frac{\partial g}{\partial y} \end{pmatrix} \tag{4.3}$$

where $\partial f/\partial x$, for example, is the *partial derivative* of f with respect to x evaluated at the equilibrium point, treating y as a constant. We now subtract λ from each *principal diagonal element* (the diagonal from upper left to lower right) and set the *determinant* of the resulting matrix to zero

$$\det \begin{pmatrix} a - \lambda & b \\ c & d - \lambda \end{pmatrix} = 0 \tag{4.4}$$

where $a = \partial f/\partial x, \ldots$, leading to the *characteristic* (or *indicial*) *equation*

$$(a - \lambda)(d - \lambda) - bc = 0 \tag{4.5}$$

whose two solutions

$$\lambda = \frac{a + d}{2} \pm \frac{1}{2}\sqrt{(a - d)^2 + 4bc} \tag{4.6}$$

may be positive or negative, but also *complex* if the quantity under the square root is negative. A complex number has two parts, $\lambda = A + Bi$,

[2]The term 'Jacobian' was coined by the eccentric and gifted English mathematician James J. Sylvester (1814–1897) in 1852 in honor of the German mathematician Carl Gustav Jacob Jacobi (1804–1851).

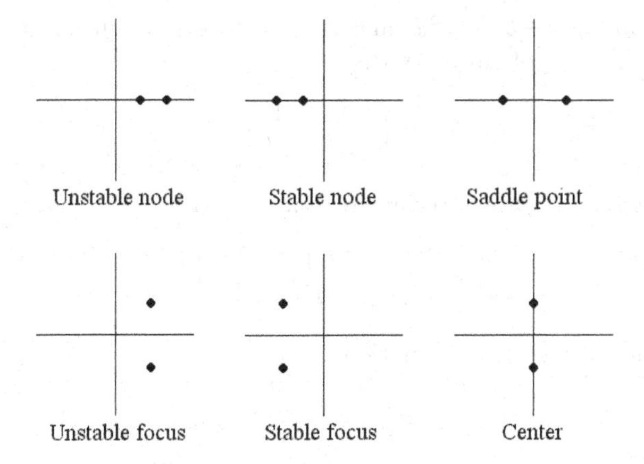

Fig. 4.2 Eigenvalues for two-dimensional flows in the complex λ plane.

where $i = \sqrt{-1}$. A and B are called the *real* and *imaginary* parts of λ, respectively.

Just as real numbers can be thought of as lying along a *line*, complex numbers reside in a *plane*. We can plot the solutions as points in this *complex plane* (called an *Argand diagram*[3]), in which the horizontal axis (A) is the real part of λ and the vertical axis (B) is the imaginary part of λ. Figure 4.2 shows the possibilities for two-dimensional flows. The complex solutions λ are called the *eigenvalues*[4] (or *characteristic values* or *characteristic exponents*). In two dimensions, they are either both real ($B = 0$) or form a *complex conjugate* pair ($A \pm Bi$). If all the eigenvalues lie off the imaginary axis (have $A \neq 0$), as they do for all the cases except the center, then the system is called *hyperbolic*. If any eigenvalues have zero real parts, then the system is *nonhyperbolic*. The eigenvalues move around in the complex plane as the system parameters change, and a bifurcation occurs whenever an eigenvalue crosses the $A = 0$ axis or moves onto or off the $B = 0$ axis (see Chapter 7).

4.3 Damped harmonic oscillator revisited

As an example of the application of these ideas, we return to the damped harmonic oscillator in §3.6 whose equations can be written

$$\frac{dx}{dt} = v \tag{4.7}$$

$$\frac{dv}{dt} = -bv - \omega^2 x \tag{4.8}$$

[3]The geometric representation of complex numbers was championed by Jean-Robert Argand (1768–1822), a French bookkeeper and amateur mathematician.

[4]The term 'eigenvalue' derives from the German 'eigenwerte' (own value) introduced by the German mathematician David Hilbert (1862–1943) in 1904.

Thus $f = v$ and $g = -bv - \omega^2 x$, and there is a single equilibrium point at $x^* = v^* = 0$. The Jacobian matrix is

$$J = \begin{pmatrix} 0 & 1 \\ -\omega^2 & -b \end{pmatrix} \tag{4.9}$$

and the eigenvalues λ are determined from

$$\det \begin{pmatrix} -\lambda & 1 \\ -\omega^2 & -b-\lambda \end{pmatrix} = 0 \tag{4.10}$$

leading to the characteristic equation

$$\lambda^2 + b\lambda + \omega^2 = 0 \tag{4.11}$$

whose solution

$$\lambda = -\frac{b}{2} \pm \frac{1}{2}\sqrt{b^2 - 4\omega^2} \tag{4.12}$$

is identical to eqn (3.25) except for a change of notation ($a \to \lambda$).

With appropriate combinations of b and ω, all the cases in Fig. 4.2, except the saddle point, occur. In particular, for $b > 2\omega$ (overdamped), we have $B = 0$, and for $b < 2\omega$ (underdamped), we have $\lambda = A \pm Bi$. For $b = 2\omega$ (critically damped), the eigenvalues coincide at $\lambda = -b/2$. Roughly speaking, the real part of the eigenvalue is the growth (or damping) rate ($A \simeq b$), and the imaginary part is the frequency ($B \simeq \omega$). If all the eigenvalues have negative A, then the equilibrium is stable, and it is an *attractor* or *sink* with a decay rate limited by the eigenvalue with the smallest $|A|$. If all the eigenvalues have positive A, then the equilibrium is unstable, and it is a *repellor* or *source* with a growth rate dominated by the eigenvalue with the largest A.

In §3.2 we assumed a solution of the form e^{at} and had to explain the meaning when a turned out not to be a real number. In the more general eigenvalue description, we assume from the outset that λ is complex and make good use of its real and imaginary parts.

4.4 Saddle points

Saddle points have a special importance in nonlinear dynamics. Because they both attract and repel, they provide a stretching and folding that often produces chaos. Furthermore, in high-dimensional systems they are the most common type of equilibrium.

Perhaps the simplest two-dimensional saddle point is the one in Fig. 4.1 reproduced in more detail in Fig. 4.3 and given by

$$\frac{dx}{dt} = y \tag{4.13}$$

$$\frac{dy}{dt} = x \tag{4.14}$$

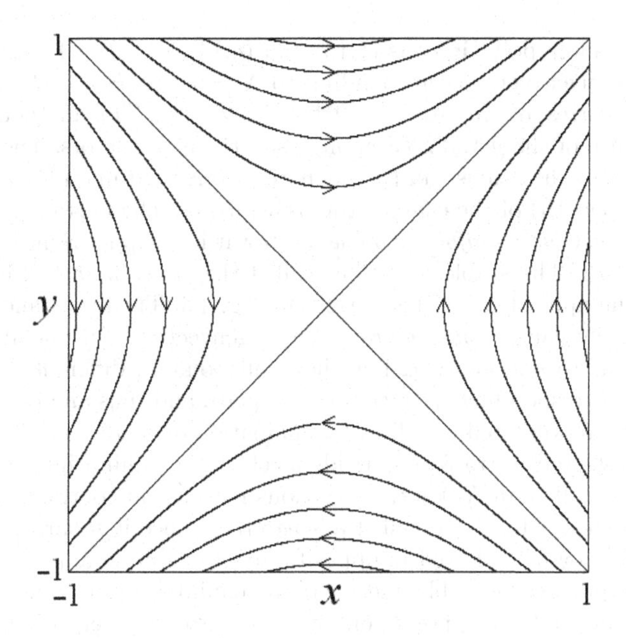

Fig. 4.3 Phase portrait for the saddle point of the system in eqns (4.13) and (4.14).

Thus $f = y$ and $g = x$, and there is a single equilibrium point at $x^* = y^* = 0$. The Jacobian matrix is

$$J = \begin{pmatrix} 0 & 1 \\ 1 & 0 \end{pmatrix} \tag{4.15}$$

and the eigenvalues λ are determined from

$$\det \begin{pmatrix} -\lambda & 1 \\ 1 & -\lambda \end{pmatrix} = 0 \tag{4.16}$$

leading to the characteristic equation $\lambda^2 - 1 = 0$ with solution $\lambda = \pm 1$ as indicated in the upper right of Fig. 4.2.

The fact that one eigenvalue is negative and the other positive means there is an attracting direction and a repelling direction, but the eigenvalues do not identify these directions. For that purpose, we calculate the *eigenvector* $\mathbf{R}$ (with components R_x and R_y) corresponding to each eigenvalue, such that $J\mathbf{R} = \lambda\mathbf{R}$. In two dimensions, this leads to a matrix equation

$$\begin{pmatrix} a & b \\ c & d \end{pmatrix} \begin{pmatrix} R_x \\ R_y \end{pmatrix} = \lambda \begin{pmatrix} R_x \\ R_y \end{pmatrix} \tag{4.17}$$

which is a fancy way of writing

$$aR_x + bR_y = \lambda R_x \tag{4.18}$$
$$cR_x + dR_y = \lambda R_y \tag{4.19}$$

For the saddle point in eqns (4.13) and (4.14), $a = d = 0$ and $b = c = 1$. Thus the eigenvector corresponding to $\lambda = 1$ has $R_x = R_y$, and the eigenvector corresponding to $\lambda = -1$ has $R_x = -R_y$. This method always gives the ratio of the y and x components of the eigenvectors. The angle θ with respect to the x axis can be calculated from $\theta = \tan^{-1}(R_y/R_x)$. The magnitude (length) of the eigenvectors is arbitrary. The unstable direction (called the *unstable manifold* or *outset*) is toward the upper right and lower left in Fig. 4.3. The stable direction (called the *stable manifold* or *inset*) is toward the upper left and lower right in Fig. 4.3. The insets and outsets are collectively called *outstructures*. The eigenvectors will not always be perpendicular to one another, but they will point in different directions (unless the corresponding eigenvalues are equal), and thus any point in the plane can be represented as a linear combination of them.

The stable and unstable manifolds meet at the equilibrium point and form what is called an *X-point*, for obvious reasons (in contrast to the *O-point* at a center). It is also called a *separatrix*, since it separates regions in which the flow is deflected in opposite directions upon approaching the equilibrium point. The stable and unstable manifolds always pass through a saddle point and cause points on either side to have very different trajectories. These manifolds are important because they organize the phase space into distinct regions. It appears that two trajectories cross at an X-point, but they do not because the flow always goes to zero as the point is approached along the inset. Note that the eigenvectors are only defined for real eigenvalues. We will not usually need the eigenvectors, which is fortunate since the calculation is tedious in higher dimensions and is usually done numerically.

4.5 Area contraction and expansion

Imagine a collection of initial conditions (x_0, y_0) at $t = 0$ distributed inside a small parallelogram whose sides are oriented parallel to the two eigenvectors ($\mathbf{R}_1$ and $\mathbf{R}_2$) as in Fig. 4.4. The area of this parallelogram is $A_0 = R_1 R_2 \sin\theta$, where θ is the angle between the vectors $\mathbf{R}_1$ and $\mathbf{R}_2$ and R is the magnitude of $\mathbf{R}$.

After a time t, the sides of the parallelogram will have changed to $Re^{\lambda t}$, and the area will be $A = R_1 e^{\lambda_1 t} R_2 e^{\lambda_2 t} \cos\theta = A_0 e^{(\lambda_1+\lambda_2)t}$. Hence, we are led to the useful result

$$\frac{1}{A}\frac{dA}{dt} = \lambda_1 + \lambda_2 \tag{4.20}$$

which says that the sum of the eigenvalues is the fractional rate of area expansion (or contraction if the sum is negative) of a cluster of initial conditions. Note that this sum is always a real number since complex eigenvalues always occur in complex conjugate pairs ($A \pm Bi$). The result extends to higher dimensions, where the sum of the eigenvalues is the rate of (hyper)volume expansion. This result gives new meaning to an *attractor*

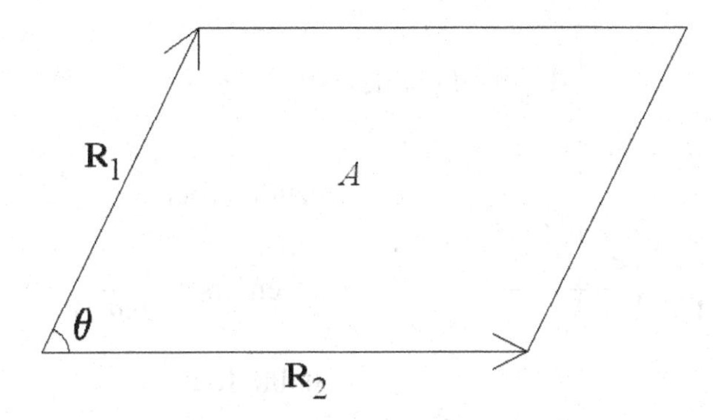

Fig. 4.4 Parallelogram of initial conditions.

where all eigenvalues are negative, a *repellor* where all are positive, and a *saddle point* where some are positive and others negative.

From eqn (4.6) it follows that the sum of the eigenvalues is the sum of the principal diagonal elements (called the *trace*) of the Jacobian matrix

$$\lambda_1 + \lambda_2 = \text{trace}(J) = \frac{\partial f}{\partial x} + \frac{\partial g}{\partial y} \tag{4.21}$$

This quantity is the *divergence* of the flow vector. Note also that the *product* of the eigenvalues is the determinant of the Jacobian matrix

$$\lambda_1 \lambda_2 = \det(J) = \frac{\partial f}{\partial x}\frac{\partial g}{\partial y} - \frac{\partial f}{\partial y}\frac{\partial g}{\partial x} \tag{4.22}$$

Both the trace and determinant of J are real numbers since complex eigenvalues occur in complex conjugate pairs.

Another way to visualize the various equilibria in two dimensions is to plot them on a diagram in which the axes are trace(J) and det(J), as in Fig. 4.5. For the damped harmonic oscillator in eqns (4.7) and (4.8), trace(J) = $-b$ and det(J) = ω^2, and the diagram thus indicates the behavior in the $\omega^2 b$ plane. The border between the foci and nodes in this plot is given by trace(J) = $\pm 2\sqrt{\det(J)}$, and points on this curve are critically damped. Like the centers along the horizontal axis (where trace(J) = 0), points on the vertical axis (where det(J) = 0) are also special because they imply a whole line of equilibria. These special cases are relatively uncommon and are said to be *structurally unstable* because a slight change in the equation will qualitatively change the solution (see §6.9).

Do not confuse *structural* instability with *dynamical* (or *Lyapunov*) instability. The former involves perturbations to the form of the equations, whereas the latter involves perturbations of the initial conditions. Structural stability is a property of the whole system of equations. The existence of a center implies structural instability. Structurally unstable equations are

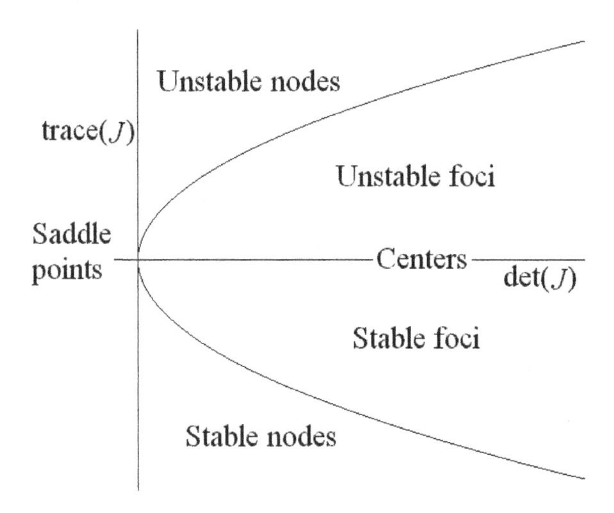

Fig. 4.5 Classification equilibrium points in two-dimensional systems.

rarely good models of nature. An undamped harmonic oscillator is structurally unstable because the topology of the solution changes when even the smallest damping is added.

4.6 Nonchaotic three-dimensional attractors

In two dimensions, the situation is relatively simple. For *linear* systems, there is only one equilibrium point with characteristics described in the previous section. For *nonlinear* systems, there may be many equilibrium points, but each can be examined by linearizing about the equilibrium using the Jacobian matrix of partial derivatives in eqn (4.3). Nonlinear two-dimensional systems such as the van der Pol equation in §3.8 can also have *limit cycles*, in which case there must be an area-expanding region (inside the loop) and an area-contracting region (outside the loop).

In *three dimensions*, there are more types of equilibrium points. There are also limit cycles and two new kinds of attractors, *tori* and *strange attractors*. In this section we discuss nonchaotic attractors and defer the discussion of strange attractors to §4.8 on chaotic dissipative flows.

4.6.1 Equilibrium points

Three-dimensional equilibria are analyzed in the same way as the two-dimensional equilibria in §4.2. The Jacobian matrix of partial derivatives has $3 \times 3 = 9$ elements. The determinant of a 3×3 matrix is

$$\det \begin{pmatrix} a & b & c \\ d & e & f \\ g & h & j \end{pmatrix} = aej + bfg + cdh - ceg - bdj - afh \tag{4.23}$$

The eigenvalues λ are determined analogously to eqn (4.5), leading, in three dimensions, to a cubic equation

$$\lambda^3 - (a+e+j)\lambda^2 + (ae+aj+ej-cg-bd-fh)\lambda$$
$$-aej-bfg-cdh+ceg+bdj+afh=0 \qquad (4.24)$$

There is a standard method for solving cubic equations of the form $f(\lambda)=0$ (see §B.9), but it is unwieldy. There are three solutions, which are either all real, or one real and the other two a complex conjugate pair. A relatively simple numerical solution begins by calculating one of the real roots using the *Newton–Raphson*[5] *method* (sometimes simply called *Newton's method* or *Newton's rule*). This method starts with an initial guess ($\lambda=0$ will usually suffice), and improves the guess using $\bar{\lambda}=\lambda-f(\lambda)/f'(\lambda)$, where $f'=df/d\lambda$. The new value $\bar{\lambda}$ is then inserted back into the formula to improve the estimate. The iterates rapidly converge to λ_1. We then write $f(\lambda)=(\lambda-\lambda_1)(A\lambda^2+B\lambda+C)$ and calculate the coefficients A, B, and C by equating powers of λ. The other two roots, λ_2 and λ_3, are determined by solving the quadratic equation $A\lambda^2+B\lambda+C=0$.

This method leads to the combinations in Fig. 4.6. Special cases where the eigenvalues coincide or are nonhyperbolic ($A=0$) have been omitted, since they are not *generic*.[6] Note that half of the eight possibilities are saddle points, having both stable ($A<0$) and unstable ($A>0$) eigenvalues. The saddle points are identified by an *index* of either 1 or 2 to specify the number of eigenvalues with $A>0$. Geometrically, the index is the dimension of the outset (the unstable manifold).[7] For example, a spiral saddle with index 2 has a radial flow toward the equilibrium point along the inset line, and then an outward spiral in the outset plane. A node can be considered an index-0 equilibrium, and a repellor an index-3 equilibrium in three dimensions. For an index-0 equilibrium, the stable manifold is the entire basin of attraction. This nomenclature is not universal; for example, the 'spiral saddle' is sometimes called a *saddle focus* or *saddle node*. In higher dimensions, new types of saddle points occur, such as *crossed spirals*, and so on.

It is evident why saddle points are the most common equilibrium in high-dimensional systems. Since a system with D dimensions has D eigenvalues, the likelihood that they all lie on the same side of the imaginary axis approaches zero as D increases. Because they both attract and repel, saddle points are especially conducive to chaos, which should therefore be common in high-dimensional systems.

[5]Joseph Raphson (1648–1715) was an English mathematician and contemporary of Isaac Newton (1643–1727) and one of the few people Newton allowed to see his mathematical papers. Raphson wrote a biography of Newton called *History of Fluxions*, published in 1715 after his death. In 1690 Raphson published the approximation method Newton had derived in 1671 but was not published until 1736.

[6]A system is 'generic' if it does the 'typical' things and avoids the exceptional things that occur with zero probability.

[7]The term 'index' is used in a slightly different sense in *index theory*. Also some authors use the dimension of the *inset* to define the index. To avoid ambiguity, specify *stability index* or *instability index*.

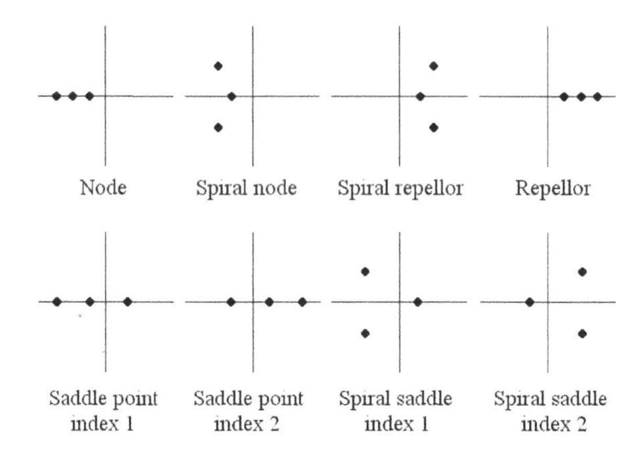

Fig. 4.6 Eigenvalues for three-dimensional hyperbolic flows in the complex λ plane.

4.6.2 Limit cycles

Not surprisingly, limit cycles can also occur in three dimensions. In fact, we already encountered one with the driven damped harmonic oscillator in §3.7 where the torus became progressively thinner, approaching a single closed loop.

A simple example of a three-dimensional *autonomous* system with a limit cycle is

$$\frac{d^3x}{dt^2} + a\frac{d^2x}{dt^2} - \left(\frac{dx}{dt}\right)^2 + x = 0 \tag{4.25}$$

Written as three first-order equations, eqn (4.25) becomes

$$\frac{dx}{dt} = y \tag{4.26}$$

$$\frac{dy}{dt} = z \tag{4.27}$$

$$\frac{dz}{dt} = -az + y^2 - x \tag{4.28}$$

There is a single equilibrium point at $x^* = y^* = z^* = 0$ with a characteristic equation $\lambda^3 + a\lambda^2 + 1 = 0$. For $a > 0$ the origin is a spiral saddle with index 2. For most values of a and initial conditions, the trajectory spirals outward to infinity.

However, there is a narrow range of a over which some initial conditions approach a limit cycle, as in Fig. 4.7, which projects the trajectory onto the xy plane. As a is reduced from 2.14, the limit cycle splits into a more complicated one that loops twice before repeating. Such behavior cannot occur in two dimensions because the trajectory needs the third dimension to cross underneath itself without intersecting, but in higher dimensions it is very common. The splitting continues in a series of period doublings

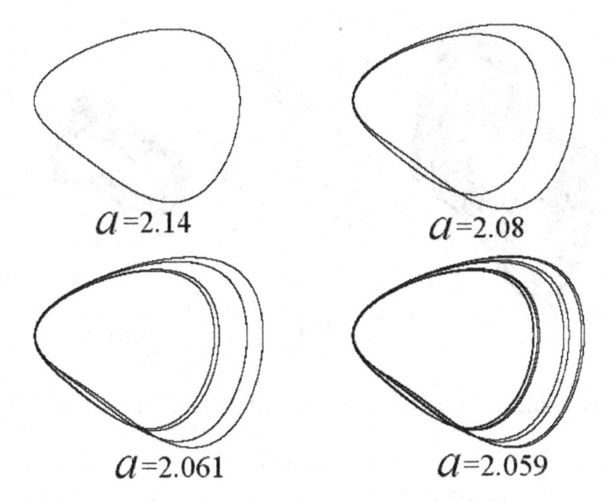

Fig. 4.7 Period doublings of a limit cycle.

reminiscent of the logistic equation. If you cannot guess what happens with a further decrease in a, jump ahead to §4.8.4.

4.6.3 Tori

In three dimensions, the trajectory of a nonlinear system can also attract to a *torus*, similar to the invariant torus for the linear undamped driven harmonic oscillator in §3.5, but with a basin of attraction. One way to produce such an attractor is a sinusoidally-driven limit cycle. The driven (or forced) van der Pol equation with $b = 1$

$$\frac{d^2x}{dt^2} + (x^2 - 1)\frac{dx}{dt} + x = A \sin \Omega t \tag{4.29}$$

is such an example (Cartwright and Littlewood 1945, Hayashi 1964).

If the driving amplitude A is sufficiently small, then the two motions decouple, and the trajectory attracts to a torus, transiting the long way with frequency Ω and the short way with frequency $\omega = 0.9429\ldots$. The trajectory projected onto the xy plane for $A = 0.9$ and $\Omega = 0.5$ is on the left of Fig. 4.8. The trajectory is quasiperiodic and consists of two incommensurate frequencies, just like the invariant torus in §3.5, but it is structurally stable in contrast to the invariant torus.

A Poincaré section through the torus produces a closed loop resembling a limit cycle, as shown on the left of Fig. 4.9, but the motion jumps around the loop rather than moving continuously. Such an object is called a *drift ring*. Taking a Poincaré section of a torus is a good way to detect incommensurate frequencies. Their section will be a ring (an infinite period); otherwise there will be a finite number of dots, indicative of the period. A Poincaré section of an attractor has a dimension one less than the attractor dimension. A torus can also be a repellor if its drift ring is unstable.

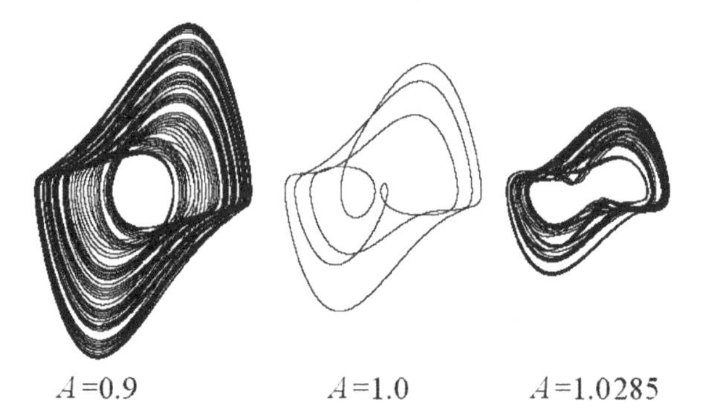

Fig. 4.8 Three solutions for the driven van der Pol oscillator with $b = 1$ and $\Omega = 0.5$.

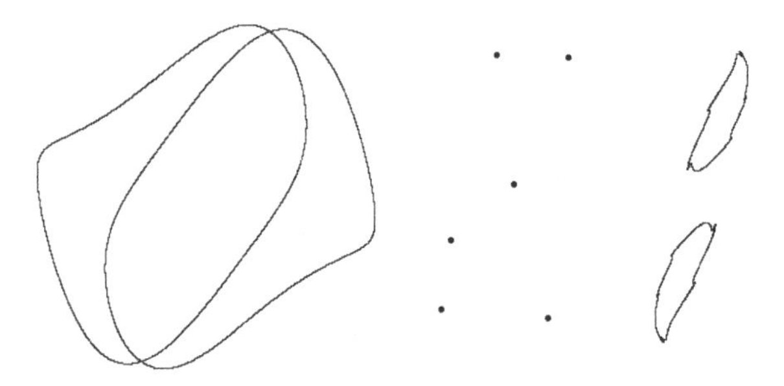

Fig. 4.9 Poincaré sections for Fig. 4.8 at $\sin \Omega t = 0$.

For larger A, the limit cycle locks to the drive frequency, producing a more complicated limit cycle with a frequency that is a rational multiple of Ω. The variation of resonant frequency with amplitude in a nonlinear oscillator is the mechanism that allows locking over a range of drive frequencies. Such a case with $A = 1$ and $\Omega = 0.5$ is in the center of Fig. 4.8 and Fig. 4.9, where the dots have been enlarged to improve visibility. Sometimes only the points that cross the Poincaré section in a particular direction are plotted, for example where $\Omega t \pmod{2\pi} = 0$, and there must be half as many of them. Thus this case is a period-3 orbit sampled twice around the torus (at 0° and 180° where $\sin \Omega t = 0$).

Evidently, a small change in A (or Ω) can qualitatively alter the solution. The conditions under which locking occurs in the driven van der Pol oscillator are shown in Fig. 4.10. Note the *Arnold* (often written *Arnol'd*) *tongues* or *horns* (Arnold 1983) near where the frequency ratio is rational. The tongues at $\Omega/\omega = 1$ and $\Omega/\omega = 3$ are especially prominent. This figure may not be accurate in detail since infinite computing is required to distinguish a torus from a limit cycle with a very long period. A small portion of the cases labeled 'torus' are actually chaotic (Mettin *et al.* 1993).

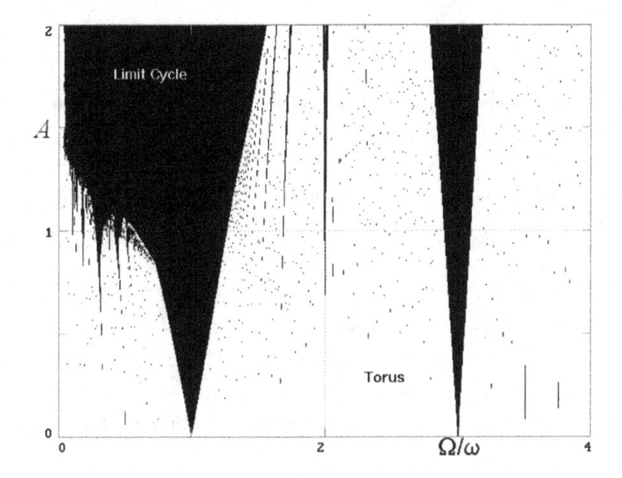

Fig. 4.10 Regions of attraction for a driven van der Pol oscillator.

Van der Pol and van der Mark (1927) reported locking for over 200 frequencies in a driven vacuum tube oscillator described by the van der Pol equation. Cumming and Linsay (1988) observed over 300 Arnold tongues in a periodically modulated relaxation oscillator using operational amplifiers.

This behavior is generic for driven limit cycles as well as for systems in which both frequencies are generated internally, such as two mutually coupled van der Pol oscillators with different natural frequencies. In the latter case, the frequencies 'pull' one another toward a common *compromise frequency*. This frequency is generally not the midpoint between the two frequencies, but is shifted by an amount that depends on the respective coupling strengths. Examples of such *entrainment* are two pendulum clocks mounted on the same wall that tick in perfect synchrony because vibrations of the wall couple their motion (Pippard 1989),[8] the synchronous flashing of fireflies (Buck 1988) and chirping of crickets, and synchronism of circadian (from 'circa diem', about one day), menstrual, and other biological cycles (Winfree 1980, Glazier and Libchaber 1988, Glass and Mackey 1988, Strogatz and Stewart 1993).

4.7 Stability of two-dimensional maps

The stability of a three-dimensional flow is best determined by analyzing its two-dimensional Poincaré map. Although it is often difficult to find an algebraic expression such as eqns (3.37) and (3.38) for the map corresponding to a given Poincaré section, the procedure for determining its stability is straightforward and parallels the method for two-dimensional flows in §4.2.

[8]The first reported observation of clock synchronization was made in 1665 by the Dutch physicist and astronomer Christiaan Huygens (1629–1695) in a letter written to his father Constantin (Huygens 1673).

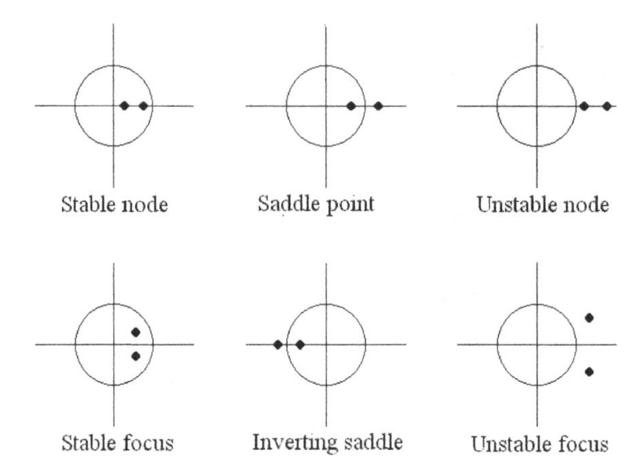

Fig. 4.11 Eigenvalues for two-dimensional maps in the complex λ plane.

First, find the fixed points (X^*, Y^*) of the map. Then construct the Jacobian matrix of partial derivatives evaluated at the fixed point. Subtract λ from the principal diagonal elements, and equate the determinant of the resulting matrix to zero to get the characteristic equation whose solutions are the eigenvalues. A two-dimensional map will have two such eigenvalues, which are either both real or a complex conjugate pair. The eigenvalues of a map are also called *Floquet multipliers* or *characteristic multipliers*.

The stability is determined by the absolute value of the eigenvalues $|\lambda|$, as in Fig. 4.11. If $|\lambda| < 1$ for all eigenvalues, then the fixed point is stable. If $|\lambda| > 1$ for any of the eigenvalues, then the fixed point is unstable. Stability is lost when one of the eigenvalues crosses the unit circle in the complex plane. The reason is that the solution is of the form $X_n = X_0\lambda^n$, rather than $x = x_0 e^{\lambda t}$ as was the case for flows where the sign of λ determined the stability (see Table 3.1). Note that the eigenvalues of a map are dimensionless numbers, whereas the eigenvalues of a flow have units of inverse time.

Two-dimensional Poincaré maps can be used to analyze the stability of limit cycles in three dimensions. If the Poincaré section has a node (two real eigenvalues on the same side of the unit circle), the resulting trajectory is a *nodal cycle*, either stable or unstable depending on the stability of the node. If the Poincaré section has a focus (a complex conjugate pair of eigenvalues), then the resulting trajectory is a *spiral cycle* or *focus cycle*. If the Poincaré section is a saddle point (eigenvalues span the unit circle), then the resulting trajectory is a *saddle cycle* and is always unstable.

4.8 Chaotic dissipative flows

The final and most interesting three-dimensional flow is chaotic and approaches a *strange attractor*. Strange attractors are so important and central to the study of chaos that the whole of Chapter 6 is devoted to them.

Here we whet the appetite by showing some important examples of dissipative chaotic flows. A *dissipative* mechanical system is one in which mechanical energy is converted into heat (*dissipated*) and the phase-space volume contracts, in contrast to a *conservative* system that conserves both energy and phase-space volume, to which Chapter 8 is devoted. We thus use the term *dissipative* for any state-space contracting system. Only dissipative systems have attractors, although there can be some preferential accumulation of points in a conservative system, producing a so-called *strange accumulator* (Smith and Spiegel 1987). For a dissipative system to continue fluctuating, there must be an external source of energy either explicitly in a (nonautonomous) forcing term or implicitly through an autonomous term that pushes the system away from equilibrium. Such a system is necessarily out of equilibrium, with a continual throughput of energy.

4.8.1 Nonautonomous chaotic flows

The damped driven harmonic oscillator in §3.7 is three-dimensional and has a nonlinearity $A \sin \phi$, but it cannot exhibit chaos. The reason is that the nonlinearity only serves to drive the trajectory around the torus. In fact, the nonlinearity can be removed by adding a new variable as in eqns (3.17) and (3.18). Chaos requires a different nonlinearity. Replacing the linear restoring force with $\sin x$ gives eqn (1.1) for the damped driven pendulum, which can exhibit chaos, as shown by the Poincaré section in Fig. 1.6.

The driven van der Pol oscillator has a nonlinearity in its damping term, giving chaos in small regions of parameter space. These regions occur near the boundary where frequency locking onsets. An example is on the right of Figs 4.8 and 4.9 where the Poincaré section is not a drift ring, or even a pair of rings, but a fractal whose structure is barely evident because the chaos is weak. The window of chaos has a width of only about 0.1% in A. The strange attractor coexists with a limit cycle, and so initial conditions must be chosen carefully to observe it (try $x_0 = z_0 = 0, y_0 = 2$).

This behavior was observed experimentally by van der Pol and van der Mark (1927) who wrote

> Often an irregular noise is heard in the telephone receivers before the frequency jumps to the next lower value. However, this is a subsidiary phenomenon, the main effect being the regular frequency demultiplication.

They missed the opportunity to discover chaos some forty years before the modern era.

Another example replaces the restoring force $-kx$ in the damped driven harmonic oscillator with a nonlinear force $-k_1 x - k_2 x^3$. The cubic term preserves the symmetry of the force. The resulting equation

$$\frac{d^2x}{dt^2} + b\frac{dx}{dt} + k_1 x + k_2 x^3 = A \sin \Omega t \tag{4.30}$$

describes *Duffing's oscillator* (Duffing 1918). It is a good model of various phenomena (Virgin 2000) such as a magnetoelastic buckled beam (Moon

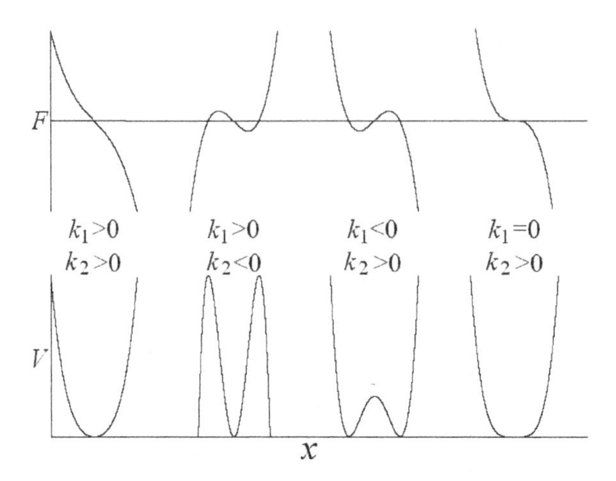

Fig. 4.12 Force and potential profiles for Duffing's oscillator.

and Holmes 1979). We can distinguish four cases, depending on the signs of k_1 and k_2. If both signs are negative, then the solutions are unbounded and will not be considered. The other three cases are shown in Fig. 4.12.

The case with $k_1 > 0$ and $k_2 > 0$ is called the *stiffening-spring* (or sometimes *hardening-spring*) case, because it would model a unit mass on a spring in which the spring gets stiffer as it is stretched or compressed. Its natural frequency of oscillation increases with amplitude. With a sufficiently large A, the system becomes *bistable* with attractors of two different amplitudes, and it exhibits *hysteresis* in which the solution depends on the past history (Jackson 2001). The onset of bistability is an example of a *cusp catastrophe.* (Thom 1975). However, the two attractors remain symmetric about the origin (for every value of x and v on the attractor, $-x$ and $-v$ is also on it) up to some value of A, at which there is a *symmetry-breaking* and the symmetric attractor splits into a pair of asymmetric attractors (for every value of x and v on one attractor, there is a corresponding value of $-x$ and $-v$ on the other).

The case with $k_1 > 0$ and $k_2 < 0$ is called the *softening-spring* case because the spring gets weaker when it is stretched or compressed. Eventually, the force reaches zero and reverses, which is not typical for a real spring, although the force will reach zero if the spring breaks. We say the equilibrium is *linearly stable*, but *nonlinearly unstable* since the force is toward the origin for small displacements but away for large displacements. It has been used to model the stability of a rolling ship (Virgin 1987). It could also model the *damped driven pendulum* (Baker and Gollub 1996) in eqn (1.1) since $\sin x \simeq x - x^3/6 + \cdots$, but it is not a good approximation if the pendulum goes 'over the top' ($|x| > \pi$).

The case with $k_1 < 0$ and $k_2 > 0$ is linearly unstable, but nonlinearly stable. The equilibrium at the origin is a repellor, but there are two nearby

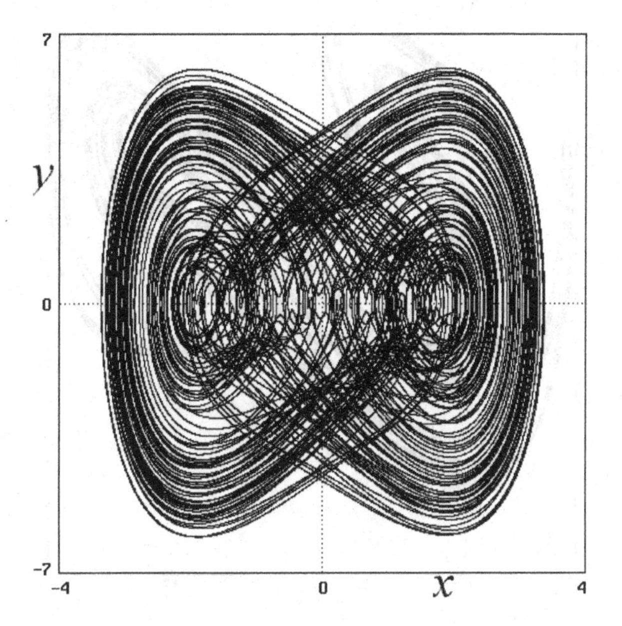

Fig. 4.13 Ueda attractor projected onto the xy plane.

attractors at $x^* = \pm\sqrt{-k_1/k_2}$. For this case it is especially instructive to consider the *potential function* $V(x) = -\int F(x)dx$ (or, equivalently, $F = -dV/dx$) which is plotted in Fig. 4.12. It provides intuition about the stability. Loosely speaking one can imagine a ball rolling on the $V(x)$ curve. Stable equilibria are at the bottom of a well, and unstable equilibria are at the top. This case is sometimes called *Duffing's two-well oscillator* for obvious reasons. In two dimensions the function $V(x, y)$ defines a surface, like the surface of the Earth, with hills and valleys. The equilibria are at the top of the hills and bottom of the valleys. The contour in the vicinity of a saddle point is shaped like a mountain pass.

All these cases exhibit chaos with appropriate parameters, but the simplest case has $k_1 = 0$. With $k_2 = 1$ and $\Omega = 1$, the resulting equation

$$\frac{d^2x}{dt^2} + b\frac{dx}{dt} + x^3 = A \sin t \tag{4.31}$$

has chaotic solutions (Ueda 1980). Nothing is lost by setting k_2 and Ω to unity, because the variables x and t can be rescaled. The trajectory with $A = 7.5$ and $b = 0.05$ projected onto the xy plane in Fig. 4.13 is a strange attractor called the *Ueda attractor*.[9]

[9]Yoshisuke Ueda first observed chaos, which he called 'randomly transitional phenomena,' on November 27, 1961 in simulations of the forced Duffing-van der Pol oscillator (the 'broken-egg' attractor) using an analog computer with vacuum tubes while he was a graduate student at Kyoto University, but the observation was so contentious that his famous professor Chihiro Hayashi suppressed its publication for many years (Abraham and Ueda 2000).

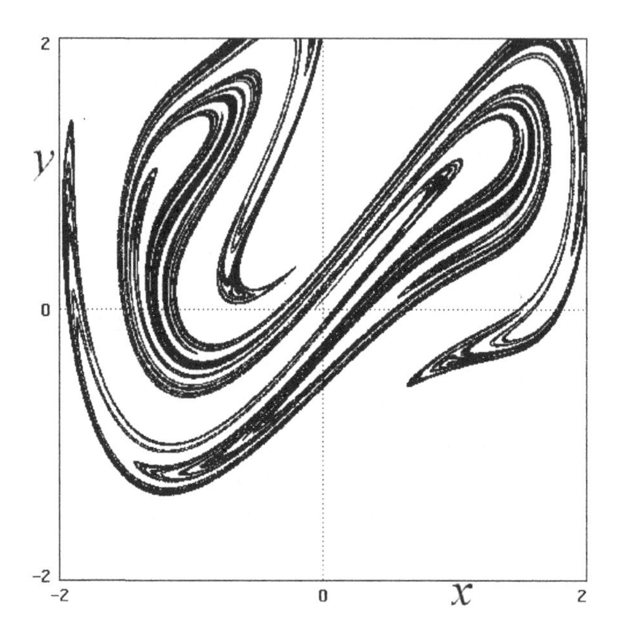

Fig. 4.14 Poincaré section of the Ueda attractor with $t \pmod{2\pi} = 0$ in the xy plane.

It is not obvious that the projection is other than a torus. In fact, for some parameters there is an attracting torus. You can show that the trajectory is chaotic and the attractor is strange by taking a Poincaré section with $t \pmod{2\pi} = 0$ as in Fig. 4.14, for example, or using the more advanced techniques described in the following chapters. Figure 4.14 was dubbed the *Japanese attractor* by Ruelle (1980).

4.8.2 Lorenz attractor

The story is now well known (Gleick 1987, Lorenz 1993) how meteorologist Edward Lorenz at the Massachusetts Institute of Technology in the early 1960s accidentally discovered sensitive dependence on initial conditions while modeling atmospheric convection on a primitive digital computer[10] when he repeated his calculation with the six-digit initial conditions rounded to three digits, naively assuming that a difference of one part in a thousand would be inconsequential. He subsequently simplified his twelve-dimensional equations to the celebrated three-dimensional *Lorenz system* (Lorenz 1963)

[10]Lorenz used a Royal-McBee LGP-30 computer with 16 kB of memory, capable of about 60 multiplications per second.

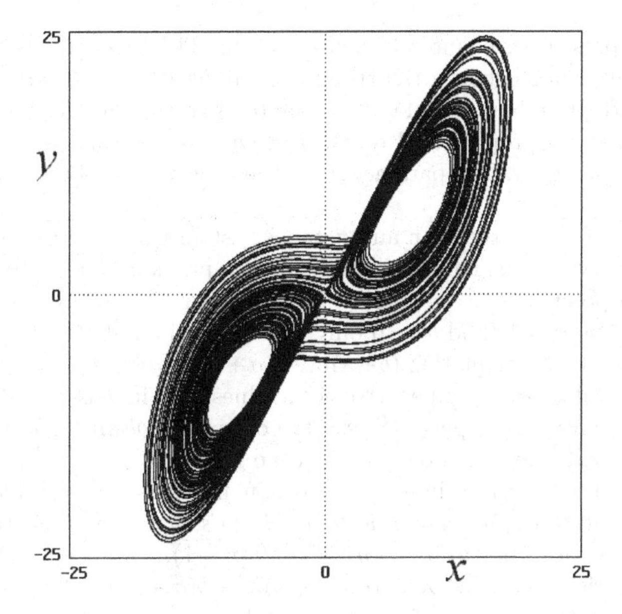

Fig. 4.15 Lorenz attractor projected onto the xy plane.

$$\frac{dx}{dt} = \sigma(y - x) \tag{4.32}$$

$$\frac{dy}{dt} = -xz + rx - y \tag{4.33}$$

$$\frac{dz}{dt} = xy - bz \tag{4.34}$$

describing the convective motion of the atmosphere heated by the ground below and cooled from above. The air rises and falls along opposite edges of long rotating cylinders. The usual parameters, $\sigma = 10$, $r = 28$, and $b = 8/3$, give chaos in which the direction of rotation occasionally changes. The attractor (sometimes called the 'butterfly attractor') for these parameters is in Fig. 4.15 projected onto the xy plane. This attractor became an emblem for early chaos researchers, and its resemblance to butterfly wings led to the metaphor first used by Lorenz in a 1972 talk entitled 'Predictability: Does the flap of a butterfly's wings in Brazil set off a tornado in Texas.'[11]

The variables x, y, and z are not spatial variables but are more abstract. The variable x is the speed of rotation, positive representing clockwise and

[11]This brief talk, presented at the 139th meeting of the American Association for the Advancement of Science in Washington, D.C. on December 29, 1972, was finally published as an appendix in Lorenz (1993). Lorenz originally used a seagull as the symbol, but the change to a butterfly was made by the session convenor, Philip Merilees. Much earlier, Ray Bradbury had written a short story 'A Sound of Thunder' in which the death of a prehistoric butterfly changed the outcome of a present-day presidential election.

negative representing counterclockwise motion. The variable y is the temperature difference between the rising and falling fluids. The variable z is the distortion from linearity of the vertical temperature profile. The parameters r and σ are proportional to the *Rayleigh number* and the *Prandtl*[12] *number*, respectively. The parameter b is the aspect ratio of the convection cylinders.

The Lorenz attractor is an autonomous system with equilibrium points whose stability can be analyzed by the methods previously described. There is one such point at $x^* = y^* = z^* = 0$ and two at $x^* = y^* = \pm\sqrt{b(r-1)}$, $z^* = r - 1$. For $r \leq 1$ (and $b > 0$) only the equilibrium at the origin exists, and its characteristic equation $(b+\lambda)[\lambda^2+(\sigma+1)\lambda+\sigma(1-r)] = 0$ predicts a stable node with three real negative eigenvalues, and its basin of attraction includes the entire xyz space. We say the origin is *globally stable* for these parameters. All convection dies out as time goes on.

At $r = 1$ the equilibrium at the origin becomes an unstable saddle point with index 1, corresponding to *Rayleigh's instability*, and the other two equilibria are born at $x^* = y^* = \pm\sqrt{b(r-1)}, z^* = r - 1$, both with the characteristic equation $\lambda^3 + (\sigma + b + 1)\lambda^2 + b(\sigma + r)\lambda + 2b\sigma(r-1) = 0$, corresponding to the onset of convection. These two equilibria are born at the origin and move away as r increases above 1.

They are stable nodes until $r = 1.34561\ldots$, whereupon they become stable spiral nodes until r reaches a value given by $\sigma(\sigma+b+3)/(\sigma-b-1) = 24.7368\ldots$ with $\sigma > b + 1$, whereupon they become unstable spiral saddle points with index 2. Trajectories spiral away from them, approaching the origin where the flow divides, throwing the trajectory back toward one of the other equilibria. For the *supercritical* value of $\sigma = 28$ used by Lorenz, all three equilibria are unstable, and chaos exists, with a globally stable attractor. Chaotic systems are locally unstable, but globally stable. In 1963 the term 'chaos' did not exist,[13] and Lorenz titled his paper 'Deterministic nonperiodic flow,' which captures some aspects of chaos.

Despite its importance, Lorenz's work went largely unnoticed for over a decade because he published in the respectable but not widely read *Journal of the Atmospheric Sciences.* Now this reference is one of the most frequently cited in the field, and the Lorenz system is often considered the prototypical autonomous chaotic flow. There is a whole book (Sparrow 1982) devoted to it. It has only recently been proved that the Lorenz system is chaotic (Tucker 1999), dispelling a lingering concern that it might be a numerical artifact. The Lorenz equations well approximate a number of other chaotic systems, including a waterwheel (Strogatz 1994), a dynamo (Cook and Roberts 1970), and a laser (Weiss and Brock 1986).

[12]Ludwig Prandtl (1875–1953) was a world-renowned German engineering professor who is considered the father of aerodynamical theory.

[13]The term 'chaos' was first used in its modern sense in the paper 'Period three implies chaos' by Li and Yorke (1975).

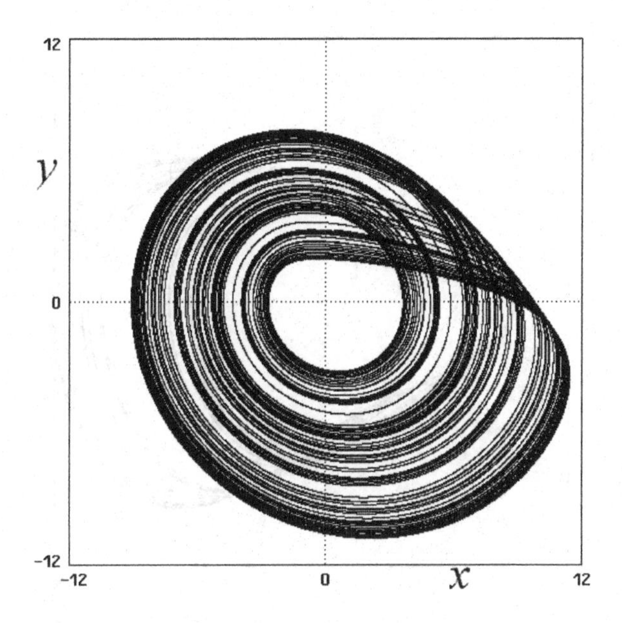

Fig. 4.16 Rössler attractor projected onto the xy plane.

4.8.3 Rössler attractor

In the 1970s, Otto Rössler, a nonpracticing medical doctor, became interested in chaos and concocted a three-dimensional autonomous system (Rössler 1976) even simpler than Lorenz's

$$\frac{dx}{dt} = -y - z \tag{4.35}$$

$$\frac{dy}{dt} = x + ay \tag{4.36}$$

$$\frac{dz}{dt} = b + z(x - c) \tag{4.37}$$

with chaotic solutions for $a = b = 0.2$ and $c = 5.7$. The *Rössler system* has the same number of terms as the Lorenz system but only a single quadratic nonlinearity (zx). Its attractor for these parameters, shown in Fig. 4.16 projected onto the xy plane, is sometimes called a *folded band*. The trajectory spirals outward close to the $z = 0$ plane and then jumps out of the plane to a large value of z and reinjects close to the origin. The system has a period-doubling route to chaos similar to the logistic map, for example, by increasing c from 2.0 to 5.7 with fixed $a = b = 0.2$.

For the parameters used by Rössler, there are two equilibria, one close to the origin ($x^* \simeq 0.0070262$, $y^* \simeq -0.035131$, $z^* \simeq 0.035131$) and one far from the attractor ($x^* \simeq 5.69297$, $y^* \simeq -28.46486$, $z^* \simeq 28.46486$). The one near the origin is a spiral saddle with index 2 and eigenvalues $\lambda \simeq -5.686975$, $0.097001 \pm 0.995193i$, and the other is a spiral saddle with

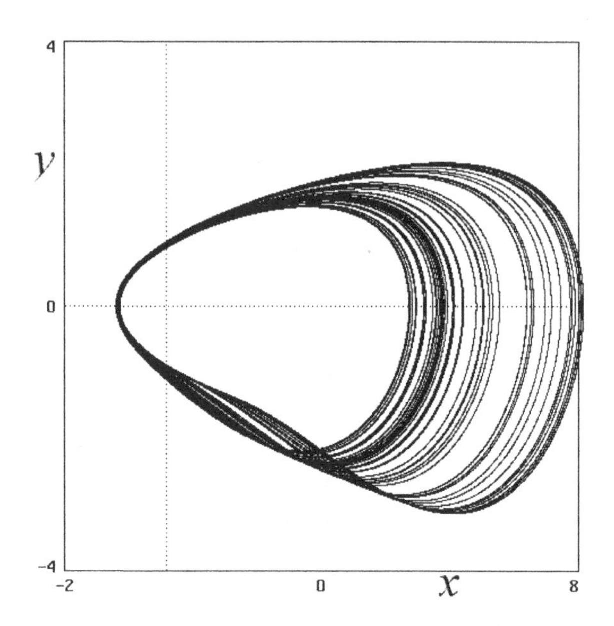

Fig. 4.17 Attractor for the simplest quadratic chaotic flow projected onto the xy plane.

index 1 and eigenvalues $\lambda \simeq 0.192983$, $-4.596 \times 10^{-6} \pm 5.428026i$. Analysis of the Rössler system is more difficult than the Lorenz system despite its simpler form.

4.8.4 Simplest quadratic chaotic flow

The Rössler system is often said to be the simplest chaotic flow, but there are many systems that are algebraically simpler (Sprott and Linz 2000), including one discovered by Rössler (1979a) himself. Perhaps the simplest autonomous dissipative chaotic flow with a quadratic nonlinearity (Sprott 1997a) is the one in eqns (4.26)–(4.28). This system has two fewer terms than the Rössler (1976) system and a single quadratic nonlinearity (y^2). It also has a single parameter (a), greatly simplifying its analysis. Chaos occurs over the narrow range $2.0168\ldots < a < 2.0577\ldots$. The attractor for $a = 2.017$ is in Fig. 4.17, projected onto the xy plane. It has a relatively small basin of attraction, so that initial conditions must be chosen carefully, such as $x_0 = y_0 = 0$, $z_0 = 1$.

The control parameter a can be considered a damping rate for the nonlinear oscillator. For all $a > 0$, the origin is a spiral saddle with index 2, and most trajectories go to infinity. As a decreases to about 2.309, a stable limit cycle is born. The limit cycle grows in size until it bifurcates into a more complicated limit cycle with two loops, which then bifurcates into four loops in a sequence of period doublings, as described in §4.6.2, until chaos finally onsets. A further decrease in a causes the chaotic attractor to

Table 4.1 Algebraically simple three-dimensional chaotic flows.

Case	Equations	Equilibria
A	$\dot{x} = y, \dot{y} = -x + yz, \dot{z} = 1 - y^2$	none
B	$\dot{x} = yz, \dot{y} = x - y, \dot{z} = 1 - xy$	f2, f2
C	$\dot{x} = yz, \dot{y} = x - y, \dot{z} = 1 - x^2$	c0, c0
D	$\dot{x} = -y, \dot{y} = x + z, \dot{z} = xz + 3y^2$	c0
E	$\dot{x} = yz, \dot{y} = x^2 - y, \dot{z} = 1 - 4x$	c0
F	$\dot{x} = y + z, \dot{y} = -x + 0.5y, \dot{z} = x^2 - z$	f1, f2
G	$\dot{x} = 0.4x + z, \dot{y} = xz - y, \dot{z} = -x + y$	f1, f2
H	$\dot{x} = -y + z^2, \dot{y} = x + 0.5y, \dot{z} = x - z$	f2, f2
I	$\dot{x} = -0.2y, \dot{y} = x + z, \dot{z} = x + y^2 - z$	f2
J	$\dot{x} = 2z, \dot{y} = -2y + z, \dot{z} = -x + y + y^2$	f2
K	$\dot{x} = xy - z, \dot{y} = x - y, \dot{z} = x + 0.3z$	f1, f2
L	$\dot{x} = y + 3.9z, \dot{y} = 0.9x^2 - y, \dot{z} = 1 - x$	n1
M	$\dot{x} = -z, \dot{y} = -x^2 - y, \dot{z} = 1.7 + 1.7x + y$	f1, f2
N	$\dot{x} = -2y, \dot{y} = x + z^2, \dot{z} = 1 + y - 2z$	f2
O	$\dot{x} = y, \dot{y} = x - z, \dot{z} = x + xz + 2.7y$	f0, f1
P	$\dot{x} = 2.7y + z, \dot{y} = -x + y^2, \dot{z} = x + y$	f2, f2
Q	$\dot{x} = -z, \dot{y} = x - y, \dot{z} = 3.1x + y^2 + 0.5z$	f1, f2
R	$\dot{x} = 0.9 - y, \dot{y} = 0.4 + z, \dot{z} = xy - z$	f2
S	$\dot{x} = -x - 4y, \dot{y} = x + z^2, \dot{z} = 1 + x$	f1, f2

grow in size, passing through infinitely many periodic windows, and finally becoming unbounded at $a \simeq 2.0168$ when the attractor grows to touch the boundary of its basin of attraction (a *crisis*).

For the range of a over which solutions are bounded, the eigenvalues are given to within about 1% by $\lambda \simeq -2.24, 0.10 \pm 0.66i$. The period of the nearly periodic solution over this range is within about 1% of 11.8, corresponding to a dominant angular frequency of $\omega \simeq 0.53$, which is about 20% lower than the linear frequency for rotation about the equilibrium at the origin ($\omega \simeq 0.66$). A bifurcation diagram for this system closely resembles the logistic map in Fig. 2.4.

Other autonomous three-dimensional chaotic flows with quadratic nonlinearities that are simpler than the Rössler system are listed in Table 4.1 (Sprott 1994a). Case A is the conservative system in eqns (3.19)–(3.21) with chaos for initial conditions $x_0 = 0, y_0 = 5, z_0 = 0$, but all the others have strange attractors for initial conditions in their basin of attraction. For brevity, the overdot denotes a time derivative, $\dot{x} = dx/dt$. The equilibrium types are denoted by f (focus), n (node), and c (center), followed by the index number.

These examples were not originally motivated by any reference to real physical systems. It was subsequently discovered that case A is a special case of the Nosé–Hoover oscillator (Hoover 1995) describing a thermostated dynamical system that had earlier been shown (Porsch *et al.* 1986) to exhibit chaos (see §8.8.1). Case B is a reduced diffusionless version of the Lorenz equations in eqns (4.32)–(4.34) studied by van der Schrier and Maas (2000).

4.8.5 Jerk systems

It is known that any ordinary differential equation can be cast in the form of a system of coupled first-order equations, but the converse does not hold in general. The system

$$\frac{d^3x}{dt^3} = f\left(\frac{d^2x}{dt^2}, \frac{dx}{dt}, x\right) \tag{4.38}$$

can be transformed into three first-order equations by substituting $y = dx/dt$ and $z = dy/dt$. In a Newtonian system, x, y, and z are the *displacement*, *velocity*, and *acceleration*, respectively, and the quantity $dz/dt = d^3x/dt^3$ is called the *jerk* (Schot 1978).[14] Equations of the form of eqn (4.38) are called *jerk equations* and are the simplest scalar differential equations that can exhibit chaos for continuous functions f since they are equivalent to three-dimensional systems.

The Lorenz and Rössler systems can be written in jerk form (Linz 1997), but the resulting equations are complicated. The Lorenz system is given by

$$\begin{aligned} &\frac{d^3x}{dt^3} + \left(1 + \sigma + b - \frac{1}{x}\frac{dx}{dt}\right)\frac{d^2x}{dt^2} \\ &+\left[b(1 + \sigma + x^2) - (1 + \sigma)\frac{1}{x}\frac{dx}{dt}\right]\frac{dx}{dt} \\ &-b\sigma(r - 1 - x^2)x = 0 \end{aligned} \tag{4.39}$$

and the Rössler system in slightly modified form is given by

$$\begin{aligned} &\frac{d^3y}{dt^3} + \left(c - \epsilon + \epsilon y - \frac{dy}{dt}\right)\frac{d^2y}{dt^2} \\ &+\left[1 - \epsilon c - (1 + \epsilon^2)y + \epsilon\frac{dy}{dt}\right]\frac{dy}{dt} + (\epsilon y + c)y + \epsilon = 0 \end{aligned} \tag{4.40}$$

where $\epsilon = a = b$ (typically 0.2).

However, all the cases in Table 4.1 with a single nonlinearity and some of those with two nonlinearities can be organized into a hierarchy of quadratic

[14]The time derivative of the jerk d^4x/dt^4 has been called a 'spasm,' 'jounce,' 'sprite,' 'surge,' or 'snap,' with its successive derivatives 'crackle' and 'pop' (Sprott 1997b).

Table 4.2 Simple chaotic quadratic jerk systems.

Model	Equation	Parameters
JD_0	$\dddot{x} = a\ddot{x} + \dot{x}^2 - x$	$a = -2.017$
JD_1	$\dddot{x} = a\ddot{x} + bx + x\dot{x} - 1$	$a = -1.8, b = -2$
JD_2	$\dddot{x} = a\ddot{x} + b\dot{x} + x^2 - 1$	$a = -0.5, b = -1.9$
JD_3	$\dddot{x} = a\ddot{x} + b\dot{x} + cx^2 + x\dot{x} - 1$	$a = -0.6, b = -3, c = 5$
JD_4	$\dddot{x} = a\ddot{x} + b\dot{x} + cx^2 + x\ddot{x} - 1$	$a = -0.6, b = -2, c = 3$
JD_5	$\dddot{x} = a\dot{x} + bx^2 + \dot{x}^2 - x\ddot{x}$	$a = 0.5, b = -1$
JD_6	$\dddot{x} = a\ddot{x} + b\dot{x} + cx^2 + d\dot{x}^2$	$a = -1, b = -1$
	$+x\ddot{x} - 1$	$c = 2, d = 2$
JD_7	$\dddot{x} = a\ddot{x} + b\dot{x} + cx^2 + d\dot{x}^2$	$a = -1, b = 1, c = 2$
	$+ex\dot{x} + x\ddot{x} - 1$	$d = -3, e = 1$

jerk equations of increasing complexity (Eichhorn *et al.* 1998) as in Table 4.2 with typical parameters $(a, b, c, \ldots)$ that give chaos. The cases JD_1 and JD_2 have been studied further by Eichhorn *et al.* (2001).

The case JD_0, corresponding to eqn (4.25), is apparently the simplest chaotic jerk system with a quadratic nonlinearity, although there are a number of systems besides eqns (4.26)–(4.28) that are equivalent to it. There are systems as simple as eqns (4.26)–(4.28) whose jerk forms are not simple quadratic polynomials, but they have small basins of attraction. Systems simpler than eqns (4.26)–(4.28), having a single quadratic nonlinearity but fewer than five terms, apparently cannot be chaotic, as shown by Zhang and Heidel (1997) for dissipative cases and by Heidel and Zhang (1999) for conservative cases. Any such examples would have no adjustable parameters, severely limiting their existence.

The simplest chaotic jerk system with a cubic nonlinearity (Malasoma 2000) appears to be

$$\frac{d^3x}{dt^3} = -a\frac{d^2x}{dt^2} + x\left(\frac{dx}{dt}\right)^2 - x \tag{4.41}$$

with $2.0278... < a < 2.0840....$ It has an infinite sequence of chaotic windows for ever smaller ranges of a, with basins of attraction that shrink in size (Malasoma 2002). This system is parity invariant under the transformation $x \to -x$ and might be the simplest such example.

The case JD_2 in Table 4.2 is of the form

$$\frac{d^3x}{dt^3} + a\frac{d^2x}{dt^2} + \frac{dx}{dt} = g(x) \tag{4.42}$$

Integrating each term reveals that this equation is a damped harmonic oscillator driven by a nonlinear memory term that involves the integral of

Table 4.3 More simple chaotic jerk systems with $a = 0.6$.

Equation	b
$\dddot{x} + a\ddot{x} + \dot{x} = \pm(b\lvert x\rvert - c)$	1.0
$\dddot{x} + a\ddot{x} + \dot{x} = -b\max(x, 0) + c$	6.0
$\dddot{x} + a\ddot{x} + \dot{x} = \pm[bx - c\,\mathrm{sgn}\,x]$	1.2
$\dddot{x} + a\ddot{x} + \dot{x} = \pm b(x^2/c - c)$	0.58
$\dddot{x} + a\ddot{x} + \dot{x} = bx(x^2/c - 1)$	1.6
$\dddot{x} + a\ddot{x} + \dot{x} = -bx(x^2/c - 1)$	0.9
$\dddot{x} + a\ddot{x} + \dot{x} = -b[x - 2\tanh(cx)/c]$	2.2
$\dddot{x} + a\ddot{x} + \dot{x} = \pm b\sin cx/c$	2.7
$\dddot{x} + a\ddot{x} + \dot{x} = \pm b\cos cx/c$	2.7
$\dddot{x} + a\ddot{x} + \dot{x} = (b - \mathrm{e}^{cx})/c$	5.0
$\dddot{x} + a\ddot{x} + \dot{x} = \{1 - b[1 + \tanh(cx)]\}/c$	13

$g(x)$. Such an equation often arises in the feedback control of an oscillator in which the experimentally accessible variable is a transformed and integrated version of the fundamental dynamical variable (Baillieul *et al.* 1980, Holmes and Moon 1983). Despite its importance and the richness of its dynamics, it has been relatively little studied (Coullet *et al.* 1979), and it does not appear to be generally known that many forms for $g(x)$ lead to chaos, some of which are listed in Table 4.3 (Sprott and Linz 2000). The last case in the table turned up in a model of the interaction of the solar wind with the Earth's ionized magnetosphere (Horton *et al.* 2001). Typical values of b that give chaos for $a = 0.6$ are shown. The value of c is arbitrary and only affects the size of the attractor.

Equation (4.42) may be a useful general model for fitting chaotic experimental data since the procedure reduces to finding a value of a and a function $g(x)$, or perhaps a more general function $g(dx/dt, x)$, that gives the best fit to the data. It can be viewed as an extension of finding a potential $V(x)$ for a damped nonlinear oscillator governed by Newton's second law

$$\frac{d^2x}{dt^2} + a\frac{dx}{dt} = -\frac{dV}{dx} \tag{4.43}$$

except that it permits chaotic solutions, which eqn (4.43) does not. Constraining a to be positive ensures that the system is dissipative with an attractor for trajectories that lie within the basin of attraction. Different forms of $g(x)$ produce a wide variety of attractors, some examples of which are in Fig. 4.18.

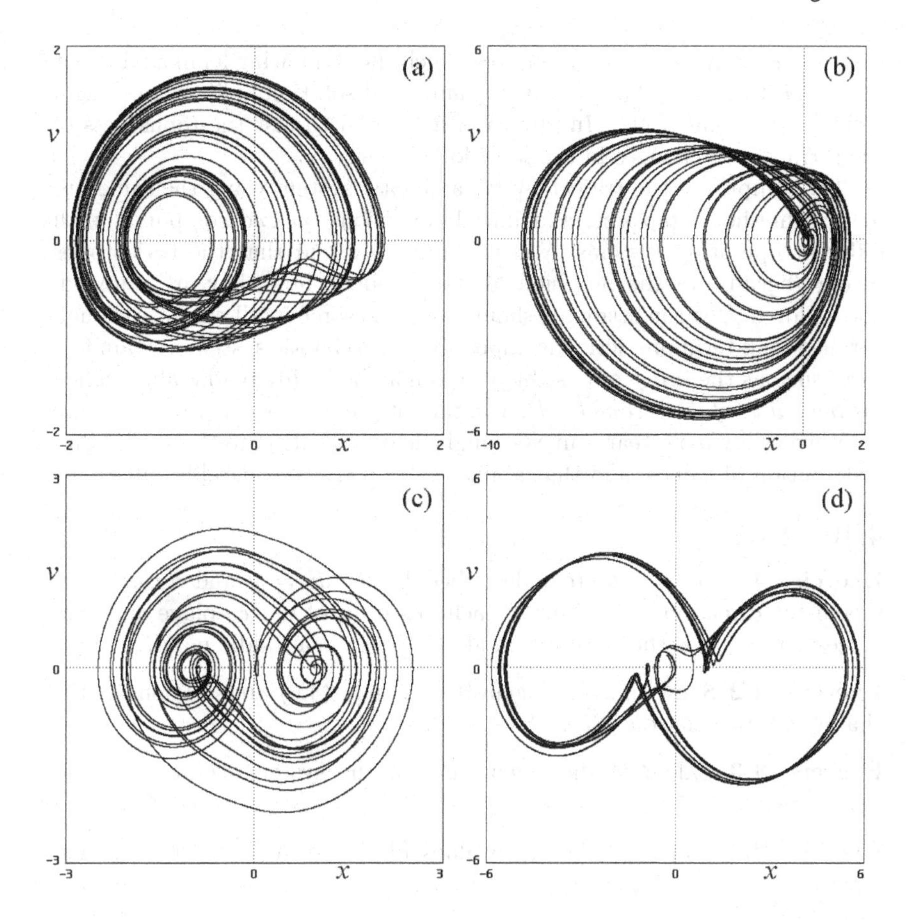

Fig. 4.18 Some strange attractors produced by eqn (4.42) with $a = 0.6$ (a) $g = 0.58(x^2 - 1)$, (b) $g = 6\max(x, 0) + 1$, (c) $g = 1.2(\operatorname{sgn} x - x)$, (d) $g = 2.7 \sin x$.

4.9 Shadowing

Since the trajectory of a chaotic system exhibits sensitive dependence on initial conditions and there are always small numerical errors in computing the trajectory, you might wonder how you can be sure these strange attractors are not just numerical artifacts. The answer is provided by the *shadowing lemma* (Grebogi *et al.* 1990), which says that although the calculated trajectory is not the real one for the chosen initial conditions, it closely resembles the real trajectory for a nearby set of initial conditions. The real trajectory thus 'shadows' the one you computed from an approximate model. The reason is that if errors push the trajectory off the attractor, it will be rapidly drawn back and will then follow one of the infinitely many trajectories on the attractor until errors push it off again. A similar result applies to experimental trajectories contaminated by noise, although the model equations are imperfect in a much more fundamental way. However, there are (nonhyperbolic) systems for which the shadowing lemma fails, but you can estimate for how long your numerical solution is likely to remain valid (Sauer *et al.* 1997). In any case, it is good to test the robustness of the attractor to changes in the iteration step size.

Try computing the trajectory for a chaotic system using the same numerical method, step size, and initial conditions on two computers with different operating systems. You will probably find that the trajectories diverge because of the precision of the numbers, how they are rounded, and so forth. But the attractor should look the same in all cases. Although detailed long-term prediction is impossible for a chaotic system, certain features such as the range of possible values may be highly predictable (called *predictability of the second kind*). You cannot predict the exact temperature on a given day next year, but you might learn to calculate the probability distribution of values, and that ability could be of considerable value.

4.10 Exercises

Exercise 4.1 For the system described by $dx/dt = y$ and $dy/dt = x$, sketch the direction of the flow in each quadrant of the xy plane and convince yourself that the equilibrium at the origin is a saddle point.

Exercise 4.2 Show that the eigenvalues in eqn (4.6) are solutions of the characteristic equation (4.5).

Exercise 4.3 Calculate the eigenvalues of the system $dx/dt = y$ and $dy/dt = x$.

Exercise 4.4 Calculate the eigenvalues of the system $dx/dt = y$ and $dy/dt = -x$.

Exercise 4.5 Find values of b and ω for the damped harmonic oscillator in eqn (4.12) that give the equilibria in Fig. 4.2, and show that a saddle point cannot occur in this system.

Exercise 4.6 Put arrows on the flows in Fig. 4.1 in each of the two possible directions, and describe the properties of each case.

Exercise 4.7 Calculate the eigenvectors for the damped harmonic oscillator in eqns (4.7) and (4.8) with $b > 2\omega$, and relate your results to the middle plot in Fig. 4.1 for which $b = 2.5\omega$.

Exercise 4.8 Calculate the eigenvalues and eigenvectors of the system $dx/dt = x + y$ and $dy/dt = -2x + 4y$.

Exercise 4.9 Prove that the trace of the Jacobian matrix is the sum of the eigenvalues and the determinant of the Jacobian matrix is the product of the eigenvalues.

Exercise 4.10 Show that a critically damped oscillator lies on the curve $\mathrm{trace}(J) = \pm 2\sqrt{\det(J)}$.

Exercise 4.11 Consider a love affair between Romeo and Juliet (Strogatz 1988, 1994) governed by the equations

$$\frac{dR}{dt} = aR + bJ \tag{4.44}$$

$$\frac{dJ}{dt} = cR + dJ \tag{4.45}$$

where R is Romeo's love (or hate if $R < 0$) for Juliet and J is Juliet's love (or hate) for Romeo at time t. Discuss the dynamics and ultimate fate for each of the sixteen combinations of romantic styles determined by the signs of a, b, c, and d.

Exercise 4.12 Find the equilibria and their eigenvalues for the *damped anharmonic oscillator*

$$\frac{d^2x}{dt^2} + b\frac{dx}{dt} + x + x^2 = 0 \tag{4.46}$$

Exercise 4.13 Find the equilibria and their eigenvalues for the *predator–prey system*

$$\frac{dx}{dt} = x - xy \tag{4.47}$$

$$\frac{dy}{dt} = -ky + xy \tag{4.48}$$

These equations are called the *Lotka–Volterra*[15] *equations* (Lotka 1920, 1925, Volterra 1926, 1931) and predict oscillations as often observed in nature, but the model is flawed because it gives a continuous family of neutrally stable cycles with a backward circulation for different initial conditions rather than an attracting limit cycle (May 1972, Gilpin 1973, Murray 1993).

Exercise 4.14 Make a plot of trace(J) versus det(J) similar to Fig. 4.5, identifying the three-dimensional equilibria in Fig. 4.6.

Exercise 4.15 Calculate the fixed points and Floquet multipliers for the *delayed logistic map*

$$X_{n+1} = AX_n(1 - Y_n) \tag{4.49}$$

$$Y_{n+1} = X_n \tag{4.50}$$

Exercise 4.16 Calculate the eigenvalues for the simple three-dimensional autonomous system in eqn (4.25) for $a = 2.14$, and show that the origin is a spiral saddle with a frequency of $\omega = 0.649\ldots$.

[15] Alfred James Lotka (1880–1949) was an Austrian-born chemist, demographer, ecologist, and mathematician who left academic science and spent most of his life working for a US insurance company. Vito Volterra (1860–1940) was an Italian mathematical physicist and later biologist who left the University of Rome in 1931 rather than take an oath of allegiance to the Fascist government.

Exercise 4.17 Find the equilibria and their eigenvalues for the Lorenz equations (4.32)–(4.34) as a function of σ, r, and b. Show that bifurcations occur at $r = 1$ and $r = \sigma(\sigma + b + 3)/(\sigma - b - 1)$.

Exercise 4.18 Show that an initial condition that starts on the z axis for the Lorenz system remains there (the axis is an *invariant line*).

Exercise 4.19 Calculate the equilibria and their eigenvalues for the *diffusionless Lorenz equations* (van der Schrier and Maas 2000)

$$\frac{dx}{dt} = -y - x \tag{4.51}$$

$$\frac{dy}{dt} = -xz \tag{4.52}$$

$$\frac{dz}{dt} = xy + R \tag{4.53}$$

which have chaotic solutions for $R = 1$ (see case B of Table 4.1).

Exercise 4.20 (difficult) Calculate the equilibria and their eigenvalues for the Rössler equations (4.35)–(4.37) as a function of a, b, and c.

Exercise 4.21 Calculate the equilibrium and its eigenvalues for the simplest quadratic chaotic flow in eqns (4.26)–(4.28) as a function of a.

Exercise 4.22 Calculate the equilibrium points for each of the equilibria in Table 4.1.

Exercise 4.23 Derive the characteristic equation for each of the equilibria in Table 4.1.

Exercise 4.24 Show that the jerk form in eqn (4.39) is equivalent to the Lorenz system in eqns (4.32)–(4.34).

Exercise 4.25 Show that the jerk form in eqn (4.40) is equivalent to the Rössler system in eqns (4.35)–(4.37).

Exercise 4.26 Show that cases F though S in Table 4.1 can be transformed into one of the jerk forms in Table 4.2.

Exercise 4.27 Find the characteristic equation for the eigenvalues of the equilibrium for the jerk system in eqn (4.42).

4.11 Computer project: The Lorenz attractor

In this project you will extend the method you developed in the previous project for numerically solving a system of differential equations to three dimensions and will use it to study the Lorenz equations, the first documented example of a flow whose limiting solution is a chaotic attractor. You may use the fourth-order, Runge–Kutta method or another suitable numerical method of your choosing. You will recreate Lorenz's discovery of the sensitive dependence on initial conditions.

The Lorenz equations are

$$\frac{dx}{dt} = \sigma(y - x) \tag{4.54}$$

$$\frac{dy}{dt} = -xz + rx - y \tag{4.55}$$

$$\frac{dz}{dt} = xy - bz \tag{4.56}$$

The values that Lorenz used for the parameters are $\sigma = 10$, $r = 28$, and $b = 8/3$.

1. Start with an arbitrary initial condition at $t = 0$ reasonably close to the attractor (but not at the equilibrium point $x^* = y^* = z^* = 0$!), and solve the Lorenz equations numerically using his parameters. Make a graph in the zy plane, and show that different initial conditions are drawn to the attractor. (See Fig. 4.15.)
2. Make a graph of x versus t over the range $25 < t < 50$, presuming that by $t = 25$ you are on the attractor.
3. Determine the exact values of x, y, and z at $t = 25$, and rerun the calculation using those values as initial conditions but rounded to three significant digits. Show that $x(t)$ for the two cases initially coincide, but that the small error eventually grows to the size of the attractor.
4. (optional) Study the behavior of the system as the parameter r is varied over the range $0 < r < 200$. You should be able to observe point attractors, limit cycles, and chaos. Look carefully in the range $99 < r < 101$, where a periodic window occurs within which there is a period-doubling cascade with the same Feigenbaum constants as in the logistic map.

5

Lyapunov exponents

Although there is no universally accepted definition of chaos, most experts would concur that chaos is the *aperiodic, long-term behavior of a bounded, deterministic system that exhibits sensitive dependence on initial conditions.* It is relatively easy to establish that a deterministic system (one arising from equations without random variables) is aperiodic and bounded, at least for time-scales over which numerical computations are feasible. It is more difficult to establish sensitive dependence on initial conditions. For that purpose, we must quantify the sensitivity. This chapter is devoted to calculation and properties of the *Lyapunov exponent*[1] (or more specifically the *spectrum* of Lyapunov exponents), whose sign signifies chaos and whose value measures how chaotic. A bounded dynamical system with a positive Lyapunov exponent is chaotic, and the exponent describes the average rate at which predictability is lost. In this chapter we assume you know the equations that produced the dynamics. The important but more difficult problem of calculating Lyapunov exponents from experimental data will be deferred to Chapter 10.

The Lyapunov exponents are closely related to the eigenvalues discussed in the previous chapter and are calculated by similar means, but there are important differences, as indicated in Table 5.1. Whereas eigenvalues are usually calculated at a point in state space, such as an equilibrium point, Lyapunov exponents are usually geometrically *averaged* along the orbit or trajectory. The Lyapunov exponents are always *real* numbers, and their associated directions are mutually *orthogonal* (perpendicular), although the directions change as the orbit moves though space. Both quantities are determined from the Jacobian matrix assuming linear local dynamics. A system with n dimensions has n Lyapunov exponents, just as it has n eigenvalues at each point.

[1]'Lyapunov exponent' (sometimes spelled 'Liapunov') was named by Oseledec (1968) for the great Russian mathematician and amateur horticulturist Aleksandr Mikhailovich Lyapunov (1857–1918) who shot himself three days after his wife of 32 years (who was also his first cousin) died of tuberculosis, leaving a note asking to be buried with her in the same grave. The origin of the concept goes back to the work of the Russian mathematician Sofya Kovalevskaya (1850–1891), who was a professor at the University of Stockholm and probably the first woman to hold such a post in Europe.

Table 5.1 Comparison of eigenvalues and Lyapunov exponents.

Eigenvalue	Lyapunov exponent
Local quantity	*Global* quantity
Constant value	*Average* value
Complex number	*Real* number
Not usually orthogonal	Mutually *orthogonal*

5.1 Lyapunov exponent for one-dimensional maps

We begin with the simplest example — a one-dimensional map $X_{n+1} = f(X_n)$ such as the logistic map in Chapter 2. Imagine two nearby initial points at X_0 and $X_0 + \Delta X_0$, respectively. After one iteration of the map, the points are separated by

$$\Delta X_1 = f(X_0 + \Delta X_0) - f(X_0) \simeq \Delta X_0 f'(X_0) \tag{5.1}$$

where $f' = df/dX$. Now define the *local Lyapunov exponent* λ at X_0 such that $\mathrm{e}^\lambda = |\Delta X_1/\Delta X_0|$, or

$$\lambda = \ln|\Delta X_1/\Delta X_0| \simeq \ln|f'(X_0)| \tag{5.2}$$

The quantity $|\Delta X_1/\Delta X_0|$ is the *local Lyapunov number*, a measure of the stretching at $X = X_0$. The absolute value ensures that the Lyapunov number is positive so that its logarithm (the Lyapunov exponent) is a real number. If $\Delta X_1/\Delta X_0$ is negative, it means the two nearby points interchange their order (the larger becomes smaller, and vice versa) upon iteration.

Calculation of the local Lyapunov numbers is very similar to calculation of the eigenvalues. In fact, for a map, the local Lyapunov numbers are the absolute values (moduli) of the eigenvalues. Knowing how the local Lyapunov exponent (or number) varies in space allows you to identify regions of an attractor with good and poor predictability for small initial errors. Nicolis *et al.* (1983) define a *nonuniformity factor* (NUF), which is the standard deviation of the distribution of local Lyapunov exponents.

To obtain the *global* Lyapunov exponent, average eqn (5.2) over many iterations

$$\lambda = \lim_{N\to\infty} \frac{1}{N} \sum_{n=0}^{N-1} \ln|f'(X_n)| \tag{5.3}$$

If the orbit is periodic, you need only average over one period once the orbit is on the attractor. The average stretch factor e^λ is the *global Lyapunov number*. Note that we have used the same symbol (λ) for eigenvalues, local Lyapunov exponents, and global Lyapunov exponents as is common in the literature, and thus the context must define which is intended. It is also common to confuse the properties, which are quite different. The global

Lyapunov exponent determines the average exponential rate of separation of two nearby initial conditions, or the average stretching of the space. A positive value signifies chaos, and a negative value implies a fixed point or periodic cycle.

The local Lyapunov exponent may vary widely and even be negative or infinite for a chaotic system as with the logistic map at $X = 0.5$. Furthermore, the orbit sometimes remains in a local region of space for a long time before eventually sampling a different region (*intermittency*), or it may eventually settle onto a periodic orbit or fixed point or escape to infinity (*transient chaos*). In such cases, long computations are required to avoid incorrect results (see §7.5.2). No matter how long you calculate, you can never be completely sure of your numerical result.

We now consider some examples of one-dimensional maps where explicit values of λ can be deduced.

5.1.1 Binary shift map

A trivial example is the binary shift map in §2.5.4 with $f(X) = 2X \pmod 1$ and $f'(X) = 2$ everywhere. Thus the Lyapunov exponent is $\lambda = \ln 2 = 0.693147181\ldots$. If you use base-2 logarithms instead of base-e, the Lyapunov exponent is simply $\lambda = 1$, and the units are bits (factors of two) per iteration.

Beware that the literature contains both base conventions (and others), sometimes without specifying which is being used. Factors of two (bits) are also called *shannons* (Sh), factors of e are called *natural units* (nats), and factors of 10 are called *Hartleys* (Hart), so that 1 bit = 1 Sh = $0.693147180559945\ldots$ nat = $0.301029995663981\ldots$ Hart. These units are used in information and communication theory and are dimensionless quantities like radians.

In this book, we will use 'log' generically for the logarithm in any base. Often the base chosen does not matter, such as where the ratio of two logarithms is used, or where the choice is arbitrary. When base-e logarithms are required, we will use the symbol 'ln', and when another base (a) is required, we will use the symbol '$\log_a$'.

5.1.2 Tent map

A slightly less trivial example is the tent map in eqn (2.15) with $f(X) = A\min(X, 1 - X)$ and $|f'(X)| = |A|$ everywhere. The Lyapunov exponent is $\lambda = \ln|A|$. This value is positive for $|A| > 1$ and negative for $|A| < 1$ in agreement with the bifurcation diagram in Fig. 2.14, confirming that fixed and periodic points become unstable when $|f'(X)| > 1$ at $X = X^*$. For $A = 2$, just before the solution becomes unbounded, the Lyapunov exponent is $\lambda = \ln 2$ as with the binary shift map.

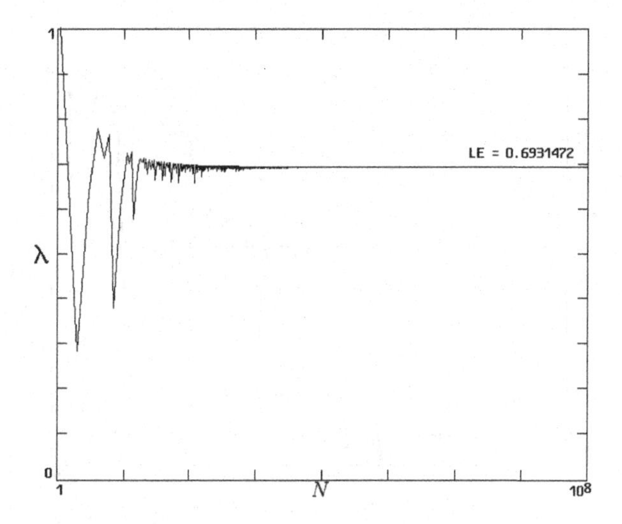

Fig. 5.1 Convergence of the Lyapunov exponent for the logistic map with $A = 4$.

5.1.3 Logistic map

More challenging is to calculate the Lyapunov exponent for the logistic map where $f(X) = AX(1 - X)$ and $f'(X) = A(1 - 2X)$. This calculation can be done numerically using eqn (5.3) which becomes

$$\lambda = \lim_{N \to \infty} \frac{1}{N} \sum_{n=0}^{N-1} \ln |A(1 - 2X_n)| \tag{5.4}$$

At $X = 0.5$ (the peak of the parabola), the logarithm is minus infinity. Thus you should use at least double precision when performing the calculation to resolve the singularity at this point. Discarding values near $X = 0.5$ skews the results, but including them makes the convergence very slow, since modest positive values of the logarithm are offset by occasional but important large negative ones. The map is stretched mostly near its ends ($X = 0$ and $X = 1$) where the local Lyapunov exponent is $\ln |A|$ and strongly compressed near its middle ($X = 0.5$) where the local Lyapunov exponent is $-\infty$. However, the average in eqn (5.4) is finite for almost all values of A.

Figure 5.1 shows the convergence of the Lyapunov exponent with increasing N for the logistic map with $A = 4$. The value after $N = 10^8$ iterations agrees with the expected value to seven significant digits. Figure 5.2 shows the variation of the exponent with A for a million iterations at each value of A. As expected, the exponent is positive where the bifurcation diagram in Fig. 2.4 indicates chaos and negative where it indicates periodicity, including the periodic windows. Since there are infinitely many such periodic windows, it is not obvious that the regions of chaos have a finite measure, but a rigorous proof was provided by Jakobson (1981).

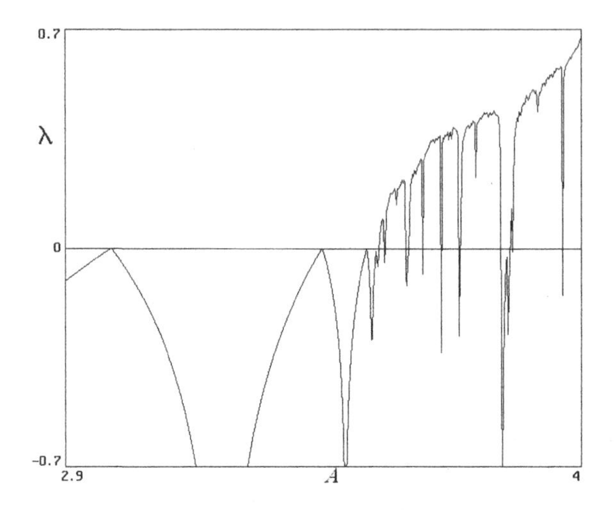

Fig. 5.2 Lyapunov exponent for the logistic map.

Table 5.2 Superstable orbits through period-6 for the logistic map.

Period	Approximate A values
1	2.0 (exact value)
2	$\sqrt{5}+1 = 3.2360679774997897\ldots$
3	3.8318740552833156
4	3.4985616993277015, 3.9602701272211526
5	3.738914912970685, 3.905706469831283, 3.990267046973701
6	3.62755753, 3.84456878, 3.93753644, 3.97776642, 3.99758311

The exponent is zero at each bifurcation point where the solution is on the verge of instability. Between each zero is a value of A at which the Lyapunov exponent is $-\infty$. These *superstable* orbits (or *supercycles*) occur when one of the iterates of the periodic orbit occurs at $X = 0.5$ where $f' = 0$, and they attract initial conditions in their basin of attraction faster and faster as the solution is approached. There is an infinite number of these points even within the region $3.57 < A < 4$ where the dynamics are primarily chaotic (see §6.9). The superstable orbits through period 6 are shown in Table 5.2. These orbits are used to calculate the Feigenbaum number (see §2.3.4) since their fast convergence makes them easier to identify than the bifurcation points where convergence is slow, and they have the same limiting ratio (Briggs 1991). They lie 'midway' between the birth (where $f' = 1$) and the death (where $f' = -1$) of each stable cycle.

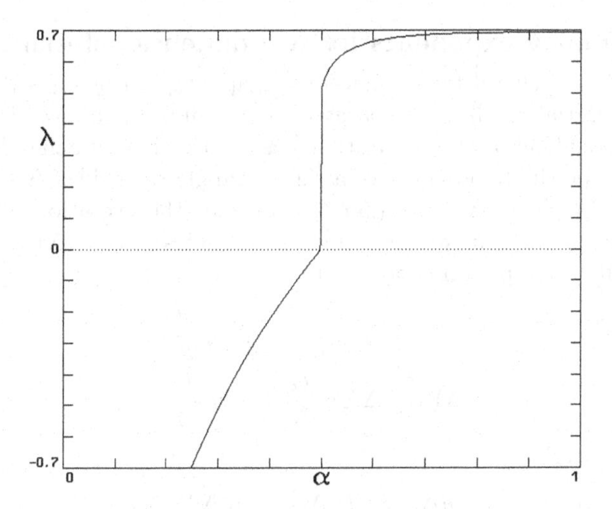

Fig. 5.3 Lyapunov exponent for general symmetric maps with $A = 1$.

For $A = 4$, the Lyapunov exponent for the logistic map can be calculated analytically using the probability distribution function in eqn (2.11)

$$\lambda = \int_0^1 P(X) \ln |f'(X)| dX = \frac{1}{\pi} \int_0^1 \frac{\ln |4(1 - 2X)|}{\sqrt{X(1 - X)}} dX \qquad (5.5)$$

It is not obvious that an average along an orbit and a weighted average over space give the same result, but the proof follows from the topological transitivity (see §2.4). Substituting $X = \sin^2(\pi Y/2)$, gives

$$\lambda = \int_0^1 \ln |4 \cos \pi Y| dY = \ln 2 \qquad (5.6)$$

The result is the same as for the tent map at $A = 2$ and for the binary shift map.

5.1.4 Other one-dimensional maps

Other unimodal maps have similar Lyapunov exponents. For example, the sine map with $A = 1$ in eqn (2.14) where $f(X) = \sin \pi X$ and $f'(X) = \pi \cos \pi X$ has $\lambda \simeq 0.689067$, half a percent smaller than $\ln 2$. For the general symmetric maps with $A = 1$ in eqn (2.16) where $f(X) = 1 - |2X - 1|^\alpha$ and $|f'(X)| = 2\alpha |2X - 1|^{\alpha - 1}$, the Lyapunov exponent is negative for $\alpha < 0.5$ and rises abruptly from $\lambda = 0$ to $\lambda = 0.5$ at $\alpha = 0.5$ and then rapidly to $\lambda = \ln 2$ at $\alpha = 1$ as in Fig. 5.3, and remains relatively constant thereafter.

The reason for the similarity is that all these cases map the unit interval back onto itself twice, and so the stretching is roughly a factor of two per iteration. The departures from this value are due to the nonuniform probability distribution or equivalently the fact that the orbit does not spend the same time in each part of the interval.

5.2 Lyapunov exponents for two-dimensional maps

Now consider a general two-dimensional map as given by eqns (3.37) and (3.38) with initial conditions separated by an infinitesimal ΔR. Unlike the one-dimensional map, the separation has a direction associated with it. Take ΔR to be the hypotenuse of a right triangle with sides ΔX and ΔY such that $(\Delta R)^2 = (\Delta X)^2 + (\Delta Y)^2$. Consider the evolution of a set of initial conditions that lie within a square with sides $\Delta X_0 = \Delta Y_0$. After one iteration, the corresponding values are

$$\Delta X_1 = \Delta X_0 \frac{\partial F}{\partial X} + \Delta Y_0 \frac{\partial F}{\partial Y} \tag{5.7}$$

$$\Delta Y_1 = \Delta X_0 \frac{\partial G}{\partial X} + \Delta Y_0 \frac{\partial G}{\partial Y} \tag{5.8}$$

and the new ΔR is

$$(\Delta R_1)^2 = (a\Delta X_0 + b\Delta Y_0)^2 + (c\Delta X_0 + d\Delta Y_0)^2 \tag{5.9}$$

where $a = \partial F/\partial X$, $b = \partial F/\partial Y$, $c = \partial G/\partial X$, and $d = \partial G/\partial Y$.

Upon iteration, the square typically distorts into a parallelogram as Fig. 5.4 shows. Note that the area of the parallelogram decreases, but it becomes successively more stretched with the direction of the stretch changing at each iteration. This *shearing* of the state space is caused by the off-diagonal terms (b and c) of the Jacobian matrix (more properly called the *monodromy matrix* since it is evaluated along the orbit rather than at a fixed point). An example of this effect with shear but *no* area contraction is Arnold's cat map (see §8.9.2).

5.2.1 Largest Lyapunov exponent

If we define the largest Lyapunov exponent as for the one-dimensional case

$$\lambda_1 = \lim_{N\to\infty} \frac{1}{N} \sum_{n=0}^{N-1} \ln |\Delta R_{n+1}/\Delta R_n| \tag{5.10}$$

some algebra leads to

$$\lambda_1 = \lim_{N\to\infty} \frac{1}{2N} \sum_{n=0}^{N-1} \ln \left[\frac{(a + bY_n')^2 + (c + dY_n')^2}{1 + Y_n'^2} \right] \tag{5.11}$$

Fig. 5.4 Evolution of a set of initial conditions for a typical two-dimensional chaotic map.

where $Y' = \Delta Y/\Delta X$ is the tangent of the direction of maximum growth (called the *tangent vector*) which evolves according to

$$Y'_{n+1} = \frac{c + dY'_n}{a + bY'_n} \tag{5.12}$$

and is independent of the initial Y'_0 after many iterations since any two initial conditions will orient themselves in the direction of maximum stretch (or minimum shrink). Note that a, b, c, and d generally depend on X_n and Y_n. Hence λ_1 is computed by iterating eqn (5.12) along with the map while computing the cumulative average in eqn (5.11). Typically the average will fluctuate and converge slowly as N increases. You can improve the convergence slightly by excluding the first few iterations from the average, thereby letting the orbit approach the attractor and the tangent vector orient along the maximally expanding direction.

5.2.2 Hénon map

As an example, consider the *Hénon map* (Hénon 1976)

$$X_{n+1} = 1 - 1.4X_n^2 + 0.3Y_n \tag{5.13}$$

$$Y_{n+1} = X_n \tag{5.14}$$

with parameters chosen to give chaos. The factor 0.3 is usually put in the second equation, but the above equivalent form has the virtue that Y is the previous value of X ($Y_n = X_{n-1}$), and so $X_{n+1} = 1 - 1.4X_n^2 + 0.3X_{n-1}$. It is always possible to write a D-dimensional map in terms of a time-delay one-dimensional map, but the number of time delays generally exceeds D (see §9.6). If the factor 0.3 were zero, then the Hénon map would reduce to a one-dimensional quadratic map with a chaotic solution having iterates in the range $-0.4 < X < 1$.

The Jacobian (monodromy) matrix for the Hénon map is

$$J = \begin{pmatrix} a & b \\ c & d \end{pmatrix} = \begin{pmatrix} -2.8X & 0.3 \\ 1 & 0 \end{pmatrix} \tag{5.15}$$

and the Lyapunov exponent is

$$\lambda_1 = \lim_{N\to\infty} \frac{1}{2N} \sum_{n=0}^{N-1} \ln\left[\frac{(-2.8X_n + 0.3Y'_n)^2 + 1}{1 + Y'^2_n}\right] \tag{5.16}$$

with $Y'_{n+1} = 1/(-2.8X_n + 0.3Y'_n)$. Figure 5.5 shows a numerical calculation of λ_1 versus N for the Hénon map with $X_0 = Y_0 = Y'_0 = 0$. About 10^4 iterations are required to get a result accurate to 1%. In fact, the error in λ_1 is of order $\delta\lambda_1 \sim 1/\sqrt{N}$, which is typical for a chaotic map, although there are exceptions (Theiler and Smith 1995). A much longer calculation ($N = 10^{11}$) gives $\lambda_1 = 0.419222 \pm 0.000003$, in good agreement with the value of $\lambda_1 = 0.4192 \pm 0.0001$ given by Grassberger and Procaccia (1984). Similar calculations for other two-dimensional chaotic maps give the results in Appendix A.

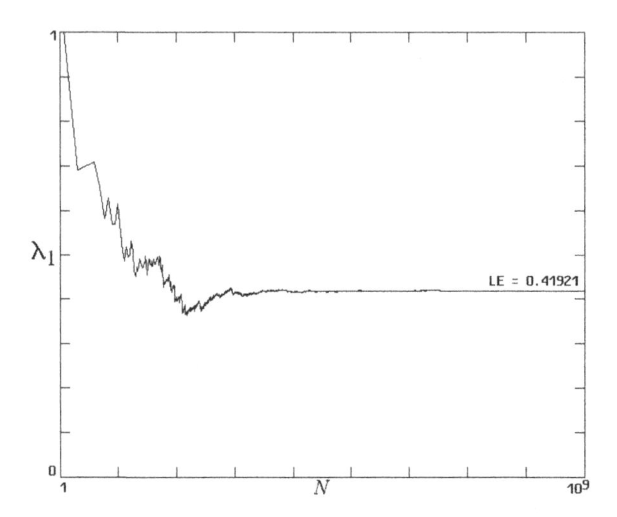

Fig. 5.5 Convergence of the largest Lyapunov exponent for the Hénon map.

5.2.3 Area expansion and contraction

For a shearless flow ($b = c = 0$), the area of the square after one iteration is $A_1 = \Delta X_1 \Delta Y_1 = a\Delta X_0 d\Delta Y_0 = adA_0$. However, as Fig. 5.4 illustrates, the shear leaves part of the rectangle empty. In fact, the empty region has area $b\Delta Y_0 c\Delta X_0$, and so the area of the remaining parallelogram is $A_1 = (ad - bc)A_0$. The quantity $ad - bc$ is the determinant of J, and thus the fractional area expansion is $A_1/A_0 = |\det(J)|$, where the absolute value makes the ratio always positive. A negative ratio means that the orientation of the points underwent a *mirror flip*, which is the two-dimensional extension of the flip that occurs in one-dimensional maps with $f'(X) < 0$.

Now describe the parallelogram in terms of a maximum expanding direction, as previously calculated, and a direction perpendicular to that whose exponent is the second Lyapunov exponent λ_2 as required for a two-dimensional map. In this rotated coordinate system, $A_1/A_0 = \mathrm{e}^{\lambda_1}\mathrm{e}^{\lambda_2} = \mathrm{e}^{\lambda_1+\lambda_2}$, from which follows

$$\lambda_1 + \lambda_2 = \ln(A_1/A_0) = \ln|\det(J)| = \ln|ad - bc| \tag{5.17}$$

This expression generalizes to invertible maps of higher dimension where the sum of the Lyapunov exponents is the logarithm of the fractional (hyper)volume expansion. The quantity $\det(J)$ generally varies along the orbit, and so eqn (5.17) gives the sum of the *local* Lyapunov exponents. The sum of the *global* exponents is obtained by averaging $\ln|\det(J)|$ along the orbit. The global area expansion cannot be positive for a bounded orbit. If the system has an attractor, then the expansion must be negative, corresponding to a contraction of initial conditions onto the attractor.

For example, the Hénon map has $\det(J) = -0.3$ independent of X and Y. Thus it has the special property of constant area contraction with

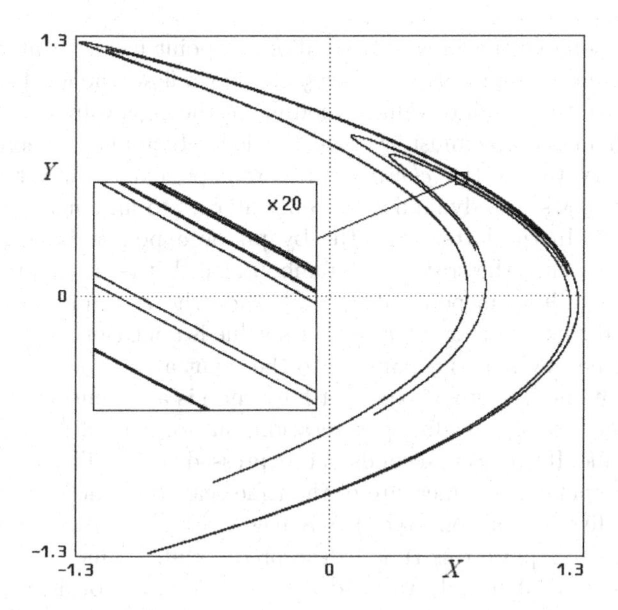

Fig. 5.6 Strange attractor for the Hénon map.

$\lambda_1 + \lambda_2 = \ln 0.3 = -1.203972\ldots$, and it is the most general quadratic map with this property. From the previous numerical calculation of λ_1, we obtain $\lambda_2 = -1.623195 \pm 0.000003$. A cluster of initial conditions will stretch out to form an infinitely long line because of the positive exponent, while collapsing to a region of zero area because of the area contraction. The line must be folded, however, to keep the orbit bounded. The resulting strange attractor in Fig. 5.6 shows the expected fractal structure in the 20-times zoomed inset. Benedicks and Carleson (1991) proved that the Hénon map has a strange attractor.

5.3 Lyapunov exponent for one-dimensional flows

To calculate the Lyapunov exponent for a one-dimensional *flow* given by $dx/dt = f(x)$, take two nearby initial conditions x_0 and $x_0 + \Delta x_0$ and advance time forward by a small Δt. The new separation is

$$\Delta x_1 = \Delta x_0 + [f(x_0 + \Delta x_0) - f(x_0)]\Delta t \simeq \Delta x_0[1 + f'(x_0)\Delta t] \tag{5.18}$$

where $f'(x_0) = df/dx$ evaluated at $x = x_0$. Define the local Lyapunov exponent λ by

$$\mathrm{e}^{\lambda \Delta t} = \Delta x_1/\Delta x_0 \simeq 1 + f'(x_0)\Delta t \tag{5.19}$$

For small $\lambda\Delta t$ where $\mathrm{e}^{\lambda\Delta t} \simeq 1 + \lambda\Delta t$, we have simply

$$\lambda = f'(x) \tag{5.20}$$

which is reasonable since a stable equilibrium point has a negative f' and an unstable one has a positive f' (see §3.1). For a flow, the local Lyapunov exponents are the absolute values (moduli) of the eigenvalues.

As with maps, you must average the local Lyapunov exponent over the trajectory to get the *global* Lyapunov exponent. However, for one-dimensional flows, the dynamics can only attract to an equilibrium point or to infinity. In the latter case, the Lyapunov exponent is positive, but uninteresting, since the trajectory is unbounded. If the trajectory attracts to a stable equilibrium point at $x = x^*$, then the Lyapunov exponent is negative and equal to $f'(x)$ at $x = x^*$. Its value is a measure of how rapidly a cluster of nearby points collapse onto the point attractor.

Note that the Lyapunov exponent for a map is a dimensionless number since it is the rate of spreading per iteration, but for flows it has dimensions of inverse time. If time is in seconds, λ is expressed as s^{-1}. The inverse of the Lyapunov exponent is a measure of the time-scale on which nearby initial conditions diverge (or converge if λ is negative) on average. However, it is dangerous to equate this time to the predictability time since the local Lyapunov exponent usually varies drastically along the orbit, as with the logistic map with $A = 4$ where it varies from $-\infty < \lambda < \log 4$. The global Lyapunov exponent is the average rate of error growth, but it is a bad idea to assume $\langle 1/\lambda \rangle$ is equal to $1/\langle \lambda \rangle$. Furthermore, the Lyapunov exponent measures only the growth of *infinitesimal* errors, and as such, it can hardly affect predictability. When the error becomes finite, the growth rate may change considerably. Often the *uncertainty doubling time* is of more interest and easier to calculate than the Lyapunov exponent (Smith 1997).

5.4 Lyapunov exponents for two-dimensional flows

Calculation of the Lyapunov exponents for a two-dimensional flow follows the method used for a two-dimensional map. In fact, you can use the Euler method (see §3.9.1) to convert the flow to a map for which $F(x, y) = x + f(x, y)\Delta t$ and $G(x, y) = y + g(x, y)\Delta t$ to derive the corresponding equations

$$\lambda_1 = \lim_{T\to\infty} \frac{1}{T} \int_0^T \frac{a + by' + cy' + dy'^2}{1 + y'^2} dt \tag{5.21}$$

$$\frac{dy'}{dt} = \frac{c + dy' - ay' - by'^2}{1 + (a + by')\Delta t} \tag{5.22}$$

$$\lambda_1 + \lambda_2 = \lim_{T\to\infty} \frac{1}{T} \int_0^T (a + d)dt \tag{5.23}$$

where $a = \partial f/\partial x, b = \partial f/\partial y, c = \partial g/\partial x$, and $d = \partial g/\partial y$. The first equation averages the local Lyapunov exponent along the trajectory. The second equation evolves the direction of the tangent vector. You might be tempted to set Δt (the step size of the numerical integration) in its denominator to zero, but it is needed to get the tangent vector 'around the corner' where $y' = \Delta y/\Delta x$ is infinite. This factor makes the singularity numerically integrable. The third equation averages the trace of the Jacobian (monodromy) matrix along the trajectory to get the average area expansion (or contraction if negative), $dA/dt/A$, which is the average divergence of the flow vector.

These equations are of somewhat limited use because of the restricted dynamics for autonomous two-dimensional flows. Solutions can attract to a point, attract to infinity, attract to a limit cycle, or circulate about a center. If they attract to a point, both exponents must be negative, and if the point is a node, there is no need to perform the average in eqns (5.21) and (5.23) since y' is given by setting $dy'/dt = 0$ in eqn (5.22) with $\Delta t = 0$ and the other quantities are evaluated at the point. If they attract to infinity, at least one exponent will usually be positive, but the solution is unbounded and hence of limited physical interest. Solutions that attract to a limit cycle have $\lambda_1 = 0$, corresponding to the direction parallel to the flow, and $\lambda_2 = a + d < 0$ in the contracting direction perpendicular to the flow. For a limit cycle, there must be a region in the xy plane (external to the limit cycle) where the area is contracting with $a + d < 0$ and a region (internal to the limit cycle) where the area is expanding with $a + d > 0$. Solutions that circulate about a center have both exponents zero since they neither contract nor expand.

5.5 Lyapunov exponents for three-dimensional flows

Only in three or more dimensions can continuous flows exhibit chaos and have a positive Lyapunov exponent without being unbounded. Calculation of the largest Lyapunov exponent in three dimensions follows the procedure for the two-dimensional case, but the equations are unwieldy because the Jacobian matrix has nine terms in general and the tangent vector is three-dimensional, requiring the solution of two additional equations, one for $y' = \Delta y/\Delta x$ and another for $z' = \Delta z/\Delta x$. Fortunately, there is an easier way, as described in the next section.

Note, however, that an autonomous three-dimensional flow has special properties that allow all three exponents to be easily determined for most systems of interest. If the system is not a point attractor, then the largest exponent cannot be negative, and hence $\lambda_1 \geq 0$. For a bounded system, the sum of the exponents cannot be positive, and hence $\lambda_3 \leq 0$. (The exponents are conventionally ordered so that $\lambda_1 \geq \lambda_2 \geq \lambda_3 \geq \cdots$.) A bounded system with $\lambda_1 > 0$ must have a zero exponent corresponding to the direction of the flow (Haken 1983a), and hence $\lambda_2 = 0$. It is easy to see that two nearby initial conditions on the same trajectory cannot separate on average since they have the same time history except for a slight time delay. Thus a calculation of λ_1 and the rate of volume expansion

$$\frac{1}{V}\frac{dV}{dt} = \lambda_1 + \lambda_2 + \lambda_3 = \langle \text{trace}(J) \rangle$$
$$= \lim_{T\to\infty} \frac{1}{T}\int_0^T \left(\frac{\partial f}{\partial x} + \frac{\partial g}{\partial y} + \frac{\partial h}{\partial z}\right) dt \quad (5.24)$$

gives all three exponents, which for a chaotic system are positive, zero, and negative, respectively.

Another important result is that for a system to have an attracting torus, there must be a region of space (inside the torus) that is volume-expanding, so that initial conditions within the torus can expand to it. For a similar reason, repellors require an expanding region of the state space surrounding them. Thus if the volume expansion or contraction is constant everywhere, as it is for the Lorenz attractor, the only possible equilibrium points are nodes or saddles (no repellors), and the only possible dynamical equilibria are limit cycles and strange attractors (no tori).

5.6 Numerical calculation of the largest Lyapunov exponent

Since Lyapunov exponents must usually be calculated numerically and the largest exponent is the most important, it is useful to develop a general numerical technique that works for any system in any dimension and that does not require explicit evaluation of the monodromy matrix. You can apply this method to any system without having to calculate the matrix of partial derivatives. In essence, you numerically evaluate the derivative along the direction of maximum expansion (or minimum contraction) and average its logarithm over the orbit (Benettin *et al.* 1980b). We consider the case of maps, since flows are converted to maps when calculated numerically. The procedure is independent of the numerical method used to solve the differential equations. In any such calculation of numerical derivatives, at least double precision should be used to reduce round-off errors.

The procedure as indicated in Fig. 5.7 is as follows:

1. Choose an initial condition $\mathbf{R}_0$ and a second initial condition separated from it by a small amount $\Delta\mathbf{R}_0$ in any direction.[2] The magnitude of the vector $\Delta\mathbf{R}$ should be much smaller than the scale on which the flow changes but a few powers of ten larger than the numerical precision. In double precision $\Delta R_0 = 10^{-10}$ will usually suffice.
2. Iterate the equations one time step for each initial condition and determine $\mathbf{R}_1$ and $\Delta\mathbf{R}_1$. You can think of the first orbit as the unperturbed one (sometimes called the *fiducial trajectory*) and the second as the perturbed one. If the spreading is small, it is slightly more efficient to advance several time steps using some criterion for the allowed $\Delta\mathbf{R}_n$, but additional bookkeeping is required.

[2]The boldface quantity $\mathbf{R}$ is a vector pointing from the origin to the point $(x, y, z, \ldots)$. The magnitude (length) of $\mathbf{R}$ is a scalar denoted by R.

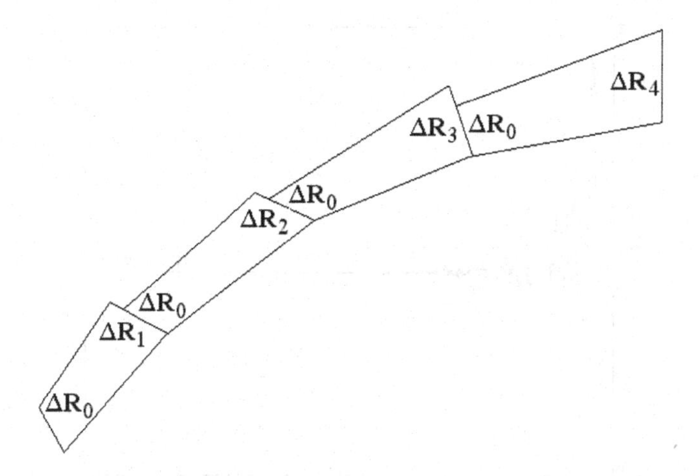

Fig. 5.7 Numerical method for calculating the largest Lyapunov exponent.

3. Change the position of the second orbit from $\mathbf{R}_1 + \Delta\mathbf{R}_1$ to $\mathbf{R}_1 + (\Delta R_0/\Delta R_1)\Delta\mathbf{R}_1$. This is the crucial step that keeps the orbits close (separation ΔR_0) while letting the direction orient to that of maximum expansion.
4. Add the quantity $\lambda_1 = \ln(\Delta R_n/\Delta R_0)$ to a running average, and loop back to step 2, exiting when the average appears to have converged. The convergence can be improved slightly by eliminating from the average some of the early iterations where the orbit is not yet on the attractor and $\Delta\mathbf{R}$ is not properly oriented.
5. If the system is a flow, divide the resulting average λ_1 by the step size h so that the units are correct.

Applying the method to the Lorenz attractor in eqns (4.32)–(4.34) with $\sigma = 10$, $r = 28$, $b = 8/3$, $h = 0.01$, and $\Delta R_0 = 1 \times 10^{-10}$ gives the result in Fig. 5.8. As with the Hénon map, the error in λ_1 is of order $\delta\lambda_1 \sim 1/\sqrt{\omega t}$, where ω is a characteristic frequency, which we take to be the linear frequency of oscillation $\omega = 12.47528\ldots$. A much longer calculation ($t = 10^8$, corresponding to $N = 10^{10}$ iterations with $h = 0.01$) gives $\lambda_1 = 0.90564 \pm 0.00003$. This result is consistent with the value of $\lambda_1 = 0.91 \pm 0.01$ given by Grassberger and Procaccia (1984). The Lorenz attractor has $\lambda_1 + \lambda_2 + \lambda_3 = \mathrm{trace}(J) = -\sigma - 1 - b = -41/3$. With $\lambda_2 = 0$, we obtain $\lambda_3 = -14.57231 \pm 0.00003$. Similar calculations for other chaotic flows give the results in Appendix A. This method has been benchmarked against the methods using analytic derivatives described earlier in this chapter with typical agreement to seven digits.

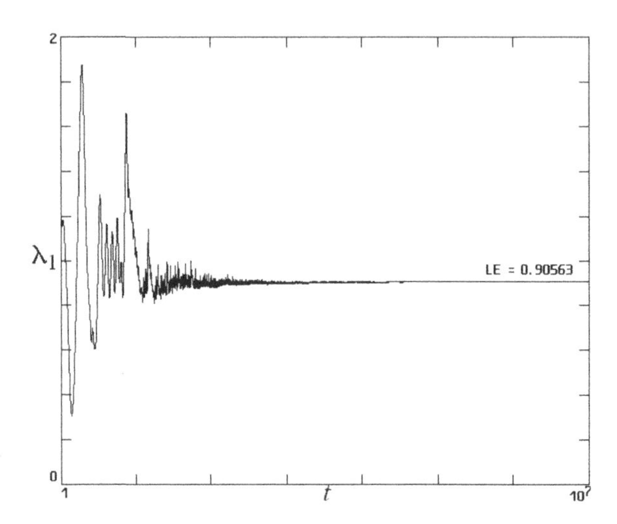

Fig. 5.8 Convergence of the largest Lyapunov exponent for the Lorenz attractor.

5.7 Lyapunov exponent spectrum in arbitrary dimension

If you only want to determine whether a system is chaotic, calculating the *largest* Lyapunov exponent suffices. If you want an estimate of the average predictability, you need all the *positive* exponents. Otherwise, you need the entire *spectrum* of exponents. For low-dimensional systems, the methods described above will often suffice. For dimensions higher than two for maps and three for flows, more general methods are needed. Such methods are beyond the scope of this book, but we outline what is involved and provide a few standard references. A survey of methods is given by Geist *et al.* (1990).

One method is to calculate the spectrum of local Lyapunov exponents at each point on the orbit and then average each exponent along the orbit (Benettin *et al.* 1980b). Alternately, multiply together the Jacobian (monodromy) matrices at all time steps along the orbit and then calculate the eigenvalues of the resulting product matrix.[3] Note, however, that matrix multiplication does not commute (Oseledec 1968). This method is conceptually equivalent to following the evolution of an initial (hyper)sphere of initial conditions, which distorts into an ellipsoid with the longest principal axis corresponding to λ_1, the next longest to λ_2, and so forth. The method is straightforward but computationally intensive. In practice, it fails with a chaotic system for two reasons. First, the individual terms of the cumulative product matrix grow exponentially. Second, the eigenvectors all tend to align in the direction of maximum growth and hence do not accurately span the space.

[3]Proof that a solution exists and is unique was first shown by Oseledec (1968) and further refined by Pesin (1977) and Katok (1980).

Both of these problems are overcome using *Gram–Schmidt reorthonormalization* (Press *et al.* 1992), which factors out a large scalar multiplier to prevent divergence and does row reduction with pivoting to retain the independence of the columns of the product matrix. For the theory, implementation, and sample numerical algorithms,[4] see Nemytskii and Stepanov (1960), Shimada and Nagashima (1979), Ruelle (1982), Eckmann and Ruelle (1985), Wolf *et al.* (1985), Geist *et al.* (1990), Habib and Ryne (1995), Christiansen and Rugh (1997), von Bremen *et al.* (1997), and Janaki *et al.* (1999). A method that does not require reorthonormalization has been proposed by Rangarajan *et al.* (1998).

5.8 General characteristics of Lyapunov exponents

The Lyapunov exponent spectrum provides additional useful information about the system. Conversely, certain geometric properties of the system imply characteristics of the spectrum. Some of these properties have already been mentioned, but we now summarize and expand on those results.

Specifically, consider an autonomous flow with four variables. Lower-dimensional cases have mostly been considered, and higher-dimensional cases are a simple extension of this example. If the four exponents sum to a positive value, then the system is unbounded since the state space is expanding, and hence it is probably not a useful physical model. If the exponents sum to zero, then the system is volume-conserving and usually structurally unstable. However, there are important *conservative* examples, such as mechanical systems with little or no friction where energy is approximately conserved as in planetary motion, to which Chapter 8 will be devoted.

Here we consider *dissipative* bounded cases where the exponents sum to a negative value, such that at least one must be negative. In such cases, the state space contracts, and the dynamics collapse onto an attractor. Transients, which are dominated by the least negative exponent, are often very important since they might destroy the circuits in your computer when you turn it on, but they correspond to flow off the attractor. However, we will focus on the long-term dynamics on the attractor, which are easier to study.

Table 5.3 summarizes the various attractors and the signs of their Lyapunov exponents. Three new ideas emerge — a *3-torus*,[5] which is a quasiperiodic system with three incommensurate frequencies, a *hyperchaotic attractor*, which has more than one expanding direction for which the standard example (Rössler 1979b) is

[4] An 'algorithm' is a set of instructions for solving a problem, the word coming from a distortion of al-Khwarizmi (*c.* 780–850), an Arab mathematician.

[5] To make a 3-torus, start with a cube that you stretch and bend until each of the three pairs of opposite faces touch one another. CAUTION: You need to go into the fourth dimension to do this.

Table 5.3 Characteristics of the attractors for a four-dimensional flow.

λ_1	λ_2	λ_3	λ_4	Attractor	Dimension
$-$	$-$	$-$	$-$	Equilibrium point	0
0	$-$	$-$	$-$	Limit cycle	1
0	0	$-$	$-$	2-torus	2
0	0	0	$-$	3-torus	3
$+$	0	$-$	$-$	Strange (chaotic)	>2
$+$	$+$	0	$-$	Strange (hyperchaotic)	>3

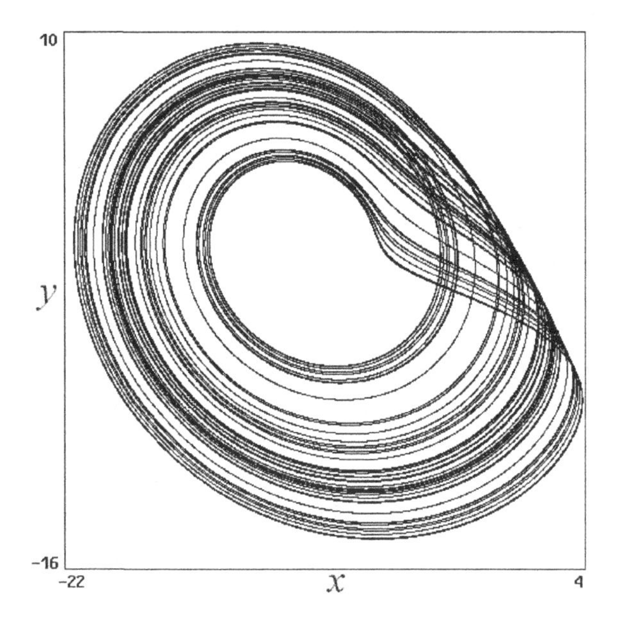

Fig. 5.9 Rössler hyperchaotic attractor projected onto the xy plane.

$$\frac{dx}{dt} = -y - z \tag{5.25}$$

$$\frac{dy}{dt} = x + 0.25y + w \tag{5.26}$$

$$\frac{dz}{dt} = 3 + xz \tag{5.27}$$

$$\frac{dw}{dt} = 0.05w - 0.5z \tag{5.28}$$

shown in Fig. 5.9 with (base-e) Lyapunov exponents 0.11, 0.02, 0, and -27 (Wolf *et al.* 1985), and the concept of *noninteger dimension*, to which we now turn.

5.9 Kaplan–Yorke (or Lyapunov) dimension

Whereas the Lyapunov exponent measures the average predictability of a dynamical system, the dimension of its attractor measures its complexity. The attractor of a system can have a dimension up to but not equal to the number of variables, and it is often significantly less, in which case its long-term behavior might be described by a much simpler model. The dimension of the attractor is a lower limit on the number of variables required to model the dynamics. The existence of simple models for complex dynamics is what motivates much of the interest in chaos.

Dimensions less than two should be readily evident in the state-space trajectory. For dimensions of two or greater, the trajectory will fill a region of the plane, and visual inspection cannot reveal its dimension. Taking a Poincaré section reduces the dimension by one and exposes attractors with dimension less than three. In principle, a slice of the Poincaré section would allow visual inspection of higher dimensions, but each successive cross-section vastly increases the computation required to generate sufficient data (Lorenz 1984a, Moon and Holmes 1985). Such *double Poincaré sections* are also called *Lorenz cross-sections*.

For these reasons, and to automate the dimension calculation, the spectrum of Lyapunov exponents is useful. If the largest exponent is zero or negative, the attractor dimension is equal to the number of zero exponents (point attractor, limit cycle, 2-torus, 3-torus, etc.). Each zero exponent implies a dimension in which the state space does not collapse. If the largest exponent is positive, the system is chaotic and will usually have a strange attractor with noninteger dimension. A chaotic *flow* will have a zero exponent in addition to the positive ones (Haken 1983a), and hence its dimension cannot be less than two. Nor can it equal two, since the trajectory would then lie on a surface and could not intersect itself. The solution for a chaotic *map* can have any dimension greater than zero, as with the one-dimensional logistic map.

Recall that the sum of the Lyapunov exponents is the long-time average rate of hypervolume expansion. Suppose the first D of these exponents (in descending order) sum to zero. Then the D-dimensional hypervolume will neither expand nor contract, and the trajectory will be D-dimensional. The largest such D for which $\lambda_1 + \lambda_2 + \cdots + \lambda_D \geq 0$ is the *topological dimension*. The next higher dimension ($D+1$) is the minimum dimension in which the attractor can exist. In this dimension, the hypervolume is contracting, and thus the attractor dimension must lie between D and $D+1$. One definition of a *fractal* is an object whose space-filling dimension exceeds its topological dimension as is the case for a strange attractor.

A fractional dimension for such an attractor can be defined by simple interpolation

$$D_{\mathrm{KY}} = D + \frac{1}{|\lambda_{D+1}|} \sum_{j=1}^{D} \lambda_j \tag{5.29}$$

This dimension is called the *Kaplan–Yorke dimension* or *Lyapunov dimension*. Kaplan and Yorke (1979) originally conjectured that it is equivalent to the capacity dimension[6] (see §12.2), but Ledrappier and Young (1985) proved that it is equal to the information dimension (see §13.2.3) for two-dimensional maps, although there are counter-examples in higher dimensions. Its relation to other geometrical measures of the dimension has been discussed by Farmer *et al.* (1983) and Young (1984). One implication of this result is that the exponents with $j > D + 1$ have little significance to the attractor, which is fortunate since they are hard to calculate. The Kaplan–Yorke dimension usually converges faster than the Lyapunov exponents from which it is derived, suggesting that the attractor dimension is a more robust quantity than the Lyapunov exponents.

A dissipative two-dimensional chaotic map has $D_{\text{KY}} = 1 - \lambda_1/\lambda_2$. For example, the Hénon map has $D_{\text{KY}} \simeq 1.258269$, which is reasonable since Fig. 5.6 is more like a line ($D = 1$) than a surface ($D = 2$). A three-dimensional chaotic flow has $\lambda_2 = 0$ and $D_{\text{KY}} = 2 - \lambda_1/\lambda_3$. For example, the Lorenz attractor has $D_{\text{KY}} \simeq 2.06215$, which is reasonable since the butterfly wings are nearly planar and the entanglement of the trajectory is restricted to a small Cantor-like region near the z axis. Most Poincaré sections of the Lorenz attractor are indistinguishable from line segments except at very high magnification. The Rössler hyperchaotic attractor in eqns (5.25)–(5.28) has $D_{\text{KY}} = 3 - (\lambda_1 + \lambda_2)/\lambda_4 \simeq 3.005$. The Kaplan–Yorke dimensions of other common chaotic systems are given in Appendix A.

Given that the Kaplan–Yorke dimension is an interpolation, you might think of fitting the sum of the first D exponents to a $(D - 1)$-degree polynomial and finding the dimension from its zero crossing. For a three-dimensional chaotic system with $\lambda_2 = 0$ the result of fitting a parabola to the three exponents gives

$$D_\Sigma = 1.5 + 0.5\sqrt{1 - 8\lambda_1/\lambda_3} \tag{5.30}$$

For the Lorenz attractor, the result is $D_\Sigma = 2.1118$, which is further than D_{KY} from the value calculated in other ways (see Chapter 12). This method seems to give worse results for attractors such as the Lorenz and Rössler whose dimension is close to an integer, but it may improve other cases.

5.10 Precautions

Before leaving the subject of Lyapunov exponents it is useful to mention a few potential difficulties that can lead to errors and even cause misidentification of an attractor.

1. Be sure the orbit calculation gives sensible results. If you have a mathematical or programming error, use too large a step size, or choose initial conditions outside the basin of attraction, the orbit may not be

[6] A similar suggestion was made independently by Mori (1980).

the intended one. There may be multiple attractors with complicated basin boundaries, including *riddled basins* (Alexander *et al.* 1992, Sommerer and Ott 1993) where every point in the basin of one attractor is arbitrarily close to a point in the basin of another. The best precaution is to observe the orbit graphically while the Lyapunov exponent is calculated. At least test the variables to see if they are going to infinity or settling to a static equilibrium.

2. Be sure to calculate long enough for the orbit to reach the attractor, orient in the direction of maximum expansion, and adequately sample the attractor. A real-time plot is useful to show the approach to the attractor and how densely the orbit has filled it, and a plot of the cumulative average exponent will show whether convergence has occurred and how many digits are significant.
3. For attractors with a strongly contracting region, such as the logistic map at $X = 0.5$, you might encounter separations ΔR_1 smaller than can be represented or logarithms of numbers too near zero. You should test for such underflows and increase ΔR_0 or the calculation precision, since ignoring them can give a false positive exponent.
4. Vary the initial conditions, step size h, initial orbit displacement ΔR_0 and orientation, and number of iterations N to see that the calculated exponent is insensitive to the chosen values. Repeating the calculation many times with different initial conditions gives a measure of the precision through the distribution of estimates. Variation of h will help detect systems with slower than exponential growth of their separation.
5. Since it is difficult to distinguish numerically a small positive exponent from a zero exponent, supplement the calculation with other methods such as Poincaré sections and power spectra to test for periodicity.

Even with these precautions, there are systems whose Lyapunov exponents are undefined or difficult or impossible to calculate. These include cases where the functions are discontinuous so that the Jacobian matrix has infinities, cases where the variables are discrete, and cases where the Lyapunov exponent is unmeasurably large, such as with maps used to generate *pseudo-random numbers*. For such cases, other analysis methods may be required.

5.11 Exercises

Exercise 5.1 Calculate the Lyapunov exponent for the *decimal shift map* $X_{n+1} = 10X_n \pmod 1$.

Exercise 5.2 Show that, in contrast to the logistic map, the tent map does *not* have periodic windows within the chaotic region $1 < A < 2$.

Exercise 5.3 Show that the Lyapunov exponent for the logistic map is zero at the bifurcation points.

Exercise 5.4 Derive an expression for the Lyapunov exponent of the logistic map in terms of A for the region $0 < A < 3$ where the attractor has a stable fixed point.

Exercise 5.5 Derive an expression for the Lyapunov exponent of the logistic map in terms of A for the region $3 < A < 3.44948\ldots$ where the attractor has a stable 2-cycle.

Exercise 5.6 Show that $A = 2$ gives a period-1 supercycle for the logistic map, and calculate the iterates of $X_0 = 0.1$ until they converge to the fixed point to within the precision of your calculator.

Exercise 5.7 Show that $A = 1 + \sqrt{5} = 3.23606797\ldots$ gives a period-2 supercycle for the logistic map, and calculate the two corresponding values of X^*.

Exercise 5.8 Evaluate the integral in eqn (5.5) to show that the Lyapunov exponent of the logistic map at $A = 4$ is $\lambda = \ln 2$.

Exercise 5.9 Calculate the Lyapunov exponent for the general symmetric map in eqn (2.16) for $A = 1$ over the range $0 < \alpha < 0.5$ where the map has a single stable fixed point.

Exercise 5.10 Show that the largest Lyapunov exponent for a general two-dimensional map is given by eqn (5.11) and that the tangent vector evolves according to eqn (5.12).

Exercise 5.11 Show that the Hénon map in eqns (5.13) and (5.14) can alternately be written

$$X_{n+1} = 1.4 + 0.3Y_n - X_n^2 \tag{5.31}$$

$$Y_{n+1} = X_n \tag{5.32}$$

Exercise 5.12 Show that the Lyapunov exponents for two decoupled tent maps

$$X_{n+1} = A\min(X_n, 1 - X_n) \tag{5.33}$$

$$Y_{n+1} = B\min(Y_n, 1 - Y_n) \tag{5.34}$$

are $\lambda_1 = \ln\max(|A|, |B|)$ and $\lambda_2 = \ln\min(|A|, |B|)$ and that $\lambda_1 + \lambda_2 = \ln|AB|$.

Exercise 5.13 Calculate the Jacobian matrix for each of the chaotic maps in Appendix A.

Exercise 5.14 Calculate the Lyapunov exponent for the logistic differential equation in eqn (3.2) as a function of the parameter a.

Exercise 5.15 Derive eqns (5.21)–(5.23) for the Lyapunov exponents of a two-dimensional flow.

Exercise 5.16 Find the region of the xy plane that is area expanding and the region that is area contracting for the van der Pol oscillator.

Exercise 5.17 Verify that the Lyapunov exponents for a two-dimensional flow around a center with eqns (3.4) and (3.5) are $\lambda_1 = \lambda_2 = 0$.

Exercise 5.18 Find the Lyapunov exponents for the overdamped harmonic oscillator in eqn (3.23) with $\omega = 1$ as a function of the damping constant b.

Exercise 5.19 Verify that the Lorenz system has a constant volume expansion and thus cannot have an attracting torus for any choice of parameters or initial conditions.

Exercise 5.20 Calculate the Jacobian matrix for each of the chaotic flows in Appendix A.

Exercise 5.21 Calculate the sum of the Lyapunov exponents for those three-dimensional flows in Table 4.1 where its value is constant in space, and indicate which cases require a numerical average along the trajectory.

Exercise 5.22 What can you deduce about the properties of an attractor for a four-dimensional system whose Lyapunov exponents are $+, 0, 0, -$, respectively?

Exercise 5.23 Calculate the Kaplan–Yorke dimension of the Rössler hyperchaotic system in eqns (5.25)–(5.28).

Exercise 5.24 Calculate the Kaplan–Yorke dimension of a system whose Lyapunov exponents are $0.5, 0.1, 0, -0.3$, and -0.6.

Exercise 5.25 By fitting a parabola to the sum of the first D Lyapunov exponents for a three-dimensional chaotic flow and calculating its zero crossing, show that the resulting dimension is given by eqn (5.30).

Exercise 5.26 Calculate the dimension D_Σ for each autonomous dissipative flow in §A.5 using the parabolic fit in eqn (5.30).

Exercise 5.27 Estimate the spurious positive Lyapunov exponent for the nonlinear oscillator $d^2x/dt^2 = -x^3$ with $x_0 = 1$, $v_0 = 0$, and $h = 0.01$.

Exercise 5.28 Calculate the Lyapunov exponent for a linear congruential generator $X_{n+1} = AX_n + B \pmod{C}$, used to generate pseudo-random numbers, where A, B, and C are arbitrary large integers.

5.12 Computer project: Lyapunov exponent

In this project you will develop a numerical test for chaos in the solution of a system of equations. This test will consist of estimating the largest Lyapunov exponent. If the solution is bounded and has a positive value of the largest Lyapunov exponent, then it is generally considered to be chaotic. This test allows you to automate the detection of chaos while scanning a range of parameters in a numerical experiment and to quantify the sensitivity to initial conditions. It is a good way to distinguish between quasiperiodicity and chaos.

1. Calculate numerically the largest Lyapunov exponent for the Hénon map

$$X_{n+1} = 1 - aX_n^2 + bY_n \tag{5.35}$$

$$Y_{n+1} = X_n \tag{5.36}$$

with the usual parameters of $a = 1.4$ and $b = 0.3$, using the procedure in §5.6. Be careful to advance X and Y simultaneously, *not* sequentially (i.e., do not accidentally use the value of X_{n+1} determined from the first equation above instead of X_n when you evaluate Y_{n+1} from the second equation above). Iterate the equations a few dozen times to be sure the solution is on the attractor, and then calculate the average logarithmic rate of separation of two nearby orbits separated by a distance ΔR_0

$$\lambda_1 = \lim_{N\to\infty} \frac{1}{N} \sum_{n=1}^{N} \ln |\Delta R_n / \Delta R_0| \tag{5.37}$$

Be sure R_0 remains small compared with the size of the attractor and that it is oriented along the direction of fastest growth. You should get a value close to 0.419 per iteration (using base-e logarithms). (Note that the other Lyapunov exponent is easily obtained from $\lambda_1 + \lambda_2 = \ln |b|$.)
2. Make a plot of λ_1 versus N to ensure that your estimate of the exponent has settled down to a unique value.
3. Make a plot of λ_1 versus a over the range of $0 < a < 2$. You should see a period-doubling route to chaos ($\lambda_1 = 0$ at each bifurcation point) and periodic windows (regions of negative λ_1) within the chaotic region, similar to the case for the logistic equation.
4. (optional) Use different nonlinear functions $F(X, Y)$ and $G(X, Y)$ of your choosing and try to discover some new examples of chaotic maps.

6 Strange attractors

More often than not, the orbit or trajectory of a chaotic system will be drawn to a small region of state space as time evolves, whereupon it moves in a deterministic but unpredictable manner on a fractal object called a *strange attractor.*[1] Strange attractors are the usual geometric manifestation of chaos. They were originally called 'strange' because of their fractal structure. The terms 'strange attractor' and 'chaotic attractor' are often used interchangeably, according to whether the interest is in their geometrical or dynamical properties, although some authors make a distinction between them.

It is useful to discuss their properties and to examine examples of them so that you will recognize them when they occur in experimental data. With a sufficiently large collection of such objects, we can address certain statistical questions, such as how common is chaos, what are the most probable values of Lyapunov exponent and dimension of various chaotic systems, and what are the most common routes to chaos? Since strange attractors can have arbitrarily high dimension, we discuss visualization techniques, which leads to consideration of their aesthetic qualities. The methods used to discover and display strange attractors represent a potential new art form as well as a way to discover new dynamical behaviors and to visualize experimental and numerical data.

6.1 General properties

Like so many terms in nonlinear dynamics, there is no universally accepted definition of a strange attractor, but most such objects of practical interest have the following properties, many of which are shared by other attractors and some of which have been previously mentioned:

1. It is a *limit set* as time goes to infinity. It is called an *omega set* in contrast to an *alpha set*, which is the limit set as time goes to *minus* infinity. (Alpha and omega are the first and last letters of the Greek alphabet.) Note that it takes an infinite time for an arbitrary initial condition to reach the attractor, but it is usually approached

[1]The term 'strange attractor' was first used in print by Ruelle and Takens (1971), but neither author call recall who coined it (Ruelle 1991).

very closely in a few times the inverse of the least negative Lyapunov exponent.

2. It is an *invariant set* . Any orbit or trajectory that starts on it stays on it for all time.
3. It is *bounded*. It does not stretch to infinity but can be enclosed within a region of finite (hyper)volume, but see §13.7 for some exceptions. It is contained within a *basin of attraction* that may stretch to infinity but often has a finite and sometimes fractal boundary.
4. It is a *set of measure zero* in the state space. If the state space is two-dimensional, the attractor will have a dimension less than two and hence zero area. If you were to throw darts at it, you would never hit it. Neither could you see it, just as you cannot see a line in a plane unless you draw it with a finite width, in which case it is not really a line.
5. It is a *fractal*. Chapter 11 will more fully describe fractals, but in brief they are *self-similar* objects with structure on all size scales and usually a noninteger dimension.
6. It is *dense in periodic orbits*. Every point on the attractor is arbitrarily close to one of these orbits, but they comprise a set of measure zero, and they are all unstable. Most of these orbits have extremely long periods.
7. It is *transitive*. This means that if you start almost anywhere on the attractor (other than one of the periodic orbits), the dynamics will take you arbitrarily close to every other point on the attractor. The nonperiodic orbits are *dense*, which means that every point on the attractor is arbitrarily close to one.
8. It is *measure invariant*. This means that every transitive orbit spends the same fraction of its time in a given region of the attractor after infinite time, but the measure may not be uniform (not the same everywhere on the attractor). No matter where you put the drop of cream in your coffee, stirring chaotically will eventually produce a uniform density of cream. Almost no orbit is special; almost all are typical. The exceptions are the periodic orbits, but they are a set of measure zero on the attractor, albeit infinitely numerous.
9. It is *indecomposable* (or *ergodic*). This means that it is not made of smaller attractors that come close or even touch one another. Some authors (e.g., Ruelle 1989) distinguish attractors from the more general *attracting sets*, which are produced by a cloud of initial conditions and can be decomposed.
10. It is *structurally stable*. Although most attractors violate the strict requirement that there be a nonvanishing neighborhood in parameter space that gives topologically equivalent attractors, as a practical matter, they usually remain intact for most perturbations of the coefficients or the addition of other small terms. As a result, their structure

is not sensitive to small numerical errors, although the sequence of points on the attractor visited by an orbit usually is. Structurally stable systems are sometimes called *robust*, *coarse*, or *rough*, although the term 'robust' also has a more restricted meaning (see §6.9).

11. It is usually *chaotic*. This means that most nearby initial conditions separate exponentially on average. There are exceptions (Grebogi *et al.* 1984, Ditto *et al.* 1990), such as the logistic map at the accumulation point, which is a strange attractor (a Cantor set), but with zero Lyapunov exponent, and certain quasiperiodically driven oscillators (Romeiras and Ott 1987, Hunt and Ott 2001). Typical trajectories on a *strange nonchaotic attractor* separate exponentially in finite time intervals but eventually converge (Pikovsky and Feudel 1995, Lai 1996). Not all chaotic systems, such as the logistic map with $A = 4$ and the conservative cases in Chapter 8, have strange attractors.
12. It is *aesthetically appealing*. While this is not a mathematical property, it is one of their most evident characteristics and the reason many people are interested in them.

Requiring all these conditions to be satisfied is an overly restrictive definition of a strange attractor, but they encompass most of their common and useful properties.

6.2 Examples

Several strange attractors have already been described, such as the two-dimensional Hénon map in Fig. 5.6 and the three-dimensional chaotic flows in Figs 4.13–4.18. To illustrate the variety of strange attractors that can arise from a simple mapping, consider the system

$$X_{n+1} = a_1 + a_2 X_n + a_3 Y_n + a_4 X_n^2 + a_5 X_n Y_n + a_6 Y_n^2 \tag{6.1}$$

$$Y_{n+1} = X_n \tag{6.2}$$

which is a generalization of the Hénon map. Various choices of the parameters a_1 to a_6 give the attractors in Fig. 6.1 with Kaplan–Yorke dimensions from 1.244 to 2.0. Each of these cases has 2.5×10^5 iterations after the initial transient has decayed from an initial condition of $X_0 = Y_0 = 0.05$.

The parameters are like settings on a combination lock, most giving uninteresting solutions, but with countless others producing strange attractors, nearly all different. The information needed to reproduce these images can be stored very compactly. If each parameter is allowed 256 different values, only six bytes (48 bits) are needed to describe each case, and yet there are $256^6 \simeq 3 \times 10^{14}$ combinations, a significant fraction of which give strange attractors. And this is just one relatively simple map! While it is extremely difficult to predict the shape of the attractor for a given set of parameters, it is even more difficult, and perhaps impossible, to determine the equations that produced a given strange attractor. In the language of

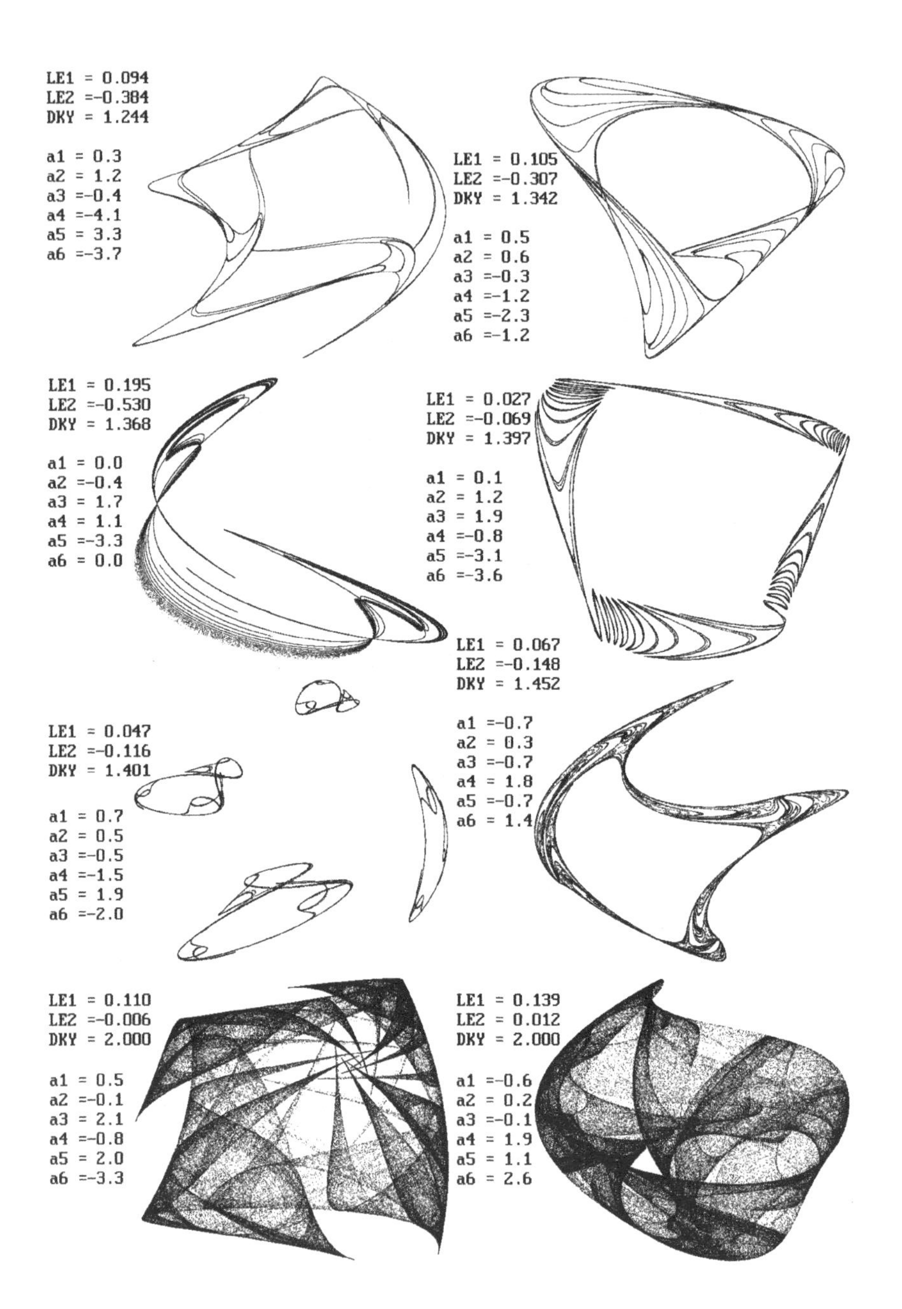

Fig. 6.1 Sample strange attractors from the simple two-dimensional quadratic map in eqns (6.1) and (6.2).

complex systems, we say that the attractor is an *emergent property* of the underlying chaotic complexity.

The lower two cases in Fig. 6.1 are arguably not attractors since their dimension is 2.0. They are not dissipative systems, and their Lyapunov exponents have a positive sum, the average of $\ln|a_3 + a_5X + 2a_6Y|$ along the orbit. In fact, the one in the lower right has both exponents positive. These systems are bounded, not because of area contraction, but because they are *noninvertible* (many points have multiple preimages) and folded so that two or more regions map onto a single region, thereby reducing the area even while being stretched. The folding is evident in the images. Such cases are a two-dimensional generalization of the chaotic logistic map, which is also noninvertible and folded. If $a_6 = 0$ in eqn (6.1), then the map is invertible, and all the bounded chaotic solutions are area-contracting with $D_{\mathrm{KY}} < 2$. In such systems, the initial condition completely defines not only the *future* states of the system, but also the *past* behavior. Invertible maps are especially important because Poincaré sections of autonomous flows are invertible since the trajectories cannot intersect.

6.3 Search methods

You may wonder how the strange attractors in Fig. 6.1 or Table 4.1 were found from the enormous combination of parameters, most of which do not give chaos, especially since the chaotic regions may be small and not contiguous or even nearby. This problem arises frequently when you have a mathematical model for some aperiodic phenomenon and want to know if it has chaotic solutions for some choice of parameters.

Fortunately, computers can solve this problem by calculating the Lyapunov exponent for many randomly selected parameters. Such a method cannot prove the absence of chaos, but it can demonstrate its existence. Finding a chaotic solution points you to regions of parameter space that are physically interesting and where other chaotic solutions are likely. Another approach is to maximize the largest Lyapunov exponent or the Kaplan–Yorke dimension. The Kaplan–Yorke dimension often converges faster and is independent of the time-scale. However, there is not usually a continuous uphill path from your initial guess to the maximally chaotic solution, but a search in the neighborhood of your best solution will often be rewarded.

In principle, each parameter can range from minus to plus infinity, but the model often dictates the sign and relevant range. In the worst case, you do not know what range to explore. If you limit the search to a region $a_{\mathrm{min}} < a < a_{\mathrm{max}}$, you risk missing interesting solutions, but if you make the range too large, most orbits are unbounded and the search is very time consuming.

For example, the logistic map has periodic solutions (including period-1 fixed points) for $-1.56994\ldots < A < 3.56994\ldots$, chaos over most of the range $-2 < A < -1.56994\ldots$ and $3.56994\ldots < A < 4$, and unbounded solutions for $A < -2$ and $A > 4$. The Hénon map has two parameters

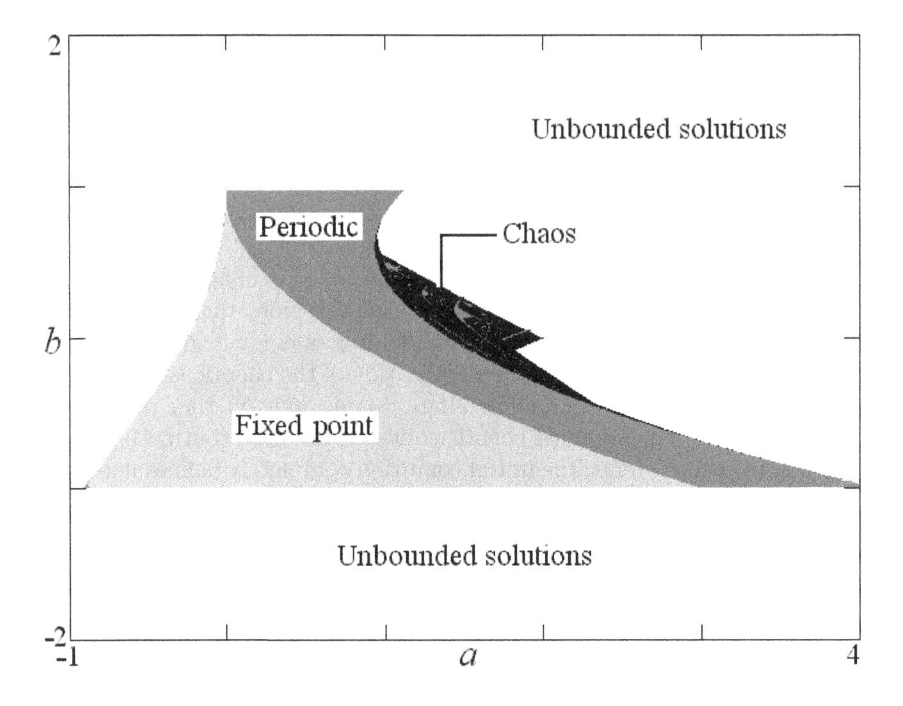

Fig. 6.2 Regions of various dynamical behaviors for the Hénon map.

(usually taken as $a = 1.4, b = 0.3$) and similar behavior as in Fig. 6.2 (Gallas 1993). The region of bounded solutions occupies an island in the ab plane with chaos confined to a small beach on the northeast side of the island. On that beach are numerous (in fact, infinitely many) ponds of periodicity.

Chaotic solutions typically have at least some of the parameters of order unity and the solution on the verge of being unbounded. If the coefficients are all too small, the linear terms dominate, and the only likely solutions are fixed points. If they are too large, the nonlinear terms dominate, and most solutions are unbounded. Unbounded solutions can be detected quickly by calculating the absolute value of one or more of the variables and moving on if it is too large. Stable fixed points can usually be detected by testing whether successive iterates are identical to some precision, although there is always the risk that the point is weakly unstable. Hence it is better to test not just for a value that approximately repeats but one whose deviation is decreasing with time. Periodic cycles can be similarly identified, but there is a point of diminishing returns for large periods. It is usually simpler to detect fixed-point and periodic solutions by their nonpositive Lyapunov exponent. Quickly discard solutions with large negative exponents, and only iterate many times for ambiguous cases, which is much easier for maps than for flows where zero Lyapunov exponents are common.

It is sometimes helpful to choose initial conditions randomly to ensure that at least some are in the basin of attraction. However, if the equations have enough parameters, the same attractor will occur at multiple positions, and so an arbitrary initial condition will usually lie in one of their basins. It is best to avoid initial conditions at the origin, which would be an equilibrium point for many simple systems, and to avoid initial conditions too far from the origin, which usually give unbounded solutions.

It is often useful to choose parameters and initial conditions from a random distribution over the whole range $-\infty < a < \infty$, but with most values of order unity. You can choose the coefficients from a normal (Gaussian) distribution (see §9.5.2), or from a Gaussian distribution raised to some odd power if you want a longer tail. Three other simple functions with this property are

$$f_1 = \tan[\pi(r - 0.5)] \tag{6.3}$$

$$f_2 = \ln\left(\frac{r}{1-r}\right) \Big/ \ln 3 \tag{6.4}$$

$$f_3 = \frac{6r - 3}{8r(1-r)} \tag{6.5}$$

where r is a random number uniform in the interval $0 < r < 1$. Each function is antisymmetric about $r = 0.5$, infinite at $r = 0$ and $r = 1$, and has half its values in the range $-1 < f < 1$. These functions nonlinearly map the unit interval onto the whole space of real numbers, but they are not iterated maps, and the choice is a matter of taste. Improvements may result from multiplying the functions by a factor to change the range over which most values occur. You can modify the formulas to limit the range of values, such as by replacing the π in f_1 with 3.12159... to give $-100 < f_1 < 100$. The values produced by f_1 have a *Cauchy distribution*

$$P(f_1) = \frac{1}{\pi(1 + f_1^2)} \tag{6.6}$$

which is the distribution of the *ratio* of two independent and identically distributed random numbers with mean zero.

If you are trying to maximize the Lyapunov exponent or some other parameter, try using $a = a_{\text{best}} + f\delta a$, where a_{best} is the value of a for the best solution found, and δa is the size of the region being explored, which is slowly decreased with each trial, using some schedule such as 99% of the previous value of δa. If you find a better solution, you might increase δa by a factor like two to keep the region from shrinking too quickly while improvements are being made.

6.3.1 Simplest piecewise linear chaotic flow

After finding chaotic solutions, you could repeat the search with small coefficients set to zero to simplify the model, or with coefficients rounded to convenient values (like unity), which also tests the structural stability of

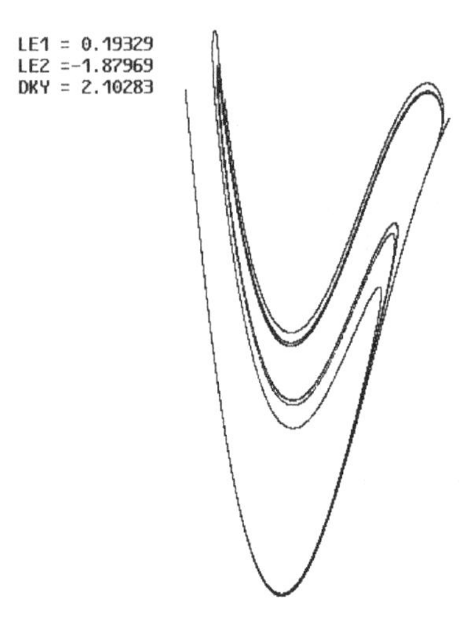

Fig. 6.3 Poincaré section in the xy plane for the driven van der Pol oscillator.

the solution. This method led to the discovery of the algebraically simplest chaotic flow with a quadratic nonlinearity in §4.8.4 and its piecewise linear counterpart (Linz and Sprott 1999)

$$\frac{d^3x}{dt^3} + 0.6\frac{d^2x}{dt^2} + \frac{dx}{dt} - |x| + 1 = 0 \tag{6.7}$$

Such piecewise linear systems are interesting because they lend themselves to accurate electronic circuit implementation using diodes and operational amplifiers, as described in §1.5.3.

6.3.2 Chaotic driven van der Pol oscillator

As another example, the method was applied to the driven van der Pol oscillator

$$\frac{d^2x}{dt^2} + a_1(x^2 - 1)\frac{dx}{dt} + x = a_2 \sin a_3 t \tag{6.8}$$

for which chaos exists in small regions of parameter space as shown in Figs 4.8 and 4.9. The largest Lyapunov exponent was found near $a_1 = 3, a_2 = 5$, and $a_3 = 1.788$, where the Lyapunov exponents are $(0.1932, 0, -1.8789)$ and the Kaplan–Yorke dimension is $D_{\mathrm{KY}} \simeq 2.1028$. The corresponding Poincaré section in the xy plane at $a_3 t \pmod{2\pi} = 0$ in Fig. 6.3 shows the expected fractal structure.

Shaw (1981) reported chaos in this system with parameters $a_1 \simeq 3.16$, $a_2 \simeq 13.1$, and $a_3 \simeq 3.79$, where the Lyapunov exponents are $(0.1352, 0, -2.9883)$ and the Kaplan–Yorke dimension is $D_{\mathrm{KY}} \simeq 2.0452$. He also suggested that chaos is more easily observed if the sinusoidal drive term is in the velocity equation rather than the acceleration equation and gave parameters for chaos

$$\frac{dx}{dt} = v + 0.25 \sin 1.57t \tag{6.9}$$

$$\frac{dv}{dt} = 10(0.1 - x^2)v - 0.7x \tag{6.10}$$

His five-parameter case can be rewritten with three parameters, rounded to convenient values as in Appendix A. Such a forcing term is unusual for a mechanical system, but it might be realized in an electrical or chemical system. Its attractor has a particularly interesting topological structure as described by Thompson and Stewart (1986). A variant of the driven van der Pol oscillator in eqn (6.8) with the x term replaced by x^3 also exhibits chaos (Ueda and Akamatsu 1981).

6.4 Probability of chaos

An interesting question is whether chaos is the rule or the exception in nature. More tractable is determining the probability of chaos in simple mathematical models that are sufficiently general to represent a wide range of phenomena. For example, the logistic map is chaotic for about 13% of the A values that give bounded solutions, and the Hénon map is chaotic for about 6% of the bounded region of the ab plane.

6.4.1 Quadratic maps and flows

Changing eqn (6.2) to the same form as eqn (6.1) gives the most general two-dimensional quadratic map, and some portion of the resulting twelve-dimensional parameter space has chaotic solutions. If the parameters are chosen uniformly in the interval $-a_{\max} < a < a_{\max}$, then they fill a twelve-dimensional hypercube. For $a_{\max}$ sufficiently large (greater than about 2), most of the solutions are unbounded and hence not physical. The small portion that are bounded have chaotic solutions for $11.10 \pm 0.36\%$ of the cases (Sprott 1993b). Other polynomial nonlinearities give similar results. Thus chaos is not the rule, but neither is it rare in low-dimensional iterated maps.

This application of the *Monte-Carlo method*[2] (Shreider 1966) can be extended to higher-dimensional quadratic maps and flows (Sprott 1993b) with the result in Fig. 6.4. Curiously and counter-intuitively, the maps become less chaotic as the dimension increases, whereas the flows become more chaotic, crossing at $D \simeq 8$ where the probability is about 2%.

6.4.2 Artificial neural networks

Similar studies were done by Albers *et al.* (1998) using *artificial neural networks* of the form

[2]The Monte-Carlo method is named after the famous gambling casino in Monaco and dates from about 1944. It was proposed by Joseph Edward Mayer (1904–1983) as a tool for calculating the behavior of liquids and independently by Stanislaw Marcin Ulam (1909–1984) as a general procedure.

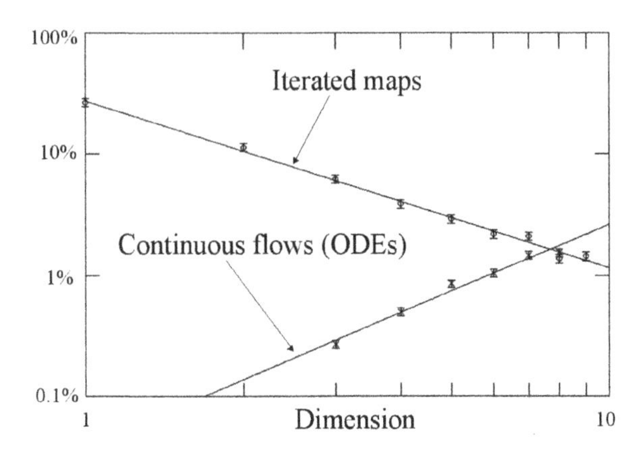

Fig. 6.4 Probability of chaos in quadratic maps and flows of various dimensions.

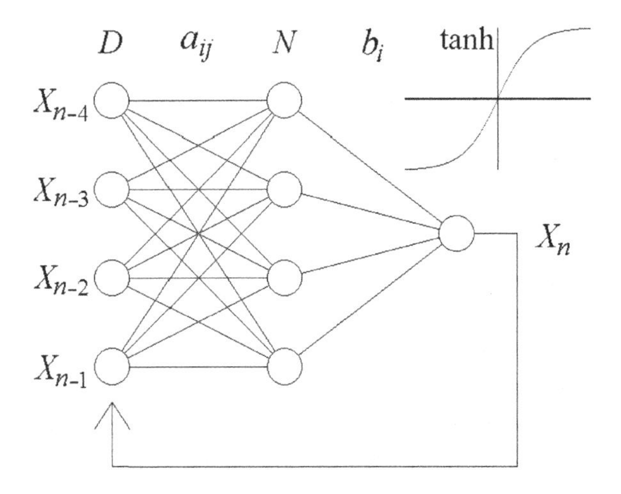

Fig. 6.5 Artificial neural network architecture.

$$X_n = \sum_{i=1}^{N} b_i \tanh\left(a_{i0} + \sum_{j=1}^{D} a_{ij} X_{n-j}\right) \tag{6.11}$$

where N is the number of 'neurons' and D is the dimension (number of time lags) as in Fig. 6.5. This *single-layer, feedforward network* is just an iterated map with a hyperbolic tangent *squashing function* that guarantees X is bounded. With N sufficiently large, it can approximate any measurable function $f(X_{n-1}, X_{n-2}, \ldots)$ arbitrarily closely (Hornik 1989, Hornik *et al.* 1990). The coefficients b_i were chosen from a random distribution uniform over $0 \le b_i < 1$ but rescaled so that their squares sum to N, and the coefficients a_{ij} (*connection strengths* or *weights*) were chosen from a random Gaussian distribution (see §9.5.2) with zero mean and standard deviation s.

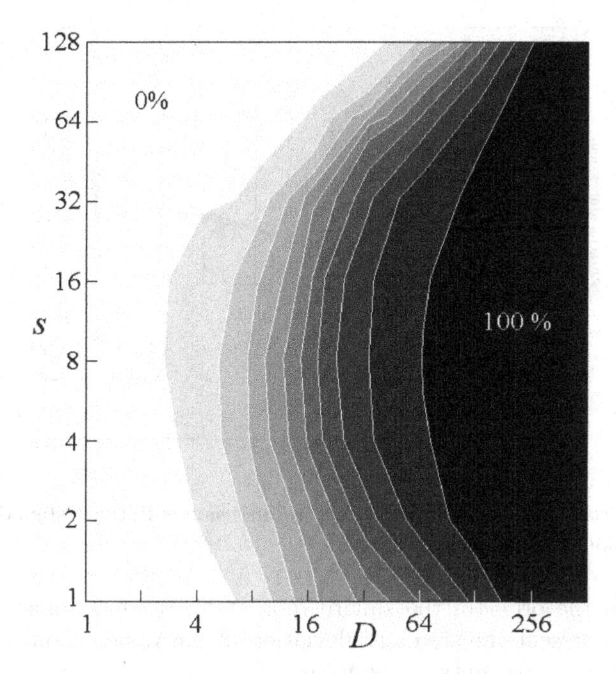

Fig. 6.6 Probability of chaos in artificial neural networks of various dimensions and connection strengths.

The percentage of chaotic solutions as a function of D and s for $N = 8$ in Fig. 6.6 approaches 100% for intermediate values of s at a dimension of about 100, consistent with the quadratic flows but in striking contrast to the quadratic maps. For small s, $\tanh x \simeq x$, and the network is approximately linear everywhere, precluding chaos. For large s, $\tanh x \simeq \pm 1$, and the network cannot be chaotic because there are only a finite number of states. At intermediate s, the probability of chaos increases with N, reaching about 50% when ND is of the order of a few hundred with $s = 8$.

6.5 Statistical properties

With a large collection of strange attractors, other statistical properties can be studied, such as how the attractor dimension, Lyapunov exponent, and routes to chaos vary with the system dimension.

6.5.1 Attractor dimension

Using the method in the previous section, 3840 chaotic maps and 2240 chaotic flows with polynomial nonlinearities were produced for systems up to nine dimensions (Sprott 1994b). The correlation dimensions (see §12.3) of the resulting attractors were estimated. The values are relatively insensitive to the degree of the polynomial, but the average dimension increases with the dimension of the system, as in Fig. 6.7. The attractor dimension is

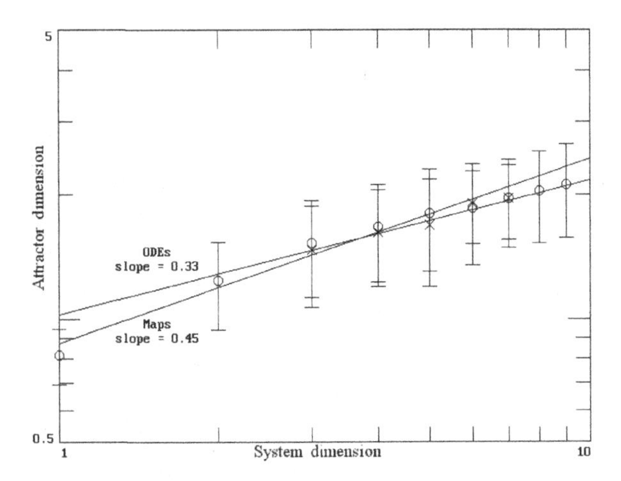

Fig. 6.7 Variation of the average attractor dimension with the system dimension for chaotic polynomial maps and flows.

typically on the order of the square root of the system dimension. The error bars represent the standard deviation of the values about the mean rather than an uncertainty in the mean.

6.5.2 Lyapunov exponent

The average value of the largest Lyapunov exponent was calculated for the same collection of strange attractors as in the previous section, with the results in Fig. 6.8. As with the dimension, the degree of the polynomial does not matter much, but the Lyapunov exponents are spread over a much wider range, and the average decreases approximately as the inverse of the system dimension. The error bars represent the standard deviation of the values about the mean rather than an uncertainty in the mean, and the difference in the slopes may not be statistically significant.

Thus we conclude that high-dimensional systems with polynomial nonlinearities tend to be only weakly chaotic. Similar results were obtained with the artificial neural networks in eqn (6.11), but the reason there may simply be that these particular higher-dimensional systems average over more past time steps, slowing their dynamical evolution.

6.5.3 Routes to chaos

It is likely that some routes by which a stable equilibrium becomes chaotic are yet to be identified. Those that result from local (rather than global) bifurcations can be categorized into three basic types: *period-doubling*, *quasiperiodic*, and *intermittent* (see Chapter 7). Period doubling has been described in connection with the logistic map, but more generally occurs in maps and flows where stable points become unstable in a series of *flip bifurcations* and subharmonic generation. Quasiperiodicity has been described

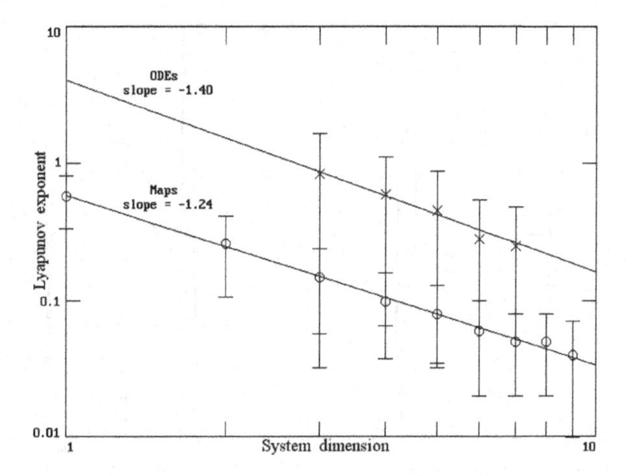

Fig. 6.8 Variation of average largest Lyapunov exponent with system dimension for chaotic polynomial maps and flows.

in connection with the van der Pol oscillator where a spiral node becomes unstable and gives birth to a limit cycle in a *Hopf bifurcation* (Marsden and McCracken 1976), which can then become a torus by a *secondary Hopf* (or *Neimark–Sacker*[3]) bifurcation after which the torus usually develops kinks and becomes chaotic (Ruelle and Takens 1971, Newhouse *et al.* 1978). Intermittency occurs to the left of the period-3 window in the logistic map and is frequently characterized in higher-dimensional systems by a *saddle-node bifurcation* (also called a *tangent bifurcation* or a *fold bifurcation*) in which a saddle point and a stable node coalesce and annihilate one another, producing an orbit that has periods of chaos interspersed with periods of regular oscillation (Pomeau and Manneville 1980).

A simple map that exhibits intermittency is the *tangent map*

$$X_{n+1} = \tan X_n \tag{6.12}$$

with X in radians and whose time series is in Fig. 6.9. Strogatz (1994) calls this map a 'nasty mess.' Hybrid routes involving combinations of these fundamental types are common.

The quasiperiodic route to chaos is historically important. Before strange attractors were known, the accepted theory of turbulence was due to Landau[4] (1944) who imagined the fluid contained many oscillators with incommensurate frequencies that were successively excited as more energy

[3]Hopf proved the bifurcation theorem for flows, but the corresponding bifurcation for maps was proved independently by Neimark (1959) and Sacker (1965) and hence is also called the *Neimark–Sacker bifurcation*.

[4]Lev Davidovich Landau (1908–1968) was a famous Russian theoretical physicist who won the Nobel Prize (1962) in Physics for studies of liquid helium but in the same year was involved in a car accident after which he lived another six years, never regaining his creativity.

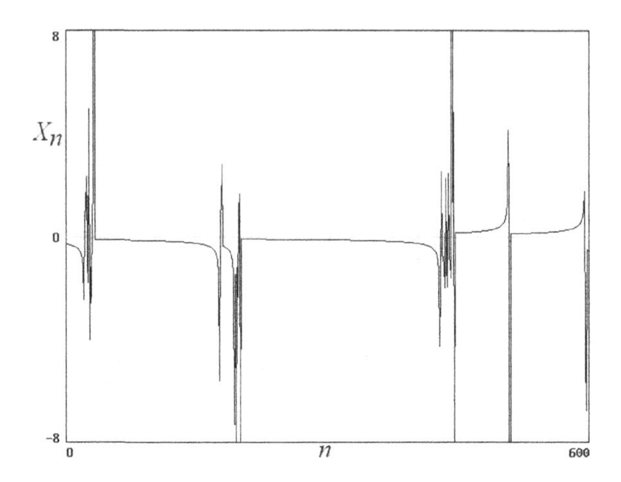

Fig. 6.9 Intermittency in the tangent map.

is added to the system (Landau and Lifshitz 1959). Thus the dynamics progress from a stable equilibrium to a limit cycle, to a 2-torus, to a 3-torus, and so forth, until the frequency spectrum is essentially continuous. The German mathematician Eberhard Hopf independently developed a similar theory (Hopf 1948). However, Peixoto (1962) showed that tori are stable only with dimensions less than three. Ruelle and Takens (1971) proposed an alternate route in which the two-torus becomes a strange attractor.[5] Independently, Gollub and Swinney (1975) and Swinney and Gollub (1978) used spectrum analysis to analyze the onset of turbulence in a *Taylor–Couette system*[6] containing water between two concentric counter-rotating cylinders, but they failed to confirm the Landau theory. Ruelle heard about their results and provided the correct interpretation for this important experiment that suggested turbulence was a form of chaos. He also predicted that oscillating chemical reactions should produce strange attractors (Ruelle 1973), and the prediction was confirmed by Roux *et al.* (1980, 1983) using the BZ reaction (Rinzel *et al.* 1984). The Landau route to turbulence has not been observed experimentally and is now considered to be incorrect. Strange attractors have been found in weakly turbulent Taylor–Couette flows (Brandstater and Swinney 1987).

With a sufficiently large collection of numerical chaotic systems, the probability of various routes can be estimated by varying some bifurcation parameter and seeing which eigenvalues become unstable. Albers *et al.* (1998) performed such a study using artificial neural networks as de-

[5]The Ruelle–Takens paper, like Feigenbaum's original paper on the period-doubling route to chaos, was initially rejected for publication (Ruelle 1991).

[6]Geoffrey Ingraham Taylor (1886–1975) was an English mathematical physicist who improved the apparatus devised by the French hydrodynamicist Maurice Couette (1858–1943).

scribed in §6.4.2 with the standard deviation s as the bifurcation parameter. Although it is difficult to identify the complete sequence of bifurcations leading to chaos, the one in which the fixed point first becomes unstable was found to be about equally divided among the three types (flip, Hopf, and saddle-node) at low dimension, but in the limit of large dimension, the Hopf was overwhelmingly most common, suggesting that the quasiperiodic route to chaos is generic in high-dimensional systems as suggested also by Doyon *et al.* (1993). This result is reasonable, given that the eigenvalues of a random high-dimensional system are spread throughout the complex plane, with relatively few on the real axis. The actual route may be much more complicated, involving multiple Hopf, inverse-Hopf and other bifurcations, with regions of chaos interspersed with regions of quasiperiodicity. However, three-tori (three incommensurate frequencies) are also sometimes observed (Bergé *et al.* 1986).

6.6 Visualization methods

A simple two-dimensional plot such as Fig. 6.1 suffices for displaying attractors of two-dimensional maps but is inadequate in higher dimensions. In particular, flows are chaotic only in dimensions of three or more, and their strange attractors have dimension greater than two. Thus additional visualization methods are required (Sprott 1993c). With all these methods, you should perform at least a few hundred iterations before plotting to ensure that the orbit is close to the attractor and to determine the size of the attractor so that the border can be chosen appropriately.

6.6.1 Projections onto a plane

The simplest method is to plot two of the variables, ignoring the others, as was done in Figs 4.13–4.18. This method amounts to projecting the attractor onto a plane, as would occur with the shadow of an object illuminated by a point source of light. If the attractor has a dimension of two or greater, then the shadow will be two-dimensional, and only its boundary will show structure. There are infinitely many such planes upon which it can be projected, and some will reveal the structure of the attractor better than others. It may also help to limit the number of iterations to reveal the density of points for a map or the shape of the trajectory for a flow. You can enhance the effect by plotting each pixel[7] in a shade of gray that depends on the number of times it has been visited by the orbit.

6.6.2 Poincaré sections

The structure of a flow can often be revealed in a Poincaré section (also called a *surface of section*) such as Figs 1.6, 4.9, 4.14, and 6.3. This method reduces the dimension of the attractor by one. The dimension of the

[7]A *pixel* (short for 'picture element') is the smallest dot that can be displayed on the computer screen or printed page.

Poincaré section is usually independent of the section taken, as long as it includes the attractor. The reason is that there is a simple mapping from one section to another that preserves the topology, albeit with stretching and distortion. There are infinitely many such sections, and some are more revealing than others. The map produced in this way (called a *Poincaré map*) has a subset of the same dynamics as the corresponding flow, including Lyapunov exponents and bifurcation behavior, except that it is missing the zero Lyapunov exponent corresponding to the direction of the flow. The other Lyapunov exponents are the same as for the corresponding flow except multiplied by the average time between successive intersections.

For a periodically-driven nonautonomous flow, it is natural to take a section at a constant phase of the drive, called a *stroboscopic view*. It represents what would be seen if a strobe lamp synchronized to the drive frequency illuminated the trajectory. This is most easily done when you choose the numerical integration time step as some integer sub-multiple of the drive period. For autonomous flows, plot the value of two of the variables when a third has a particular value such as zero. Or, plot two of the variables when one of the derivatives is zero, or equivalently when one of the variables reaches a maximum or minimum.

Another way is to plot the maximum value of one of the variables versus its previous maximum. In fact, most deterministic rules that impose a constraint on the system suffice unless the resulting map contains no points. However, do not try to plot the position at arbitrary constant time increments for a nonautonomous flow, since the result will be a uniform sampling of the whole attractor rather than a cross-section of it.

In the numerical calculation of a flow, identify or define a variable whose value is zero in the chosen Poincaré section. Test each iteration to see if this value is negative before the iteration and positive after (or vice versa), in which case you can interpolate to get the values of the other variables at the estimated time the plane was crossed by the trajectory. A linear interpolation will usually suffice unless the step size is unduly large, in which case you can fit a higher-degree polynomial to a few points in the temporal vicinity of the crossing.

For more careful work, such as where you want to zoom in on a portion of the Poincaré section to see its fractal structure, you can recalculate the time interval during which the zero crossing occurs with successively smaller time steps to locate the point with arbitrary precision. Even better is to use the method of *inverse interpolation* to calculate the time step that puts the trajectory exactly on the Poincaré section and then determine its location in one time step (Hénon 1982).

Maps do not have Poincaré sections, but you can still take a thin slice though their attractor (Lorenz 1984a, Moon and Holmes 1985). The fine-scale structure will be washed out unless the slice is extremely thin, in which case many iterations are required. Since a Poincaré section of a flow is a map, this method can in principle be applied successively to reduce the dimension by integers until the structure becomes apparent if sufficient data are available. However, maps produced in this way will not generally inherit the dynamical properties of the original map or flow.

6.6.3 Colors and gray scales

Another method is to plot two of the variables with a third represented by a color or shade or gray. This method does not accurately portray the attractor, but it can reveal structure not otherwise evident, and it makes a more interesting image. Many books and Web sites have such color images, but no examples are given here for reasons of economy.

For systems with more than three dimensions, you can combine these methods with others to reveal structure that would not otherwise be apparent. For example, you can take a Poincaré section of a five-dimensional system and plot two of the variables in a color chosen by a third variable from a rainbow palette with an intensity proportional to a fourth variable. You can also use three of the variables to control independently the red, green, and blue components of the color for each point. Such extreme methods are probably more visually appealing than scientifically enlightening.

6.6.4 Illumination and shadows

There are sophisticated *ray-tracing* methods (Watkins and Sharp 1992) for making photo-realistic images of three-dimensional objects even using only shades of gray. In essence they postulate one or more sources of illumination with a light intensity that decreases with some power of the distance from the source (typically $1/r^2$) and follow a collection of light rays from each source, plotting a point wherever the ray intersects the object or the background wall or floor. Shadows are naturally produced by this computationally intensive method.

You can obtain reasonable results using a similar but much simpler method. Suppose the object is projected onto the xy plane at a height z above the plane. Add a constant to z so that its minimum is zero, making the object appear to rest on the plane. Then plot each point with an intensity proportional to z as if the illumination were close above the object. Before plotting each point, test to see that its z value is greater than the point currently plotted at that position so that a near portion of the attractor will correctly occlude a more distant portion. It is more realistic to plot the object in a light shade of the color against a darker background (at $z = 0$).

As an enhancement, plot a shadow point in a darker shade of the background for each point on the attractor displaced down and to the right (as is the customary convention for a light source over your left shoulder) by an amount proportional to z if the shadow falls on the background at $z = 0$. You can make the shadow less harsh by adding some randomness to the position at which each shadow point is plotted, simulating a more diffuse light source.

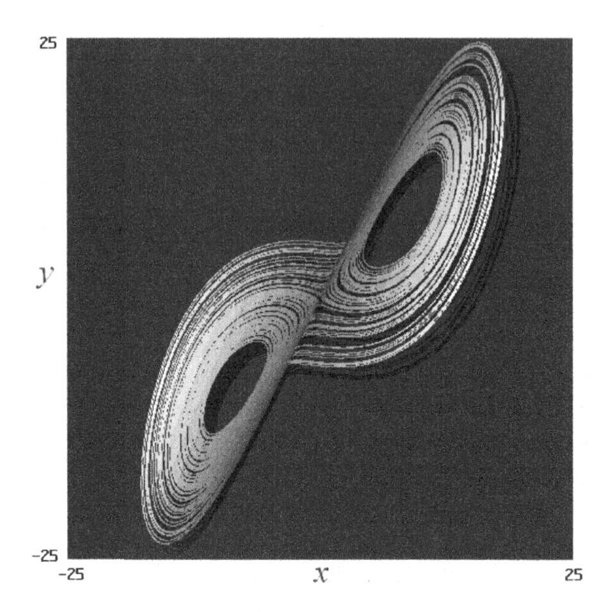

Fig. 6.10 Lorenz attractor with illumination and shadow.

This method is not precise since it does not, for example, include the shadow of one portion of the object on another portion. With more computing you can include that effect as well by tracing the ray from each point on the attractor down and to the right until it intersects a point whose z value is greater than that of the ray, and then plotting that point in a darker shade of the color that exists at that point. The method nonetheless gives reasonable results with as few as sixteen gray shades, as Fig. 6.10 shows for the Lorenz attractor.

6.6.5 Anaglyphs

A more realistic method for visualizing a third dimension uses binocular vision, projecting a different image into each eye as would occur for an actual three-dimensional object.[8] One such method is to plot two overlapping images displaced by a small amount proportional to z, one in red and the other in blue, and to view the image with red/blue glasses (Sprott 1992). Such images, called *anaglyphs*, have been used in comic books and movies. The displacement is not critical, but its maximum value should be about 10% the width of the image, which is the ratio of the distance between your eyes to the viewing distance. The length of your arm is about ten times the distance between your eyes, and so this method correctly projects an object with depth equal to its width when held at arm's length.

[8]The concept of the binocular stereogram dates back to Socrates (469–399 BC), and the first stereograms were produced by the English physicists, Sir Charles Wheatstone (1802–1875) and Sir David Brewster (1781–1868).

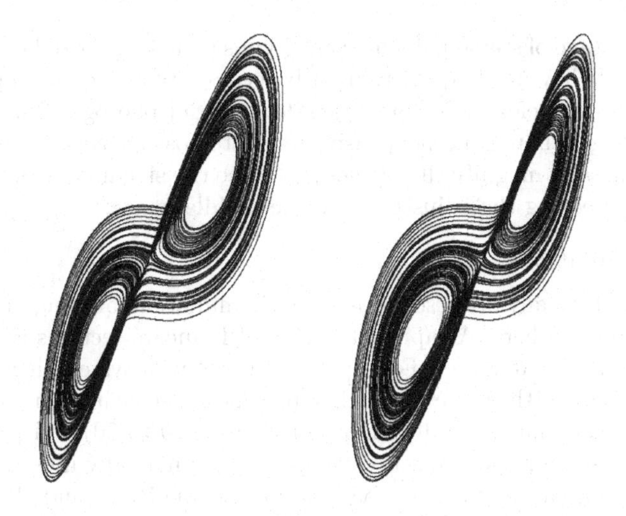

Fig. 6.11 Lorenz attractor displayed as a stereo pair.

You might add a constant to z so its median value is zero to make it lie partly above the plane and partly below. In such a case, the images will overlap near $z = 0$, and such points should be plotted as black or white as required to contrast with the background. Reversing the colors of either the image or glasses, or changing the background from white to black, reverses the sense of the z axis, which sometimes improves the view. Note that color-blindness should not hinder the depth perception since it is not necessary (indeed, not desirable) to perceive the individual colors, but only to respond to their intensities.

6.6.6 Stereo pairs

You can achieve binocular vision without special glasses by plotting the images side-by-side in the same color without overlapping as a *stereo pair*. The individual images should be small enough that they can be separated by a distance no more than the separation of your eyes (about 6.5 cm for most people). An example showing the Lorenz attractor is in Fig. 6.11.

Hold the book directly in front of your face at normal reading distance and exactly horizontal, and gaze into the distance until you see three images. Concentrate on the one in the middle, which should appear in three dimensions, while trying to ignore the two-dimensional ones on each side. Some people find it easier to cross the eyes, focusing on a point halfway to the page (perhaps placing your finger or a pencil there to help with focusing), in which case the three images should float up off the page with the middle one in three dimensions. Such *short-focus viewing* reverses the sense of z from the usual *free-viewing*, but it is less relaxing. Short-focus viewing allows the images to be farther apart than the eye separation, permitting much larger images.

An advantage of stereo pairs over anaglyphs is that they can be rendered in full color. They are frequently used by geologists and cartographers to determine terrain elevation from *hyperstereo* aerial photographs. You can view the images though an inexpensive hand stereoscope containing prisms that separate and magnify the images and force the side images out of your field of view, easing the adjustment to the middle image.

6.6.7 Animations

You can combine any of these methods with animation, although it is hard to show examples here. A rapid succession of Poincaré sections allows you to 'fly though' the attractor. For a periodically-driven system, it is natural to use the phase of the drive as the time parameter, since it will repeat after one period, and thus a small number of frames (20 to 30) will produce a nearly smooth animation that cycles for ever, dramatically illustrating the stretching that causes the sensitivity to initial conditions and the folding that produces the fractal structure.

For an autonomous system, the best animation is probably a series of views from different angles that periodically repeat, giving the illusion of rotation. Some axes of rotation are better than others for revealing the structure of the attractor. When combined with color, variable illumination, and shadows, the effect can be striking.[9]

6.7 Unstable periodic orbits

Strange attractors are dense in periodic orbits, all unstable. For some purposes, it is useful to identify those orbits with small periods (Lathrop and Kostelich 1989, So *et al.* 1996, Schmelcher and Diakonos 1997). Basic ergodic properties such as dimension, Lyapunov exponents, and entropy can be expressed in terms of the periodic orbits (Eckmann and Ruelle 1985, Auerbach *et al.* 1988, Cvitanović 1988, Grebogi *et al.* 1988). Their nature and behavior provide a means to classify strange attractors (Gilmore and Lefranc 2002). Furthermore, many of the schemes for controlling (suppressing) chaos (Ott *et al.* 1990, Shinbrot *et al.* 1993, Ott and Spano 1995) use feedback to keep the system on such an orbit. Sometimes it is desirable to use feedback to keep the orbits *unstable* ('anti-control' of chaos).

Suppose you want to find all orbits with periods less than N. Iterate the equations until the solution is on the attractor. Then record the state-space position $X_0, Y_0, \ldots$ and iterate N more times, testing whether each iterate is within some small ϵ of the initial condition. If it is, you have presumably found an orbit with a period equal to that number of iterations. If not, then use the state-space position after N iterations as the new initial condition and continue searching. Eventually, all the orbits with period up to N will emerge. With flows, the procedure is the same, except that the period is the

[9] See http://sprott.physics.wisc.edu/fractals/animated/ for examples of animated strange attractors.

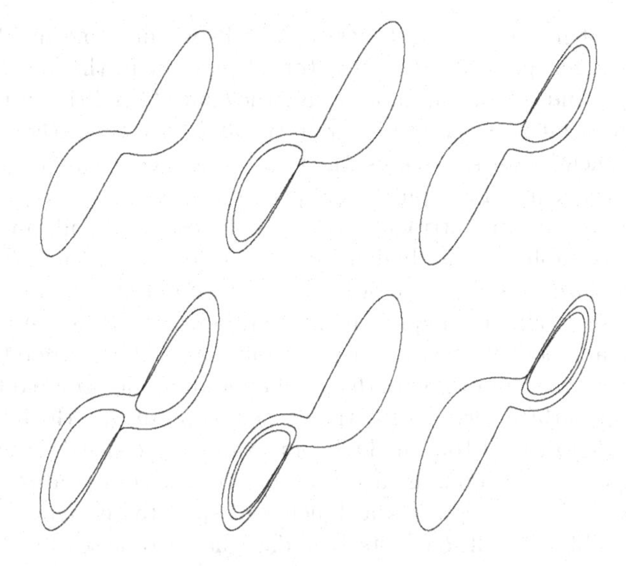

Fig. 6.12 First six unstable periodic orbits for the Lorenz attractor.

number of iterations times the step size, which should be chosen relatively small.

As an example, Fig. 6.12 shows the first six periodic orbits for the Lorenz attractor with the standard parameters (see §4.8.2), projected onto the xy plane (Mirus and Sprott 1999). These were produced with a fourth-order Runge–Kutta step size of $h = 0.0004$, $\epsilon = 0.01$, and $N = 3.1/h$, giving periods up to $Nh = 3.1$ in the units in which time is measured in the Lorenz equations. The case in the upper left has a period of $1.559\ldots$, and the other two upper ones have periods of $2.306\ldots$. The case in the lower left has a period of $3.084\ldots$, and the other two lower ones have periods of $3.024\ldots$. The orbits occur in symmetric pairs due to the symmetry of the Lorenz equations with respect to a sign change in x and y. These periodic orbits form a kind of 'skeleton' of the attractor, sometimes called the 'Cheshire set,'[10] since they are normally invisible because of their instability.

6.8 Basins of attraction

Every attractor is enclosed within a *basin of attraction*, which may stretch to infinity as with the Lorenz attractor. The basin of attraction comprises all the preimages of the attractor. You can find the basin by iterating the equations with many initial conditions, while testing whether the orbit has found the attractor. Because of the transitivity, it usually suffices to

[10]The Cheshire cat was a character in Lewis Carroll's *Alice's Adventures in Wonderland* (1865) who vanished, leaving behind only its broad grin. Lewis Carroll was the pen name of Charles Lutwidge Dodgson (1832–1898), an eccentric and withdrawn mathematician, writer, and pioneer photographer, who was born in Cheshire, England, and lectured on mathematics at Oxford from 1855 to 1881.

declare a point in the basin if it eventually falls within some small ϵ of an arbitrarily chosen point on the attractor. Points not in the basin usually go to infinity, and so you can move to another initial condition if the orbit reaches some arbitrary distance far from the attractor. Since the growth is usually exponential for such cases, the test can use a very large value (10^6 or more) without much computational penalty.

If there are multiple attractors, then both tests may fail, and so you will need to establish some 'bailout' condition for the maximum number of iterations after which you move on to a new initial condition. Points near the basin boundary require many iterations to categorize and thus also require a bailout. In fact, the regions outside the basin are often color coded or gray scaled according to the number of iterations required for them to cross some surface surrounding the attractor, producing the interesting patterns in Chapter 14. Too small a bailout value will conceal the structure near a boundary, and too large a value will slow the computation.

The basin of attraction has a dimension equal to the number of dynamical variables m, although its boundary has a topological dimension of $m - 1$ and may be a fractal with a noninteger dimension greater than $m - 1$ (McDonald *et al.* 1985). To facilitate visualization, the examples in Fig. 6.13 as described below have $m = 2$.

6.8.1 Hénon map

The Hénon map has already been described. Its basin boundary is relatively smooth and nearly touches the attractor in several places, as is typical for strange attractors and other chaotic systems such as the logistic map with $A = 4$.

6.8.2 Lozi map

The *Lozi map* (Lozi 1978)

$$X_{n+1} = 1 - 1.7|X_n| + 0.5Y_n \tag{6.13}$$

$$Y_{n+1} = X_n \tag{6.14}$$

is a piecewise linear variant of the Hénon map with a strange attractor (Misiurewicz 1980) and a similar basin, except that its boundary is also piecewise linear.

6.8.3 Tinkerbell map

An example of a strange attractor with a fractal basin boundary is the *Tinkerbell map* (Nusse and Yorke 1994)

$$X_{n+1} = X_n^2 - Y_n^2 + 0.9X_n - 0.6Y_n \tag{6.15}$$

$$Y_{n+1} = 2X_nY_n + 2X_n + 0.5Y_n \tag{6.16}$$

Its boundary also nearly touches the attractor at several places and fits rather tightly around the attractor.

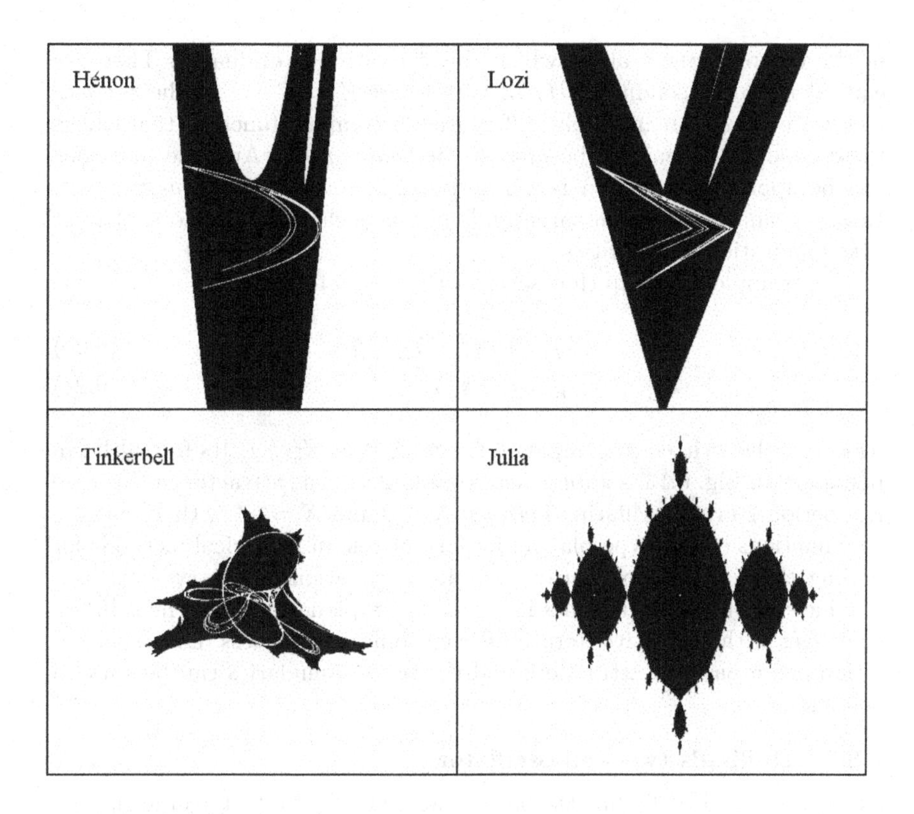

Fig. 6.13 Sample basins of attraction for various attractors.

6.8.4 Julia sets

There is no obvious relation between the shape of the attractor and the shape of its basin. However, two-dimensional maps usually have fractal basin boundaries if they satisfy the *Cauchy–Riemann equations*[11] (Arfken 1985)

$$\frac{\partial F}{\partial X} = \frac{\partial G}{\partial Y} \tag{6.17}$$

$$\frac{\partial F}{\partial Y} = -\frac{\partial G}{\partial X} \tag{6.18}$$

or, equivalently

$$\frac{\partial^2 F}{\partial X \partial Y} = -\frac{\partial^2 G}{\partial X \partial Y} \tag{6.19}$$

These equations ensure that the derivative df/dZ of a complex function $f(Z) = F(X,Y)+iG(X,Y)$ with $Z = X+iY$ is independent of the direction

[11]The Cauchy–Riemann equations are named after the prolific French mathematician Augustin Louis Cauchy (1789–1857) and the German mathematician Georg Friedrich Bernhard Riemann (1826–1866) whose Riemannian geometry was the foundation of general relativity.

in the complex plane along which the derivative is evaluated. They are equivalent to the condition $\partial f/\partial \bar{Z} = 0$, where $\bar{Z} = X - iY$ is the *complex conjugate* of Z. An infinitely differentiable complex function that obeys these equations is said to be *analytic* or *holomorphic*. Analytic functions can be approximated by a power series, $f(x) = a_0 + a_1 x + a_2 x^2 + \cdots$. However, analytic maps apparently do not have chaotic attractors, at least with quadratic nonlinearities.

An example of a map that satisfies these equations is

$$X_{n+1} = X_n^2 - Y_n^2 - 1 \tag{6.20}$$

$$Y_{n+1} = 2X_n Y_n \tag{6.21}$$

which can be written in complex form as $Z_{n+1} = Z_n^2 - 1$. Its fractal basin boundary in Fig. 6.13 is a *Julia set* (see §11.2.4). The attractor in this case is a period-2 orbit oscillating between $X = 0$ and $X = -1$ with $Y = 0$.

Functions of this type play an important role in numerical methods for finding the complex solutions of nonlinear equations. Such systems usually have multiple fixed points in the complex plane to which the solution converges with successive iterations from some initial guess. Each such solution has a basin of attraction, and the basin boundaries can be smooth or fractal.

6.8.5 Duffing's two-well oscillator

Another method for finding the basin of attraction is to start a large number of trajectories at different positions near the attractor (but not on it!) and run time backwards. To illustrate the idea, consider Duffing's autonomous two-well oscillator (see §4.8.1)

$$\frac{d^2x}{dt^2} + b\frac{dx}{dt} - x + x^3 = 0 \tag{6.22}$$

Running time backwards means replacing d/dt with $-d/dt$, which in this case only changes the sign of b. The system has two equilibria at $x^* = \pm 1$ and $v^* = 0$, where $v = dx/dt$. The attractors become repellors for $b < 0$, forcing trajectories that start nearby to spiral outward in alternating bands, as in Fig. 6.14. As the damping is reduced ($b \to 0$), the basins become more tightly intertwined, producing another kind of sensitive dependence on initial conditions, not related to chaos. With three equilibria, the basins can become even more intertwined with a fractal boundary (see §14.7).

The same method can be applied to chaotic flows, but in that case the state space is at least three-dimensional, and it is harder to visualize the basin, which is a cavity of nonzero (hyper)volume, except by taking Poincaré sections. Since this chapter is supposed to concern *strange* attractors, we will defer further discussion of basin boundaries to Chapter 14.

Fig. 6.14 Basins of attraction for the equilibria in Duffing's autonomous two-well oscillator with $b = 0.1$.

6.9 Structural stability and robustness

For an attractor to be structurally stable, it must preserve its topology in the presence of arbitrary small perturbations to the *form* of its equations. The topology is preserved only if the attractor dimension is constant or changes continuously. Although it is easy to demonstrate structural *instability*, it is usually impossible to prove structural *stability* except in special cases (Robbin 1971), since you cannot test all possible perturbations. However, most strange attractors appear to be stable to most perturbations in numerical experiments.

For example, consider linear perturbations to the Lorenz equations

$$\frac{dx}{dt} = 10(y - x) \pm \epsilon x \pm \epsilon y \pm \epsilon z \tag{6.23}$$

$$\frac{dy}{dt} = -xz + 28x - y \pm \epsilon x \pm \epsilon y \pm \epsilon z \tag{6.24}$$

$$\frac{dz}{dt} = xy - \tfrac{8}{3}z \pm \epsilon x \pm \epsilon y \pm \epsilon z \tag{6.25}$$

in which the $\pm$ signs can occur in $2^9 = 512$ different combinations, not all of which give unique solutions. Figure 6.15 shows the largest Lyapunov exponent λ_1 for 10^4 Lorenz attractors perturbed with various random combinations of the signs and values of ϵ in the range $0 < \epsilon < 1$. Positive values of λ_1 indicate a strange attractor (dimension greater than 2), and negative values indicate a point attractor (dimension of 0). Values of ϵ smaller than about 0.1 always seem to give strange attractors with similar exponents, suggesting structural stability. Quadratic perturbations give similar results, albeit in a smaller neighborhood. These numerical tests certainly do not constitute a proof of structural stability even for the Lorenz attractor.

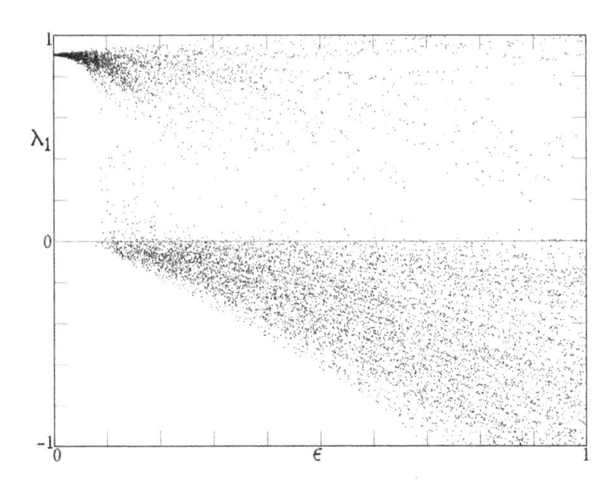

Fig. 6.15 Lyapunov exponents for randomly perturbed Lorenz attractors, showing structural stability for $\epsilon \lesssim 0.1$.

In addition to the difficulty of testing all possible perturbations, there may be narrow windows of nonchaotic attractors for arbitrarily small ϵ, but their probability typically goes to zero as $\epsilon \to 0$ (Barreto *et al.* 1997), and you are unlikely to see them in experiments with even a small amount of noise (Schaffer *et al.* 1986). The logistic map for $A = 4 - \epsilon$ with $\epsilon \ll 1$ is a simple such example with features shared by more complicated systems such as the Rössler attractor (see §4.8.3) whose return maps have smooth maxima. These systems typically have a dense set of stable periodic windows for any range of parameter values (Graczyk and Swiatek 1997), which means that a carefully chosen but arbitrarily small adjustment to a parameter will cause the chaotic system to become periodic.

Systems that retain their topology in the presence of small perturbations to their *parameters* (as opposed to the *form* of the equations) are called *robust* (Banerjee *et al.* 1998). Robustness is a necessary but not sufficient condition for structural stability, and it is a desirable property for experimental models. All structurally stable systems are robust, but not the converse.

The logistic map can be made robust in the region just below $A = 4$ by adding a small ($\delta \ll 1$) component of the tent map

$$f(X) = \frac{A}{1 + 4\delta}[X(1 - X) + \delta(1 - |2X - 1|)] \tag{6.26}$$

to avoid having $f'(X) = 0$ for any $0 < X < 1$. Figure 6.16 shows the Lyapunov exponent for eqn (6.26) with $\delta = 0.01$ and $\delta = 0$ for $3 < A < 4$. Even a δ this small is sufficient to remove nearly all the periodic windows and the superstable orbits (see §5.1.3). The period-3 window also closes

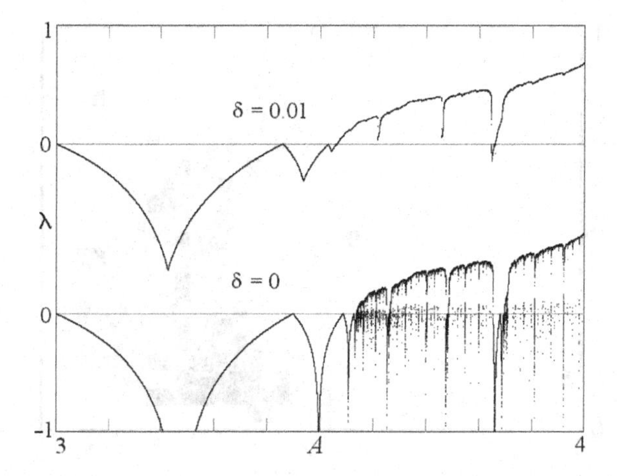

Fig. 6.16 Logistic map made robust using eqn (6.26) with $\delta = 0.01$ (top) compared with $\delta = 0$ (bottom).

when δ reaches $0.01611406\ldots$ where the last stable period-3 orbit occurs for $A = 3.81966876693728\ldots$.

While it is apparently true that *most* attractors survive *most* perturbations, it is unlikely that a *particular* strange attractor will survive *all* perturbations. You are not likely to be troubled by this problem in numerical experiments with dissipative systems unless you have intentionally chosen parameters to be near a bifurcation point. Conservative systems described in Chapter 8 are another matter, however.

6.10 Aesthetics

Strange attractors in state space have considerable visual appeal, presumably because they resemble natural patterns in real space produced by chaotic processes. An interesting question is whether their aesthetics relate to measurable quantities such as fractal dimension and Lyapunov exponent. If so, it suggests an automatic method for generating computer art.

6.10.1 Aesthetic evaluation

In one study (Sprott 1993a), 7500 strange attractors as in Fig. 6.1 for two-dimensional quadratic maps were evaluated by eight volunteers including three artists and five scientists. The attractors were chosen randomly from a large collection and displayed sequentially on the computer screen. Each volunteer evaluated about a thousand cases on a scale of one to five according to their aesthetic appeal.

The average evaluation is plotted in Fig. 6.17 as a function of the largest Lyapunov exponent[12] (λ_1) and fractal dimension (D) with a gray scale in

[12]In this study, the Lyapunov exponent was evaluated in base-2 so that the numbers correspond to the loss rate of information in bits per iteration.

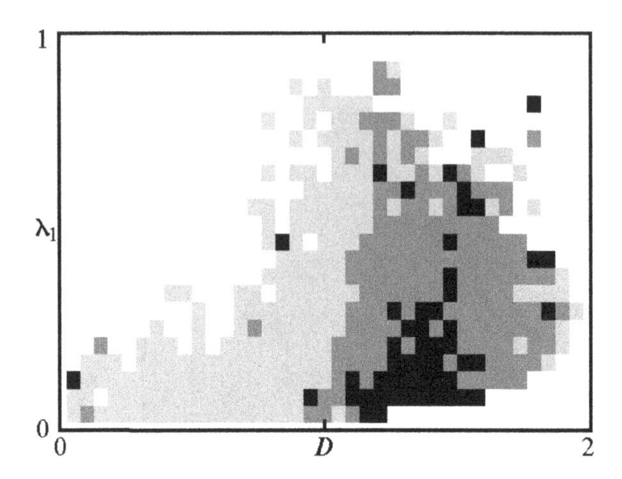

Fig. 6.17 Aesthetic evaluation of 7500 attractors of various dimensions and Lyapunov exponents.

which the darker shades are preferred. Individual preferences differ slightly (Aks and Sprott 1996), but all subjects preferred attractors with dimension between about 1.1 and 1.5 and a Lyapunov exponent between zero and about 0.3. Some of the most interesting cases have Lyapunov exponents below about 0.1 bits per iteration.

The dimension preference is not surprising since many natural objects have dimensions in this range. The Lyapunov exponent preference is harder to understand but suggests that strongly chaotic systems are too unstructured to be appealing. For the 443 cases that were rated five (best) by the subjects, the average correlation dimension was $D = 1.30 \pm 0.20$, and the average Lyapunov exponent was $\lambda_1 = 0.21 \pm 0.13$ bits per iteration. Hence a computer can be taught to critique its own art and select patterns likely to appeal to a human. Since artificial neural networks can produce strange attractors and can be easily trained, they are a natural vehicle for producing automated computer art (Sprott 1998).

6.10.2 Computer art

To produce true computer art, you do not want to be restricted to two-dimensional monochromatic maps but will want to use color and other visualization methods discussed in §6.6 for higher dimensions. A simple way to add dimensions is to include with eqns (6.1) and (6.2) equations of the form[13] $Z_{n+1} = Y_n, \ldots$, which is equivalent to adding more time delays $Z_n = X_{n-2}, \ldots$. An example of such an attractor is on page viii. The search methods work in higher dimensions, and there are many nonlinear functions $X_{n+1} = f(X_n, Y_n, Z_n, \ldots)$ to explore (Sprott 1993c). You can also use chaotic maps to produce music (see §9.4.7), since the iterates combine

[13] The Z here is a real third dimension (X, Y, Z), and not a complex variable.

Fig. 6.18 A sample strange attractor symmetric icon.

determinism with unpredictability, as is common in music (Voss and Clarke 1978, Pressing 1988, Meloon and Sprott 1997).

6.10.3 Symmetries

One problem with strange attractors as objects of art is that they are often too unstructured. You can constrain the model equations so that the solutions have a symmetry (Stewart and Golubitsky 1992, Field and Golubitsky 1992), but such constraints limit the range of allowed equations and complicate the mathematics. It is simpler to choose the equations arbitrarily and introduce the symmetry when the solution is plotted. You can do this by replicating the pattern several times and distorting the individual images to fit within pie-shaped slices like petals of a flower (Sprott 1996). It is difficult to do justice to the technique without color, but Fig. 6.18 shows an example of such an image, called a *symmetric icon.*[14]

Symmetries play an important role in nature and in the mathematical theories that describe it (Weyl 1952). Heavenly objects such as stars and planets have a spherical symmetry, and many earthbound objects such as humans have a bilateral symmetry. Flowers, starfish, diamonds, and snowflakes have symmetries with respect to rotations through specific angles. Symmetries with respect to changes in scale characterize the fractals in Chapter 11. Symmetries of the laws of physics with respect to displacements in space, time, and rotations lead to the conservation of linear momentum, energy, and angular momentum, respectively.[15] The ubiquity of symmetries in nature probably accounts for their aesthetic appeal in computer art.

[14]See http://sprott.physics.wisc.edu/fractals/icons/ for examples of colored symmetric icons.

[15]The connection between symmetries and conservation laws was first noted by Emmy Amalie Noether (1882–1935), a gifted female German mathematician who suffered obstinate discrimination but finally secured a faculty position with the help of David Hilbert.

6.11 Exercises

Exercise 6.1 Show that the Hénon map is invertible, and derive the inverted map.

Exercise 6.2 Find the condition under which eqns (6.1) and (6.2) have fixed-point solutions, and derive a formula for the fixed points (X^*, Y^*) when the condition is satisfied.

Exercise 6.3 Find the fixed points for the Hénon map in eqns (5.13) and (5.14), and locate them on the attractor in Fig. 5.6.

Exercise 6.4 Calculate the eigenvalues of the fixed points for the Hénon map in eqns (5.13) and (5.14).

Exercise 6.5 Show that the general two-dimensional map in eqns (6.1) and (6.2) is invertible for $a_6 = 0$, and derive the inverted map.

Exercise 6.6 Find the preimages of the point (X, Y) for $a_6 \neq 0$ in eqns (6.1) and (6.2).

Exercise 6.7 Show that the functions in eqns (6.4)–(6.6) are antisymmetric about $r = 0.5$, infinite at $r = 0$ and $r = 1$, and have half their values in the range $-1 < f < 1$ for r uniform in the range $0 < r < 1$.

Exercise 6.8 Calculate the probability distribution functions for the functions in eqns (6.4)–(6.6).

Exercise 6.9 Calculate the equilibria and their eigenvalues for the simplest piecewise linear chaotic flow in eqn (6.7).

Exercise 6.10 Show by a linear transformation of $x = \alpha x'$ and $t = \beta t'$, that the most general form of the driven van der Pol equation

$$\frac{d^2x}{dt^2} + a_1(x^2 - a_4)\frac{dx}{dt} + a_5 x = a_2 \sin a_3 t \tag{6.27}$$

can be reduced to the form in eqn (6.8) in which $a_4 = a_5 = 1$ so that the parameter space is three dimensional.

Exercise 6.11 Show that eqns (6.9) and (6.10) can be reduced to the form of the Shaw van der Pol oscillator in §A.4.3 and that the parameters are approximately consistent.

Exercise 6.12 Show that an appropriate choice of the a's and b's in the artificial neural network in eqn (6.11) with $N = 2$ and $D = 1$ maps the interval $0 < X < 1$ back onto itself in a manner similar to the logistic map for $A = 4$.

Exercise 6.13 Compare the dimensions and largest Lyapunov exponents for the maps and flows in Appendix A with the predictions of Figs 6.7 and 6.8.

Exercise 6.14 Suppose you want to plot a return map for the maximum value of the variable x in a chaotic flow and you have three points x_1, x_2, and x_3 along the trajectory, with x_2 larger than x_1 and x_3. By fitting a parabola to the three points, derive an expression for the maximum x in terms of x_1, x_2, and x_3.

Exercise 6.15 By linear interpolation, derive an expression for a point (X, Y) in the Poincaré section of a three-dimensional flow at $z = 0$ if the trajectory crosses the $z = 0$ plane during the time step bounded by (x_1, y_1, z_1) and (x_2, y_2, z_2).

Exercise 6.16 Show that an anaglyph is correctly rendered for a spherical object if the horizontal displacement of the images is equal to the horizontal size of the image times the ratio of the separation of the eyes to the viewing distance.

Exercise 6.17 Derive a more accurate formula for the position on the computer screen of the two image points in an anaglyph of a point on an attractor located at (x, y, z), including the fact that more distant points are closer together in the field of view.

Exercise 6.18 Derive a set of equations showing where to plot a point (x_p, y_p) on the computer screen for a point (x, y, z) from an attractor in the nth frame of an animation that makes the attractor appear to rotate about its z axis with N frames per rotation $(1 \leq n \leq N)$.

Exercise 6.19 Show that the difference in the periods of the first two unstable periodic orbits for the Lorenz attractor $(2.306 - 1.559 = 0.747)$ is approximately equal to the linear period of oscillation about one of the unstable equilibrium points.

Exercise 6.20 Find the fixed points for the Lozi map in eqns (6.13) and (6.14).

Exercise 6.21 Find the fixed points for the Tinkerbell map in eqns (6.15) and (6.16).

Exercise 6.22 Show that the two-dimensional map in eqns (6.20) and (6.21) for the Julia set has two unstable fixed points and a stable 2-cycle.

Exercise 6.23 Derive the Cauchy–Riemann equations by requiring that df/dZ is the same when evaluated along the real and imaginary axes.

Exercise 6.24 Show that the Hénon, Lozi, and Tinkerbell maps do not satisfy the Cauchy–Riemann equations, but that eqns (6.20) and (6.21) for the Julia set do.

Exercise 6.25 Show that the Cauchy–Riemann equations constrain the coefficients of eqn (6.1) to preclude chaotic solutions.

Exercise 6.26 Show that the most general quadratic map in two dimensions that satisfies the Cauchy–Riemann equations is

$$F = a_1 + a_2X + a_3Y + a_4(X^2 - Y^2) + 2a_5XY \tag{6.28}$$

$$G = a_6 + a_2Y - a_3X - a_5(X^2 - Y^2) + 2a_4XY \tag{6.29}$$

Exercise 6.27 Show that the logistic map perturbed with a tent map in eqn (6.26) with $A = 4$ maps the unit interval ($0 < X < 1$) onto itself and that the derivative dF/dX is nowhere zero.

6.12 Computer project: Hénon map

In this project you will examine the properties of the Hénon map, which is probably the most studied example of a two-dimensional chaotic map. You will observe the self-similar (fractal) structure of its strange attractor. The Hénon map is a two-dimensional generalization of the one-dimensional logistic equation, and was proposed by Hénon as a model of the Poincaré section that arises from a solution of the Lorenz equations. The Hénon map is given by the following equations

$$X_{n+1} = 1 - aX_n^2 + bY_n \tag{6.30}$$

$$Y_{n+1} = X_n \tag{6.31}$$

The usual choice for the parameters a and b that give chaotic solutions are $a = 1.4$ and $b = 0.3$.

1. Using the parameters above, choose initial conditions X_0 and Y_0 somewhere in the basin of attraction, and make a plot of X_n versus n to verify that the solution is bounded and apparently chaotic.
2. Start with two nearby initial conditions, and show that their separation increases rapidly as expected for a chaotic system.
3. Make a plot of several thousand values of Y_n versus X_n after discarding a sufficient number of initial iterates to ensure that the plotted points lie on the (strange) attractor.
4. Expand your plot above in the vicinity of the double lines that you should have seen, and convince yourself that what look like individual lines are (upon magnification) really pairs of lines, each of which are pairs of lines, ad infinitum.
5. (optional) Make a plot of the basin of attraction (those initial values of X_0 and Y_0 that do not escape to infinity) for the Hénon map with the parameters given above. Overlay the plot of the basin onto the plot of the attractor.

7 Bifurcations

A bifurcation is a qualitative change in the dynamical behavior of a system or the topological structure of its phase portrait as one or more parameters pass through a critical value. Any point in parameter space where the dynamical system is structurally unstable is a *bifurcation point*, and the set of all such points is a *bifurcation set*. This set may contain infinitely many points but usually has zero measure.

Bifurcations are important because they provide strong evidence of determinism in otherwise seemingly random systems, especially if the parameter can be repeatedly changed back and forth across the critical value with a consistent response of the system. Some systems exhibit *hysteresis*, in which the bifurcation occurs at different values of the parameter depending on the direction in which it is changed. It is often important to know where these bifurcations occur so that systems can be designed to avoid undesirable effects (Chen *et al.* 2000). There are dozens of different bifurcations and probably some not yet discovered. The terminology is not precise or universal and is still evolving, and many bifurcations have several names. Bifurcation theory has its origins in the work of Poincaré, but it is not yet a completed subject.

There are a number of ways to classify bifurcations:

1. **Maps and flows**: Both *discrete-time* and *continuous-time* systems can have bifurcations. Some bifurcations occur in both systems and have the same or similar names; others are unique to one or the other.
2. **Dimension**: The *dimension* of the system is the number of dynamical variables. Some bifurcations occur only when the dimension exceeds a minimum value. It suffices to examine each bifurcation in the minimum dimension in which it occurs, since nothing fundamentally new is added by the extra dimensions.
3. **Codimension**: We will only consider *codimension-1* bifurcations in which only a single parameter is changed with others held constant. With higher codimension, two or more parameters have to be adjusted, and such cases are more difficult to analyze and are less readily observed in experiments.
4. **Local and global**: Bifurcations in which equilibrium points appear, disappear, or change their stability are called *local*. Those that involve

Table 7.1 Summary of local codimension-1 bifurcations.

Bifurcation	Type	Normal Form
Fold	1-D flow	$\dot{x} = \mu \pm x^2$
Transcritical	1-D flow	$\dot{x} = \mu x \pm x^2$
Pitchfork	1-D flow	$\dot{x} = \mu x \pm x^3$
Hopf	2-D flow	$\dot{r} = r(\mu \pm r^2), \dot{\theta} = 1$
Fold	1-D map	$X_{n+1} = \mu + X_n \pm X_n^2$
Flip	1-D map	$X_{n+1} = -(1+\mu)X_n \pm X_n^3$
Transcritical	1-D map	$X_{n+1} = (1+\mu)X_n \pm X_n^2$
Pitchfork	1-D map	$X_{n+1} = (1+\mu)X_n \pm X_n^3$
Neimark–Sacker	2-D map	$Z_{n+1} = (1-\mu)Z_n \pm Z_n^2 \bar{Z}_n$

the entire orbit or trajectory are called *global* and are usually harder to analyze. The distinction is not always obvious, since there are bifurcations with both characteristics.

5. **Continuous and discontinuous**: In a *continuous bifurcation* (also called *subtle* or *supercritical*), an eigenvalue becomes stable or unstable. In a *discontinuous bifurcation* (also called *catastrophic* or *subcritical*), eigenvalues appear or vanish. A discontinuous bifurcation that lacks hysteresis is called *explosive*[1] (Smale 1967).

There are many ways to organize the discussion of bifurcations. We will begin with local bifurcations (Wiggins 1990, Crawford 1991) in flows, then in maps, and finally discuss global bifurcations (Wiggins 1988, Kuznetsov 1995). Only the most common and important codimension-1 types will be described. Table 7.1 summarizes the important local bifurcations and their properties. For brevity, the overdot denotes a time derivative, $\dot{x} = dx/dt$.

7.1 Bifurcations in one-dimensional flows

Since local bifurcations are determined by the properties of equilibrium points, it is customary and convenient to assume the equilibrium is at $x = 0$ and to represent the flow near the equilibrium by

$$\frac{dx}{dt} = f(x, \mu) \tag{7.1}$$

where μ is the bifurcation parameter, chosen such that the bifurcation occurs at $\mu = 0$. You can always redefine the variables to satisfy these conditions. We consider *normal forms* (Kahn and Zarmi 1997), which are

[1] The term 'explosive' is also used for instabilities that grow faster than exponential such that they reach infinity in a finite time.

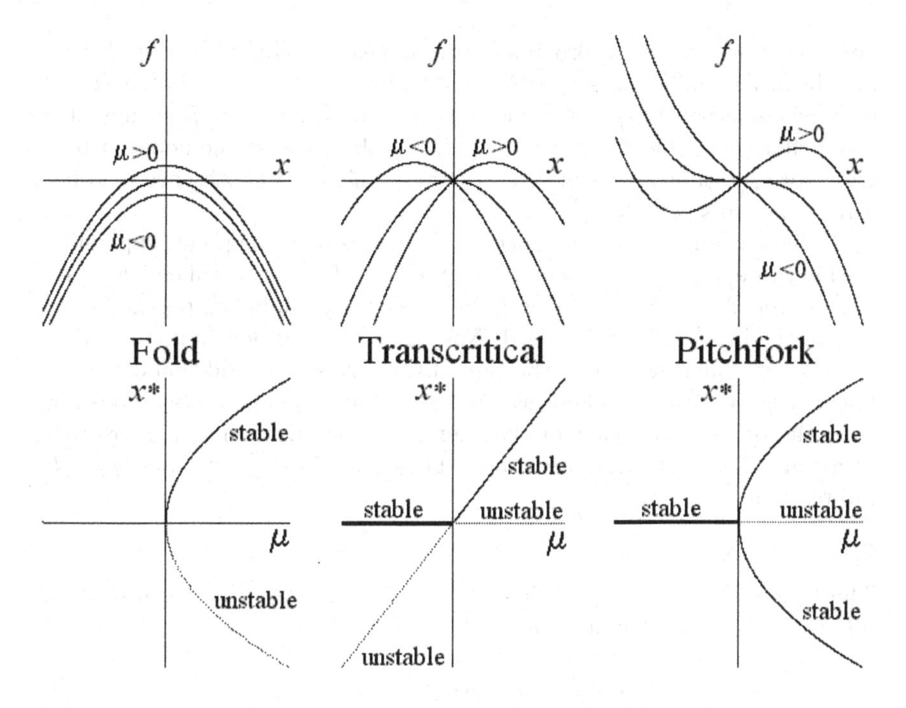

Fig. 7.1 Bifurcations of one-dimensional flows.

the simplest polynomial representations for each type of bifurcation. You can think of them as Taylor series expansions of the flow in the vicinity of $x = 0$ and $\mu = 0$. The existence and nature of these forms are a consequence of *center manifold theory* (Wiggins 1990). The *center manifold* is the subspace of the dynamical system in which the eigenvalues have zero real parts, in contrast to the *stable manifold* where they are negative and the *unstable manifold* where they are positive (see §4.4).

7.1.1 Fold

The basic mechanism for creating and destroying equilibrium points is the *fold*, with a normal form

$$f = \mu - x^2 \tag{7.2}$$

There are various ways to visualize its behavior. The upper left of Fig. 7.1 shows f versus x for negative, zero, and positive values of μ. Also shown is the equilibrium value of x (the value for which $f = 0$) versus μ, with stable equilibria indicated by solid lines and unstable ones with dotted lines. The shape of the lower plot accounts for the name 'fold.' It is also called a *tangent bifurcation* because it occurs when the upper plot is tangent to the x axis. The fold is the most fundamental bifurcation in nonlinear dynamics.

For $\mu < 0$ the solution of eqn (7.2) is imaginary for $f = 0$, and there is no real equilibrium. The function f is negative everywhere, and trajectories are pushed toward minus infinity, although you should remember that the normal form is only an approximation valid near the equilibrium, and there may be other equilibria. At $\mu = 0$ two equilibrium points are born at $x = 0$ and then separate for $\mu > 0$. One of these nodes ($x^* = -\sqrt{\mu}$) is unstable (a repellor), and the other ($x^* = +\sqrt{\mu}$) is stable. The stable node attracts trajectories within its basin of attraction, which in the absence of other attractors is $0 < x < \infty$.

In higher dimensions, the unstable branch is a saddle point, explaining the alternate name *saddle-node bifurcation*. The fold is also called a *blue-sky bifurcation* (Abraham and Shaw 1988), since the equilibrium points appear 'out of the blue' as μ is increased. Note that an equivalent normal form is $f = \mu + x^2$, differing only in the sign of the terms. The addition of a term linear in x to either of these forms destroys the symmetry but does not alter the qualitative behavior. Furthermore, the addition of higher-order terms in x does not alter the results. Thus the fold is a *structurally stable* bifurcation.

7.1.2 Transcritical

The logistic differential equation in §3.2 is an example of a *transcritical bifurcation* with a normal form

$$f = \mu x - x^2 \tag{7.3}$$

It has two equilibrium points ($x^* = 0, \mu$) whose stability switches at $\mu = 0$, corresponding to $a = 0$ in eqn (3.2), as in the center of Fig. 7.1. For $\mu < 0$, the equilibrium at $x^* = 0$ is stable, and for $\mu > 0$, the one at $x^* = \mu$ is stable.

The transcritical bifurcation is structurally unstable, since even the smallest constant term ϵ on the righthand side of eqn (7.3)

$$f = \epsilon + \mu x - x^2 \tag{7.4}$$

will either remove the bifurcation (if $\epsilon > 0$) or convert it into a fold (if $\epsilon < 0$) as in the top of Fig. 7.2. Such bifurcations are called *imperfect*, and the small constant ϵ is the *imperfection parameter*. In one-dimensional flows, the fold is generic, and hence all bifurcations are folds unless made otherwise by the existence of some special symmetry. As with the fold, an equivalent normal form is $f = \mu x + x^2$, differing only in the sign of the x^2 term.

7.1.3 Pitchfork

Even less generic is the *pitchfork bifurcation* whose normal form

$$f = \mu x - x^3 \tag{7.5}$$

lacks both a constant and quadratic term, and has only odd powers of x. It occurs in systems with precise antisymmetry $f(x) = -f(-x)$, for which the

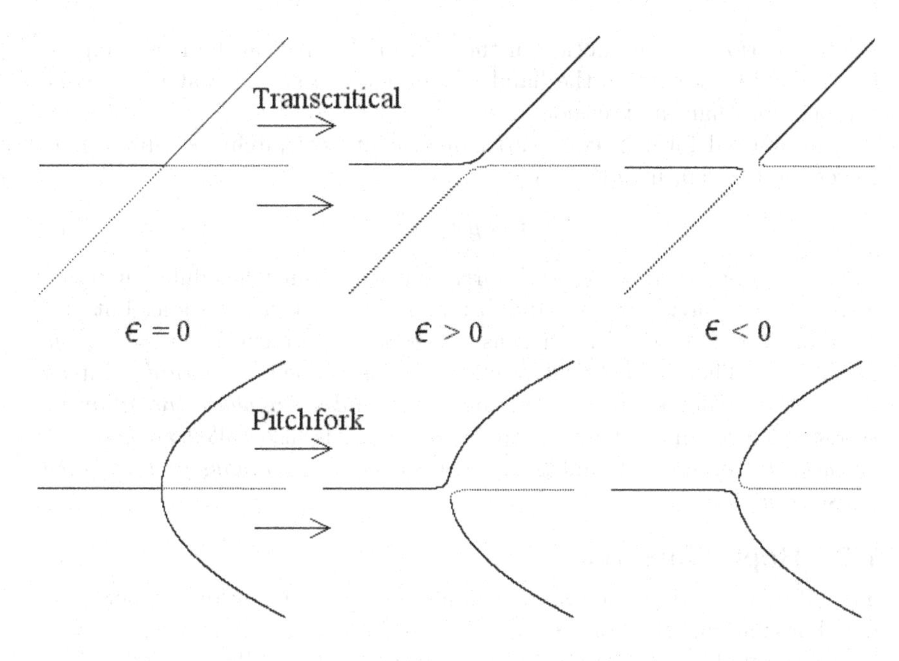

Fig. 7.2 Imperfect bifurcations.

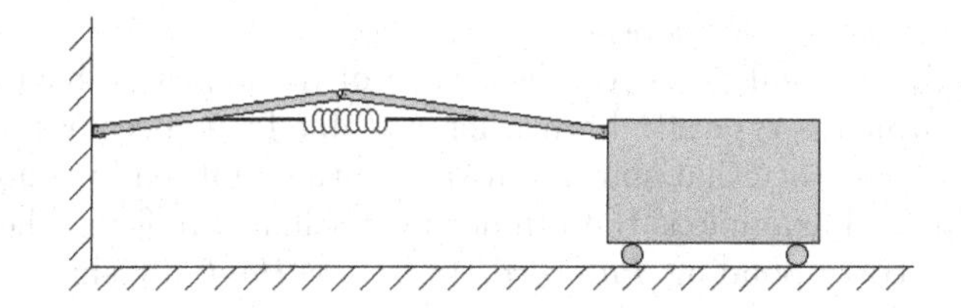

Fig. 7.3 Cart on a spring with a hinged handle undergoes a pitchfork bifurcation.

dynamical equation $dx/dt = f(x)$ is *invariant* under a change of variable x to $-x$. For example, a vertical column supporting a heavy weight can buckle and fail either to the right or left with equal probability. A cart attached to a hinged, massless handle with a spring at the hinge, as in Fig. 7.3, undergoes a pitchfork bifurcation when the oscillation amplitude is lowered to the point where the hinge is either always up as shown or always down (Golubitsky and Schaeffer 1985). The name 'pitchfork' comes from the lower-right plot in Fig. 7.1, which shows a stable equilibrium at $x^* = 0$ for $\mu < 0$ that becomes unstable for $\mu > 0$ giving birth to two new stable branches with $x^* = \pm\sqrt{\mu}$. It is also called a *cusp* bifurcation.

The pitchfork bifurcation is structurally unstable, since even the smallest imperfection parameter ϵ on the righthand side of eqn (7.5)

$$f = \epsilon + \mu x - x^3 \tag{7.6}$$

will remove the bifurcation, as in the bottom of Fig. 7.2. For example, the bifurcation for the motion of the cart in Fig. 7.3 can be made imperfect by adding weight to the handle or by using a spring that is stiffer in compression than in extension.

The normal form in eqn (7.5) is a *supercritical* pitchfork bifurcation. Reversing the sign in eqn (7.5)

$$f = \mu x + x^3 \tag{7.7}$$

leads to a *subcritical* pitchfork bifurcation, with somewhat different properties. The behavior of the equilibrium at $x^* = 0$ is the same, but the pitchfork points to the left and has two *unstable* branches at $x^* = \pm\sqrt{-\mu}$ for $\mu < 0$. The supercritical pitchfork is also called a *forward* bifurcation and is analogous to a *continuous* or *second-order phase transition* in statistical mechanics. The subcritical pitchfork is also called an *inverted* or *backward* bifurcation and is analogous to a *discontinuous* or *first-order phase transition*.

7.2 Hopf bifurcation

The bifurcations described above can also occur in dimensions higher than one. For example, the Lorenz attractor has a pitchfork bifurcation at $r = 1$ where the node at the origin becomes an unstable saddle and gives birth to two new nodes that become spirals at $r = 1.34561\ldots$ and eventually become the centers of the lobes.

In two-dimensional flows, the new type of dynamical behavior is the *limit cycle*, which is typically born when a stable focus becomes unstable, producing a trajectory that spirals outward until stabilized by a nonlinearity. The standard example is the van der Pol oscillator in §3.8. The normal form for the corresponding *Poincaré–Andronov–Hopf* (or simply *Hopf*[2]) bifurcation (Marsden and McCracken 1976) is

$$\frac{dx}{dt} = -y + x[\mu - (x^2 + y^2)] \tag{7.8}$$

$$\frac{dy}{dt} = x + y[\mu - (x^2 + y^2)] \tag{7.9}$$

which can be rewritten more compactly in polar coordinates as

$$\frac{dr}{dt} = r(\mu - r^2) \tag{7.10}$$

$$\frac{d\theta}{dt} = 1 \tag{7.11}$$

where $r = \sqrt{x^2 + y^2}$ and $\theta = \tan^{-1}(y/x)$. For $\mu < 0$, the focus at $r = 0$ is stable, but for $\mu > 0$ the focus at $r = 0$ is unstable, and a stable circular

[2]Eberhard Hopf (1902–1983) was an Austrian mathematician, trained in Germany, who was on the MIT faculty before returning to Germany in 1936, where he remained during World War II, after which he returned to the US in 1947, became a US citizen, and spent his remaining years at the University of Indiana.

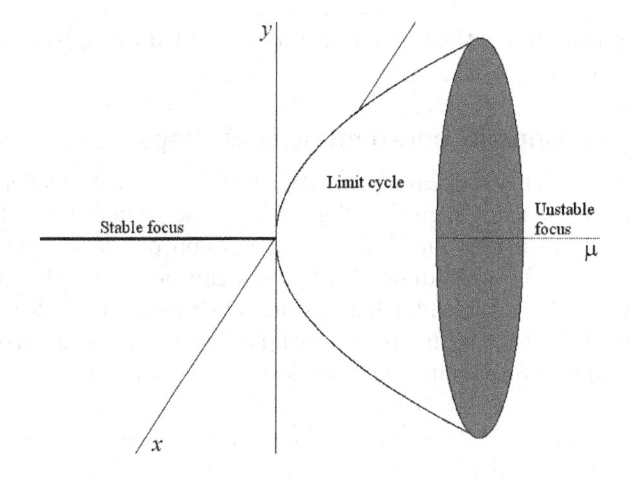

Fig. 7.4 Hopf bifurcation.

limit cycle with $r = \sqrt{\mu}$ and angular frequency $\omega = d\theta/dt = 1$ exists. At $\mu = 0$, the eigenvalues, which are a complex conjugate pair, cross the imaginary axis, and the sinusoidal oscillation changes from decay to growth. The dimension of the attractor changes from zero (a point) to one (a closed loop). The similarity of Fig. 7.4 to the pitchfork bifurcation is evident.

As with the pitchfork, the normal form in eqn (7.10) is a *supercritical* (continuous) Hopf bifurcation (the limit cycle is born with zero radius). The Hopf bifurcation can also be *subcritical* with a radial normal form

$$\frac{dr}{dt} = r(\mu + r^2) \tag{7.12}$$

in which case a stable focus at $r = 0$ and an unstable limit cycle at $r = \sqrt{-\mu}$ for $\mu < 0$ coalesce and annihilate at $\mu = 0$. The trajectory then spirals outward to infinity unless another nonlinearity, such as a term $-r^5$ in eqn (7.12), limits the growth. This bifurcation is *discontinuous* or *catastrophic* and exhibits *hysteresis*. Once the large amplitude oscillations onset, they cannot be eliminated by reducing μ to zero. A subcritical Hopf bifurcation occurs in the Lorenz attractor at $r = 24.7368\ldots$.

The *Hopf bifurcation theorem* (Casti 2000) predicts when a stable limit cycle exists in a two-dimensional flow. Suppose the eigenvalues are $\lambda = A \pm Bi$, where A and B are differentiable functions of the bifurcation parameter μ. Then a stable limit cycle will exist if

$$\frac{dA}{d\mu} > 0 \tag{7.13}$$

$$A(0) = 0 \tag{7.14}$$

$$B(0) \neq 0 \tag{7.15}$$

Under some conditions, the theorem can be extended to higher dimensions (Sacker 1965).

7.3 Bifurcations in one-dimensional maps

Discrete-time systems also have bifurcations, which mostly parallel those in flows and have the same name. An important class of map is the Poincaré map derived from a flow and having a dimension one less than the corresponding flow. Thus two-dimensional flows can be studied by analyzing their one-dimensional maps. Such maps provide a means to understand the fate of limit cycles that bifurcate. As with flows, we assume the fixed point is at $X = 0$ and approximate the map near the fixed point by

$$X_{n+1} = f(X_n, \mu) \tag{7.16}$$

where μ is the bifurcation parameter chosen such that the bifurcation occurs at $\mu = 0$.

7.3.1 Fold

The fold is one mechanism for creating and destroying fixed points in maps. Its normal form is

$$f = \mu + X - X^2 \tag{7.17}$$

A fixed point exists when $f = X$ or $X^* = \pm\sqrt{\mu}$. No such points exist for $\mu < 0$, but two fixed points are born at $\mu = 0$, one stable and the other unstable. This bifurcation occurs when the function $f(X)$ is tangent to the 45° line, and is thus also called a *tangent bifurcation*. An equivalent normal form is $f = \mu + X + X^2$, differing only in the sign of the X^2 term. The addition of higher-order terms in X does not alter the results, and thus the fold is a *structurally stable* bifurcation.

A familiar example of the fold is the birth of the stable period-3 cycle in the logistic map at $A = 1 + \sqrt{8} = 3.82842712\ldots$. Figure 7.5 shows $f(X)$ for three values of A near the bifurcation point, indicating how the new stable 3-cycle is born.

As a Poincaré section of a flow, imagine a limit cycle in a plane with a straight line also in the plane but that does not intersect the limit cycle. As the limit cycle grows, it touches the line, and a pair of fixed points are born and separate as the cycle continues growing. In this case, the bifurcation of the map does not correspond to a bifurcation in the flow, since the bifurcation value depends on the where the line was drawn, and nothing special happens when it touches the line.

For values of A slightly below the critical value, *intermittency* as in Fig. 7.6 (see also Fig. 6.9) is observed in which short bursts of chaos are embedded in long regions of nearly periodic behavior as the orbit slowly works its way through the narrow channel between the $f(X)$ curve and the 45° line (Bergé *et al.* 1986), sometimes called *bottlenecking*. Intermittency

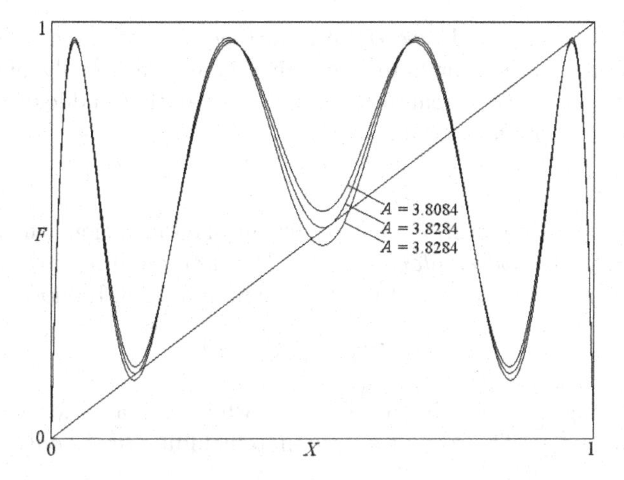

Fig. 7.5 Tangent bifurcation in the logistic map.

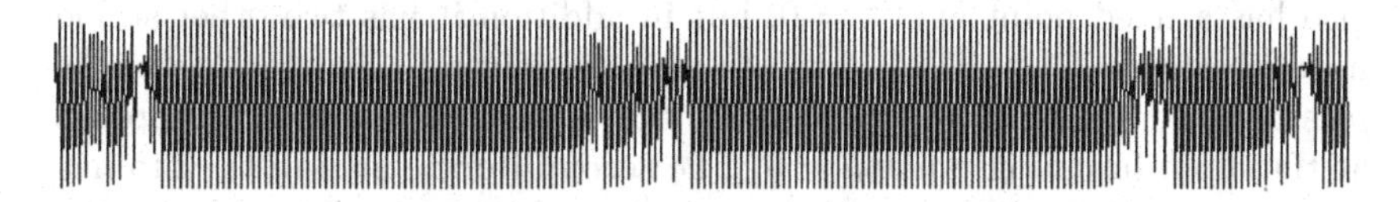

Fig. 7.6 Intermittency in the logistic map with $A = 3.8284$.

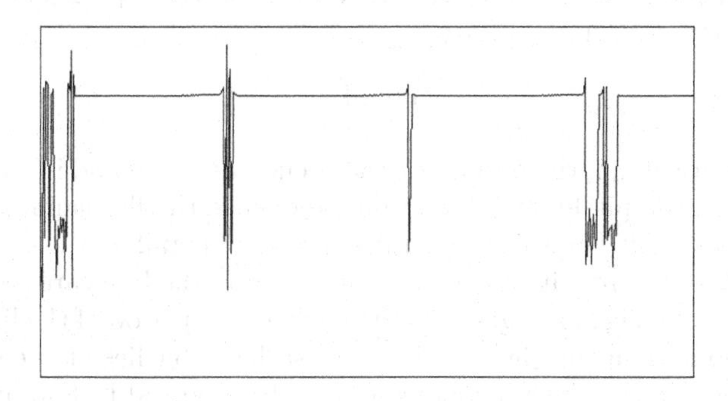

Fig. 7.7 Intermittency in the Lorenz attractor with $r = 166.065$.

also occurs in flows, for example, in Duffing's oscillator (Trickey and Virgin 1998) and in the Lorenz system with $166 \simeq r \simeq 166.2$ (Pomeau and Manneville 1980) and may be a source of the '$1/f$ noise' often observed in physical systems (Schuster 1995). Figure 7.7 shows the time between successive $x = 0$ crossings for the Lorenz attractor with $r = 166.065$.

The *type I* intermittency associated with the tangent bifurcation is only one of three types that have been identified in the vicinity of local bifurcations (Hilborn 2000). *Type II* intermittency is associated with the subcritical Hopf bifurcation, and *type III* intermittency is associated with inverse period doubling (see below). Intermittency can be especially puzzling in experiments where an apparently periodic oscillation suddenly becomes chaotic for no apparent reason.

7.3.2 Flip

Another common bifurcation that occurs in one-dimensional maps is the *flip*, also called the *period-doubling* or *subharmonic* bifurcation, described in some detail in the context of the logistic map in §2.3. Its normal form is

$$f = -(1 + \mu)X + X^3 \tag{7.18}$$

There is a single fixed point at $X^* = 0$, which is stable for $\mu < 0$ and unstable for $\mu > 0$. The second-iterate map including terms only up to X^3

$$f[f(X)] \simeq (1 + \mu)^2 X - (1 + \mu)(2 + 2\mu + \mu^2)X^3 \tag{7.19}$$

also has a fixed point at $X^* = 0$, but in addition it has two other stable fixed points at $X^* \simeq \pm\sqrt{\mu}$, for μ small, constituting a stable 2-cycle. As μ is reduced toward zero from above, the 2-cycle shrinks and disappears at the flip bifurcation point $\mu = 0$. The second iterate map of the flip has a pitchfork bifurcation since there are two stable values. The first-iterate map oscillates between these two values.

The normal form in eqn (7.18) is a *supercritical* flip. There is also a *subcritical* flip with normal form

$$f = -(1 + \mu)X - X^3 \tag{7.20}$$

In the subcritical flip, the period-2 cycle occurs for $\mu < 0$ and is unstable by analogy with the pitchfork bifurcation. Note that the flip cannot occur in a one-dimensional *flow* since the trajectory would stall at the equilibrium point if it tried to cross it. It can and does occur in the Poincaré section of a limit cycle in dimensions higher than two where the period of the limit cycle suddenly doubles, as in Fig. 4.7. The new stable orbit lies at the edge of a *Möbius strip* whose width increases with μ. It is typical to find an infinite succession of flips with a finite accumulation point beyond which chaos onsets (see §2.3), although a finite number of bifurcations followed by the same number of reverse bifurcations without leading to chaos sometimes occurs.

Both the fold and the flip in maps are structurally stable. The fold has an eigenvalue (df/dX in one dimension) of $+1$, and the flip has an eigenvalue of -1.

7.3.3 Transcritical

Entirely analogous to the transcritical bifurcation in flows is the transcritical bifurcation in one-dimensional maps with normal form

$$f = (1+\mu)X - X^2 \tag{7.21}$$

There is a fixed point at $X^* = 0$ and another at $X^* = \mu$ whose stability exchange at $\mu = 0$. The standard example is the logistic map in §2.3 with $A = 1$. This bifurcation is structurally unstable since it is destroyed by a small imperfection similar to Fig. 7.2. An alternate form is $f = (1+\mu)X + X^2$, differing only in the sign of the X^2 term.

7.3.4 Pitchfork

Similarly, the pitchfork bifurcation occurs in one-dimensional maps with a normal form

$$f = (1+\mu)X - X^3 \tag{7.22}$$

There is a fixed point at $X^* = 0$ that becomes unstable for $\mu = 0$, where two stable fixed points are born with $X^* = \pm\sqrt{\mu}$ similar to Fig. 7.1. The standard example is the second-iterated logistic map at any of the period-doubling points such as $A = 3$. This bifurcation is also structurally unstable since it is destroyed by a small imperfection similar to Fig. 7.2. An alternate form is $f = (1+\mu)X + X^3$, differing only in the sign of the X^3 term.

7.4 Neimark–Sacker bifurcation

As with flows, the dynamics become richer when maps of dimension greater than one are considered. The one-dimensional bifurcations just described also occur in higher dimensions. The main new behavior is the formation of a *drift ring* when the map is a Poincaré section of a quasiperiodic flow. This bifurcation occurs when a limit cycle in a flow becomes unstable, giving birth to a torus. If the Poincaré section is chosen appropriately, then the map corresponding to the limit cycle contains a single fixed point whose stability is analyzed as described in §4.7. When a torus is born in the flow, the eigenvalues of the map are a complex conjugate pair with $|\lambda| = |\bar{\lambda}| = 1$.

By analogy with the Hopf bifurcation, the birth of a torus in a flow is called a *secondary Hopf*, and the corresponding appearance of a drift ring in the map is called a *Neimark–Sacker bifurcation* (Neimark 1959, Sacker 1965). The normal form for the Neimark–Sacker bifurcation is rather complicated and depends on subtleties involving rational and irrational numbers not evident in the previous discussion (Wiggins 1990). For the irrational case, the normal form is

$$f = (1-\mu)Z \pm Z^2\bar{Z} \tag{7.23}$$

where Z is complex and $\bar{Z}$ is its complex conjugate. This bifurcation has both a supercritical and a subcritical form.

An example of a Neimark–Sacker bifurcation is the *delayed logistic map*

$$X_{n+1} = AX_n(1 - X_{n-1}) \tag{7.24}$$

at $A = 2$ (Aronson *et al.* 1983). This system has a trivial fixed point at $X^* = 0$ and a nontrivial one at $X^* = 1 - 1/A$ whose eigenvalues are

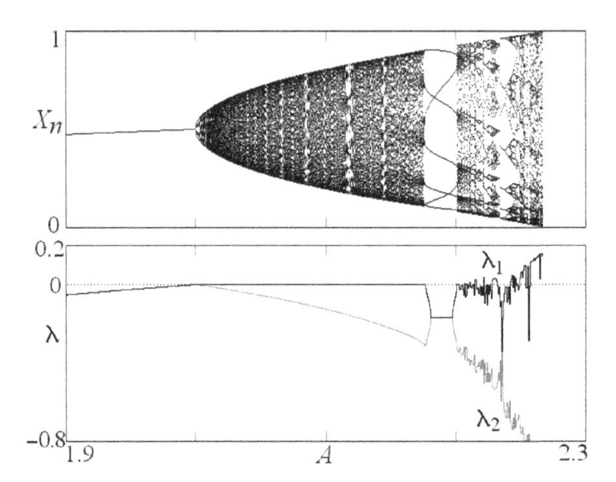

Fig. 7.8 Bifurcation diagram and Lyapunov exponents for the delayed logistic map.

$\lambda = \frac{1}{2} \pm \sqrt{\frac{5}{4} - A}$. For $A > \frac{5}{4}$, the eigenvalues are complex, and they cross the unit circle when $A = 2$. Quasiperiodic behavior exists over the range $2 < A < 2.176398\ldots$, followed by a period-7 window and a band of chaos near $A = 2.27$, after which the orbit becomes unbounded at $A = 2.271012\ldots$ as in Fig. 7.8. In such a bifurcation diagram it is difficult to distinguish chaos from quasiperiodicity, but the plot showing the Lyapunov exponents indicates quasiperiodicity in the region where λ_1 remains at zero. The Lyapunov exponents for $A = 2.1$ are $\lambda \simeq 0, -0.1267672$ and for $A = 2.27$ are $\lambda \simeq 0.18312, -1.24199$.

Two-dimensional maps arising from three-dimensional flows can also exhibit chaos, although the bifurcations leading to chaos are usually a sequence of period doublings rather than the destruction of a torus. In dimensions higher than three, the quasiperiodic route to chaos is more common (see §6.5.3).

7.5 Homoclinic and heteroclinic bifurcations

The theory of global bifurcations is much less developed than the theory of local bifurcations. Global bifurcations typically occur when an attractor touches its basin of attraction or when a limit cycle touches an equilibrium point or, more generally, when the stable and unstable manifolds intersect. The most important examples of the latter type are the *homoclinic tangency* where the stable and unstable manifold of a saddle point (see §4.4) touch one another and the *heteroclinic tangency* where the stable manifold of one equilibrium point touches the unstable manifold of another. These manifolds are important because they are boundaries between regions where nearby trajectories diverge in different directions as they approach the saddle point. When the manifolds intersect, nearby tra-

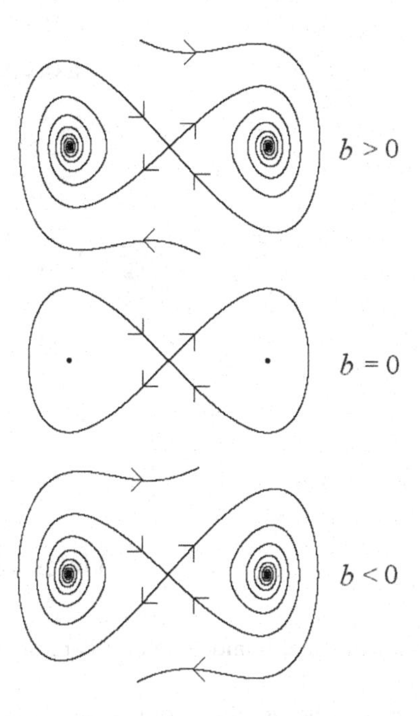

Fig. 7.9 Two-dimensional homoclinic ($b = 0$) and heteroclinic ($b \neq 0$) trajectories.

jectories are repeatedly repelled by and attracted to the equilibrium points, producing chaos.

Homoclinic and heteroclinic *connections* can occur in dissipative systems, but they are ubiquitous in conservative systems. Note that two stable manifolds or two unstable manifolds cannot intersect because the direction of flow would then not be unique at the intersection. A homoclinic connection cannot occur in a linear system, because the stable and unstable manifolds are straight and cannot bend as required to make them touch, and it requires a saddle point since it must have both a stable and unstable manifold.

7.5.1 Autonomous Duffing's oscillator

Homoclinic and heteroclinic connections occur in the autonomous, two-well Duffing's oscillator

$$\frac{d^2x}{dt^2} + b\frac{dx}{dt} - x + x^3 = 0 \tag{7.25}$$

where b is the damping constant (friction), which we take to be either positive or negative, but small. This two-dimensional system has three equilibrium points at $x^* = 0, \pm 1$ and $y^* = dx/dt = 0$. The one at the origin is a saddle point, and the other two are stable foci for $b > 0$ and unstable foci for $b < 0$. Figure 7.9 shows the phase portraits for three values of b. For $b \neq 0$ there is a heteroclinic connection between the origin and the

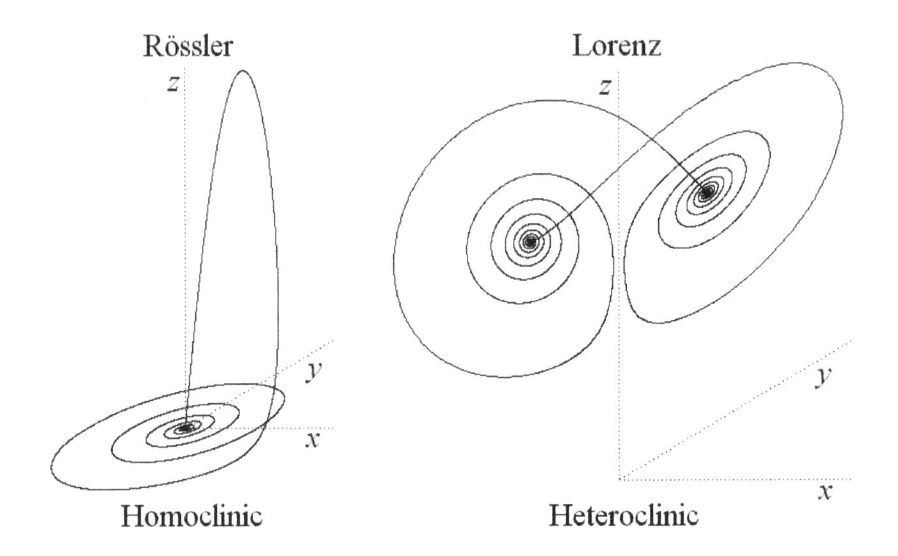

Fig. 7.10 Three-dimensional homoclinic and heteroclinic trajectories.

foci, and for $b = 0$ there is a homoclinic connection between the stable and unstable manifolds of the saddle point at the origin. In this example, the heteroclinic connection is structurally stable, and the homoclinic connection is structurally unstable. It appears that the homoclinic trajectories cross, but they do not, since their period is always infinite and they stall as they approach the equilibrium point. There is a *homoclinic bifurcation* at $b = 0$.

7.5.2 Rössler and Lorenz attractors

In higher dimensions, such bifurcations are even more common. Figure 7.10 shows a homoclinic connection in the Rössler attractor given by eqns (4.35)–(4.37) with $a = b = 0.2$ and $c \simeq 10.25505$ and a heteroclinic connection in the Lorenz attractor given by eqns (4.32)–(4.34) with $\sigma \simeq 15.39947, b = 8/3$, and $r = 78$. These trajectories were found using a *shooting method*, starting at various points on the unstable manifold near the equilibrium point and adjusting the parameters so that the trajectory returned as close as possible to the equilibrium on its next return. These cases are not unique. For example, the Lorenz attractor, as in Fig. 7.11, also has a double *homoclinic* connection at $\sigma = 10, b = 8/3$, and $r \simeq 13.92656$ where the saddle point at the origin is connected back to itself by two trajectories that loop around the respective lobes, and a double *heteroclinic* connection at $\sigma = 10, b = 8/3$, and $r \simeq 24.0579$ where the saddle point at the origin connects to the off-axis spiral saddle points, marking the abrupt onset of a chaotic attractor.

The bifurcation typically occurs when the stable and unstable manifolds

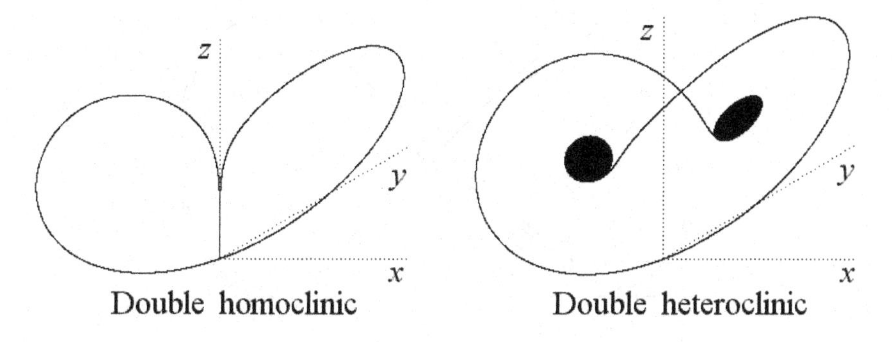

Fig. 7.11 Double homoclinic and heteroclinic trajectories in the Lorenz attractor.

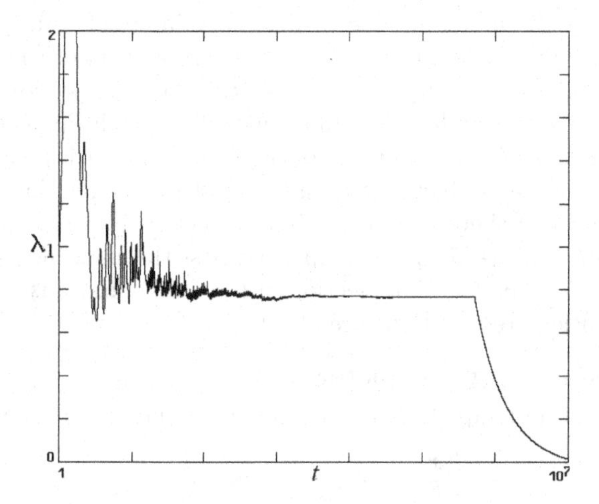

Fig. 7.12 Calculated Lyapunov exponent for the Lorenz attractor with $r = 23.9$, just before a heteroclinic bifurcation, showing transient chaos.

are tangent to one another at the *homoclinic* (or *heteroclinic*) *point* in what is called a *homoclinic* (or *heteroclinic*) *tangency.* Prior to the bifurcation, the manifolds approach closely but do not touch, and *transient chaos* is often observed in which the motion is apparently chaotic for a long time before eventually settling to a nonchaotic attractor. Figure 7.12 shows that the calculated Lyapunov exponent for the Lorenz attractor (see §5.6) with $r = 23.9$ initially converges to a constant value and then suddenly falls to zero after nearly ten million cycles as the trajectory attracts to one of the off-axis stable equilibrium points. Such chaotic transients typically occur near a global bifurcation, signaling the onset of chaos.

Beyond the bifurcation point, the manifolds intersect *transversely* (at a nonzero angle), and chaos typically occurs. Whereas tangencies are structurally unstable, transverse intersections are structurally stable. It can be

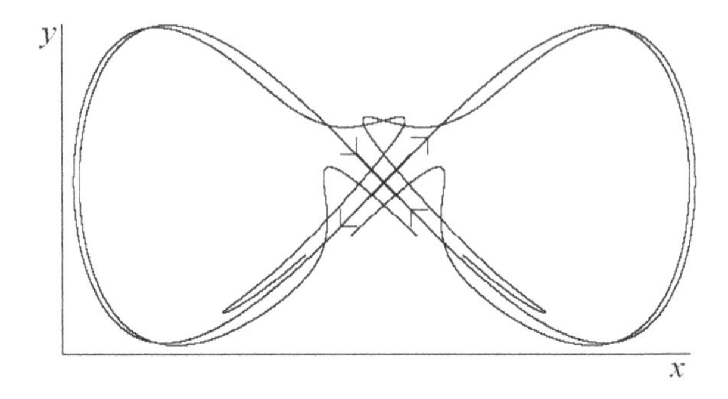

Fig. 7.13 First stages of a homoclinic tangle in the driven Duffing's oscillator.

shown that if the manifolds intersect once, they intersect infinitely many times. To see why this is so, construct a Poincaré section of the flow. For a three-dimensional flow with saddle points, the Poincaré section is a two-dimensional map with saddle points independent of which Poincaré section is taken as long as the section intersects the flow. Each such point has images that lie along the outset of the saddle and preimages that lie along the inset. However, if there is a homoclinic trajectory in the flow, the corresponding homoclinic orbit in the map must pass through infinitely many points, all of which lie at an intersection of the stable and unstable manifold, a result anticipated by Poincaré (1921).

7.5.3 Driven Duffing's oscillator

Now consider the undamped, *driven* Duffing's oscillator (see §4.8.1)

$$\frac{d^2x}{dt^2} - x + x^3 = A \sin \Omega t \tag{7.26}$$

whose state space is three-dimensional. For $A = 0$, this system has a double homoclinic connection of the saddle point at the origin ($x^* = 0, y^* = dx/dt = 0$) as in Fig. 7.9. For $A \neq 0$ but small, the trajectory lies near the surface of a torus and has a Poincaré section in the xy plane that is a perturbed version of the original trajectory. In particular, following the inset and outset of the equilibrium at the origin with $A = 0.05$ and $\Omega = 2$ gives a map for $\Omega t \pmod{2\pi} = 0$ as in Fig. 7.13. For clarity, only a portion of each manifold, showing the first few intersections, has been shown, but it is clear that they oscillate about the unperturbed orbit with an amplitude that grows as time advances both forward and backwards. The infinity of oscillations, called a *homoclinic tangle*, leads to an infinite Cantor set of intersections, and the amplitude growth suggests sensitive dependence on initial conditions. Note that these curves are not trajectories, but sets of points whose images and preimages lie on the unstable and stable manifolds, respectively.

The stretching and folding of the phase space is apparent and is an example of a *Smale horseshoe* (Smale 1967) as in Fig. 7.14. The horseshoe

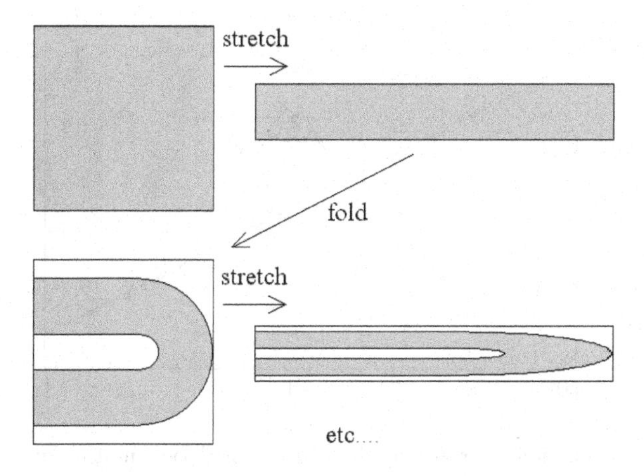

Fig. 7.14 Smale horseshoe map.

map takes a rectangular region and stretches it horizontally by a factor $a > 1$ while compressing it vertically by a factor $b < 1$. The resulting rectangle is then folded into a horseshoe shape, and the process is repeated over and over as with Danish pastry or the taffy-pulling machine in Fig. 1.7. For $ab < 1$, the area contracts, producing a strange attractor. The stretching accounts for the repulsion of orbits on the small scale, and the folding allows attraction on the large scale.

Since the homoclinic tangle is structurally stable, it also occurs when a small damping term bdx/dt is added to the lefthand side of eqn (7.26). Using a perturbation method due to Melnikov (1963), Wiggins (1990) has shown that transverse intersections occur in the damped, driven Duffing's oscillator provided

$$b < \frac{3\pi\Omega A}{\sqrt{8}} \cosh \frac{\pi\Omega}{2} \tag{7.27}$$

Numerical experiments confirm the presence of chaos when this condition is satisfied.

This is a rare example of a case in which the condition for the onset of chaos can be calculated analytically. In fact, Sil'nikov (1965) has proved that a three-dimensional flow with a homoclinic connection to a spiral saddle point with eigenvalues $\lambda, \alpha \pm i\omega$ is chaotic for a range of parameters near the bifurcation point if $|\lambda| > |\alpha|$. Under some conditions, the results can be generalized (Mees and Sparrow 1987), and similar behavior occurs near heteroclinic points (Lau and Finn 1992).

7.6 Crises

The final class of global bifurcation that we will discuss is called a *crisis*. It occurs when a chaotic attractor collides with an unstable periodic orbit or its basin of attraction (Grebogi *et al.* 1982, 1983). We can distinguish three types of crises — a *boundary crisis* in which the attractor touches its basin

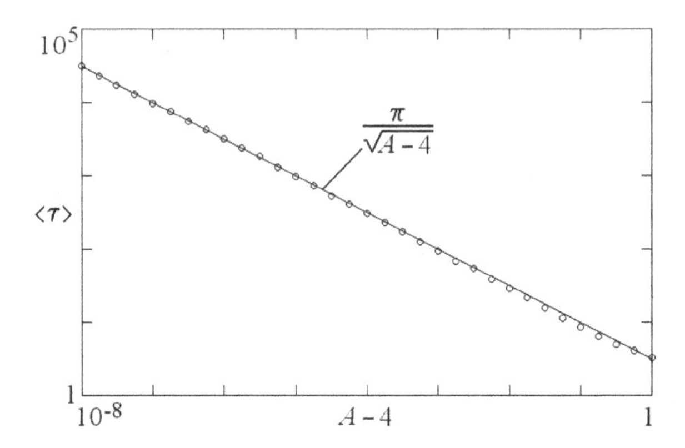

Fig. 7.15 Duration of transient chaos in the logistic equation with $A > 4$.

boundary, an *interior crisis* in which the attractor touches a periodic orbit within its basin, and an *attractor merging crisis* in which two or more attractors simultaneously touch a periodic orbit on the basin boundary that separates them. We will consider a simple example of each case in a one-dimensional map.

7.6.1 Boundary crisis

A system undergoes a boundary crisis when its attractor collides with the basin of attraction separating it from another coexisting attractor. After the crisis, the attractor is destroyed and converted into a nonattracting chaotic saddle. The trajectory wanders in the vicinity of the saddle for a time before asymptotically approaching the other attractor, producing transient chaos. These transients are especially long-lived (*superpersistent*) and have been observed in coupled chaotic electrical oscillators (Zhu *et al.* 2001).

A simple and familiar example of a boundary crisis is the logistic map with $A = 4$. For $A < 4$ the attractor spans the range $\frac{A^2}{4}(1 - \frac{A}{4}) < X < \frac{A}{4}$, and its basin spans the range $0 < X < 1$ with an unstable fixed point at $X^* = 0$. For $A = 4$ the attractor touches its basin boundary, and for $A > 4$ orbits are attracted to $-\infty$. However, for $A = 4 + \epsilon$, with ϵ small and positive, the orbits do not escape until X happens to fall in the range $0.5 - 0.25\sqrt{\epsilon} < X < 0.5 + 0.25\sqrt{\epsilon}$ (see §2.3.7) near the peak of the parabola.

The probability that this will happen can be calculated using eqn (2.11), with the result that the average number of chaotic iterations is given by $\langle\tau\rangle \simeq \pi/\sqrt{\epsilon}$. A numerical experiment with 4000 randomly chosen initial conditions in Fig. 7.15 confirms the result. In general, $\langle\tau\rangle$ scales as $\epsilon^{-\gamma}$, where γ is the critical exponent of the crisis. The value of $\gamma = 1/2$ holds for any one-dimensional map with a quadratic maximum, but higher dimensional systems typically have $\gamma > 1/2$. The probability that the resulting transient chaos will last at least τ iterations for a collection of random ini-

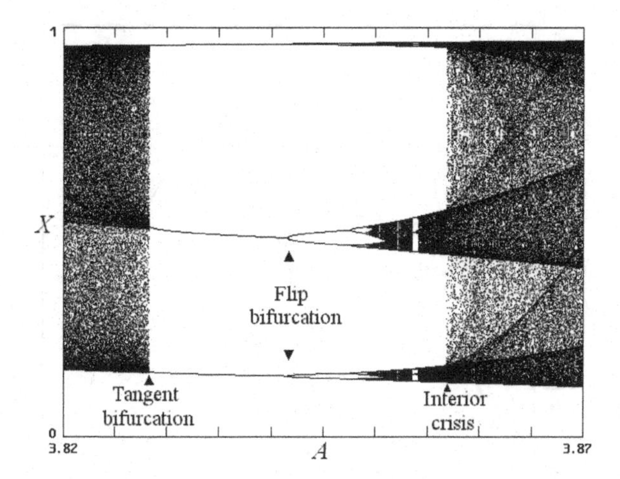

Fig. 7.16 Interior crisis in the period-3 window of the logistic map.

tial conditions is $P(\tau) \simeq e^{-\tau/\langle\tau\rangle}$. Viewed from the opposite direction, the boundary crisis is another route to chaos through chaotic transients that become progressively longer as the bifurcation point at $\epsilon = 0$ is approached.

Boundary crises are very common in chaotic systems. The Hénon map has one at $a \simeq 1.08$ (Grebogi *et al.* 1983).

7.6.2 Interior crisis

In an interior crisis, the chaotic attractor collides with an unstable periodic orbit or limit cycle *within* its basin of attraction. When the collision occurs, the attractor suddenly expands in size but remains bounded. An example occurs in the periodic windows of the logistic map (see Fig. 2.4). The period-3 window expanded in Fig. 7.16 shows three miniature bifurcation diagrams that are similar to the large one in Fig. 2.4. Just as with the boundary crisis at $A = 4$, these smaller regions suddenly expand at $A \simeq 3.8568$ to fill a single wider interval.

This same behavior occurs in all the periodic windows within the logistic map, including the miniature windows within the larger windows. The reason is that to the left of each such window is a tangent bifurcation (or fold) that gives birth to both a stable and unstable orbit. The stable orbit is evident in the diagram, but the unstable one remains nearby and eventually collides with the attractor as it expands. This orbit then repels the iterates to regions that were not previously accessible.

The behavior in the vicinity of the interior crisis (Grebogi *et al.* 1983) is similar to the boundary crisis at $A = 4$. To the right of the crisis point, *crisis-induced intermittency* (Grebogi *et al.* 1987) is observed in which the orbit switches intermittently from the narrow bands to the broader region. The average time spent in the narrow bands increases inversely with the square root of the deviation of A from the crisis point, going to infinity at

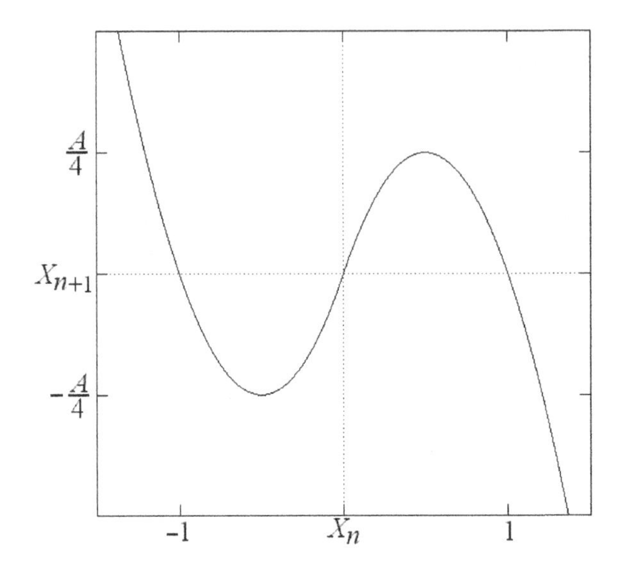

Fig. 7.17 Antisymmetric logistic map.

the crisis point (Hilborn 2000). This behavior is evident from the density of points in Fig. 7.16. Interior crises occur frequently in maps and flows and have been observed in electrochemical experiments (Krischer *et al.* 1991).

7.6.3 Attractor merging crisis

An attractor merging crisis occurs in the *antisymmetric logistic map*

$$X_{n+1} = AX_n(1 - |X_n|) \tag{7.28}$$

shown graphically in Fig. 7.17. $X > 0$ gives the familiar logistic map, but for $X < 0$ there is a similar inverted map. The attractors for these maps grow larger as A increases, until they simultaneously collide with the fixed point $X^* = 0$ at $A = 4$. Thus two separate attractors merge into one at $A = 4$ and remain bounded until $A = 2 + \sqrt{8} = 4.82842712\ldots$, where a boundary crisis occurs. If the orbit is initially confined to one of the maps, the region occupied by it suddenly doubles in size at the crisis point, as in Fig. 7.18.

For values of A only slightly above 4, intermittency occurs as the orbit moves about one of the attractors for a long time before finding its way to the other where it then spends a long time before moving back at apparently random intervals. Hysteresis also occurs, since reducing the value of A is likely to trap the orbit in the other attractor when the crisis point is crossed in the opposite direction.

Some systems, such as the one-dimensional maps

$$X_{n+1} = A \sin X_n \tag{7.29}$$

$$X_{n+1} = AX_n \sin X_n \tag{7.30}$$

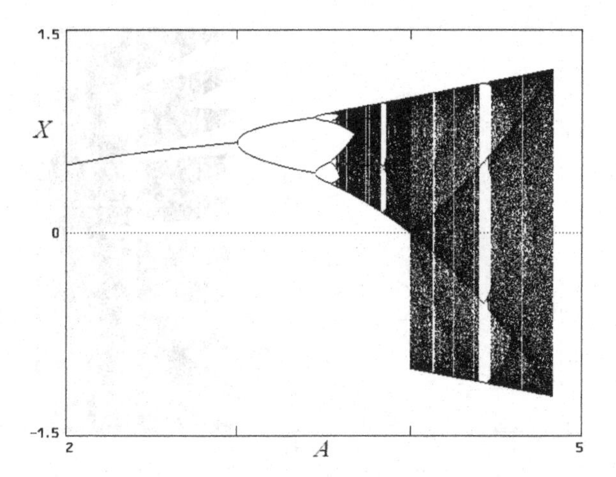

Fig. 7.18 Attractor merging crisis in the antisymmetric logistic map at $A = 4$.

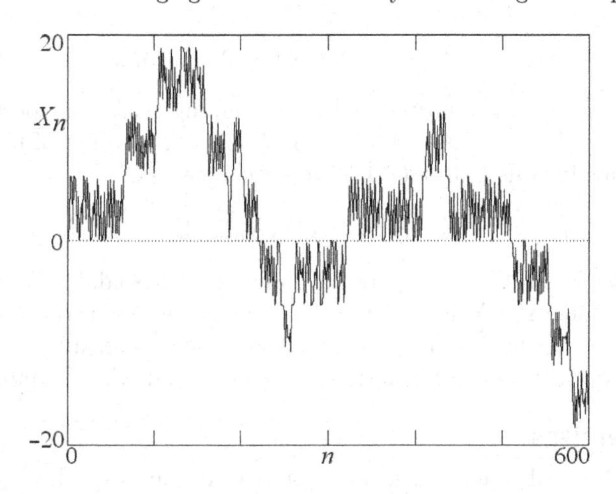

Fig. 7.19 Intermittency producing an apparent random walk in eqn (7.31) with $A = 4.61$.

$$X_{n+1} = X_n + A \sin X_n \tag{7.31}$$

$$X_{n+1} = X_n + \sin AX_n \tag{7.32}$$

have infinitely many attractors that can either merge simultaneously or sequentially as A increases. Just above the bifurcation point, a form of intermittency occurs where the orbit remains in one basin for a long time before switching to an adjacent basin, producing diffusion of the orbits along the X axis, as Fig. 7.19 shows for eqn (7.31) with $A = 4.61$ (Geisel and Nierwetberg 1982). This system is called the *process equation* by Kauffman and Sabelli (1998) and is a variant of the *sine–circle map* (Arnold 1965)

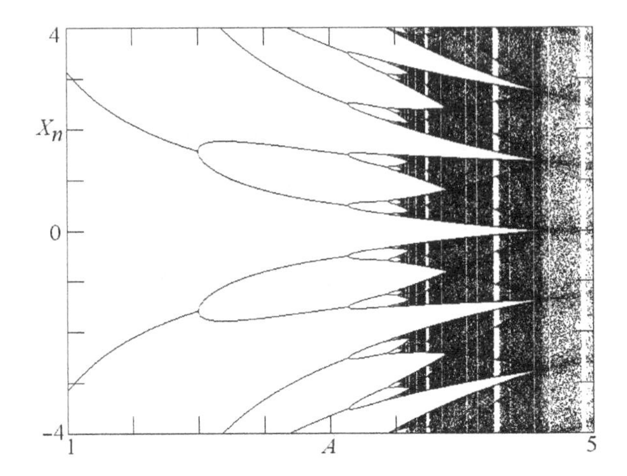

Fig. 7.20 Multiple attractor merging crisis at A= 4.6033... in eqn (7.32).

$$X_{n+1} = X_n + \Omega - \frac{K}{2\pi} \sin 2\pi X_n \pmod 1 \tag{7.33}$$

Figure 7.20 shows the multiple attractor merging crisis that occurs when $1/A = -\cos\sqrt{A^2-1}$ or $A = 4.6033388487517\ldots$ for eqn (7.32).

The maps in eqns (7.31)–(7.33) are special cases of

$$X_{n+1} = X_n + F(X_n) + \Omega \tag{7.34}$$

where $F(X)$ is an arbitrary periodic function (sinusoidal in these cases) with odd symmetry $F(X) = -F(-X)$ and Ω is a bias term (zero in the first two cases), which exhibits intermittency and deterministic diffusion for an appropriate choice of parameters (Barkai and Klafter 1997).

7.7 Exercises

Exercise 7.1 Calculate the location and eigenvalues of the equilibrium points for the one-dimensional flow in eqn (7.2) with $\mu \geq 0$.

Exercise 7.2 Calculate the location and eigenvalues of the equilibrium points for the one-dimensional flow in eqn (7.3).

Exercise 7.3 Show that the system $dy/dt = a \ln y + y - 1$ has a transcritical bifurcation, and find relations between x and y and between μ and a that reduces it to the normal form in eqn (7.3).

Exercise 7.4 Calculate the location and eigenvalues of the equilibrium points for the one-dimensional flow in eqn (7.5).

Exercise 7.5 Calculate the location and eigenvalues of the equilibrium points for the one-dimensional flow in eqn (7.7).

Exercise 7.6 Sketch the bifurcation diagram for the system $dx/dt = ax - \sin x$ for $a > 0$, and classify all the bifurcations that occur.

Exercise 7.7 Calculate the location and eigenvalues of the equilibrium points for the two-dimensional flow in eqns (7.8) and (7.9).

Exercise 7.8 Show that eqns (7.10) and (7.11) for the Hopf bifurcation are equivalent to eqns (7.8) and (7.9).

Exercise 7.9 Find the curves in ab space for which Hopf bifurcations occur in the biased van der Pol oscillator $d^2x/dt^2 + b(x^2 - 1)dx/dt + x = a$.

Exercise 7.10 Calculate the stable limit cycle for the radial normal form $dr/dt = r(\mu + r^2 - r^4)$, and show that the subcritical Hopf bifurcation exhibits hysteresis.

Exercise 7.11 (difficult) Show that the third-iterated logistic map in the vicinity of $A = 1 + \sqrt{8}$ has a fold bifurcation with a normal form as in Fig. 7.1.

Exercise 7.12 (difficult) Show that the average duration of the periodic regions in a map of the form $f = \mu + X + X^2$ that exhibits intermittency scales as $1/\sqrt{\mu}$. (Hint: Estimate the number of iterations required for the orbit to transit the narrow gap between the function $f(X)$ and the 45° diagonal and use a scaling argument as suggested by Guckenheimer and Holmes (1990).)

Exercise 7.13 Show that eqn (7.19) is the second iterate of the normal form of the flip bifurcation in eqn (7.18) for $|X| \ll 1$ and that the fixed point has $X^* \simeq \pm\sqrt{\mu}$ and is stable for μ small and positive.

Exercise 7.14 Show that Ricker's population model $X_{n+1} = AX_n e^{-X_n}$ (Ricker 1954) has a fixed point at $X^* = \ln A$ that undergoes a flip bifurcation at $A = e^2 = 7.38905609\ldots$.

Exercise 7.15 Show that the logistic map at $A = 1$ has the normal form in eqn (7.21), and show that $\mu = 0$ corresponds to $A = 1$.

Exercise 7.16 Show that the second-iterated logistic map at $A = 3$ has the normal form in eqn (7.22), and show that $\mu = 0$ corresponds to $A = 3$.

Exercise 7.17 Calculate the eigenvalues for the nontrivial fixed point of the delayed logistic map in eqn (7.24).

Exercise 7.18 Show that the homoclinic trajectory in the system $dy/dt = x - x^3, dx/dt = y$ is given by $y^2 = x^2 - x^4/2$.

Exercise 7.19 Show that the heteroclinic trajectory in the system $dy/dt = -x + x^3, dx/dt = y$ is given by $y^2 = (1 - x^2)^2/2$.

Exercise 7.20 Show that the heteroclinic trajectory in the frictionless pendulum $dy/dt = -\sin x, dx/dt = y$ is given by $y^2 = 2(1 + \cos x)$. Note that the trajectory is *homoclinic* if plotted on the surface of a cylinder with circumference 2π.

Exercise 7.21 Show that the attractor for the logistic map is confined to the range $\frac{A^2}{4}(1-\frac{A}{4}) < X < \frac{A}{4}$.

Exercise 7.22 Show that the average duration of transient chaos in the logistic map is $\langle\tau\rangle \simeq \sqrt{A-4}/\pi$ for $0 < A-4 \ll 1$.

Exercise 7.23 Calculate the value of A for which 10% of the orbits in the logistic map exhibit transient chaos for greater than 1000 iterations.

Exercise 7.24 Show that an interior crisis occurs in the logistic map at $A \simeq 3.8568$.

Exercise 7.25 Show that a boundary crisis occurs in the antisymmetric logistic map in eqn (7.28) when $A = 2+\sqrt{8}$.

Exercise 7.26 Calculate the values of A at which an attractor merging crisis and a boundary crisis occurs in the *cubic map*[3] $X_{n+1} = AX_n(1-X_n^2)$.

Exercise 7.27 Calculate the values of A at which attractor merging crises occur in the sine map $X_{n+1} = A\sin X_n$.

Exercise 7.28 Calculate the values of A at which attractor merging crises occur in the map $X_{n+1} = AX_n \sin X_n$.

Exercise 7.29 Calculate the value of A in the process equation $X_{n+1} = X_n + A\sin X_n$, at which an attractor merging crisis occurs and the orbit becomes unbounded.

7.8 Computer project: Poincaré sections

In this project you will learn to make Poincaré sections using a driven, damped, nonlinear oscillator as an example. All real oscillators are nonlinear if driven to sufficiently large amplitude, and thus this example is really very general. The Poincaré section allows you to distinguish quasiperiodic from chaotic behavior and to observe the fractal structure of a strange attractor whose dimension is between two and three, as is the case for dissipative chaotic flows in three dimensions.

The equations you will use are the Ueda (1979) equations, which are a restricted form of Duffing's (1918) oscillator (see §4.7.1) and model a mass on a stiffening spring in which the restoring force is proportional to $-x^3$ rather than the usual $-x$ as in Hooke's law

$$\frac{dx}{dt} = y \tag{7.35}$$

$$\frac{dy}{dt} = -by - x^3 + A\sin t \tag{7.36}$$

[3]Other forms referred to as 'the cubic map' include $f(X) = AX - X^3$, $f(X) = AX(1-X)^2$, $f(X) = A|1-X|(1-X)^2$, $f(X) = AX^3-(A-1)X$, $f(X) = AX+(1-A)X^3$, and $f(X) = A(1-|2X-1|^3)$, as well as various two-dimensional cases.

Use the usual method of converting the equations to autonomous form by changing $\sin t$ to $\sin z$ and adding

$$\frac{dz}{dt} = 1 \tag{7.37}$$

1. Use the fourth-order Runge–Kutta (see §3.9.4) or another suitable method to solve the equations above in their chaotic regime. Values of $b = 0.1$ and $A = 12$ should produce a chaotic solution. As usual, calculate for a while to be sure the trajectory has reached the attractor, and then show a plot of $x(t)$ that suggests chaotic behavior.
2. Make a Poincaré section plot of y versus x for $t \pmod{2\pi} = 0$ (at constant phase of the drive term). Try to accumulate at least a few thousand points. This may take a while if you have a slow computer or compiler, but it is worth the wait.
3. Repeat 2 above using $b = 0.2$ and $b = 0.4$. Since b is the damping (friction), try to draw a general conclusion about the effect of increasing the damping in a chaotic system.
4. (optional) With $b = 0.1$, vary A over the range $9 < A < 14$, and see what kind of bifurcations you can observe in the Poincaré sections. For $A = 9.8$ you should see a period-3 orbit. The period-3 motion should bifurcate to chaos around $A = 10$ and then should become periodic again with transient chaos beyond $A = 13.3$ (Ueda 1979).

8 Hamiltonian chaos

Most previous examples have been *dissipative* systems in which friction or some equivalent process causes a set of initial conditions within the basin of attraction to collapse onto an attractor. When the dissipation is small, the time required for this to happen is long, and it is natural to ask what happens when the dissipation goes to zero. Mathematically, such a situation is *nongeneric*, and the equations describing it are *structurally unstable*, but there are many physical situations in which it is nonetheless a reasonable approximation. These cases have some conserved quantity, such as mechanical energy or angular momentum. Hence such systems are called *conservative*. Conservative systems arise naturally in the *Hamiltonian*[1] formulation of classical (Newtonian) mechanics, and hence they are also called *Hamiltonian* systems. Hamiltonian mechanics is an old, beautiful, and much studied field about which much has been written and to which we cannot do justice here.

The classic example of such a system is planetary motion, made conservative by the fact that the planets are relatively rigid and move through a near vacuum. However, there are small tidal forces and interactions with the solar wind that dissipate energy after millions of years and cause effects such as the locking of the Moon's rotation with its revolution around the Earth. Another example is the motion of charged particles (such as protons and electrons) in magnetic fields. The magnetic field exerts a force perpendicular to the motion causing the particles to move in circles around the magnetic field line, and so it does not change the energy. However, when such particles accelerate as they must when they move in circles, they radiate electromagnetic waves and slowly lose energy. Incompressible, inviscid, fluid flow is conservative and is a reasonable description of some liquids such as water. Magnetic fields can be treated as the conservative flow of 'magnetic flux.' Conservative systems also arise in nonlinear optics, quantum mechanics, and statistical mechanics.

Since conservative systems do not have attractors, there is no need to wait for an initial transient to decay when calculating the orbit or trajectory. On the other hand, the absence of an attractor means that every

[1]'Hamiltonian' is named after the Irish mathematician and child prodigy, Sir William Rowan Hamilton (1805–1865), who was fluent in ten languages by the age of twelve.

initial condition potentially has a different dynamical behavior. Some initial conditions may produce chaos, whereas others nearby may be periodic, and the regions may be intertwined in complicated ways. Since there is a conserved quantity, the dynamics for a given initial condition take place in a lower-dimensional subspace of the full space of dynamical variables. However, this subspace is an integer, unlike the noninteger dimension of strange attractors. Numerical solutions of conservative systems must be done carefully to ensure that the conserved quantities remain constant. If you want a more extensive treatment of Hamiltonian chaos, see Ozorio de Almeida (1988) or Seimenis (1994).

8.1 Mass on a spring

In §3.4 we discussed the mass on a spring as an example of a simple harmonic oscillator. The starting point was Newton's second law (force = mass × acceleration) and Hooke's law for the force exerted by a spring with spring constant (stiffness) k. The energy consists of kinetic ($mv^2/2$) and potential ($kx^2/2$) and is conserved (constant in time). The resulting motion of the mass is periodic with angular frequency $\omega = \sqrt{k/m}$, and the trajectory in xv phase space is an ellipse.

An alternate starting point is to define a conserved function $H(p, x)$, called the *Hamiltonian*, in which p is a variable *canonical* to x, chosen so that the equations

$$\frac{dx}{dt} = \frac{\partial H}{\partial p} \tag{8.1}$$

$$\frac{dp}{dt} = -\frac{\partial H}{\partial x} \tag{8.2}$$

are satisfied. In this case, the Hamiltonian $H = p^2/2m + kx^2/2$ is the total energy, and $p = mv$ is the momentum. The above equations thus imply $dx/dt = v$ and $dv/dt = -kx/m$, which are identical to eqns (3.7) and (3.8).

To show that H is conserved, take its time derivative

$$\frac{dH}{dt} = \frac{\partial H}{\partial p}\frac{dp}{dt} + \frac{\partial H}{\partial x}\frac{dx}{dt} = \frac{dx}{dt}\frac{dp}{dt} - \frac{dp}{dt}\frac{dx}{dt} = 0 \tag{8.3}$$

which implies that H must be independent of time and hence constant. Viewed in the opposite direction, conservation of H implies eqns (8.1) and (8.2). If energy flows into or out of the system (such as by friction), the Hamiltonian then has an explicit time dependence, and the considerations in this chapter do not apply.

8.2 Hamilton's equations

We can generalize the above example to a system of N particles moving in D spatial dimensions by calculating a conserved Hamiltonian $H(\mathbf{p}, \mathbf{q})$, where $\mathbf{q}$ is a vector whose components are the ND individual components of position q_i and $\mathbf{p}$ is a vector whose components are the ND individual components of the canonical momenta p_i. These components then satisfy *Hamilton's equations*

$$\frac{dq_i}{dt} = \frac{\partial H}{\partial p_i} \tag{8.4}$$

$$\frac{dp_i}{dt} = -\frac{\partial H}{\partial q_i} \tag{8.5}$$

You will recognize this as a dynamical system with $2ND$ variables or state-space dimension, usually called *phase space*. For a Hamiltonian system with no explicit time dependence of H, the state space always has an even number of dimensions. Each point in this space describes a unique system whose past and future can be calculated from the Hamiltonian. For a given value of the Hamiltonian (total energy), the motion is constrained to a subspace of dimension $2ND - 1$ (an odd integer). We say that the energy is a 'constant of the motion.'

Recall from eqn (5.24) that the fractional rate of volume expansion for a flow is given by the trace of the Jacobian matrix, which for a Hamiltonian system is

$$\frac{1}{V}\frac{dV}{dt} = \sum_{i=1}^{ND}\left(\frac{\partial f_i}{\partial q_i} + \frac{\partial g_i}{\partial p_i}\right) = \sum_{i=1}^{ND}\left(\frac{\partial^2 H}{\partial q_i \partial p_i} - \frac{\partial^2 H}{\partial p_i \partial q_i}\right) = 0 \tag{8.6}$$

where $f_i = dx_i/dt$ and $g_i = dp_i/dt$. Thus a Hamiltonian system not only conserves energy, but it also conserves phase-space volume (in a hyperspace with $2ND$ dimensions). Furthermore, the density of a cluster of initial conditions must remain constant (*Liouville's theorem*[2]) even while the phase space stretches in some directions and contracts in others in a process called *filamentation*. We will consider a volume-conserving system to be Hamiltonian even when there is no well-defined energy or obvious Hamiltonian function.

8.3 Properties of Hamiltonian systems

We can now enumerate the properties of a Hamiltonian system, some of which have been mentioned and others follow as a consequence.

1. It is *dissipationless*. There is no friction or other effect to remove energy from the system.
2. It is *conservative*. The energy or some equivalent quantity is the same at all times. This quantity is a constant of the motion.
3. It involves a *Hamiltonian function* H that is constant and whose partial derivatives satisfy Hamilton's equations.

[2]Liouville's theorem was published by the French mathematician, Joseph Liouville (1809–1882) in 1838.

4. It has *two dimensions* (variables) for each degree of freedom (spatial dimension).
5. The dynamics are *confined to a hypersurface* (a constant-energy surface) with a dimension one less than the dimension of the system.
6. It has *two zero Lyapunov exponents* in the full space of variables, one corresponding to the direction of the flow and the other perpendicular to the constant-energy surface.
7. It typically contains neutrally stable *centers* and *saddle points* with *homoclinic* or *heteroclinic trajectories.*
8. It has *no attractors* and *no basin of attraction*, although the invariant measure may not be uniform. Alternately, you can think of the attractor and its basin as coincident.
9. It has *no transients.* There is no need to wait for transients to decay when determining the long-term dynamics.
10. It is *time-reversible.* Replacing t with $-t$ does not change the form of the equations or the region accessible to the solution.
11. It is *recurrent.* The trajectory or orbit eventually returns arbitrarily close to the initial condition, which implies that it will do so infinitely many times if you wait long enough (the *Poincaré recurrence theorem*).
12. It is *volume conserving.* The phase-space volume occupied by a cluster of initial conditions is constant, as is the density of such points (Liouville's theorem). The flow is *incompressible* (like water).
13. The *Lyapunov exponents occur in equal and opposite pairs*, and hence they *sum to zero.* This property results from the time reversibility.
14. Chaotic flows *require at least two degrees of freedom* so that the dynamics are at least three-dimensional (and at least one nonlinearity).
15. The dynamics occur in a *space of integer dimension*, although the space can be a *fat fractal* with infinitely many holes but finite measure (see Chapter 11), and the measure may not be uniform (see Chapter 13).
16. The orbit typically contains *accumulators* (Smith and Spiegel 1987), which are chains of small islands where the orbit spends a long time before emerging and wandering into another region of state space.

The remainder of this chapter will describe several important and interesting examples of chaotic Hamiltonian systems, both in flows and in maps.

8.4 Simple pendulum

A simple pendulum consists of a mass m suspended by a rigid, massless rod of length L free to swing without friction along the arc of a circle under the influence of a downward gravitational acceleration g, as in Fig. 8.1. If x is the angle of the pendulum from the downward equilibrium (in radians) and p is the corresponding angular momentum $p = mL^2dx/dt$, then the

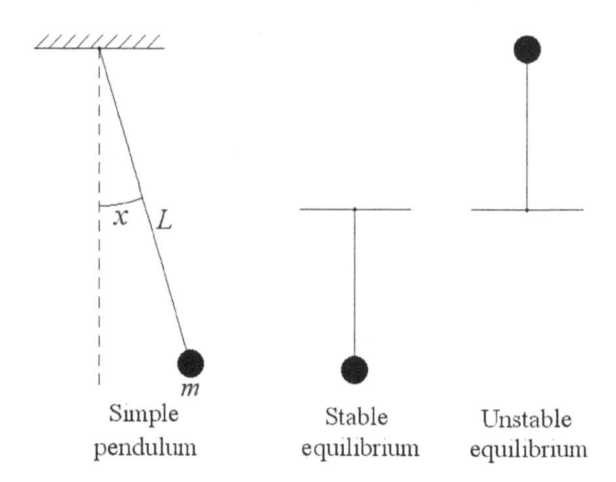

Fig. 8.1 Simple pendulum with mass m suspended by a rigid massless rod of length L.

total energy is given by the Hamiltonian $H = p^2/2mL^2 + mgL(1 - \cos x)$, where the potential energy is arbitrarily taken as zero when $x = 0$. Note that adding a constant to the Hamiltonian has no effect since Hamilton's equations, as in eqns (8.4) and (8.5), involve only derivatives of H. The pendulum has equilibria at $x = n\pi$ for all integer values of n $(0, \pm 1, \pm 2, \ldots)$. Equilibria with even n are stable, and those with odd n are unstable.

Applying Hamilton's equations leads to the equations of motion

$$\frac{dx}{dt} = v \tag{8.7}$$

$$\frac{dv}{dt} = -\omega^2 \sin x \tag{8.8}$$

where $v = dx/dt$ is the angular velocity of the mass and $\omega = \sqrt{g/L}$. The phase-space area expansion $\partial f/\partial x + \partial g/\partial v$ is zero as expected. For $|x| \ll 1$, $\sin x$ is approximately equal to x, and eqns (8.7) and (8.8) reduce to the equations for the simple harmonic oscillator in eqns (3.7) and (3.8). For a given value of total energy E, the trajectory lies on a curve in phase space given by

$$v = \pm\sqrt{2}\omega\sqrt{C + \cos x} \tag{8.9}$$

where $C = E/mgL - 1$. Note that for $E = 0$, C is -1, and the only solution of eqn (8.9) is $\cos x = 1$ and $v = 0$, corresponding to the stable equilibrium in Fig. 8.1.

Contours for other values of C are shown in Fig. 8.2. Stable equilibria are indicated by O-points (*elliptic points* or *centers*), and unstable equilibria are indicated by X-points (*hyperbolic* or *saddle points*) through which the separatrix passes. The trajectories near the O-points are elliptical and those near the X-points are hyperbolic. Hamiltonian systems cannot have

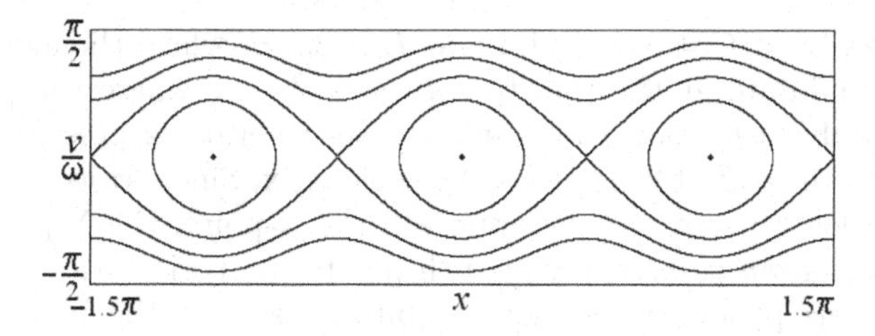

Fig. 8.2 Phase portrait for the simple pendulum.

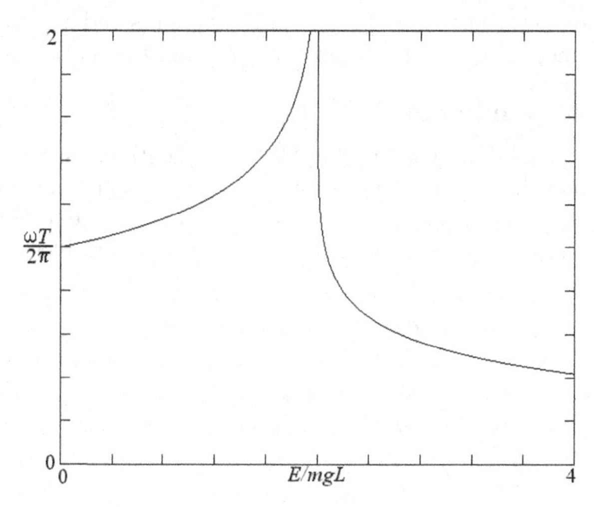

Fig. 8.3 Variation of the period T of a simple pendulum versus total energy E.

nodes or foci. The separatrix is a heteroclinic connection (or homoclinic if the phase portrait is wrapped onto a cylinder). Trajectories on the separatrix correspond to motion that just brings the pendulum to rest at the top, where it remains fixed. Trajectories near the O-points correspond to small amplitude oscillations (called *librations*). Trajectories outside the separatrix correspond to *rotations* that take the pendulum 'over the top,' either in the clockwise ($v < 0$) or counterclockwise ($v > 0$) direction. Reversing the sign of time reverses the trajectory but leaves the phase portrait unchanged.

The period of the pendulum for small oscillations is nearly constant[3] at $T \simeq (2\pi/\omega)(1+E/8mgL)$. At higher energy, it varies as in Fig. 8.3, ranging

[3]The approximate *isochronicity* of the pendulum was discovered by the Italian physicist and astronomer Galileo Galilei (1564–1642) supposedly while timing a swinging chandelier with his pulse in the cathedral at Pisa in 1583 (Seeger 1966). Like the story of Galileo dropping masses off the leaning tower of Pisa, this story is probably mythical since the chandelier was purchased after his death (Frautschi *et al.* 1986). The period for small oscillations was calculated by Christiaan Huygens (1629–1695) in 1673 and for large oscillations by Leonhard Euler (1707–1783) in 1736.

from $T = 2\pi/\omega$ at $E = 0$, to infinity at $E = 2mgL$ where the motion stalls with the pendulum at the top, to zero as $E \to \infty$ where the pendulum spins infinitely fast. As a consequence, every period is present for some energy, with the widest range near the separatrix. Since trajectories on the separatrix slow to zero as they approach the X-point, the X-point is also called a *stagnation point*. Only at such points can trajectories intersect.

The simple pendulum cannot exhibit chaos since the phase space is only two-dimensional. There are two Lyapunov exponents, one of which is zero, corresponding to the direction of the flow. The fractional rate of area expansion is the sum of the Lyapunov exponents and is zero. Thus the second Lyapunov exponent must also be zero, and the motion is periodic.

8.5 Driven pendulum

Now add a sinusoidal drive term $A \sin \Omega t$ to the righthand side of eqn (8.8). In this case the Hamiltonian has an explicit time dependence and thus is not conserved since energy flows into and out of the system. However, we can include the drive in the system by adding two variables y and w to obtain

$$\frac{dx}{dt} = v \tag{8.10}$$

$$\frac{dv}{dt} = -\omega^2 \sin x + Ay \tag{8.11}$$

$$\frac{dy}{dt} = w \tag{8.12}$$

$$\frac{dw}{dt} = -\Omega^2 y + Ax \tag{8.13}$$

The last two equations in the absence of the Ax term produce a sinusoidal oscillation $y \propto \sin \Omega t$. The Ax term, which is negligible if the oscillation amplitude is sufficiently large, makes the resulting four-dimensional system Hamiltonian with $H = v^2/2 + w^2/2 + \omega^2(1 - \cos x) - Axy + \Omega^2 y^2/2$. The trajectories lie on three-dimensional constant-energy hypersurfaces and can thus be chaotic. The energy sloshes back and forth between the nonlinear oscillator in eqns (8.10) and (8.11) and the linear oscillator in eqns (8.12) and (8.13). Initial conditions can be chosen so that the linear oscillator has a very large energy and thus is little perturbed by the Ax term, and a large energy can be imparted to the nonlinear oscillator even with a small A. Think of the linear oscillator as a massive flywheel that can absorb or provide energy with little effect on its motion. For example, replace the motor in Fig. 1.5 with a large, low-friction flywheel.

For $A \ll 1$, the nonlinear oscillator is only weakly coupled to the linear oscillator, and yet the perturbation can have a big effect on the nonlinear oscillator since it will be resonant at some value of the energy. If the resonance occurs near the separatrix where a heteroclinic connection occurs, then chaos is expected even for small A. Figure 8.4 shows a Poincaré section

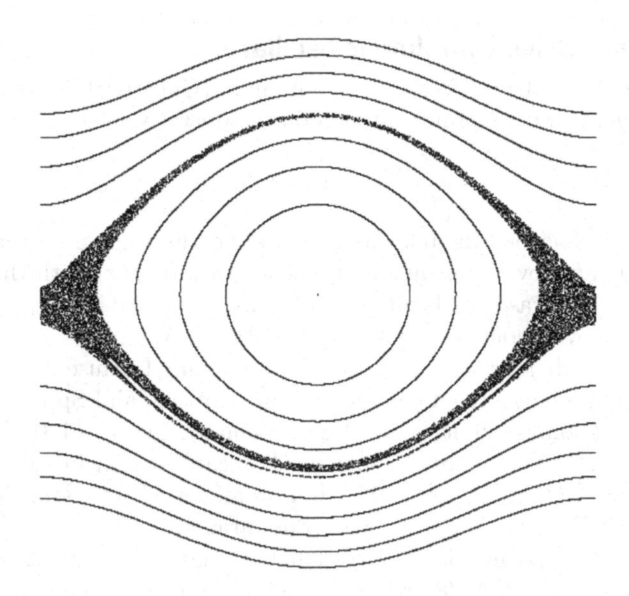

Fig. 8.4 Poincaré section of a weakly driven frictionless pendulum $\ddot{x} + \sin x = 0.01 \sin 0.5t$, showing chaos near the separatrix.

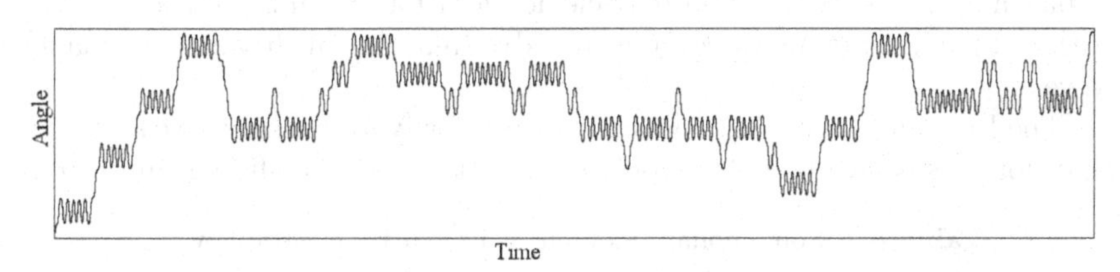

Fig. 8.5 Intermittency for a weakly driven frictionless pendulum $\ddot{x} + \sin x = 0.01 \sin 0.5t$.

in the xv plane for $\omega = 1, A = 0.01$, and $\Omega = 0.5$ at $t \pmod{2\pi} = 0$ with chaos evident near the separatrix with heteroclinic tangles. In such a case, the trajectory is confined to a small region of phase space and thus looks nearly periodic, but the period differs greatly from cycle to cycle depending on how close the trajectory is to the separatrix.

This chaotic region is called a *chaotic sea*,[4] in contrast to the larger islands of quasiperiodicity. Trajectories within the sea wander throughout it and are precluded from entering the islands. Chaotic trajectories near the island boundaries have long periods of near periodicity and hence exhibit intermittency, as in Fig. 8.5. As A increases, the sea expands, eventually enveloping the O-point for $A \simeq 1$.

[4]Much of the older literature calls it a *stochastic sea*, but the term 'stochastic' should be reserved for truly random processes (Brown and Chua 1999).

8.6 Other driven nonlinear oscillators

The example of a driven frictionless pendulum (Chirikov 1979) is one case of a more general class of nonautonomous conservative systems given by

$$\frac{d^2x}{dt^2} + f(x) = A\sin\Omega t \tag{8.14}$$

The method described in §6.3 was used to find the value of Ω that maximizes the Lyapunov exponent for $A = 1$ for various $f(x)$ with the results in Table 8.1. All cases had initial conditions of $x_0 = 0$ and $\dot{x}_0 = 0$. The other Lyapunov exponents are $\lambda_2 = 0$ and $\lambda_3 = -\lambda_1$.

The case with $f(x) = x^3$ is especially interesting because it may be the simplest sinusoidally-driven chaotic system (Gottlieb and Sprott 2001). It is the dissipationless limit of the Ueda oscillator in eqn (4.31). It has a single parameter Ω. Figure 8.6 shows a Poincaré section in the xv plane for $\Omega = 1.88$ at $\Omega t \pmod{2\pi} = 0$ with a prominent chaotic sea. Figure 8.7 shows how its largest Lyapunov exponent varies with Ω.

Another interesting class of system that has been extensively studied (Chernikov *et al.* 1988) replaces the $\sin\Omega t$ term in eqn (8.14) with $\sin(kx - \Omega t)$ to represent a nonlinear oscillator driven by a plane wave moving in the x direction with speed Ω/k, as is common in many problems in plasma physics, optics, and particle acceleration. Even the linear system with $f(x) = \omega^2 x$ exhibits complicated dynamics when driven by a plane wave.

The Lyapunov exponent converges more slowly for conservative systems than for dissipative ones because the trajectory is space-filling and often

Table 8.1 Some nonautonomous conservative chaotic flows.

Equation	λ_1
$\ddot{x} + \sin x = \sin 0.50t$	0.163
$\ddot{x} + x^3 = \sin 1.88t$	0.097
$\ddot{x} + x^5 = \sin 2.19t$	0.163
$\ddot{x} + x^7 = \sin 2.32t$	0.198
$\ddot{x} + x^9 = \sin 2.58t$	0.230
$\ddot{x} + x^{11} = \sin 2.79t$	0.242
$\ddot{x} + x^3 - x = \sin 1.87t$	0.164
$\ddot{x} + x\|x\|^{-1/2} = \sin 5.57t$	0.123
$\ddot{x} + x\|x\| = \sin 1.61t$	0.051
$\ddot{x} + x\|x\|^3 = \sin 2.00t$	0.139
$\ddot{x} + \sinh x = \sin 1.54t$	0.014
$\ddot{x} + \tanh x = \sin 0.27t$	0.009

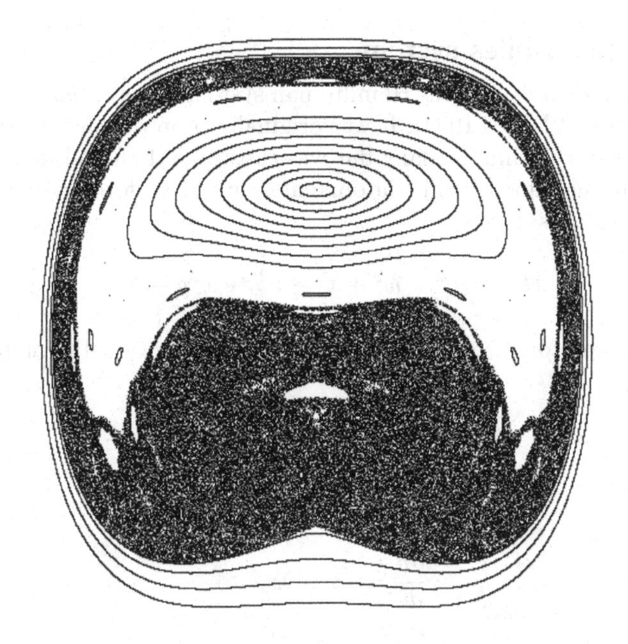

Fig. 8.6 Poincaré section of the system $\ddot{x} + x^3 = \sin 1.88t$.

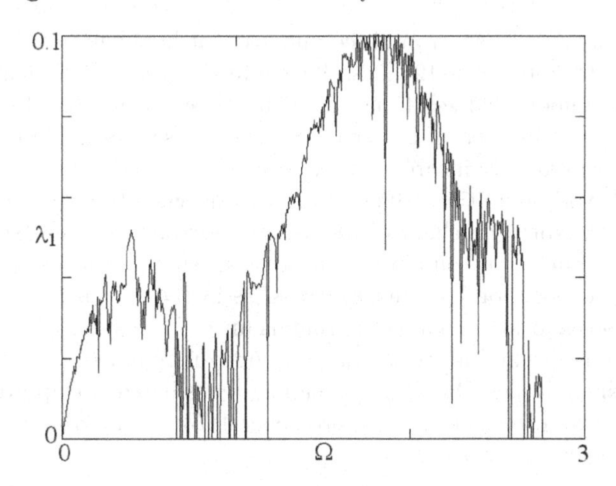

Fig. 8.7 Lyapunov exponent versus Ω for the system $\ddot{x} + x^3 = \sin \Omega t$.

remains in one region of space for a relatively long time before exploring other accessible regions. In fact, under some conditions, the trajectory will diffuse throughout an unbounded region of phase space as the drive term adds and removes energy from the system. A collection of particles obeying such dynamics undergoes *chaotic heating* (usually improperly called *stochastic heating*). The Lyapunov exponent fluctuates along with the energy.

8.7 Hénon–Heiles system

Another chaotic autonomous Hamiltonian system is the *Hénon–Heiles system* (Hénon and Heiles 1964). It was originally proposed as a model of the motion of a star within a galaxy but was subsequently found to also model a particular one-dimensional triatomic molecule (Lunsford and Ford 1972). Its Hamiltonian is

$$H = \tfrac{1}{2}(v^2 + w^2 + x^2 + y^2) + x^2 y - \tfrac{1}{3}y^3 \tag{8.15}$$

Applying Hamilton's equations gives the four-dimensional dynamical system

$$\frac{dx}{dt} = v \tag{8.16}$$

$$\frac{dy}{dt} = w \tag{8.17}$$

$$\frac{dv}{dt} = -x - 2xy \tag{8.18}$$

$$\frac{dw}{dt} = -y - x^2 + y^2 \tag{8.19}$$

The dynamics occur on a three-dimensional hypersurface of constant energy $E = H$. For $E < 1/12$, all initial conditions have either quasiperiodic solutions or unbounded solutions. As E increases, some initial conditions give chaos, and the area of the chaotic sea increases rapidly for $E \gtrsim 1/9$. Figure 8.8 shows a Poincaré section in the yw plane for $E = 1/8$ and $x = 0$ with various initial values of x in the range $-0.5 < x_0 < 0.5$ and $y_0 = z_0 = 0$. Note the chain of five islands surrounding the large island at the right center. Surrounding each of these islands is a separatrix with heteroclinic connections producing chaos, as in the driven pendulum. For larger E, the chaotic sea floods the islands until E reaches $1/6$, whereupon no islands remain, and all trajectories become unbounded.

The Hénon–Heiles Hamiltonian can be considered an approximation with $|x| \ll 1$ and $|y| \ll 1$ to the more general *Toda Hamiltonian* model of triatomic molecules (Ford 1975)

$$H = \tfrac{1}{2}(v^2 + w^2) + \tfrac{1}{3}(\mathrm{e}^{y+\sqrt{3}x} + \mathrm{e}^{y-\sqrt{3}x} + \mathrm{e}^{-2y}) - 1 \tag{8.20}$$

Curiously, the Toda system has a second invariant besides energy (Hénon 1969) and hence cannot have chaotic solutions since the motion is restricted to a two-dimensional surface (Ford *et al.* 1973). Thus complexity of the Hamiltonian is not always an indication of the complexity of the dynamics.

Furthermore, if you replace the y^2 term in eqn (8.19) with $-y^2$ and define two new variables, $\psi = x - y$ and $\chi = x + y$, then the equations separate into two independent second-order equations

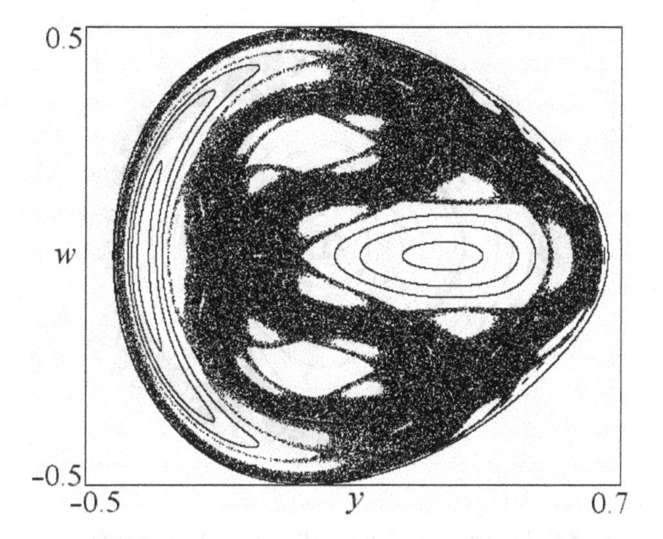

Fig. 8.8 Poincaré section for the Hénon–Heiles system with $E = 0.125$.

$$\frac{d^2\chi}{dt^2} = -\chi - \chi^2 \tag{8.21}$$

$$\frac{d^2\psi}{dt^2} = -\psi + \psi^2 \tag{8.22}$$

In such a case, we say the system is *integrable*, since a solution in terms of elliptic functions can be found by integrating each equation separately and chaos cannot occur. All Hamiltonian systems with one degree of freedom (two dynamical variables) are integrable. A system with N degrees of freedom is integrable if there are N constants of the motion (energy, angular momentum, etc.). Integrable systems cannot exhibit chaos, but most high-dimensional systems are nonintegrable and hence potentially chaotic.

8.8 Three-dimensional conservative flows

Conservative flows also occur in autonomous dynamical systems for which there is no obvious Hamiltonian or for which the Hamiltonian is not used to derive the equations. These systems nonetheless conserve phase-space volume and are time-reversible. To exhibit chaos, they must be at least three-dimensional.

8.8.1 Nosé–Hoover oscillator

Perhaps the simplest such example is the *Nosé–Hoover oscillator* (case A in Table 4.1) given by eqns (3.19)–(3.21) (Nosé 1991, Hoover 1995). Most initial conditions produce trajectories that lie on invariant tori, but some (such as $x_0 = 0, y_0 = 5$, and $z_0 = 0$) give chaos. The various regions are shown in the Poincaré section in Fig. 8.9 in which points are plotted where the trajectory punctures the $z = 0$ plane for different initial conditions.

Fig. 8.9 Poincaré section for the Nosé–Hoover oscillator.

The rate of volume expansion $dV/dt/V = \langle \text{trace}(J) \rangle = \langle z \rangle$ is not obviously zero, but numerical calculations suggest that it is. You can confirm the conservative nature of this system by noting that the equations are invariant to a change of t to $-t$, by defining new variables equal to $-x$ and $-z$.

However, time reversibility does not guarantee that a system is conservative. A counter-example is case D in Table 4.1 which has a symmetric strange attractor/repellor pair that switch roles when time is reversed, as you can demonstrate by defining new variables equal to $-x$ and $-z$. Its rate of volume expansion $dV/dt/V = \langle \text{trace}(J) \rangle = -\langle x \rangle$ does not average to zero. In fact, the two attractors lie entirely on opposite sides of the $x = 0$ plane.

8.8.2 Labyrinth chaos

There are many other examples of chaotic three-dimensional conservative flows such as the simple and elegant case

$$\frac{dx}{dt} = \sin y \tag{8.23}$$

$$\frac{dy}{dt} = \sin z \tag{8.24}$$

$$\frac{dz}{dt} = \sin x \tag{8.25}$$

which Thomas (1999) calls 'labyrinth chaos,' for a reason to be evident shortly. It has infinitely many equilibria at $x^* = \pm\pi l, y^* = \pm\pi m$, and $z^* = \pm\pi n$, with l, m, and n arbitrary integers.

Wrapping the system onto a 3-torus with period 2π in each direction gives only eight equilibria. Most initial conditions produce chaotic trajec-

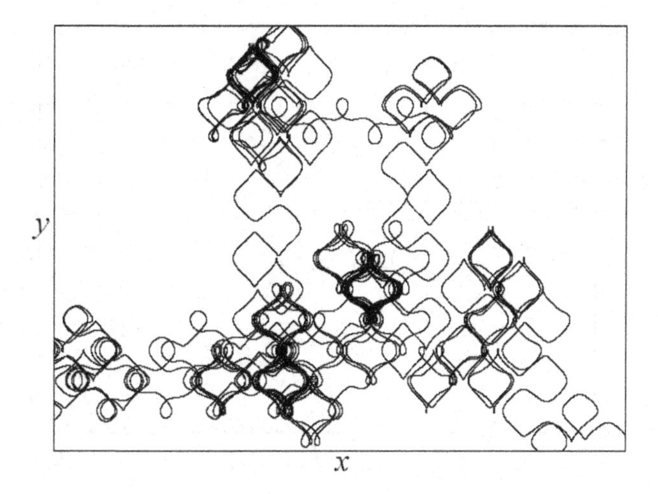

Fig. 8.10 Projection of Thomas' labyrinth chaos onto the xy plane.

tories that wander over the three-dimensional hypersurface of the torus, although there are quasiperiodic trajectories that lie on a two-dimensional submanifold of the torus.

More interesting is to view the trajectory in three-dimensional (x, y, z) *cartesian space*.[5] Then the equilibria are laid out on an infinite three-dimensional lattice like the atoms in a crystal. The trajectory wanders throughout this lattice (or labyrinth) as in Fig. 8.10.

The trajectory in this space is unbounded, but it diffuses slowly. Figure 8.11 shows how the *root mean square* (rms) distance $d = \langle (x - x_0)^2 + (y - y_0)^2 + (z - z_0)^2 \rangle^{1/2}$ for 1000 random initial conditions increases in time. The rms distance for distances greater than the lattice spacing ($d = 2\pi$) increases more slowly than linearly in time, but faster than the square root of time that would be expected for a purely diffusive process. This is an example of *fractional Brownian motion* described in §9.4.6, albeit produced by a deterministic rather than random process.

This labyrinth is a special case of the more general, *cyclically symmetric system*

$$\frac{dx}{dt} = f(x, y, z) \tag{8.26}$$

$$\frac{dy}{dt} = f(y, z, x) \tag{8.27}$$

$$\frac{dz}{dt} = f(z, x, y) \tag{8.28}$$

which has chaotic solutions for many choices of the nonlinear function f. For example, Thomas (1999) has examined the dissipative cases

[5]Cartesian (or rectangular) coordinates, named after the French philosopher and mathematician René Descartes (1596–1650), have straight, perpendicular axes.

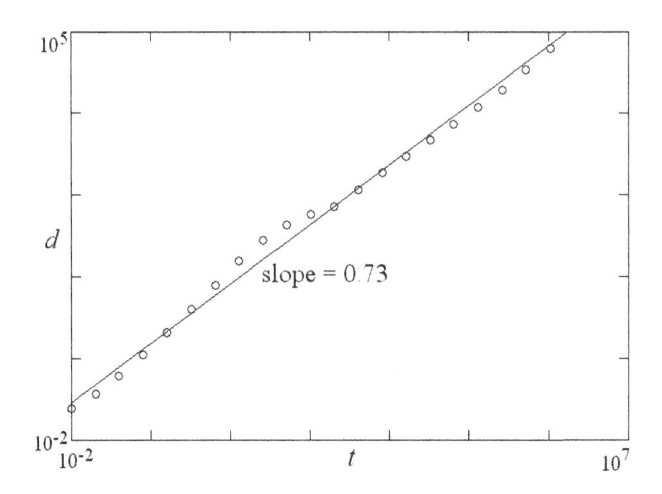

Fig. 8.11 Diffusion of trajectories in Thomas' labyrinth.

$$f(x, y, z) = -0.18x + \sin y \tag{8.29}$$

and

$$f(x, y, z) = -0.3x + 1.1y - y^3 \tag{8.30}$$

and Arne Dehli Halvorsen (unpublished) proposed the case

$$f(x, y, z) = -ax - 4y - 4z - y^2 \tag{8.31}$$

with chaos for $a = 1.27$. The symmetry gives attractors with special beauty for these dissipative examples.

8.8.3 Conservative jerk systems

Gottlieb (1996) pointed out that the chaotic Nosé–Hoover oscillator in eqns (3.19)–(3.21) can be written in jerk form as

$$\frac{d^3x}{dt^3} = -\left(\frac{dx}{dt}\right)^3 + \frac{d^2x}{dt^2}\left(x + \frac{d^2x}{dt^2}\right) \Big/ \frac{dx}{dt} \tag{8.32}$$

Although this is not a particularly elegant example of a chaotic conservative jerk system, it demonstrates their existence. A more thorough search for such systems using the method in §6.3 turned up the simple examples in Table 8.2. These cases are chaotic for the indicated initial conditions but have relatively small Lyapunov exponents. They have not been carefully optimized, and some cases may be only transiently chaotic, albeit with very long transients.

Table 8.2 Some chaotic conservative jerk systems.

Equation	$(x_0, \dot{x}_0, \ddot{x}_0)$
$\dddot{x} = -0.25\dot{x} - 0.03x + \cosh x - 1$	$(0, 0.01, 0)$
$\dddot{x} = -6.66\dot{x} \pm (\lvert x\rvert - 1)$	$(-0.26, 0, -0.03)$
$\dddot{x} = -4.09\dot{x} \pm (x - x^3)$	$(0, 0, -0.55)$
$\dddot{x} = -1.96\dot{x} \pm 0.59(x - \max(x, 0) + 1)$	$(-0.94, -0.42, 0)$
$\dddot{x} = -2.26\dot{x} \pm 0.04(x - \min(x, 0) - 1)$	$(-0.48, 9.61, -0.74)$
$\dddot{x} = -0.26\dot{x} \pm (9.52x - \sinh x)$	$(0, -0.01, 0.02)$
$\dddot{x} = -4.34\dot{x} \pm x + x^2$	$(-0.01, 0, -0.01)$
$\dddot{x} = -0.41\dot{x} \pm 3.69x - \cosh x + 1$	$(0, 0.11, -0.16)$

8.9 Symplectic maps

An inherent difficulty in studying chaotic Hamiltonian flows is that the numerical methods may not precisely conserve the invariants. As a result, the long-term trajectory may settle into a periodic motion or become unbounded, whereas the real system does not. To counter this difficulty, special *symplectic* methods (Rowlands 1991, Sanz-Serna 1994) have been developed to analyze problems such as the stability of the Solar System. In fact, a whole branch of mathematics, called *symplectic geometry*, has emerged for such studies. The Runge–Kutta method (see §3.9.4) is not inherently symplectic, but under appropriate conditions, the leap-frog method (see §3.9.2) is, even when it is of lower order.

One way to circumvent this difficulty is to analyze a map constructed to approximate a Poincaré section of the corresponding flow, and to arrange that this map is precisely conservative. A general two-dimensional map such as eqns (3.37) and (3.38) is area-preserving if $|\det(J)| = |ad - bc| = 1$. Then the main concern is round-off error, which can be minimized by computing in at least double precision (64-bit).

8.9.1 Hénon area-preserving quadratic map

Hénon (1969) showed that the most general area-preserving quadratic map with a stable fixed point at the origin ($X^* = Y^* = 0$) is

$$X_{n+1} = X_n \cos\alpha - (Y_n - X_n^2)\sin\alpha \tag{8.33}$$

$$Y_{n+1} = X_n \sin\alpha + (Y_n - X_n^2)\cos\alpha \tag{8.34}$$

with $0 < \alpha < 2\pi$. Figure 8.12 with $\cos\alpha = 0.24$ and $X_0 = 0.6, Y_0 = 0.13$ shows the narrow region of chaos near the separatrix that surrounds five quasiperiodic islands.[6] At the center of each island is an elliptic fixed point,

[6]For a given $\cos\alpha$, the quantity $\sin\alpha$ is best calculated from $\sin\alpha = \sqrt{1 - \cos^2\alpha}$.

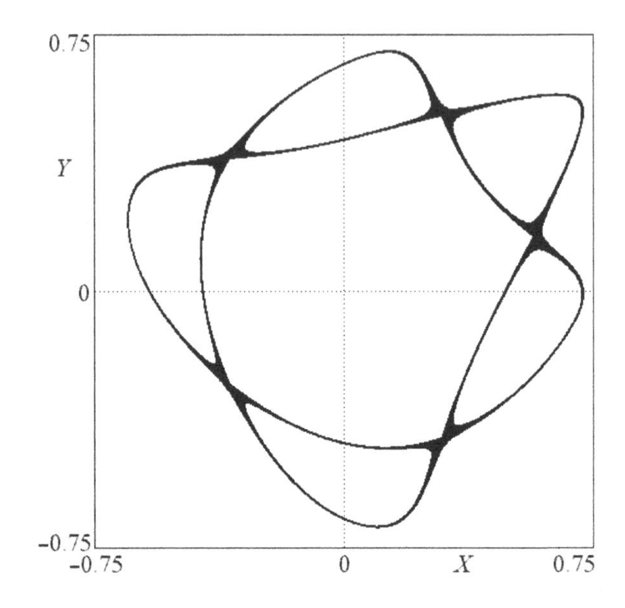

Fig. 8.12 Hénon area-preserving map with $\cos\alpha = 0.24$.

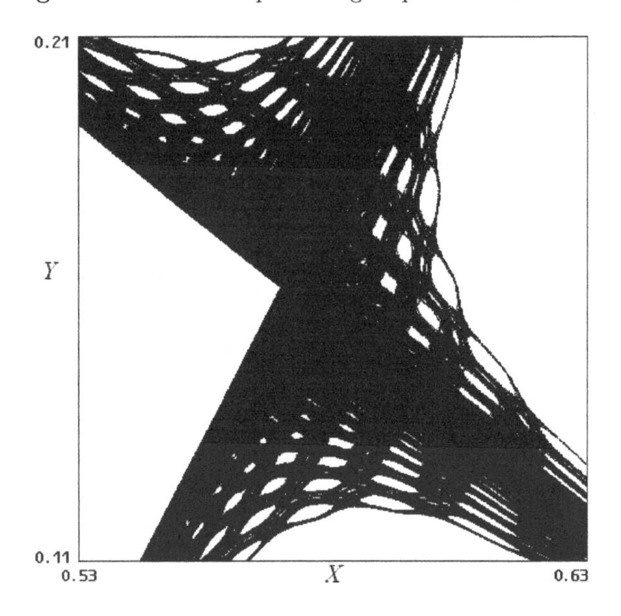

Fig. 8.13 Expanded view of Fig. 8.12 showing the island structure around the fixed point.

and between each island is a hyperbolic fixed point. Within the chaotic sea is an archipelago of islands of various sizes and periodicities, shown expanded in Fig. 8.13. Within each island are nested drift rings (not shown),

Fig. 8.14 First five iterations of the Arnold cat map.

indicating quasiperiodicity in the corresponding Hamiltonian flow.

For larger values of $\cos\alpha$, the chaotic region shrinks, and for smaller values it expands, with most orbits escaping to infinity. There are infinitely many bifurcations, with new islands and their accompanying separatrices appearing and disappearing with small changes of α. The Hénon area-preserving quadratic map is invertible and time-reversible. In fact, Engel (1955) proved that the time inverse of any polynomial area-preserving map is also a polynomial area-preserving map.

8.9.2 Arnold's cat map

A very different kind of symplectic map, whose analysis is relatively simple, is *Arnold's cat map* (Arnold and Avez 1968), sometimes called the *Anosov map* (Anosov 1962, 1963)

$$X_{n+1} = X_n + Y_n \pmod 1 \tag{8.35}$$

$$Y_{n+1} = X_n + 2Y_n \pmod 1 \tag{8.36}$$

which can be considered a two-dimensional generalization of the tent map (see §2.5.2). The name 'cat' comes from the outline of a cat that was used for the initial conditions in the original reference, a more photo-realistic version of which is in Fig. 8.14. It is suggestive of an analog video image that has lost its horizontal synchronization.

This map is structurally stable and hyperbolic since the stable and unstable manifolds are straight and perpendicular to one another. It has a fixed point at the origin and infinitely many periodic points at rational

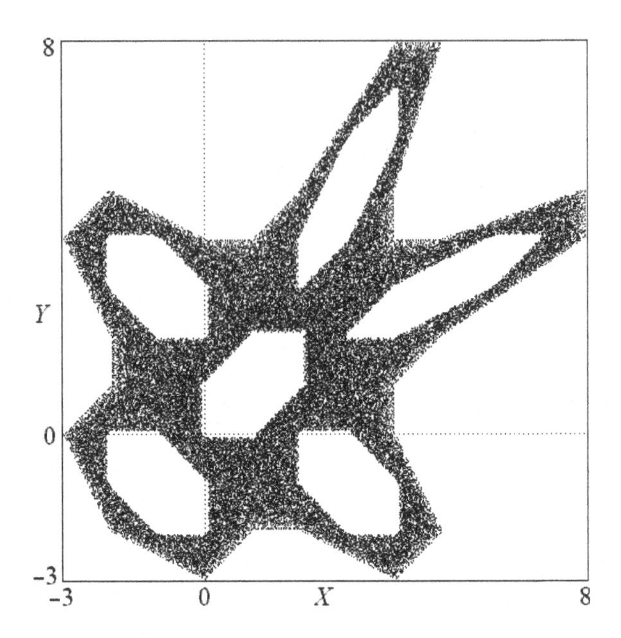

Fig. 8.15 Gingerbreadman map.

values of X and Y, each of which is either a homoclinic or heteroclinic point. All irrational initial conditions spread throughout the unit square with uniform probability after many iterations, eventually coming arbitrarily close to every point in the square, producing *mixing*. The (mod 1) operation means that the map lies on the surface of a torus with circumference 1 (rather than 2π) in each direction. This system is highly chaotic with Lyapunov exponents $\lambda = \pm \ln\left[\frac{1}{2}(3+\sqrt{5})\right] = \pm 0.96242365\ldots$. For rational values of X and Y, as is necessarily the case for an image with finitely many pixels, the map is periodic, and the picture of the cat will reappear after some large number of iterations.

8.9.3 Gingerbreadman map

Another area-preserving map

$$X_{n+1} = 1 + |X_n| - Y_n \tag{8.37}$$

$$Y_{n+1} = X_n \tag{8.38}$$

was proposed by Devaney (1984). For an initial condition such as $X_0 = 0.5, Y_0 = 3.7$, the orbit wanders chaotically throughout a region, as in Fig. 8.15. Its shape suggested the name *gingerbreadman map*. There are six hexagonal islands with periodic orbits. The point (1, 1) is a fixed point. Other points in the interior hexagon have period 6 and remain confined within the region. Orbits in the other five hexagons circulate among them with a unique period-5 orbit at $(-1,3), (-1,-1), (3,-1), (5,3)$, and (3, 5), and all other orbits of period 30.

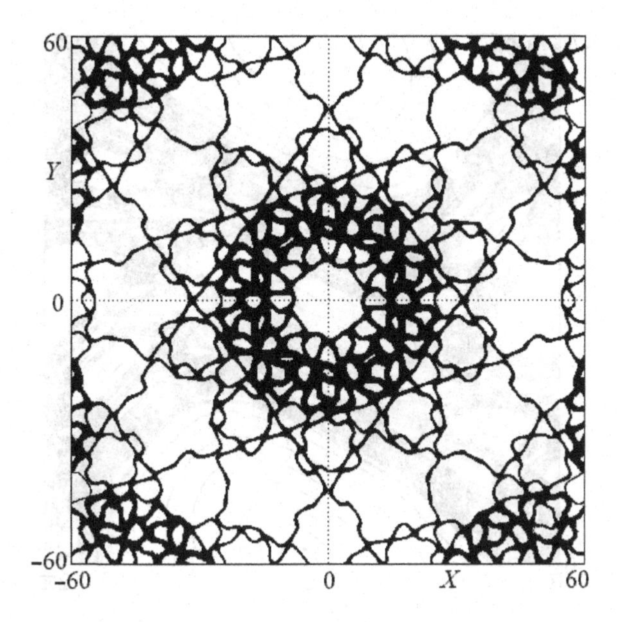

Fig. 8.16 Chaotic web map.

8.9.4 Chaotic web maps

Another chaotic conservative system similar to the Hénon area-preserving map but with periodicity is the *chaotic web map*, usually improperly called the *stochastic web map* (Chernikov *et al.* 1988)

$$X_{n+1} = X_n \cos\alpha - (Y_n + k \sin X_n) \sin\alpha \tag{8.39}$$

$$Y_{n+1} = X_n \sin\alpha + (Y_n + k \sin X_n) \cos\alpha \tag{8.40}$$

with $\alpha = 2\pi/m$ and $m = 1, 2, 3, \ldots$. It models a harmonic oscillator with periodic impulsive 'kicks.' The quantity k is a measure of the strength of the kick, and m is its periodicity. The intricate spider-web structure in Fig. 8.16 with $m = 5$ and $k = 0.8$ explains the name, also called an *Arnold web*.

For $m = 4$ the phase plane contains an infinite web whose shape is nearly a square grid. The web exists for any k, but the thickness goes to zero for $k \to 0$. For $m = 3$ and $m = 6$ the web forms a hexagonal grid. For all other values of m ($\neq 1, 2, 3, 4, 6$), a nonperiodic web is formed with so-called *quasicrystal-type symmetry*. The web tiles the plane with a symmetry slightly spoiled by the finite thickness of the web. The orbit moves throughout the web in a process called *Arnold diffusion* (Arnold 1964). The diffusion is extremely slow for small k but exists for all k. Such webs are common in higher-dimensional conservative systems.

8.9.5 Chirikov (standard) map

The most widely studied symplectic map is the *standard map* (Lichtenberg and Liebermann 1992)

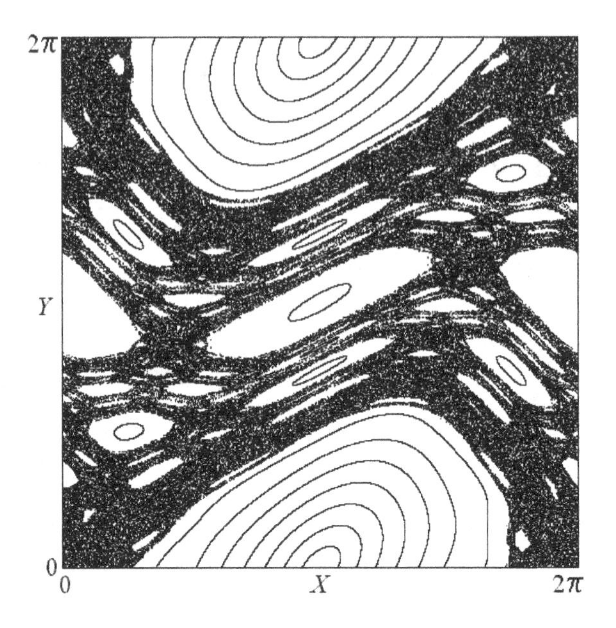

Fig. 8.17 Chirikov map with $k = 1$.

$$X_{n+1} = X_n + Y_{n+1} \pmod{2\pi} \tag{8.41}$$

$$Y_{n+1} = Y_n + k \sin X_n \pmod{2\pi} \tag{8.42}$$

so named by Chirikov (1979) and thus also called the *Chirikov map* or *Chirikov–Taylor map*. It was originally devised to model the interaction of a plasma with a magnetic field (Sinclair *et al.* 1970), but it models a variety of other nonlinear dynamical processes such as the ball on an oscillating floor in §1.3, the motion of charged particles in accelerators (Jowett *et al.* 1985), linear chains of atoms (Aubry 1983), and the periodically kicked rotor[7] (Jensen 1987). It can also be derived by applying the leap-frog method (see §3.9.2) to the simple pendulum in eqns (8.7) and (8.8) with a step size that is not small. The map is most easily calculated by first advancing Y and then X using the new value of Y. The Y_{n+1} in eqn (8.41) is not a misprint.

The parameter k measures the strength of the nonlinearity. For $k = 0$, the value of Y is conserved, and the orbit advances in constant increments of $\Delta X = Y_0$. When $Y_0/2\pi$ is rational, the motion is periodic, and when it is irrational (the usual case), the motion is quasiperiodic (a drift ring). All the fixed and periodic points of the Chirikov map are either saddle points or centers.

For $k > 0$ there is a saddle point at (0, 0) that is always unstable and a center at $(\pi, 0)$ for $k < 4$. The unstable solution gives a layer of chaos that grows with k, eventually engulfing the entire plane. The isolated chaotic regions at small k merge into a single connected region at $k \simeq 0.971635406$

[7]A rotor is a pendulum without gravity.

when the two lowest-order resonances overlap (the *Chirikov condition*), giving *global chaos* (Greene 1979). For sufficiently large values of k, such as $k = 80/9$, no islands remain. Figure 8.17 shows an intermediate case with $k = 1$ for a variety of initial conditions.

A variant of the Chirikov map, called the *dissipative standard map*, in which the Y_n term in eqn (8.42) is replaced by bY_n has been studied by Schmidt and Wang (1985). For $b = 1$, this variant reduces to the Chirikov map, but for $|b| < 1$, the map is area-contracting and has Lyapunov exponents that sum to $\log |b|$. It has chaotic solutions and a strange attractor for a wide range of b and k such as $b = 0.1$ and $k = 8.8$.

8.9.6 Lorenz three-dimensional chaotic map

The above symplectic maps are all two-dimensional. We conclude with an example of a three-dimensional symplectic map that Lorenz (1993) cites as the simplest

$$X_{n+1} = X_n Y_n - Z_n \tag{8.43}$$

$$Y_{n+1} = X_n \tag{8.44}$$

$$Z_{n+1} = Y_n \tag{8.45}$$

or even more simply

$$X_{n+1} = X_n X_{n-1} - X_{n-2} \tag{8.46}$$

Such a system is analogous to the jerk representation of a flow (see §4.8.5). It is computationally efficient since it requires only one multiplication and one subtraction per iteration, and so it is easy to explore various regions at high resolution.

The quantity $Q = X_n^2 + X_{n-1}^2 + X_{n-2}^2 - X_n X_{n-1} X_{n-2}$ is conserved, and thus the orbit lies on a two-dimensional surface in the three-dimensional space. Small values of Q give nearly spherical surfaces, with the surfaces becoming increasingly distorted as Q increases, extending to infinity for $Q > 4$.

To observe the structure, Lorenz recommends choosing Q between 1 and 2 with small initial values of X_1 and X_2. Then solve the quadratic equation involving Q to find X_3. The other root will be X_0. More simply, try values $X_2 = 0.5, X_1 = 0.5$, and $X_0 = -1$, which give $Q = 1.75$. Projecting the orbit onto the $X_n X_{n-1}$ (or XY) plane gives Fig. 8.18. With higher resolution, you should see chains of loops of ever smaller size and 'perhaps something unexpected,' as Lorenz teasingly suggests.

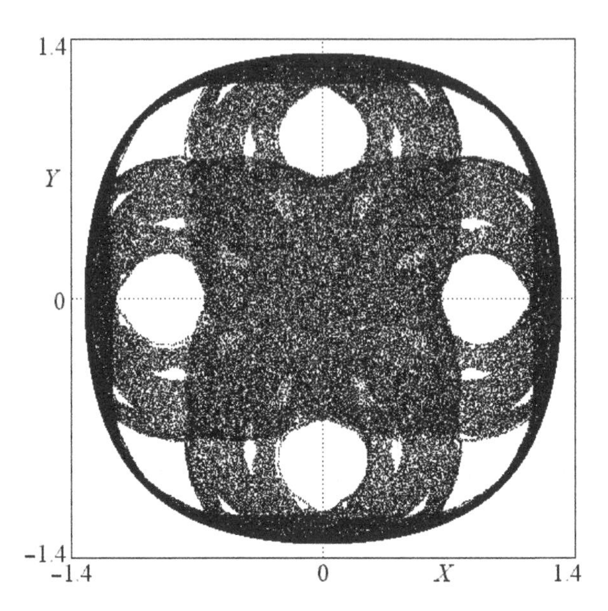

Fig. 8.18 Lorenz three-dimensional symplectic map.

8.10 KAM theory

Suppose you have an integrable system with N degrees of freedom ($2N$ dynamical variables) and a Hamiltonian H_0. There must be N constants of the motion, and the trajectory for different initial conditions must be quasiperiodic and confined to a set of nested N-tori. Now add a small nonintegrable perturbation ϵH_1, so that the total Hamiltonian is $H = H_0 + \epsilon H_1$, and consider what happens to the tori. Kolmogorov (1958), Arnold (1963), and Moser (1973) showed that for small ϵ, most of the tori survive with only a slight distortion. Although the proof of the *KAM theorem* (Arnold 1978) requires $\epsilon < 10^{-48}$ in the case of planetary motion,[8] in practice the result holds for $\epsilon \lesssim 1$.

However, with $N = 2$, each torus contains trajectories with two independent frequencies. If these frequencies are commensurate (their ratio is rational), a resonance occurs between the two motions. At these resonances are saddle points (also called *hyperbolic points*) with homoclinic or heteroclinic trajectories and tangles leading to destruction of the torus and chaotic trajectories. These trajectories are bounded by *KAM tori* that form barriers, preventing them from wandering throughout phase space (producing 'soft' chaos or 'nonglobal chaos'). Such barriers presumably explain why the Earth has remained at approximately the same distance from the Sun, as evidenced by the existence of liquid water, since its formation five billion years ago, despite perturbations by Jupiter and the other planets.

[8]Professor Ian Percival of the University of London has described this value as comparable to the 'gravitational pull of a micro-organism in Australia.'

As ϵ increases, the tori are destroyed one-by-one, leaving only the one with a frequency ratio equal to the *golden mean*[9]

$$G = \cfrac{1}{1+\cfrac{1}{1+\cfrac{1}{1+\cfrac{1}{1+\cdots}}}} = \frac{\sqrt{5}-1}{2} = 0.61803398\ldots \tag{8.47}$$

which is the 'most irrational' of the irrational numbers (the one worst approximated by rationals and whose continued fraction converges most slowly). It too finally succumbs at $\epsilon \sim 1$, leading to global ('hard') chaos. The *Poincaré–Birkhoff theorem* (Birkhoff 1927) shows that the tori break into an alternating sequence of hyperbolic and elliptic points producing a chain of quasiperiodic islands. These points, in turn, give rise to even smaller island chains down to arbitrarily small size, forming a *fat fractal*[10] (Umberger and Farmer 1985). As parameters of the system are changed, these periodic trajectories often undergo period doubling with a Feigenbaum number (see §2.3.4) of $\delta = 8.721097\ldots$ (Benettin *et al.* 1980a). Other bifurcations, such as period quintupling, can also occur.

With more than two degrees of freedom or with a time-dependent Hamiltonian, the KAM tori do not bound the chaotic trajectories, and chaotic webs with Arnold diffusion (see §8.9.4) are the norm (Lichtenberg and Liebermann 1992), producing 'hard' or 'global' chaos.

8.11 Exercises

Exercise 8.1 Prove that if Hamilton's equations in eqns (8.4) and (8.5) are satisfied, then the Hamiltonian H is conserved ($dH/dt = 0$).

Exercise 8.2 Derive the equations of motion for the simple pendulum in eqns (8.7) and (8.8) from the appropriate Hamiltonian using Hamilton's equations of motion.

Exercise 8.3 Show that the equilibrium at $x = \pi, v = 0$ in the simple pendulum in eqns (8.7) and (8.8) is a saddle point, and calculate its eigenvalues.

Exercise 8.4 Calculate the period of the simple pendulum for $E/mgL = 0.1, 1, 2,$ and 4. The answer will be in terms of elliptic integrals whose values you can find in mathematical handbooks (Abramowitz and Stegun 1965).

Exercise 8.5 Show that eqns (8.10)–(8.13) follow from the Hamiltonian $H = v^2/2 + \omega^2(1 - \cos x) - Axy + \Omega^2 y^2/2 + z^2/2$.

[9]Some authors define the golden mean as $1/G = 1 + G = 1.61803398\ldots$.

[10]The term 'fat fractal' was coined by Farmer (1985) for a Cantor set with finite Lebesgue measure and defined more precisely by Grebogi *et al.* (1985). Mandelbrot (1983) calls these sets 'dusts with positive volume.'

Exercise 8.6 Show that the Hénon–Heiles Hamiltonian in eqn (8.15) leads to eqns (8.16)–(8.19).

Exercise 8.7 Show that the Toda Hamiltonian in eqn (8.20) is of the same form as the Hénon–Heiles Hamiltonian in eqn (8.15) in the limit $|x| \ll 1$ and $|y| \ll 1$.

Exercise 8.8 Verify that eqns (8.21) and (8.22) follow from the Hénon–Heiles system in eqns (8.16)–(8.19) with the sign changed in the y^2 term in eqn (8.19), and find the solution.

Exercise 8.9 Show that the Nosé–Hoover oscillator in eqns (3.21)–(3.23) has a rate of volume expansion given by $dV/dt/V = \langle z \rangle$.

Exercise 8.10 Show that the Nosé–Hoover oscillator in eqns (3.21)–(3.23) is time reversible.

Exercise 8.11 Show that case D in Table 4.1 is time reversible and has a rate of volume expansion given by $dV/dt/V = -\langle x \rangle$.

Exercise 8.12 Show that the jerk form in eqn (8.32) is equivalent to the Nosé–Hoover oscillator in eqns (3.21)–(3.23).

Exercise 8.13 Show that the second-order Runge–Kutta method when applied to the simple pendulum in eqns (8.7) and (8.8) is not symplectic (energy-conserving), but that the leap-frog method is, at least when averaged over a period of the motion.

Exercise 8.14 Verify that the Hénon area-preserving quadratic map in eqns (8.31) and (8.32) is area-preserving.

Exercise 8.15 Find the inverse mapping for the Hénon area-preserving quadratic map in eqns (8.33) and (8.34).

Exercise 8.16 Find the conditions under which the generalized Hénon map in eqns (6.1) and (6.2) is area-preserving.

Exercise 8.17 Calculate the eigenvalues and eigenvectors for Arnold's cat map in eqns (8.35) and (8.36), and show that the stable and unstable manifolds are perpendicular to one another but that the unstable manifold makes an angle with respect to the x axis whose tangent is the inverse *golden mean* $\frac{1}{2}(1+\sqrt{5}) = 1.61803398\ldots$.

Exercise 8.18 Find the two period-2 orbits of Arnold's cat map in eqns (8.35) and (8.36).

Exercise 8.19 Verify the existence of a period-5 orbit in the gingerbreadman map of eqns (8.37) and (8.38).

Exercise 8.20 Show that the chaotic web map in eqns (8.39) and (8.40) is area-preserving.

Exercise 8.21 Derive the Chirikov map in eqns (8.41) and (8.42) using the leap-frog method in §3.9.2 to approximate the simple pendulum in eqns (8.7) and (8.8) with a step size that is not small.

Exercise 8.22 Show that the Chirikov map in eqns (8.41) and (8.42) is area-preserving.

Exercise 8.23 Calculate the fixed points for the Chirikov map in eqns (8.41) and (8.42), and analyze their stability.

Exercise 8.24 Find values of X^* and Y^* that give a 2-cycle for the Chirikov map in eqns (8.41) and (8.42).

Exercise 8.25 Find values of X^* and Y^* that give a 3-cycle for the Chirikov map in eqns (8.41) and (8.42) with $k = 0$.

Exercise 8.26 Show that the dissipative standard map has Lyapunov exponents that sum to $\log |b|$.

Exercise 8.27 Show that Lorenz's three-dimensional symplectic map in eqn (8.46) conserves the quantity $Q = X_n^2 + X_{n-1}^2 + X_{n-2}^2 - X_n X_{n-1} X_{n-2}$.

Exercise 8.28 Show that the continued fraction representation of the golden mean in eqn (8.47) satisfies $G + 1 = 1/G$ whose positive solution is $G = 0.61803398\ldots$.

8.12 Computer project: Chirikov map

In this project you will examine the most famous area-preserving, two-dimensional map, called the *standard map* or *Chirikov map*. It exhibits most of the phenomena that are observed in chaotic Hamiltonian systems such as the driven, frictionless pendulum, the ball bouncing elastically on an oscillating floor, and the toroidal magnetic field in the presence of small perturbations.

The Chirikov map is given by

$$X_{n+1} = X_n + Y_{n+1} \pmod{2\pi} \tag{8.48}$$

$$Y_{n+1} = Y_n + k \sin X_n \pmod{2\pi} \tag{8.49}$$

Note that unlike the Hénon map in eqns (5.13) and (5.14), the Y in the first equation is really Y_{n+1}, not Y_n, and so the equations can be iterated sequentially starting with the second. Note also that the mod operator does not work properly for nonintegers and negative numbers in some programming languages, but that you can implement it with statements such as the following:

IF $X \geq 2\pi$ THEN $X = X - 2\pi$
IF $X < 0$ THEN $X = X + 2\pi$

1. With $k = 1$, start with 25 different initial conditions $X_0 = Y_0 = 2\pi i/25$ for $i = 1$ to 25 and plot a few thousand iterates for each case on the same plot. (Compare your result for $k = 1$ with Fig. 8.17.) You should see that some initial conditions produce periodic orbits, while others seem chaotic.
2. Repeat the above with $k = 0.5$ and with $k = 2$.
3. Find a region of one of your plots that contains an interesting structure, and zoom in on it (using appropriate and more closely spaced initial conditions) to show that the structure appears to persist on all size scales. You may need to increase the number of iterations as you did with the Hénon map in §6.12.
4. (optional) Select a thousand or more initial points within a small circle in the chaotic region (the 'chaotic sea') and follow them forward for a few iterations, showing that they stretch but that the area occupied by the points remains constant in accordance with Liouville's theorem. Estimate the largest Lyapunov exponent from the rate of stretching. You may also want to apply the method described in §5.2 or in §5.6 to the calculation of the largest Lyapunov exponent.

9

Time-series properties

The previous chapters typically began with equations whose properties and solutions were to be examined. These equations were devised to model a physical process or exhibit a particular behavior. The main lesson is that simple equations can have complicated solutions, including sensitive dependence on initial conditions, strange attractors with fractal structure, irregular basins of attraction, and intricate bifurcation sequences.

Armed with this experience, we now tackle the inverse problem. When you observe complicated behavior in nature, you seek a simple underlying cause. With only experimental data, you ask whether the dynamics are deterministic and chaotic or nondeterministic and random. In an extreme case, you might have only a single sequence of measurements at successive times (a *time series*). Thus we move from a theoretical to a more experimental approach in this and successive chapters. Be warned that the analysis of chaotic data is much less advanced than the theory of chaotic systems.

How you proceed depends on your goal. Usually you want to know more than simply whether there is hidden determinism in your data. If you are an economist, meteorologist, or gambler, you might be interested in *prediction*. If you are a cryptographer or communications engineer, you might want to extract a deterministic signal from a noisy background (*noise reduction*). If you are a researcher, you might be interested in better *insight and understanding* of the underlying dynamics. Finally, you might want to *control* the system, using small perturbations (the butterfly effect) to alter its outcome (Chen and Dong 1993, Shinbrot *et al.* 1993).

Time-series analysis is an old and venerable field about which much has been written. It is not our goal to review this large body of work but rather to focus on those aspects that follow from an understanding of chaos. New analysis techniques have been developed specifically to characterize chaotic systems, and the availability of powerful computers has brought these methods to the desktops of most scientists and researchers.

However, you should proceed with caution. Chaotic time-series analysis is still more art than science, and there are few sure-fire methods. You need a battery of tests, and conclusions are seldom definitive. New tests are constantly being developed. It is easy to fool yourself into thinking that a random system is chaotic, and the literature is full of false claims. Finally the question 'Is it chaos?' may be too simplistic since there is a hierarchy of dynamical behaviors with varying amounts of determinism.

9.1 Hierarchy of dynamical behaviors

Moon (1992) lists seven classes of motion in nonlinear deterministic systems to which several more classes of randomness have been added in the *hierarchy of dynamical behaviors* in Table 9.1. At the simplest level are regular motions such as two-body planetary motion, a mass on a spring, and a pendulum, as well as quasiperiodic cases such as the tides and the times of moonrise and moonset. At the opposite extreme is true randomness of which the only true examples may be atomic processes governed by quantum mechanics, which is an inherently probabilistic theory. In between are various classes of dynamics with increasing randomness and unpredictability. In practice, systems often exhibit a combination of behaviors.

If you graph the average daily temperature at some location on Earth over several years, as in Fig. 9.1, you are struck by the strong annual periodicity superimposed on an apparently random fluctuation.[1] There would also be a strong 24-hour periodicity if the temperature were sampled more frequently rather than averaged over a day, giving quasiperiodicity. The temperature trajectory would thus lie near the surface of a torus with an irrational winding number of approximately 365.242 (the average number

Table 9.1 Hierarchy of dynamical behaviors.

Class	Examples
Regular predictable	planets, clocks, tides
Regular unpredictable	coin toss
Transient chaos	pinball machine
Intermittent chaos	logistic map at $A = 3.8284$
Narrow-band chaos	Rössler attractor
Broad-band low-dimensional chaos	Lorenz attractor
Broad-band high-dimensional chaos	neural networks
Correlated (colored) noise	random walk
Pseudo-randomness	computer 'random' numbers
Random noise	radioactivity, radio 'static'
Combinations of the above	most real-world data

[1]The time series for this and other examples in this book can be found at http://sprott.physics.wisc.edu/chaostsa/data/.

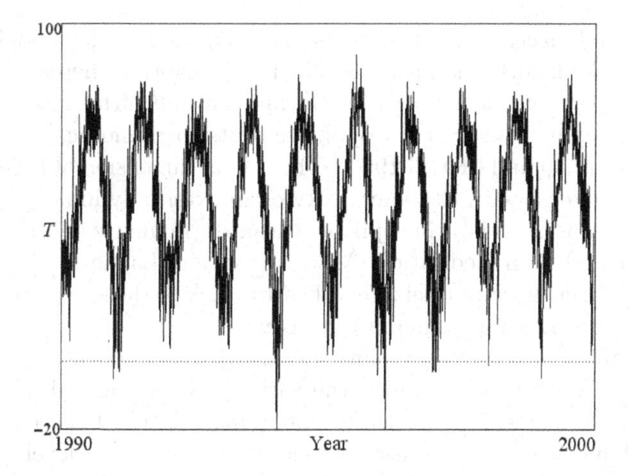

Fig. 9.1 Average daily temperature (°F) in Madison, Wisconsin, over a decade.

of days in a year). The deviations from this surface are the challenging and presumably most interesting aspects of the dynamics.

9.2 Examples of experimental time series

Meteorological observations provide good examples for time-series analysis. In addition to temperature, data are available for precipitation, wind speeds and direction, barometric pressure, cloud cover, and many other quantities for thousands of locations around the world often at hourly intervals, sometimes extending back over a hundred years. On a longer time-scale are climate variations as determined by geological core samples. Global warming and the concentration of carbon dioxide, ozone, and other gases and atmospheric particulates are of special interest. Other geological examples include tidal levels and seismic waves that might be useful for predicting earthquakes.

Extraterrestrial examples include the light output from Cepheid variable stars and the wobble of stars indicating extra-solar planets. Sunspots foretell magnetic storms on Earth, disrupting radio communications and electrical power transmission and posing radiation hazards to astronauts, and possibly other phenomena (Stetson 1937).

Financial records include the trading prices of individual stocks, market averages, and exchange rates. Much current interest in time-series analysis is motivated by market prediction, where the detection of even a miniscule component of determinism can yield lavish rewards (Brock *et al.* 1991).

Biological examples include the population of plant and animal species including humans, physiological records such as electrocardiograms[2] (ECGs

[2]The electrocardiogram has a long history perhaps beginning in 1842 when the Italian physicist Carlo Matteucci (1811–1868) showed that an electric current accompanies each heart beat.

or EKGs) and electroencephalograms[3] (EEGs), and epidemiological data (diseases). The healthy heart is thought to be chaotic, whereas a fibrillating heart is random, leading to death, and an unhealthy heart is nearly periodic, implying that it is unresponsive to its surroundings (Kaplan and Cohen 1990, Poon and Merrill 1997). The fractal dimension of EEG records appears to increase with the complexity of an externally imposed task and to decrease prior to and during epileptic seizures (Lehnertz and Elger 1998), quiet sleep, and deep meditation (Mayer-Kress and Layne 1987). The ability to predict the onset of a seizure or heart attack or the spread of a disease would have considerable humanitarian benefit.

Although we use the term 'time series,' the independent variable need not be time. With a core sample, the variable is distance, although there is presumably a linear or at least monotonic relation between the two. You could measure the variation of elevation above sea level along the equator, where distance is the independent variable not associated with time, although the two would be equivalent if you traveled around the Earth at a constant speed.

More abstract is the sequence of letters in a written document, notes in a musical composition, or bases in a DNA molecule. In such cases, there is not an independent variable at all, but just an ordered list. Other ordered lists include the sequence of times between heartbeats (*interbeat*, or more generally, *interspike interval*) or drips from a leaking faucet (see §1.4), in which case the *dependent* variable, rather than the *independent* variable, has units of time.

Psychological experiments (Guastello 2001) provide a wealth of ordered lists, such as the history of reaction time on successive trials of some task, the interval between flips of the *Necker cube*[4] (Aks and Sprott 2003), as in Fig. 9.2, or other reversible figures, and movements of the eye when performing a visual search or processing a visual stimulus (Aks *et al.* 2002). Interesting experiments involve asking a subject to perform a periodic task such as tapping a foot or finger at regular intervals to emulate a metronome, or attempting to emulate a random process such as the clicks of a Geiger counter. Subjects can be asked to write a list of 'random numbers' or to generate a random sequence of bits (0 or 1) or to click a mouse at random points on a line or in a circle or square on a computer screen. The resulting data might be useful in testing models of cognition and mental processes.

In some of these cases, such as eye movement or clicking randomly on the computer screen, the dependent variable is not a scalar but a vector

[3]Electrical signals from animals were discovered by the English physiologist Richard Caton (1842–1926) in 1879 and later used in human brains by the German psychiatrist Hans Berger (1873–1941) in 1929. The signals were first shown to be chaotic by Babloyantz and Destexhe (1986).

[4]The illusion is named after Louis Albert Necker (*c.* 1800–1861), a Swiss crystallographer who saw his line drawings of crystals spontaneously reverse in perspective (Necker 1832) and who died as a recluse after becoming disenchanted with European political life.

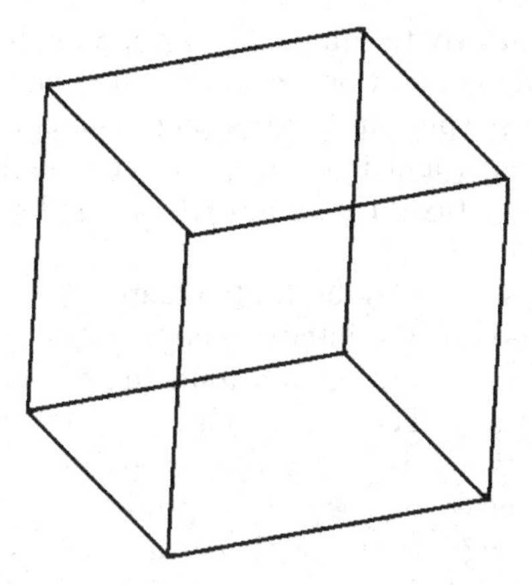

Fig. 9.2 The Necker cube repeatedly flips front to back as you stare at it.

with two or more components. In fact, in most experiments it is possible and desirable to record simultaneously more than one quantity. We will for the present assume the dependent variable is a scalar and will return later to discuss how additional information might be effectively used (see §12.7.5).

9.3 Practical considerations

The cleanest forms of data are produced by numerically iterated maps. Subject to computer limitations, you can produce a record of arbitrary length and precision, and there is no issue of choosing an appropriate sample rate since every iteration is recorded. Furthermore, you know much about the system since the equations are known. Only slightly less desirable is a time series produced by the numerical solution of a system of differential equations. The numerical method is an approximation and introduces iteration as well as round-off errors (see §3.9), and it is necessary to choose time steps both in the integration of the equations and in the interval between data samples, or you might sample the data at nonuniform times such as when the time derivative is zero (a local maximum or minimum). In either case, you can choose an optimal variable or record multiple variables.

With experimental data, you must often convert an analog signal into digital form to facilitate processing. Such *A-to-D converters* have inherent limitations in accuracy (number of bits of precision and calibration errors), bandwidth, sample rate, and total number of data points. Furthermore, any instrument used to collect data inevitably introduces some noise.

You do not usually know the fundamental dynamical variables or even how many there are. Even if you knew, you might not have experimental access to them or your access may be indirect through some form of filtering or operation such as integration or differentiation. You cannot even be sure the process is described by a finite set of ordinary differential equations. You must decide how frequently to sample the data. Sampling too infrequently will miss important dynamics that occur on fast time-scales, or worse will lead to *aliasing* in which spurious low-frequency components are introduced by the sampling process, and it might generate a uselessly short record. Sampling too frequently taxes the memory limits and slows the calculation for many of the tests.

Finally, the time series may be contaminated, such as having missing or spurious data points, or the interval between samples may not be uniform. There are always measuring and rounding errors when representing an analog signal digitally. The system may not be stationary (such as a 'bull' stock market), which is to say that there are important dynamical components at frequencies too low to be captured by the finite data record, or you may be observing a transient that is not representative of the final state of the system. Any finite time series will lead to some uncertainty in your conclusions.

9.4 Conventional linear methods

Most conventional time-series analysis methods (Box *et al.* 1994, Chatfield 1996) implicitly assume the data come from a linear dynamical system, perhaps with many degrees of freedom and some added noise. Thus the variation is assumed to be a superposition (summation) of sine waves or exponentials that grow or decay. If there are many such terms, the time series can look quite erratic, but it cannot be chaotic.

9.4.1 Stationarity

The exponential terms typically lead to *nonstationarity* (Priestley 1981, 1988) in which some characteristic of the time series such as the *mean* (average value)[5]

$$\langle X \rangle = \frac{1}{N} \sum_{n=1}^{N} X_n \tag{9.1}$$

or *standard deviation* (root-mean-square deviation)

$$\sigma = \sqrt{\frac{1}{N-1} \sum_{n=1}^{N} (X_n - \langle X \rangle)^2} \tag{9.2}$$

changes in time, perhaps approaching infinity for $N \to \infty$. Nonstationarity is a property of the process, not the data, and arises when the mechanism producing the data changes in time. However, a time series too short to capture the slowest variations of the measured quantity produces the same effect.

[5]We will use the convention that the N elements of the time series X_n are numbered from $n = 1$ to N, rather than the equally reasonable 0 to $N - 1$.

The quantity under the square root in eqn (9.2) is called the *variance*, and hence the standard deviation is the square root of the variance and measures the 'width' of the distribution.[6] A simple test is to calculate these quantities for the first half ($1 \leq n \leq N/2$) and second half ($N/2+1 \leq n \leq N$) of the data. It is then a matter of judgement whether the inevitable difference is significant. If the means of the two halves differ by more than a few *standard errors* for each half, then stationarity is almost certainly a problem. The standard error is the standard deviation divided by the square root of the number of data points $\sigma/\sqrt{N}$.

Since the standard deviation involves the *second moment* of the distribution because of the square in eqn (9.2), it is sensitive to points far from the mean (e.g., 'outliers'), and in some cases it may not even converge as the number of points increases. More robust is the *average deviation*

$$\text{average deviation} = \frac{1}{N}\sum_{n=1}^{N} |X_n - \langle X \rangle| \tag{9.3}$$

which provides much the same information as the standard deviation but gives less importance to any outliers. A time series whose first two moments (mean and variance) are constant is said to exhibit *weak stationarity*, but such a condition is generally insufficient for analyzing a chaotic system.

Higher moments of the data can be examined, but they tend to be even less robust, and their time independence still does not guarantee stationarity for a chaotic system. The third moment, called the *skewness*

$$\text{skewness} = \frac{1}{N}\sum_{n=1}^{N} \left[\frac{X_n - \langle X \rangle}{\sigma}\right]^3 \tag{9.4}$$

is a measure of the symmetry of the data about the mean. The fourth moment, called the *kurtosis*

$$\text{kurtosis} = \frac{1}{N}\sum_{n=1}^{N} \left[\frac{X_n - \langle X \rangle}{\sigma}\right]^4 - 3 \tag{9.5}$$

measures the peakedness of the distribution relative to a normal (Gaussian) distribution (hence the -3 term). A distribution with positive kurtosis is called *leptokurtic* (or *fat-tailed*), and a distribution with a negative kurtosis is called *platykurtic* (Press *et al.* 1992).

If the time series is a fractal, moments of the distribution may not exist in the sense that their calculated values depend on the length of the data record and may approach infinity or zero as the number of data points

[6]The $N-1$ in the denominator of the variance should be changed to N if you know $\langle X \rangle$ independent of the data, but as Press *et al.* (1992) comment, 'if the difference between N and $N-1$ ever matters to you, then you are probably up to no good anyway — e.g., trying to substantiate a questionable hypothesis with marginal data.'

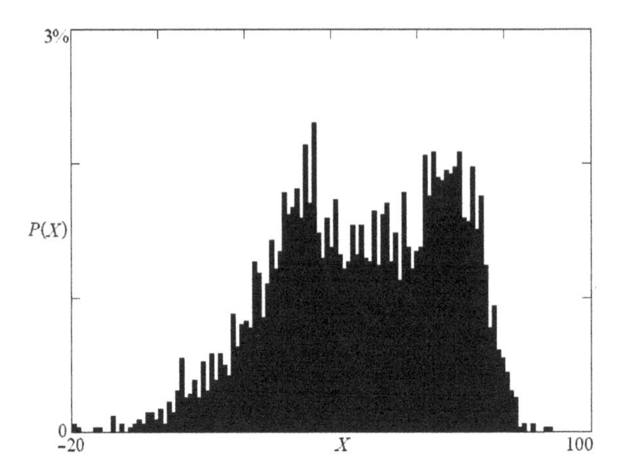

Fig. 9.3 Probability distribution of average daily temperatures in Madison, Wisconsin.

becomes large. An example is the *St Petersburg game* (Feller 1968) formulated around 1700 by Nicolas,[7] Johann, and Daniel Bernoulli. Assuming you have a 50/50 chance of winning \$2 for every \$1 that you bet, start with a bet of \$1 and double your bet each time you lose, but lower your bet to \$1 each time you win. Your *average* winnings will continue increasing without limit. If you try this at the casino, be aware that the house may limit the size of your bets even if you have the necessary infinite ante and that if you gamble long enough you will lose any finite amount (*gambler's ruin*).

9.4.2 Probability distribution

These quantities are just a few of infinitely many that characterize a distribution of values. More generally, you could plot the distribution function $P(X)$, which is the probability that X is within some ΔX of X. You can approximate $P(X)$ by partitioning the interval from the minimum to the maximum X into some convenient number of 'bins' (or 'partitions'), perhaps of the order of $\sqrt{N}$ and graphing the number of X values in each bin. With too few bins, the distribution is overly smeared out, and with too many, the number of points in each bin has large statistical fluctuations. For a time series to be *strictly stationary*, $P(X)$ must remain constant in time for all X within statistical uncertainties.

Figure 9.3 shows the distribution of temperatures corresponding to Fig. 9.1 in which the values are rounded to the nearest degree Fahrenheit and plotted with bins of 1°F width from the low (−20°F) to the high temperature (91°F). This distribution is *bimodal* with peaks at 36°F and 67°F as

[7]Nicolas Bernoulli (1687–1759) was Daniel and Johann's cousin and part of an abundant family of Swiss scientists and mathematicians where there was unfortunate rivalry, jealousy, and bitterness.

is typical of a sine wave, which spends most of its time near its extrema. The largest peak (the most probable value) is called the *mode* (36°F in this case), and it will usually change with the bin width.

Also of interest is the *median* (48°F in this case), for which half the values are above and half below. The median is equal to the mean (47.2092°F in this case) for a symmetric distribution, but it can be quite different for a highly skewed distribution. It is a more robust estimator than the mean since it is less sensitive to outliers, but it is harder to calculate since it usually requires sorting the data values, which requires on the order of $N \ln N$ operations, although there are more efficient methods (Press *et al.* 1992). Note that the median winnings for the St Petersburg game in §9.4.1 is $1.

9.4.3 Detrending

The effects of nonstationarity can often be reduced by *detrending* the data. If the nonstationarity is primarily in the mean, you can subtract from the data some smooth function $f(n)$ such as a simple linear *least-squares fit* $f(n) = A + Bn$, where

$$A = \langle X \rangle - \frac{N+1}{2} B \tag{9.6}$$

$$B = \frac{6}{N(N^2-1)} \sum_{n=1}^{N} (2n - N - 1)(X_n - \langle X \rangle) \tag{9.7}$$

The first and second half of the detrended data then have zero mean. You can also fit polynomials of higher degree or a superposition of sine waves, among others.

Sometimes the trend is the interesting part of the data. For example, applying the method above to the temperature data in Fig. 9.1 gives $A \simeq$ 46.39°F and $B \simeq 0.000448$°F/day, implying a temperature rise of $BN \simeq$ 1.6°F over the decade ($N = 3652$ days). Whether this result portends global warming requires further statistical analysis. You must be cautious in applying such methods. If the time series had ended half a year earlier (in mid summer), there would have been a much stronger (but spurious) warming, and if it had started half a year later (in mid summer), there would have been a strong (but spurious) cooling. In cases with a dominant periodicity, you should collect data for an integer number of periods (years in this weather example).

Another common method, where all the values are positive, such as stock prices, is to compute a new time series Y_n by taking *log first differences*

$$Y_n = \ln X_{n+1} - \ln X_n = \ln \frac{X_{n+1}}{X_n} \simeq \frac{X_{n+1} - X_n}{X_n} \tag{9.8}$$

and performing the analysis on this new series, which is approximately the *fractional* change in X. Note that the new time series is of length $N - 1$ since one data point is lost in the differencing. The assumption (or perhaps fervent hope) is that the interesting dynamics are preserved or even enhanced in the detrended data.

9.4.4 Fourier analysis

For stationary or detrended data with inherent periodicities, *Fourier analysis*[8] (also called *spectral analysis*, *frequency analysis*, or *harmonic analysis*) is useful (Newland 1993). You assume the time series can be represented by a superposition of sines and cosines of various amplitudes and frequencies. With a finite data set, the range of frequencies that can be discerned is limited on the low end to $f_0 = 1/N$ (the fundamental frequency) by the finite length of the record and on the high end to $f_c = 1/2$ (the *Nyquist frequency*) by the finite sample rate. In each case, the frequency has units of *cycles per time step*. The fundamental frequency comes from needing at least one full oscillation, and the Nyquist frequency from needing at least two data points per period to define an oscillation. The reciprocals of these frequencies are the *periods* of the corresponding oscillations. For example, the temperature data in Fig. 9.1 have measurable periods from 2 to 3652 days. Conversely, if you know the range of frequencies that are important in your experiment, then you can determine how frequently and for how long it should be sampled.

Thus with a finite data set, the best you can do is approximate X_n by

$$X_n \simeq \frac{a_0}{2} + \sum_{m=1}^{N/2} a_m \cos\frac{2\pi mn}{N} + b_m \sin\frac{2\pi mn}{N} \tag{9.9}$$

where the $N/2$ different frequencies or *harmonics*[9] ($f = mf_0$) have amplitudes

$$a_m = \frac{2}{N}\sum_{n=1}^{N} X_n \cos\frac{2\pi mn}{N} \tag{9.10}$$

$$b_m = \frac{2}{N}\sum_{n=1}^{N} X_n \sin\frac{2\pi mn}{N} \tag{9.11}$$

These equations constitute the *discrete Fourier transform* (DFT). The N coefficients a_m and b_m contain the same information as the original N data points X_n but are in the *frequency domain* rather than the *time domain*.

Sometimes you need the power in each Fourier component, called the *power spectral density*, or more simply, the *power spectrum*. Since the power

[8]Fourier analysis was developed by the French mathematical physicist and lifelong bachelor, Jean Baptiste Joseph Fourier (1768–1830), who also served as Napoleon's scientific advisor and was almost executed twice for his revolutionary activities.

[9]Traditionally, the fundamental ($m = 1$) is called the *first harmonic*, and the one with $m = 2$ is called the *second harmonic*, and so forth, although some authors call the one with $m = 2$ the 'first harmonic.'

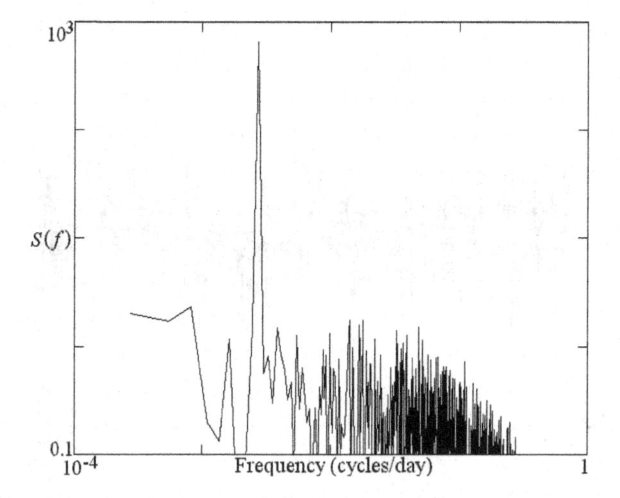

Fig. 9.4 Power spectrum of average daily temperature in Madison, Wisconsin.

is proportional to the square of the amplitude of an oscillation and since there are both sine and cosine terms 90° out of phase, the power $S(f)$ at frequency mf_0 is

$$S_m = a_m^2 + b_m^2 \tag{9.12}$$

There are various ways to normalize S_m, but since we are usually interested only in relative values, we need not be concerned with that.

Applying the formula to the temperature data in Fig. 9.1 gives the power spectrum in Fig. 9.4. Because of the wide range of frequencies and powers and because power spectra are often power laws $S(f) \propto 1/f^\alpha$, it is customary to plot the spectrum on a log–log scale (called a *Bode plot* by electrical and mechanical engineers). The strong peak with a period of 365.2 days ($m = 10, f = 1/365.2$) is evident. The dominant frequency component has $a_{10} \simeq -24.26$ and $b_{10} \simeq -7.91$, from which we can calculate a phase $\phi = \tan^{-1}(b/a) \simeq 0.315$ radians, which implies that the coldest day on average is January 18 ($365.2\phi/2\pi$ days after the beginning of the year).

One use of Fourier analysis is to remove seasonal trends in meteorological, economic, or environmental data. Suppose from the temperature data X_n we construct a new data set

$$Y_n = X_n - a_{10}\cos\frac{20\pi n}{N} - b_{10}\sin\frac{20\pi n}{N} \tag{9.13}$$

in which the annual oscillation at $m = 10$ has been subtracted. The resulting record in Fig. 9.5 shows only the fluctuations about the seasonal mean. If you suspect the periodic component is not sinusoidal, then you can include additional harmonics such as $m = 20, 30, 40, \ldots$. On the other hand, you might be interested in the seasonal variations, in which case the

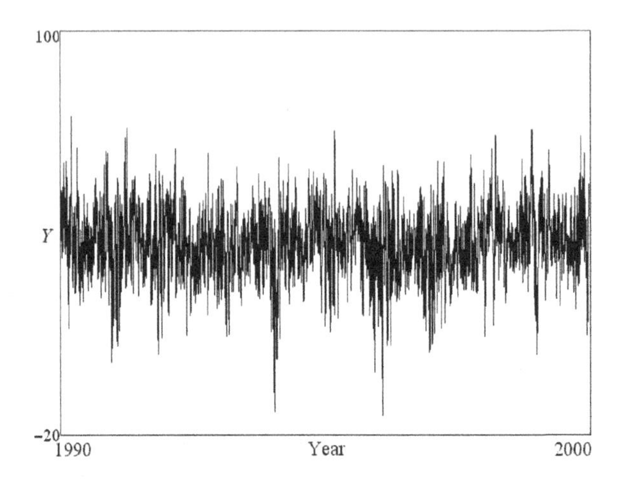

Fig. 9.5 Seasonally detrended average daily temperature in Madison, Wisconsin.

Table 9.2 Temperature predictions (°F) from Fourier analysis.

Date	Predicted	Actual	Error
Jan 1, 2000	24	35	−11
Jan 2, 2000	26	42	−16
Jan 3, 2000	36	28	+8
Jan 4, 2000	23	25	−2
Jan 5, 2000	19	16	+3
Jan 6, 2000	20	31	−11
Jan 7, 2000	28	18	+10

fluctuations are extraneous noise. Thus the method can be used for noise reduction.

Another use of Fourier analysis is prediction. To the extent that the time series is quasiperiodic, eqn (9.9) provides predictions for $n > N$ once the coefficients a_m and b_m have been determined. Since there is little evidence of additional peaks in the power spectrum, we do not expect this method to work well for short-term weather prediction. Table 9.2 shows that the predicted temperature for the first seven days of 2000 is rarely much more accurate than the 9°F standard deviation of the detrended data. The prediction is presumably better for longer-term averages where seasonal effects are dominant and for other types of quasiperiodic data such as the height of the tides. Prediction and noise reduction will be described more fully in the next chapter.

Differentiating the data (taking first differences) accentuates the high frequencies, since the derivative of $\sin \omega t$ is $\omega \cos \omega t$, where $\omega = 2\pi f$. Then high-frequency components of the power spectrum become more evident, and a broad peak with a period of about five days emerges from the temperature data. This time may be the average between the passage of successive fronts. Thus a good guess for the temperature five days from now is that it will be the same as today after correcting for seasonal changes.

Some people think the stock market consists of random noise superimposed on periodic 'cycles,' and try to predict the market accordingly. Over the long term, these traders tend not to be very successful because the expected periodicities become common knowledge and are thus suppressed by other traders who respond to them in a so-called *efficient market*.

Fourier analysis is useful for distinguishing chaos from quasiperiodicity. A quasiperiodic signal will have a finite number of sharp spectral peaks, many of which are simple harmonics (integral multiples) of one another. A chaotic signal will have a continuous (or 'broadband') spectrum, perhaps with some embedded peaks. Unfortunately, nondeterministic (random) noise also has a continuous spectrum, and so Fourier analysis is not very useful for distinguishing chaos from noise. A chaotic signal often has structure in its spectrum, whereas noise more often has a simple power law $S \propto 1/f^{\alpha}$ without sharp peaks. However, chaotic systems just at the onset of chaos, called 'the edge of chaos' by Langton (1990), often have power-law spectra (Manneville 1980, Bak 1996, Dimitrova and Yordanov 2001), and it is possible to produce noise with a narrow frequency spectrum using, for example, frequency-selective filters. An abrupt change in the power spectrum as a control parameter is varied suggests a bifurcation in a deterministic system rather than noise.

A difficulty is that the Fourier transform implicitly assumes the longest period is N and that the record repeats for $n > N$. If the last data point X_N is very different from the first X_1, then the 'glitch' produces spurious high-frequency components in the spectrum. There are at least three ways to minimize this problem. You can truncate the time series at a value of n where $X_n \simeq X_1$ if doing so does not eliminate too much data. You can subtract from each data point a value $f(n) = A + Bn$, where

$$A = \frac{NX_1 - X_N}{N-1} \tag{9.14}$$

$$B = \frac{X_N - X_1}{N-1} \tag{9.15}$$

which is a special kind of detrending that makes the first and last points zero ($X_1 - A - B = X_N - A - BN = 0$). Finally, you can multiply the data by a 'window' function (Press *et al.* 1992) such as

$$W(n) = 1 - \left| \frac{2n - N - 1}{N+1} \right| \tag{9.16}$$

called a *Parzen window* and reminiscent of the tent map, or a *Welch window*

$$W(n) = 1 - \left(\frac{2n - N - 1}{N + 1}\right)^2 \tag{9.17}$$

which resembles the logistic map, and smoothly reduces the influence of points at the ends of the time series.

A further difficulty with the discrete Fourier transform is that the number of calculations scales as N^2. A method developed by Cooley and Tukey (1965), appropriately called the *fast Fourier transform* (FFT),[10] scales as $N \log_2 N$. The difference is immense. If you have 10^6 data points (not large by current standards) and each operation requires a microsecond, the FFT requires about twenty seconds, while the DFT requires over eleven days. Unfortunately, the FFT is cumbersome to implement, and so you may want to used a canned routine, such as in Press *et al.* (1992).

9.4.5 Autocorrelation function

The Fourier transform of the power spectrum in the time domain, according to the *Wiener–Khinchin theorem*, is given by the *autocorrelation function* (or *serial correlation function*)[11]

$$G(k) \simeq \frac{\sum_{n=1}^{N-k}(X_n - \langle X\rangle)(X_{n+k} - \langle X\rangle)}{\sum_{n=1}^{N-k}(X_n - \langle X\rangle)^2} \tag{9.18}$$

which measures how strongly on average each data point is correlated with one k time steps away. It is the ratio of the *autocovariance* to the *variance* of the data. The mean is subtracted from each data point so that $G(k) = 0$ for uncorrelated data, and $G(k)$ is normalized so that $G(0) = 1$. Note that the denominator is the value of the numerator at $k = 0$. The limits on the sums ($1 \leq n \leq N - k$) are adjusted to keep the indices within the range of available data. The correlation function is only defined at integer values of k, but it can be considered a discrete sample of the continuous $G(k)$ that would result from correlating the continuous variable $x(t)$ from which the discrete time series X_n was derived.

The autocorrelation function is a linear measure, each term of which (the *lag-k autocorrelation coefficient*) measures the extent to which X_n versus X_{n+k} is a straight line. Many nonlinear systems, such as the logistic map for $A = 4$, have no linear correlation. Uncorrelated data should have $G(k)$ within $\pm 2/\sqrt{N}$ of zero (two standard deviations) for about 95% of the k values (Makridakis *et al.* 1983).

In general, the correlation function falls from a value of 1 at $k = 0$ to zero at large k. The value of k at which it falls to $1/\mathrm{e} \simeq 37\%$ is called the

[10] Cooley and Tukey subsequently discovered that the FFT had been used by others as early as 1942, but the method was not widely known until their work.

[11] The autocorrelation function, originally called the *correlogram*, was invented by George Udny Yule (1871–1951), a Scottish physicist and statistician who established many of the principles of linear time-series analysis.

correlation time τ_c. For X_n nearly periodic, the correlation function will be a decaying oscillation, in which case τ_c is the time for the *envelope* to decay to 1/e. The power spectrum can be calculated from the correlation function by

$$S_m = G(0) + G(K)\cos\left(\frac{2\pi mK}{N}\right) + 2\sum_{k=1}^{K-1} G(k)\cos\left(\frac{2\pi mk}{N}\right) \tag{9.19}$$

where K is the maximum k, which you should usually take to be about $N/4$ (Davis 1986).

The correlation function is symmetric about $k = 0$, and so the full width is $2\tau_c$, which is a measure of how much 'memory' the system has. The reciprocal of this quantity, $0.5/\tau_c$, is an estimate of the average rate at which predictability is lost. Thus it is sometimes called the 'poor man's Lyapunov exponent,' since its value is often similar to the largest Lyapunov exponent.

9.4.6 Hurst exponent

Another way to quantify the correlation between points in a time series uses *rescaled range* (or R/S) *analysis*.[12] Suppose the values of X_n represent step sizes in a one-dimensional random walk.[13] Feller (1951) has proved that the asymptotic behavior for *any* independent random process (a *Poisson process*)[14] with finite variance is given by

$$R/\sigma = \left(\frac{\pi n}{2}\right)^{1/2} \tag{9.20}$$

where R is the *range* (of cumulative sums) given by the difference between the maximum and minimum value of

$$Y_n = \sum_{i=1}^{n}(X_i - \langle X\rangle) \tag{9.21}$$

over the first n time steps and σ is the standard deviation of X_n in eqn (9.2), which is the average step size. The convergence of R/σ is rather slow, requiring a large n to get a good result.

More generally, the ratio R/σ increases with some power of time $R/\sigma \propto n^H$, where H is the *Hurst exponent*[15] (Hurst *et al.* 1965), which measures

[12] R/S analysis was developed by Harold Edwin Hurst (1900–1978) who spent 62 years in Egypt as a British civil servant, mostly studying the long-term dependence of water levels in the Nile and other rivers and reservoirs.

[13] Sometimes the term 'random walk' is reserved for cases in which the step size is constant rather than normally (Gaussian) distributed.

[14] A Poisson process is one in which events occur randomly such as the emission of particles from a radioactive sample and is named in honor of the French mathematician Siméon Denis Poisson (1781–1840).

[15] Hurst used the symbol K for the exponent, but Benoit Mandelbrot switched to the more appropriate H.

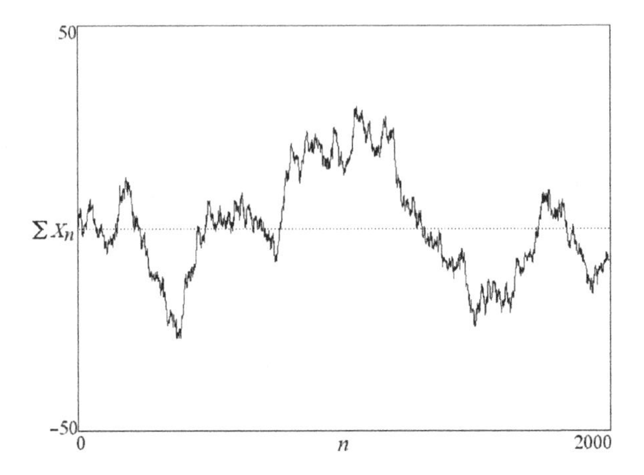

Fig. 9.6 Trace of Brownian motion from the running sum of n independent random X values with zero mean and unit variance.

the smoothness of the time series. An uncorrelated sequence of X_n values produces *Brownian motion*[16] for Y_n as in Fig. 9.6 and thus has $H = 1/2$. Other values in the range $0 < H < 1$ are called *fractional Brownian motion*, or *fBm* for short (Mandelbrot 1983), and values with $1/2 < H < 1$ are called *Lévy flights*.[17] Values with $H > 1/2$ imply *persistence* (positive correlation), where the trajectory tends to continue in its current direction and thus produces enhanced (or *anomalous*) diffusion. Values with $H < 1/2$ exhibit *antipersistence* (negative correlation), where the trajectory tends to return to the point from which it came and thus suppresses diffusion. Mandelbrot (1975) has shown that H must lie in the range $0 \leq H \leq 1$. Hurst *et al.* (1965) noted that many natural processes, such as river discharges, mud sediments, and tree rings, have $H = 0.73 \pm 0.09$ and hence have considerable long-term memory. Fractional Brownian motion is a nonstationary process. Its variance $\langle Y^2 \rangle$ increases in time without bound, which is why R/σ is used.

The Hurst exponent is also related to the fractal dimension of the Y_n versus n curve (called the *trace*) by $D = 2 - H$ (Feder 1988) and to the slope of its power spectrum by $\alpha = 2H + 1$ (Tsonis 1992). Thus the trace in Fig. 9.6 with $H = 1/2$ has a fractal dimension of 1.5, midway between

[16] *Brownian movement* (as it was originally called) was discovered by Robert Brown (1773–1858), a Scottish botanist, who in 1827 noticed the incessant motion of pollen grains suspended in water when viewed under a microscope. Brown demonstrated that the motion was physical rather than biological by boiling and freezing the water, but the explanation of collisions with randomly moving water molecules was provided by Albert Einstein in 1905, which in turn allowed the French physicist Jean Baptiste Perrin (1870–1942) to calculate the size and mass of molecules, for which Perrin received the Nobel Prize in Physics (1926).

[17] Paul Lévy (1886–1971) was a French mathematician who eventually occupied Poincaré's chair at the Paris Académie des Sciences, and was a mentor of Mandelbrot.

a line and a surface. The trace of fBm has a fractal dimension that varies from 1 for $H = 1$ to 2 for $H = 0$, and a power law that varies from $\alpha = 1$ for $H = 0$ to $\alpha = 3$ for $H = 1$. From these relations it follows that $1/f^{\alpha}$ noise has a trace with a fractal dimension $D = (5 - \alpha)/2$ for $1 \leq \alpha \leq 3$. Since Y is the integral of X and the power is proportional to X^2, the slope of the X power spectrum is $\alpha - 2$. Be careful not to confuse the X and the Y time series. Calculate R for the Y time series and σ for the X time series. In summary, the Hurst exponent can be estimated in any of the following ways

$$H = \frac{d \log(R/\sigma)}{d \log n} = 2 - D = \frac{\alpha - 1}{2} \tag{9.22}$$

These relations provide a cross check on the analysis.

To confuse matters further, the dimension of the *trajectory* (or *trail*) of fBm is $1/H$. Thus with $H = 1/2$, the dimension is two. However, the dimension cannot exceed the dimension D_E of the space in which it occurs

$$D = \min(1/H, D_E) \tag{9.23}$$

Fractional Brownian motions with $H < 0.5$ are plane filling, but with $H > 0.5$ then tend to 'wander off.' A random walk constrained to a line has dimension one and constantly intersects itself. Consequently, Brownian motion in one dimension will return infinitely many times to its starting point (although the returns are clustered in time), but in three dimensions it will never do so. In two dimensions, it will eventually return arbitrarily close to its starting point.

9.4.7 Sonification

Sometimes patterns and structure in a time series can be more readily discerned audibly than visually. This method has been called *sonification* or *audification* (Kramer 1994). It has been used to interpret data from the Voyager spacecraft as it encountered plasma waves near Jupiter and micrometeorites as it crossed the rings of Saturn. The repetitive sound of a simple periodic cycle contrasts sharply with the nonrepetitive waverings of a random or chaotic signal.

A time series can be used to produce a crude kind of music by rounding the data to a finite number of values and assigning a musical note to each value. Figure 9.7 shows the result for various kinds of colored noise, for a chaotic time series derived from the iterates of the logistic map with $A = 4$, and for a period-8 time series derived from the iterates of the logistic map with $A = 3.56$. Most listeners will agree that pink music ($\alpha = 1$) sounds more like real music, although perhaps from an alien culture. Voss and Clarke (1978) showed that most real music has a $1/f$ power spectrum. Brown music ($\alpha = 2$) is too regular, and white music ($\alpha = 0$) is too

Fig. 9.7 Music produced by colored noise, chaos, and periodicity.

erratic.[18] Chaotic music will often have a motif that repeats with slight variations, while periodic music repeats exactly. Real music has determinism, since it is possible to predict with some accuracy the next note in a musical score, but it is boring if it is too predictable, and excruciatingly boring if it is periodic.

9.5 Case study

As an example of time-series analysis, consider two time-series records, one produced by Gaussian white noise, and the other by a one-dimensional chaotic map specially constructed to mimic the random noise. We will use this example to demonstrate the limitations of conventional linear analysis and to introduce methods more suitable for chaotic signals.

9.5.1 Colored noise

There are many kinds of noise, depending on the shape of the power spectrum and the distribution of values. Noise in which the power spectrum is independent of frequency is called *white* by analogy with white light, which contains all the colors of the rainbow. White noise is *uncorrelated* since the correlation function is zero for all nonzero time lags.

Noise with a power spectrum that varies with frequency is called *correlated* (or *colored*). The noise power often varies with frequency as $1/f^\alpha$

[18] A mathematical theory of aesthetics (Birkhoff 1933), asserting that music, art, and poetry are most appealing if they are neither too predictable nor too surprising, was proposed by the American mathematician George David Birkhoff (1884–1944), who also advanced the work of Poincaré well before the advent of modern chaos.

(sometimes called *Hurst noise*). White noise has $\alpha = 0$. The case of $\alpha = 1$ is called *1/f noise*, *flicker noise*, or *pink noise* since it resembles light with more intensity in the red than in the blue. Although $1/f$ noise is common in nature and there are many models for it, its origin is often obscure. It is often used in acoustical research because it has a constant power per *octave*[19] (or any other constant logarithmic frequency interval) and thus is well matched to human audible response. The case with $\alpha = 2$ is Brownian motion and is sometimes called *brown noise*, which derives from Robert Brown and has nothing to do with color. Cases with $\alpha > 2$ are sometimes called *black noise* and might arise, for example, from a process that integrates $1/f$ noise (giving $\alpha = 3$). Less common is noise in which the power increases with frequency $\alpha < 0$, which has a primarily negative correlation function and might be called *blue noise*.

Note that a time series with a $1/f^\alpha$ power spectrum has a $1/f^{\alpha+2}$ power spectrum when integrated over time and a $1/f^{\alpha-2}$ power spectrum when differentiated with respect to time since the power is proportional to the amplitude squared of each frequency component. The noise cannot be accurately $1/f^\alpha$ for all frequencies since that would imply an infinite power when integrated from zero to infinity, but it may follow a power law over many decades. A time series whose spectrum is a power law has infinite variance and no characteristic time-scale. If recorded and played back at a different speed, it would sound the same except for a change in loudness.

9.5.2 Gaussian white noise

Noise is further characterized by the distribution of values in the time series. Even white noise has infinite variety. If the values are chosen from a *uniform* distribution of random numbers (such as 0 to 1), we call it *uniform white noise*. More typically the values have a *normal* (or *Gaussian*[20]) distribution

$$P(X) = \frac{1}{\sqrt{2\pi}} e^{-X^2/2} \tag{9.24}$$

giving the *bell curve* in Fig. 9.8. The factor of $1/\sqrt{2\pi}$ makes the integral of the probability over all X unity, and the factor of $1/2$ in the exponent makes the variance (and standard deviation) unity.

The Gaussian distribution is important because it arises in many natural processes. For example, the distribution of molecular velocities in the air you are breathing is accurately Gaussian, since that is the statistically most probable distribution. The number of heads in a large number (N) of coin

[19] An octave is a factor of two in frequency, so named because there are eight notes in the musical scale which rises by a factor of two. Actually, 'octave' is a bit of a misnomer since there are only seven distinct notes, with the eighth returning to the starting point one octave away.

[20] The Gaussian distribution is named after Johann Karl Friedrich Gauss (1777–1855), a German mathematician and physicist who contributed to electricity, magnetism, and astronomy, as well as statistics.

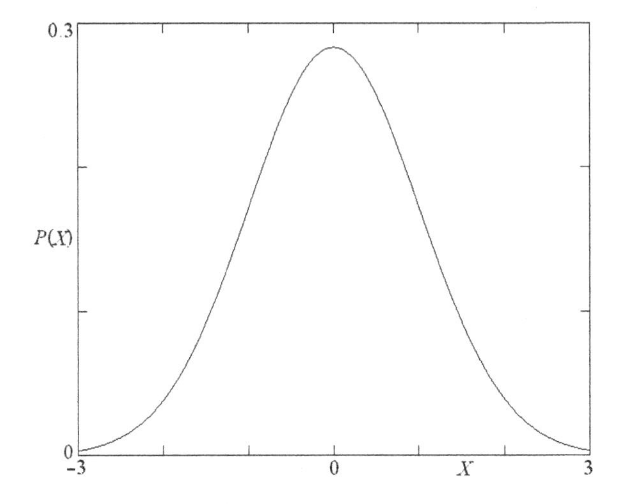

Fig. 9.8 Normal (Gaussian) probability distribution function.

tosses has a probability centered on $N/2$ with a standard deviation of $\sqrt{N}$. Statistical errors in measurements are usually assumed to be Gaussian, in which case the probability that the correct value is within one standard deviation of the mean is about 68.3%, within two standard deviations is about 95.4%, and within three standard deviations is about 99.73%. These values are called the *confidence*. Thus if you toss a coin 100 times and it comes up heads 56 times, the probability of heads is 0.56 ± 0.10 with 68% confidence, which is consistent with even odds.

It is straightforward (Press *et al.* 1992) to generate *Gaussian white noise* using

$$X = \sqrt{-2 \ln r_1} \sin 2\pi r_2 \tag{9.25}$$

where r_1 and r_2 are uncorrelated random numbers uniform in $0 < r < 1$. Another method is to use

$$X = \lim_{N \to \infty} \frac{12}{N} \sum_{n=1}^{N} r_n - 6 \tag{9.26}$$

which has a standard deviation of $\sigma = \sqrt{12/N}$. Thus if you are content with an excellent approximation, you need only sum twelve uniform random numbers and subtract six to get a normal distribution with unit variance. The numbers will be bounded to the range $-6 < X < 6$, but larger values are rare in a Gaussian distribution (one in 500 million), and their omission can be a virtue.

9.5.3 Gaussian chaotic time series

It is possible to construct a one-dimensional map with many of the same characteristics as Gaussian white noise. The inspiration comes from the general symmetric maps in Fig. 2.15 that have probability distributions progressively more peaked at their midpoint as α is reduced below one. From the Frobenius–Perron relation in eqn (2.12), it follows that

$$e^{-f^2(X)/2} = -\frac{2e^{-X^2/2}}{f'(X)} \tag{9.27}$$

and thus

$$\text{erf}(f/\sqrt{2}) = 1 - 2\,\text{erf}(X/\sqrt{2}) \tag{9.28}$$

where $\text{erf}(x)$ is the error function

$$\text{erf}(x) = \frac{2}{\sqrt{\pi}} \int_0^x e^{-t^2}\, dt \tag{9.29}$$

and f is the map function $X_{n+1} = f(X_n)$.

The implicit form in eqn (9.28) is unwieldy, and the results are very sensitive to the symmetry and precision of the map, but it can be used to devise an explicit approximation by noting

- $f(X) \to +\infty$ for $|X| \ll 1$
- $f(X) \to -|X|$ for $|X| \gg 1$
- $f(-X) = f(X)$
- $f(X) = 0$ for $X = \pm X_A = \pm 0.674489753\ldots$
- $f'(X) = \mp 1.593095481\ldots$ for $X = \pm X_A$

Equation (9.28) can be numerically integrated

$$\Delta f = -2e^{f^2/2 - X^2/2} \Delta X \tag{9.30}$$

to deduce $f(X)$ by starting at $X = X_A$, where $f(X) = 0$, and taking $\Delta X = (|X| - X_A)/M$, where M is some large integer. There is also an antisymmetric solution for $f(X)$ that guarantees $P(X)$ is symmetric about $X = 0$, but its power spectrum is not white.

A simple one-dimensional map that gives an approximate but slightly skewed Gaussian distribution is

$$X_{n+1} = \frac{1}{\sqrt[4]{|X_n|}} - \tfrac{1}{2} - |X_n| \tag{9.31}$$

although you need $X_n > -16$ to remain within the basin of attraction.

Another way to produce a deterministic Gaussian distribution is to use the tent map or binary shift map (see §2.5) to generate values of r_n for use in eqn (9.25). If you try this, use separate maps starting with different initial conditions for r_1 and r_2 to avoid correlations that drastically alter the probability distribution. The resulting map will be more complicated because it is two-dimensional, but it should accurately mimic a Gaussian distribution.

You can also produce an *approximate* Gaussian distribution with a standard deviation of π by applying the *logit transform*

$$X_n = \ln\left(\frac{r_n}{1 - r_n}\right) \tag{9.32}$$

to a time series r_n produced by the logistic map with $A = 4$. The tent map produces similar results with a standard deviation of about 1.805. The *inverse logit transform*

$$r_n = \frac{1}{e^{-X_n} + 1} \tag{9.33}$$

is useful for producing a distribution in the range $0 < r_n < 1$ from a Gaussian or other unbounded distribution with $-\infty < X_n < \infty$.

9.5.4 Conventional linear analysis

We now apply the above analysis methods to a time series consisting of Gaussian white noise produced by eqn (9.25) and to one derived from the one-dimensional map in eqn (9.28). In each case, a record of 32 000 points was produced. A segment of $X(t)$ containing 200 points is shown at the top of Fig. 9.9. Both cases look very erratic, although the one in the upper right has discernible features such as a slow rise following a large negative value and a tendency to oscillate about an unstable fixed point at $X^* = \sqrt{2}\,\mathrm{erf}^{-1}(\frac{1}{3}) = 0.430727\ldots$, where $\mathrm{erf}^{-1}(x)$ is the *inverse error function* (the quantity whose error function is x)

$$\mathrm{erf}^{-1}(x) = \frac{\sqrt{\pi}}{2}x + \frac{\pi^{3/2}}{24}x^3 + \frac{7\pi^{5/2}}{960}x^5 + \cdots \tag{9.34}$$

The probability distribution function $P(X)$ is the same for the two cases within statistical uncertainties by design. Somewhat surprising is the relative constancy of the power spectrum $S(f)$ and the total lack of autocorrelation $G(k)$ for all time lags $k \neq 0$. A white power spectrum arises from other chaotic systems including the logistic map and the tent map. Thus we conclude that these linear methods are insufficient for distinguishing chaos from noise.

9.5.5 Surrogate data

An important lesson is that a simple chaotic system can produce a time series that passes most tests for randomness. Conversely, a purely random system with a nonuniform power spectrum (correlated noise) can masquerade for chaos. Thus you are strongly advised to test your conclusions about whether a time series is chaotic by applying the tests to *surrogate data* designed to mimic the statistical properties of your data, but with the determinism removed (Theiler *et al.* 1992). Generating surrogate data is the opposite of what we did earlier in constructing chaotic systems that resemble noise.

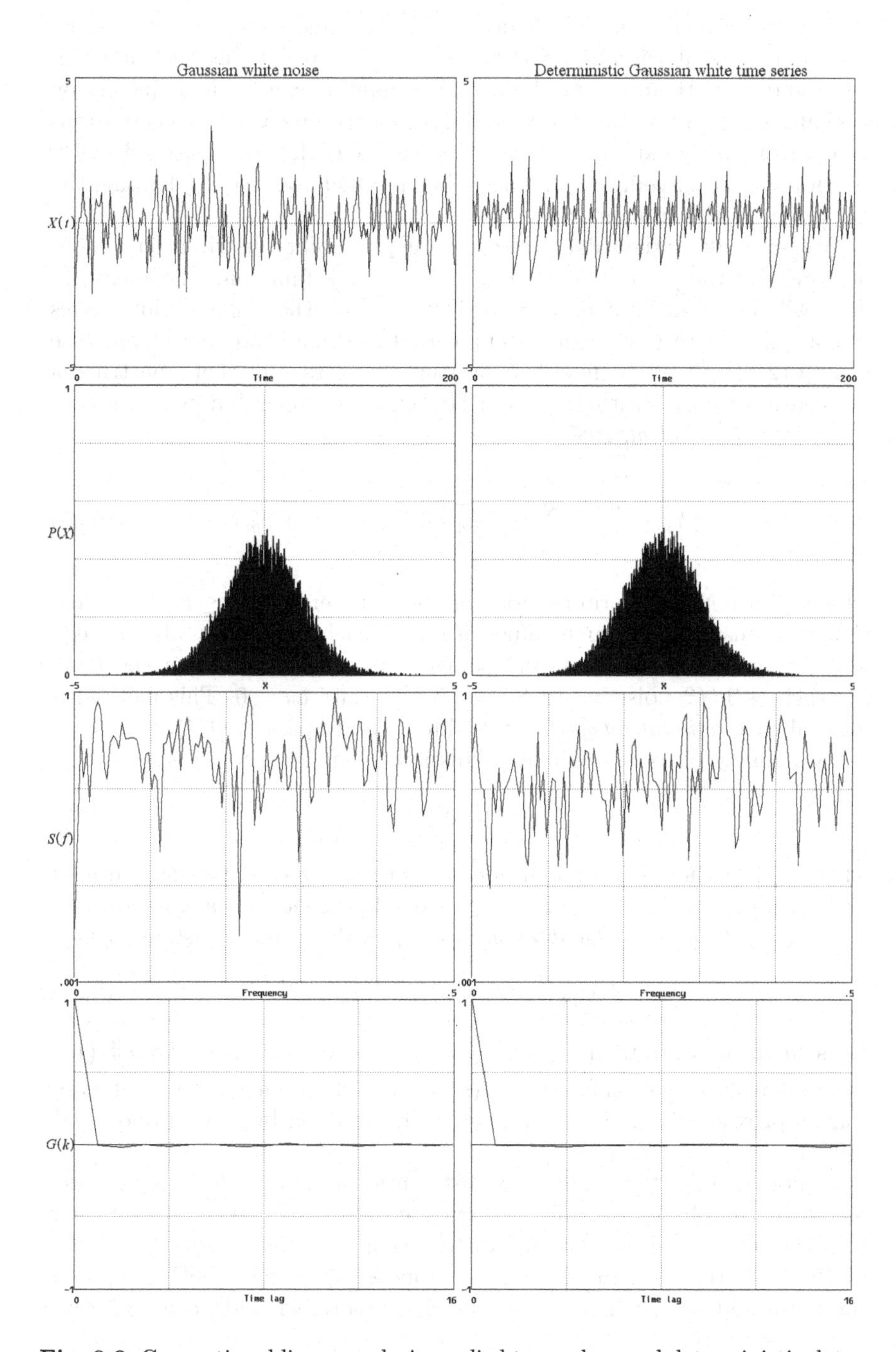

Fig. 9.9 Conventional linear analysis applied to random and deterministic data.

It is relatively easy to produce a random time series with the same probability distribution as the data. In essence you put the data values in a hat and draw them out randomly. This shuffling can be done quickly by stepping through the time series, swapping each value with one chosen randomly from anywhere in the series. On average each point is moved twice, essentially guaranteeing randomness. The method does require keeping the whole time series in memory.

While shuffling the values preserves the probability distribution, it does not preserve the power spectrum and correlation function. The surrogate data will be white and uncorrelated even when the original time series is not. To generate surrogate data with the same power spectrum, use eqn (9.12) or (9.19) to find S_m for your X_n data and then construct a surrogate series Y_n with the same Fourier amplitudes but with random phases (Osborne *et al.* 1986)

$$Y_n = \frac{a_0}{2} + \sum_{m=1}^{N/2} \sqrt{S_m} \sin 2\pi(mn/N + r_m) \tag{9.35}$$

where r_m are $N/2$ uniform random numbers chosen from $0 \le r_m < 1$. Note that you must use the same values of r_m for each n. You can also use eqn (9.35) to produce a random time series with any desired power spectrum S_m such as $1/f^\alpha$ noise where $S_m = (N/m)^\alpha$ and $a_0 = 0$. This method is also called *fractional integration*[21] (Oldham and Spainer 1974).

A simple way to approximate Gaussian $1/f$ noise ($\alpha = 1$) is

$$X_{n+1} = 0.6X_n + 0.8g_n \tag{9.36}$$

where g_n is a Gaussian random number with zero mean and unit variance such as produced by eqn (9.25) or (9.26). More generally, a discrete sampling of the *Ornstein–Uhlenbeck process* (Uhlenbeck and Ornstein 1930)

$$X_{n+1} = a_0 + a_1 X_n + a_3 g_n \tag{9.37}$$

gives surrogate data with mean $\langle X \rangle = a_0/(1 - a_1)$, variance $\sigma^2 = a_3^2/(1 - a_1^2)$, and autocorrelation function $G(k) = a_1^{|k|}$. Other one-dimensional maps that approximate a $1/f^\alpha$ power spectrum are described by Murao *et al.* (1992).

Unfortunately, the probability distribution function $P(X)$ is not preserved in the phase-randomized surrogate data, which tend to be nearly Gaussian even when the original distribution is uniform such as with the tent map. Various schemes have been proposed to preserve both the power spectrum and the probability distribution (Schreiber and Schmitz 1996).

[21]Fractional integration and differentiation were first conceived by the German mathematician and philosopher Gottfried Wilhelm Leibniz (1646–1716), who arguably invented calculus independent of Isaac Newton.

One conceptually simple method is to order by rank (smallest to largest) the values in the original data and in the phase-randomized surrogate data. Then reassign to the smallest value of the surrogate data the smallest value of the original data, and so forth through the entire record. The probability distribution functions will then be identical, but the power spectrum of the surrogate data will be only slightly altered. Better yet, find a transformation such as the logit transform in eqn (9.32) that makes your data Gaussian, generate phase-randomized surrogate data from the transformed data (which will still be Gaussian), and then inverse transform the surrogate data to recover the original distribution (Theiler *et al.* 1992). Another way to avoid these problems is to generate noise with the same autocorrelation function as the data (Schreiber 1998).

Now suppose you have developed some statistic, or worse, found one somewhere such as in the next few chapters of this book, that purports to distinguish chaos from noise. You compute the statistic for the original and surrogate data, and the values are inevitably different. You need to decide whether the difference is statistically significant. One method is to generate many surrogate data sets with different random phases. If you have the luxury of also collecting many realizations of the original data, all the better. Then compute the mean and standard error of the statistic for the data sets, and see if the value for the original data differs by more than a few standard errors from the mean. If it does, you have evidence (but not proof!) that your time series is not colored noise (Smith 1992a). This simple test may prevent you from joining the legions of others who have published false claims of chaos in experimental data.

9.5.6 Return maps

If a simple one-dimensional chaotic map produces a time series indistinguishable from Gaussian random noise by such conventional tests, what hope do you have? You must find a test that reveals the determinism that

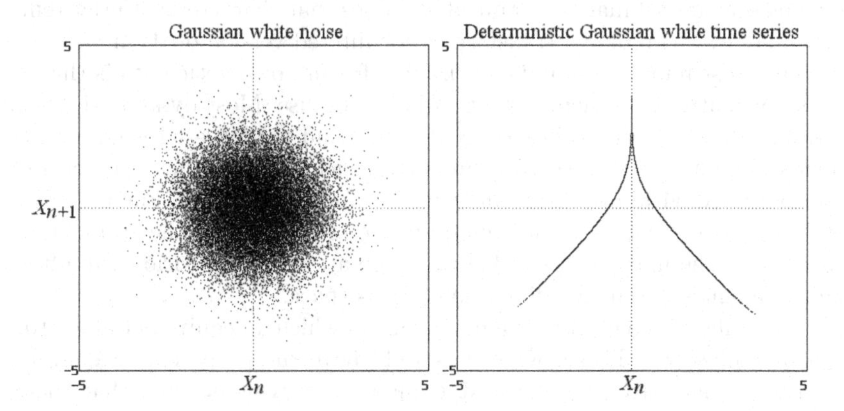

Fig. 9.10 Return maps for Gaussian random and deterministic data.

underlies chaos. The simplest such determinism would have each value depend only on its immediate predecessor. Thus a plot of X_{n+1} versus X_n should show structure as with the *return map* in the right of Fig. 9.10. By contrast the random case on the left has a return map that is a fuzzy ball. This is a particularly easy comparison for an extremely random system and a very simple deterministic system where we knew the answer in advance.

9.6 Time-delay embeddings

It is rare that a plot of each value in a time series versus its immediate predecessor reveals one-dimensional structure, even for simple chaotic systems. However, if the system is deterministic, you would expect each value to depend on some finite number of past values $X_{n+1} = f(X_n, X_{n-1}, X_{n-2}, \ldots)$ since for a rapidly sampled time series these values capture essentially the same information as a succession of time derivatives, the first few of which for small Δt are

$$\frac{df}{dt} \simeq \frac{X_n - X_{n-1}}{\Delta t} \tag{9.38}$$

$$\frac{d^2f}{dt^2} \simeq \frac{X_n - 2X_{n-1} + X_{n-2}}{(\Delta t)^2} \tag{9.39}$$

$$\frac{d^3f}{dt^3} \simeq \frac{X_n - 3X_{n-1} + 3X_{n-2} - X_{n-3}}{(\Delta t)^3} \tag{9.40}$$

$$\frac{d^4f}{dt^4} \simeq \frac{X_n - 4X_{n-1} + 6X_{n-2} - 4X_{n-3} + X_{n-4}}{(\Delta t)^4} \tag{9.41}$$

$$\frac{d^5f}{dt^5} \simeq \frac{X_n - 5X_{n-1} + 10X_{n-2} - 10X_{n-3} + 5X_{n-4} - X_{n-5}}{(\Delta t)^5} \tag{9.42}$$

We already noted in §4.8.5 that many chaotic flows can be written as a third-order ODE in a single variable (a jerk equation). Furthermore, X may be only one of many dynamical variables that characterize the system, or it may be a combined, transformed, or filtered version of them.

The most remarkable and encouraging feature of chaotic data is that it does not matter in principle what variable you use. Most dynamical properties such as the eigenvalues of fixed points and the largest Lyapunov exponent, and metric (topological) properties such as the attractor dimension are contained within (almost) any variable and its time lags. Furthermore, it is not usually necessary or even desirable to reconstruct the entire state space from the measured variable since the attractor dimension will often be much smaller than the dimension of this space.

It is sufficient to construct a new space in which an equivalent attractor can be embedded. This new space should have the properties that every point in it maps to a unique next point by the dynamics and that there is a smooth nonsingular transformation between it and the original space. Furthermore, we want this new space to be *Cartesian* (rectangular) rather than curved. The method is called *state-space reconstruction* and was proposed by Packard *et al.* (1980) and by Takens (1981). The study of how to do this embedding optimally has been called 'embedology' (Sauer *et al.* 1991). The method is rooted in the *scatter diagram* used by Yule (1927).

9.6.1 Whitney's embedding theorem

According to *Whitney's embedding theorem* (Whitney 1936), an arbitrary D-dimensional curved space can be mapped into a Cartesian space of $2D+1$ dimensions without any self intersections, hence satisfying the uniqueness requirement. If you map it into a space of D dimensions, there will in general be overlap regions of nonzero measure and hence having dimension D. If you map it into a space of $D+1$ dimensions, the overlap will lie in a subspace of dimension $D-1$, in $D+2$ it will be of dimension $D-2$, and so on until in $2D$ dimensions, the overlap will be point-like (zero-dimensional). Only in dimension $2D+1$ do the overlaps disappear. As a simple example, consider that a curved line (dimension 1) in a surface (dimension 2) almost certainly intersects itself somewhere, but almost certainly does not in a volume (dimension 3).

Sauer *et al.* (1991) generalized Whitney's theorem to fractal attractors with dimension D_F and showed that the embedding space need only have a dimension greater than $2D_F$. The proof used the capacity dimension for D_F (see §12.2), but other dimensions such as D_{KY} (see §5.9) probably suffice. Thus the penalty you pay in choosing your analysis variable arbitrarily is that you may need to embed it in a space of higher dimension than the original state space. Although it is possible for a *fractal* to be embedded in another fractal, we consider only integer embeddings.

9.6.2 Takens' delay embedding theorem

Takens (1981) proved that the time-lagged variables constitute an adequate embedding provided the measured variable is smooth and couples to all the other variables, and the number of time lags is at least $2D+1$. *Takens' delay embedding theorem* was generalized by Sauer *et al.* (1991) who showed that the dimension need only exceed $2D_F$ if there are no orbits with period Δt or $2\Delta t$ and only a finite number of orbits with higher periods, where Δt is the time lag between data samples. They call this result the *fractal delay embedding prevalence theorem*.

For many purposes, such as determining the dimension of an attractor, overlaps of zero measure are of no consequence, and the number of time lags need only exceed D_F (Eckmann and Ruelle 1985, Sauer and Yorke 1993). Of course, if you already knew D_F, you would not be calculating it! Methods for determining the optimal embedding as well as the required time delay Δt from the data will be discussed in the next chapter. Embeddings larger than two are hard to graph and larger than three are hard to visualize, but the computer has no trouble embedding and calculating in any dimension.

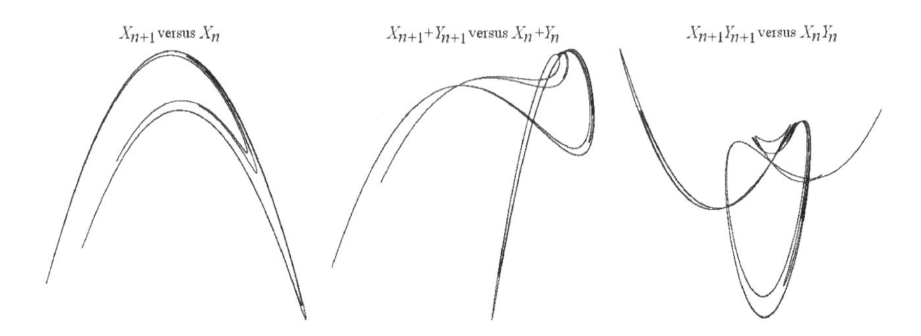

Fig. 9.11 Hénon map with various observer functions.

9.6.3 Hénon map example

The previous discussion was somewhat abstract, and so you might want an example. Consider the Hénon map in §5.2.2 as a two-dimensional chaotic dynamical system from which a time series of the X variable has been obtained. If you embed the data in a two-dimensional space as shown at the left of Fig. 9.11, the result is identical to Fig. 5.6, since in this case the Y variable is simply the previous value of X.

However, suppose the time series (*observer function* or *measurement function*) is the sum of the X and Y variables $Z_n = X_n + Y_n$, and you plot Z_{n+1} versus Z_n as in the center of Fig. 9.11. Such a measurement might come from an instrument with a slow response that averages over two time steps. The attractor is still evident with a similar structure but twisted and distorted with apparent overlaps. To untangle the overlaps, you must embed the attractor in a three-dimensional space using also Z_{n-2}.

The observer function can be more complicated and related *nonlinearly* to the variables. For example, the right of Fig. 9.11 shows the return map for the product $Z_n = X_n Y_n$. The distortion is more severe and there are more overlaps, but the structure of the attractor is preserved, and its dimension in this two-dimensional embedding space is plausibly unchanged.

9.7 Summary of important dimensions

This chapter has described many different dimensions, and thus it is useful to summarize them.

1. The *configuration space* (or *state space*) is the space of the dynamical variables, or, equivalently, the number of initial conditions that must be specified. For the Hénon map, this is a two-dimensional space with coordinates X and Y.
2. The *solution manifold* is the space of integer dimension in which the solution 'lives.' This space may be the same as the configuration space, but it might have a lower dimension, for example a Hamiltonian system where the orbit or trajectory lies on a surface of constant energy in the phase space. Another example is a limit cycle that lies on a

two-dimensional solution manifold even when it comes from a three-dimensional system of equations (see §4.6.2).

3. The *attractor dimension* is the dimension of the space filled by the solution after the initial transient has decayed. This dimension will usually be a fraction if the attractor is strange. The Kaplan–Yorke dimension in §5.9 is one estimate of this dimension, but there are infinitely many others (see §13.2). The attractor dimension for the Hénon map is about 1.26.
4. The *dimension of the observable* is the number of simultaneous quantities that are recorded. A scalar time series such as the temperature data described earlier is one-dimensional. If you had simultaneously recorded the humidity, the observable would then be a sequence of vectors with two components, each corresponding to a point in the two-dimensional observable space. Or you might have a record of wind speed and direction, which would also be two-dimensional (or even three-dimensional if the wind has a vertical component of velocity).
5. The *time-delay reconstructed state space* is the space obtained by plotting each value of the time series against some number of previous values. The dimension of this space can be chosen arbitrarily, with a value as large as the number of points in the time series N, in which case the whole data set would be represented by a single point in an N-dimensional space.
6. The *time-delay embedding space* is a reconstructed state space chosen with the minimum dimension for which the important dynamical and topological properties are preserved. For most purposes this dimension need only be the next integer larger than $2D_F$, where D_F is the attractor dimension. An embedding dimension of three suffices for the Hénon map with $D_F \simeq 1.26$. In many cases, the embedding need only be the next integer larger than D_F, as is the case for the Hénon map when the observable is X (or Y).
7. The *dimension of the time series graph* (the trace) is the fractal dimension of the $x(t)$ curve. If x is a scalar observable, then this dimension will lie between 1.0 for a very smooth curve and 2.0 for a very erratic one.

9.8 Exercises

Exercise 9.1 Write a formula for $x(t)$ with t in hours that would approximate the variation of temperature at your location including both the daily and seasonal variation.

Exercise 9.2 Find the mean and standard deviation of the time series 1, 2, 4, 5.

Exercise 9.3 Calculate the average deviation, skewness, and kurtosis of the time series 1, 2, 4, 5.

Exercise 9.4 Calculate your *average* winnings at the St Petersburg game after N turns, and show that this number approaches infinity as $N \to \infty$ but that your *median* winning is \$1.

Exercise 9.5 Use the linear least-squares fit to detrend the time series 1, 2, 4, 5.

Exercise 9.6 Use log first differences to detrend the time series 1, 2, 4, 5.

Exercise 9.7 Calculate the Fourier series representation of the time series 1, 2, 4, 5.

Exercise 9.8 Suppose you have a time series that comes from the function $x = \sin 2\pi ft$ sampled at a frequency $f_s = f_c + \Delta f = 1/\Delta t$, where f_c is the Nyquist frequency and $0 < \Delta f \leq f_c$. Show that the signal is spuriously aliased to a frequency $f_c - \Delta f$.

Exercise 9.9 Calculate the standard deviation of the actual temperature in Table 9.2, and compare the root-mean-square prediction error with this value. What do you conclude about the efficacy of this prediction method?

Exercise 9.10 Show that subtracting from each data point in a time series a value $f(n) = A + Bn$, where $A = (NX_1 - X_N)/(N - 1)$ and $B = (X_N - X_1)/(N - 1)$, makes the data consistent with one whose period is N.

Exercise 9.11 Show that integrating $1/f^\alpha$ noise from $f = 0$ to $f = \infty$ gives an infinite power.

Exercise 9.12 Show that the formula for Gaussian white noise in eqn (9.24) has the property that $\int_{-\infty}^{\infty} P(X)dX = 1$.

Exercise 9.13 Show that the formula for Gaussian white noise in eqn (9.24) has a mean of zero and a variance of 1.0.

Exercise 9.14 Show that the full width at half maximum for the normal distribution in eqn (9.24) is 2.354σ, where σ is the standard deviation.

Exercise 9.15 Use the Frobenius–Perron relation in eqn (2.12) to derive the map in eqn (9.28) that has a Gaussian probability distribution.

Exercise 9.16 Calculate values for $f'(X)$ and $f''(X)$ for the map in eqn (9.28) at $X = X_A \simeq 0.674489753$, where $f(X) = 0$.

Exercise 9.17 Describe the numerical algorithm required to determine $f(X)$ for the map in eqn (9.28) by the Euler method.

Exercise 9.18 Calculate X_A and $f'(X_A)$ for the map in eqn (9.31), and show that these values approximately satisfy the Frobenius–Perron relation in eqn (9.27) at this value of X_A.

Exercise 9.19 Show that the map in eqn (9.31) has a fixed point at $X^* = 0.3848252\ldots$, and calculate its stability.

Exercise 9.20 Show that the map in eqn (9.31) has a basin of attraction with $-16 < X_0 < \infty$.

Exercise 9.21 Using the probability distribution function for the logistic map at $A = 4$ in eqn (2.11), show that the logit transform in eqn (9.32) of the logistic map has a standard deviation of π.

Exercise 9.22 Show that eqn (9.33) is the inverse of the logit transform and that it is bounded in the range $0 < r_n < 1$.

Exercise 9.23 Show that the one-dimensional map

$$X_{n+1} = -\frac{1}{A} \ln(1 - 2e^{-AX_n}) \tag{9.43}$$

has a probability distribution function $P(X) = Ae^{-AX}$.

Exercise 9.24 Show that eqns (9.38)–(9.42) converge to the first five time derivatives in the limit $\Delta t \to 0$.

Exercise 9.25 Calculate the minimum embedding dimension that avoids overlap for an arbitrary scalar time series collected from each of the chaotic systems in Appendix A.

9.9 Computer project: Autocorrelation function

In this project you will begin to develop a set of tools for the analysis of time-series data. One of the first tests to perform on an experimental time-series record is to calculate the autocorrelation function, from which an autocorrelation time can be estimated (see §9.6). Knowledge of the correlation function is often sufficient to distinguish quasiperiodicity from chaos, and the correlation time is an important variable in the application of more advanced techniques. The correlation function is the Fourier transform of the power spectrum.

1. Numerically solve the Lorenz equations in eqns (4.32)–(4.34) with the usual parameters of $\sigma = 10, r = 28$, and $b = 8/3$. Allow enough time to elapse for the trajectory to reach the attractor and then generate and plot a series of 1000 data points representing $x(t)$ at equally spaced time intervals of $\Delta t = 0.05$. Call these values X_n for $n = 1$ to N. Hereafter you will assume that this time series is from some experiment sampled 20 times per second and you will pretend that you do not know anything else about the dynamics of the system that produced the time series.
2. Calculate and plot the autocorrelation function $G(k)$ versus k for your time series, where $G(k)$ is given approximately by

$$G(k) \simeq \frac{\sum_{n=1}^{N-k}(X_n - \langle X \rangle)(X_{n+k} - \langle X \rangle)}{\sum_{n=1}^{N-k}(X_n - \langle X \rangle)^2} \tag{9.44}$$

and $\langle X \rangle$ is the mean (average value) of X_n

$$\langle X \rangle = \frac{1}{N} \sum_{n=1}^{N} X_n \tag{9.45}$$

3. Estimate the value of k at which $G(k)$ falls to 1/e of its value at $k = 0$, and from this value find the autocorrelation time in units of seconds.
4. (optional) Plot the autocorrelation function for a time series from each of these cases:
   (a) a sine wave with many cycles
   (b) a uniform random-number generator
   (c) successive iterates of the logistic map with $A = 4$
   (d) the superposition of two incommensurate sine waves
   (e) a collection of Gaussian random numbers with zero mean

10

Nonlinear prediction and noise reduction

One of the most important applications of time-series analysis is *prediction* (or *forecasting*). This task has been highly developed over many years assuming the dynamics are governed by a linear model plus random noise. In this chapter we briefly review some linear methods and then describe predictors designed to exploit the possibility that the aperiodic fluctuations may be low-dimensional chaos. If the system is chaotic, then the sensitive dependence on initial conditions given by the largest Lyapunov exponent precludes long-term prediction, but it promises improved short-term prediction. Conversely, evidence of predictability indicates underlying determinism, and the average growth of the prediction error is a measure of the sensitivity to initial conditions.

Closely related to prediction is *noise reduction*. Any prediction method must separate the deterministic and random components of the time series, since only the deterministic part is predictable. If the deterministic part is the signal and the random part is the noise, then a new time series can be constructed with the noise reduced or removed. In practice, the separation is never perfect, and removing the noise always distorts the signal to some extent.

In some cases, prediction entails developing a global dynamical model for the data, which may illuminate the underlying mechanisms. A deterministic model is necessarily noise-free and provides an unlimited amount of data when started from an arbitrary point on the attractor. Sampling the output of a model for estimating parameters and errors is called *bootstrapping*.[1] A credible model must have a solution that resembles a skeleton of the real data, but this test is not definitive since other models may agree equally well, or better. Models, like physical theories, cannot be proved, but only falsified.

[1]The term 'bootstrapping' comes from the expression 'pulling yourself up by your own bootstraps,' which suggests getting more out of something than you put in. Similarly, 'booting a computer' launches the entire operating system with a few simple instructions.

10.1 Linear predictors

When we speak of *linear predictors,* we do not mean that the time series X_n is a linear function of n, but rather that the deterministic part can be written as a linear combination of m past values. For smooth, highly-correlated, noise-free data, you can assume X_n is a polynomial function of n and use the corresponding linear extrapolation, the first few of which are

$$X_{n+1} \simeq 2X_n - X_{n-1} \tag{10.1}$$

$$X_{n+1} \simeq 3X_n - 3X_{n-1} + X_{n-2} \tag{10.2}$$

$$X_{n+1} \simeq 4X_n - 6X_{n-1} + 4X_{n-2} - X_{n-3} \tag{10.3}$$

$$X_{n+1} \simeq 5X_n - 10X_{n-1} + 10X_{n-2} - 5X_{n-3} + X_{n-4} \tag{10.4}$$

$$X_{n+1} \simeq 6X_n - 15X_{n-1} + 20X_{n-2} - 15X_{n-3} + 6X_{n-4} - X_{n-5} \tag{10.5}$$

These formulas show one of the many uses of *binomial coefficients*. Expect disastrous results if you extrapolate more than one or two time steps into the future, especially if you use a high-order polynomial with noisy data. More generally

$$X_{n+1} = a_0 + \sum_{i=1}^{m} a_i X_{n-m+i} + r_n \tag{10.6}$$

where r_n is the random noise, assumed to have zero mean. The quantity m is the time-delay embedding dimension, and the coefficients a_i are determined by a best fit to the whole time series.

10.1.1 Autoregressive models

Equation (10.6) with assumed uncorrelated (white) noise is an *autoregressive* (AR) model of order m, written AR(m), first introduced by Yule (1927) in the study of sunspots. If the noise is correlated, you can use an *autoregressive moving average* (ARMA) model of order m, l, written ARMA(m, l), in which r_n is replaced by a linear weighted average of past values of the noise

$$X_{n+1} = a_0 + \sum_{i=1}^{m} a_i X_{n-m+i} + \sum_{j=1}^{l} b_j r_{n-l+j} \tag{10.7}$$

The case of ARMA($0, l$) is called a *moving-average* (MA) model of order l, written MA(l). These are just a few of a wide class of autoregressive models in common use (Tong 1990, Box *et al.* 1994, Shumway and Stoffer 2000, Chatfield 2000).

The solutions with $r_n = 0$ can be a superposition of growing or decaying exponentials or oscillations, but not chaotic since the model is linear. In the simplest case, the system is stationary, and the time series is a superposition of noisy sine waves whose amplitudes can be determined by the Fourier methods in §9.4.4. An example of predicting the temperature by this method was shown.

Traditionally, the task is to find values of the coefficients a_i that minimize the mean-square one-step prediction error over the entire data set (Chatfield 1988). The coefficient a_0 can be made zero by subtracting the mean from the data. The other coefficients are obtained by solving the m-dimensional linear system of equations

$$\sum_{i=1}^{m} C_{ij} a_i = \sum_{n=m}^{N-1} X_{n+1} X_{n-m+j} \tag{10.8}$$

for $j = 1$ to m, where C_{ij} is an element of the $m \times m$ *autocovariance matrix*

$$C_{ij} = \sum_{n=m}^{N-1} X_{n-m+i} X_{n-m+j} \tag{10.9}$$

It is instructive to exhibit explicitly the solution for $m = 1$

$$a_1 = \frac{\sum_{n=1}^{N-1} X_{n+1} X_n}{\sum_{n=1}^{N-1} X_n^2} \tag{10.10}$$

and for $m = 2$

$$a_1 = \frac{\sum X_n X_{n-1} \sum X_{n+1} X_n - \sum X_n^2 \sum X_{n+1} X_{n-1}}{(\sum X_n X_{n-1})^2 - \sum X_n^2 \sum X_{n-1}^2} \tag{10.11}$$

$$a_2 = \frac{\sum X_n X_{n-1} \sum X_{n+1} X_{n-1} - \sum X_{n-1}^2 \sum X_{n+1} X_n}{(\sum X_n X_{n-1})^2 - \sum X_n^2 \sum X_{n-1}^2} \tag{10.12}$$

where the unlabeled sums are from $n = 2$ to $N - 1$. For $m = 1$, the solutions can only grow or decay exponentially, but for $m > 1$ there can be oscillations whose amplitude grows or decays.

In cases where the process is not stationary or the noise has a $1/f^\alpha$ power spectrum with $\alpha > 1$, it is reasonable to fit not X_n, but $X_n - X_{n-1}$ (the *first difference*). Generically, such models are called autoregressive *integrated* moving average (ARIMA) models, since the assumption is that the observed variable is an integrated version of a stationary signal with white noise. You can take higher-order differences if necessary to obtain a stationary time series with nearly uncorrelated noise.

10.1.2 Pure sine wave example

As a simple example, apply the AR model to data from a pure sine wave, sampled once per radian, $X_n = \sin n$. After 10^8 iterations, the fit with $m = 2$ is $X_{n+1} \simeq 1.080605 X_n - X_{n-1}$, and the one-step prediction error is accurate to machine precision. The AR fit can be applied iteratively to predict many time steps ahead, and the result in Fig. 10.1 shows essentially perfect agreement within the precision of the plot for at least 15 iterations. If the data had been a superposition of two incommensurate sine waves, then the trajectory would lie on a 2-torus, and an embedding dimension of $m = 3$ would be required for a good fit.

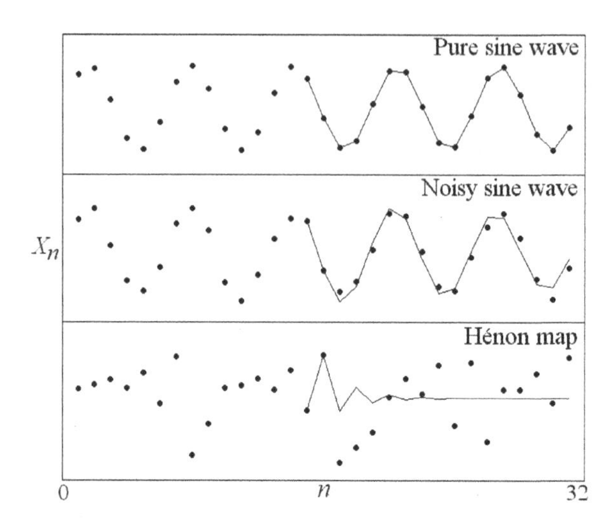

Fig. 10.1 AR model predictions for various data sets.

10.1.3 Noisy sine wave example

A more stringent test showing the usefulness of the method uses data from a noisy sine wave, $X_n = \sin n + r_n$, where r_n is an uncorrelated Gaussian random number with mean zero and standard deviation 0.1. In this and the following examples, the noise is superimposed (*additive noise*) on an otherwise deterministic signal. Such noise is an example of *measurement noise* (or *observational noise*), in contrast to *dynamical noise* (or *system noise*), which enters in a more fundamental and problematic way such as random fluctuations of the parameters. With dynamical noise, the time series is not a simple superposition of signal plus noise, but rather a signal modulated by the noise.

After 10^8 iterations, the fit with $m = 2$ is $\bar{X}_{n+1} \simeq 1.03671X_n - 0.95714X_{n-1}$. We will use $\bar{X}$ to denote the predicted value of X where it is necessary to distinguish the predicted (or noise-reduced) value from the actual value. The one-step rms prediction error normalized to the standard deviation of the data ($\sigma \simeq 1/\sqrt{2} = 0.70710678\ldots$) is $\sqrt{\langle(\bar{X}_n - X_n)^2\rangle}/\sigma \simeq 0.2457$, which is about the best you can expect for this noise level.

The multi-step prediction in Fig. 10.1 becomes progressively worse as you try to predict further into the future. In fact, the predicted oscillation is damped and will eventually vanish. It is easy to produce an undamped oscillation by setting a_1 to exactly -1.0 (rather than -0.95714), but the damping can be a virtue. Eventually the predicted oscillation will get out of phase with the actual oscillation, and the damping gracefully forces the prediction to the average rather than to a steady oscillation unrelated to the data. These are examples for which autoregressive models are expected to excel.

10.1.4 Hénon map example

Suppose instead the time series came from a noise-free nonlinear chaotic system such as the Hénon map $X_{n+1} = 1 - 1.4X_n^2 + 0.3X_{n-1}$. After 10^8 iterations, the AR fit with $m = 2$ is $\bar{X}_{n+1} \simeq 0.25692 - 0.26173X_n + 0.16583X_{n-1}$. The one-step rms prediction error normalized to the standard deviation of the data is about 0.9364, which is not much better than guessing the mean of the data. Embedding in higher dimensions ($m > 2$) does not give much improvement. Not surprisingly, the iterated multi-step prediction shown in Fig. 10.1 rapidly decays to the mean of the data. The lesson is that all linear predictors must fail for chaotic data even in the absence of noise, since the linear predictor cannot produce chaos.

10.2 State-space prediction

By now you may have despaired at ever finding a method that detects chaos in experimental data and allows prediction. An obvious extension to the AR model would include nonlinear functions such as $X_{n+1} = a_0 + a_1 X_n + a_2 X_n^2 + \cdots$ (Volterra 1959), which gives perfect predictions for noise-free polynomial maps such as the Hénon map. However, the number of coefficients for a polynomial with m variables and order l is $(m+l)!/m!l!$, which is a very large number.[2] Even worse, there are infinitely many possible nonlinear functions that could be used instead of polynomials, and optimization of the coefficients is difficult and computationally intensive. One nonlinear class is the *threshold autoregressive* (TAR) model (Tong 1983, 1990) that uses different linear AR models in different regions of state space. We will return to nonlinear modeling when we discuss neural network predictors in §10.7.

A different approach abandons model equations altogether and looks for points in the time-delay state space that are close to the present point to see how they evolved (Sugihara and May 1990). Equivalently, look for a sequence of points in the time series that most closely resembles the current sequence. This method was proposed by Lorenz (1969) for predicting the weather, but it fails for high-dimensional systems because of the paucity of close analogs. One estimate (van den Dool 1994) is that an adequate analog for the atmosphere occurs only once in 10^{30} years, considerably longer than the age of the Universe (about 10^{10} years). For a deterministic system with sufficient analogs, predictability can be good, however. A stochastic variant of the method called *random analog prediction* (RAP) selects a neighbor randomly with a probability that decreases with its state-space distance (Paparella *et al.* 1997).

10.2.1 Hénon map example

To see how state-space prediction works, take $N = 6000$ iterates of X_n for the Hénon map and then another 18 saved separately for comparison.

[2]The quantity $m!$ is the *factorial* of m given by $m! = 2 \times 3 \times 4 \times \cdots \times m$.

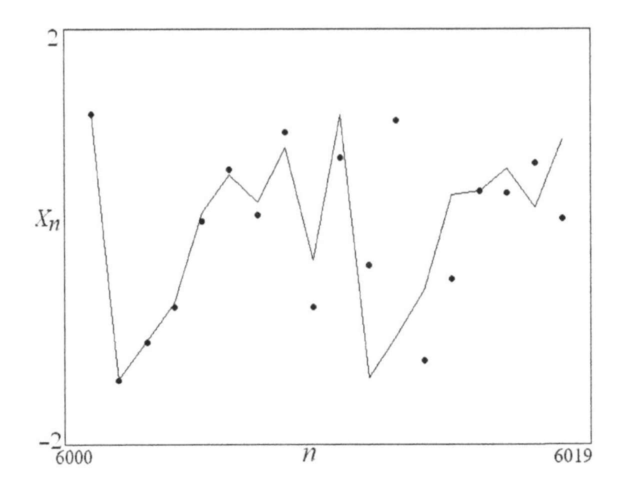

Fig. 10.2 Prediction for the Hénon map using nearest neighbors in state space.

Always use *out-of-sample* rather than *in-sample* data to validate your prediction; that is, do not test your predictor with the same data used to formulate it. The results will differ for nonstationary or *overfitted* data in which the model has fit random fluctuations. Using an embedding of $m = 2$, look at the last two points X_{N-1} and X_N, and search the time series for the points X_{j-1} and X_j with the smallest $(X_{N-1} - X_{j-1})^2 + (X_N - X_j)^2$. The *zero-order prediction* for X_{N+k} will be X_{j+k}, as in Fig. 10.2. The first few points agree well with the data, but the prediction eventually fails as it must for a system with sensitive dependence on initial conditions. Furthermore, the predicted dynamics will eventually settle into a periodic orbit, albeit usually with a large period.

The previous example is only one realization of a prediction for the Hénon map. To test its quality, repeat many times with different Hénon data sets, and plot the rms prediction error versus time. Typical results for 10^4 cases are shown in Fig. 10.3. The initial error is well below 1% and grows approximately as $e^{0.50n}$, which only slightly exceeds the rate expected from the largest Lyapunov exponent of $\lambda_1 \simeq 0.419$, which measures the average growth of a small error in the initial condition. This method has been used to estimate the sum of the positive Lyapunov exponents (Wales 1991), but it is not usually recommended because it may be sensitive to the chosen model.

10.2.2 State-space averaging

The prediction can be improved by including a cluster of nearby points rather than just the nearest one (Farmer and Sidorowich 1987) and either taking their average (zero-order approximation) or fitting them to a locally linear model (first-order approximation). For a short, noise-free data set, you only need enough points to do a proper linear interpolation. The num-

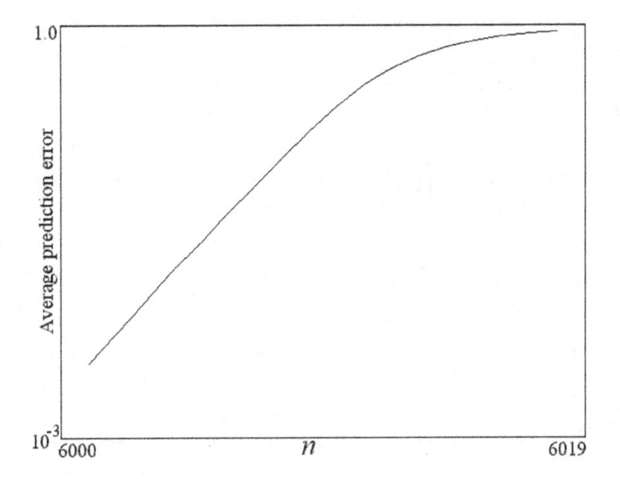

Fig. 10.3 Average prediction error for 10^4 Hénon maps.

ber required for a linear interpolation is $m+1$, since two points are needed to define a line ($m = 1$), three points are needed to define a plane ($m = 2$), and so forth. The optimal dimension for the local approximation may be smaller than the optimal global embedding dimension, may vary with location (Smith 1994), and need not be larger than the smallest integer greater than or equal to the dimension of the attractor. Higher-order polynomial approximations can also be used (Brown *et al.* 1991). For longer, noisy data, use more points so as to average over the noise fluctuations. If you know the noise level, average the prediction for all points within a cluster whose size is of the order of the rms noise level.

One way to average over a cluster of neighbors (Hegger *et al.* 1999) is to use every point in the data set, but weighted by e^{-d^2/σ_N^2}, where d is the distance of each point in the m-dimensional embedding from the last known value X_N, and σ_N is the standard deviation of the noise. The prediction is

$$X_{N+k} = \frac{\sum_{n=m}^{N-k} w_n X_{n+k}}{\sum_{n=m}^{N-k} w_n} \tag{10.13}$$

where

$$w_n = \exp\left[-\frac{1}{\sigma_N^2}\sum_{j=0}^{m-1}(X_{N-j} - X_{n-j})^2\right] \tag{10.14}$$

Other *weighting functions* (or *kernels*[3]) that fall to zero for large separations such as $w_n = 1/d^\alpha$ give similar results. Usually you do not know σ_N, and the results may be sensitive to the value chosen. The parameters m and σ_N control the quality of the fit and have to be optimized for each data set, such as by adjusting them for the best in-sample fit.

[3]A 'kernel' is a weighting function that depends only on the distance to the point.

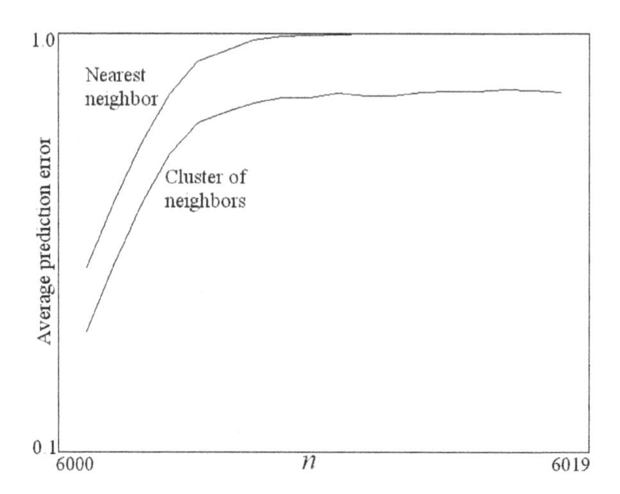

Fig. 10.4 Average prediction error for 10^4 noisy Hénon maps.

With this method, the prediction is always bounded even when predicting many time steps ahead, although the predicted orbit for chaotic data may be periodic. Alternately, use the nearest points in state space to predict the *change* ΔX rather than X. The near-term prediction may be improved, but there is then no guarantee that the orbit is bounded.

10.2.3 Noisy Hénon map example

As an example, add Gaussian white measurement noise with $\sigma_N = 0.1$ to the X_n data from the Hénon map. Figure 10.4 shows the prediction error for 18 out-of-sample time steps in 10^4 realizations of the map with different initial conditions for $N = 6000$ and $m = 2$. The figure shows the result using only the nearest neighbor and using a weighted average of all neighbors as described above. The improvement is evident but not enormous. The prediction based on a cluster of neighbors fails more gracefully by going to the average of the data after many iterations. A better method would be to fit the data in the vicinity of each point to a local linear model using least squares (Kantz and Schreiber 1997).

10.2.4 Temperature data example

A more stringent test of state-space averaging using the seasonally detrended temperature data from Fig. 9.5 with $m = 3$ and $\sigma_N = 1°\text{F}$ (round-off error) gives the predictions in Table 10.1. The rms error is about 5.5°F compared with 9.8°F for the prediction from Fourier analysis in Table 9.2. The improvement is somewhat fortuitous and depends on the choice of m and σ_N, but the state-space prediction is usually better.

Table 10.1 Temperature predictions (°F) from state-space averaging.

Date	Predicted	Actual	Error
Jan 1, 2000	33	35	−2
Jan 2, 2000	32	42	−10
Jan 3, 2000	28	28	0
Jan 4, 2000	28	25	+3
Jan 5, 2000	23	16	+7
Jan 6, 2000	26	31	−5
Jan 7, 2000	23	18	+5

10.3 Noise reduction

Closely related to prediction is noise reduction. Any method that can predict the future can predict the present at least as well, and usually much better if you also know the future.

10.3.1 Linear methods

Linear methods in the time or frequency domain are not usually recommended for reducing noise in chaotic data, because, like prediction, they tend to work poorly and can significantly distort the deterministic signal. Nevertheless, for completeness and since some processes are in fact linear, we mention here some such methods.

For oversampled data (data sampled with an unnecessarily small time step) contaminated by high-frequency noise or round-off error, try *smoothing*, which is essentially *low-pass filtering*. The simplest implementation replaces each data point with the *moving average* of its temporal neighbors

$$\bar{X}_n = \frac{1}{2m+1} \sum_{j=-m}^{m} X_{n+j} \tag{10.15}$$

where $\bar{X}_n$ is the noise-reduced approximation to X_n. The parameter m determines the smoothing, with $m = 0$ representing none at all and $m = N/2 - 1$ giving the mean of the data. Note that you lose m points from the beginning and m points from the end unless you do something special there such as reduce the value of m or use an asymmetric window. You can also use a weighted average such as $\bar{X}_n = (X_{n-1} + 2X_n + X_{n+1})/4$ and apply the formula iteratively for increased smoothing.

Rather than *averaging* the $2m + 1$ nearest points, you can take their *median* (Press *et al.* 1992). This *windowed median* is invariant to transformations of both X and time and performs better when the noise has broad tails.

If your time series is a sum of a few discrete frequencies contaminated by broadband noise, the Fourier methods in §9.4.4 work well. Identify the dominant frequencies, and construct a new time series that is a linear superposition of them. More generally, if you plot the power spectrum of your data, you may be able to discern a signal $S(f)$ riding on top of a broad background $N(f)$ that you suspect is noise. You can then construct a function

$$\phi(f) = \frac{S(f)}{S(f) + N(f)} \tag{10.16}$$

that approaches unity at frequencies where the signal dominates and approaches zero where the noise dominates. Multiply the Fourier transform of your time series by this function and inverse Fourier transform to get a noise-reduced time series. This *optimal* (or *Wiener*) *filter* works even with poor estimates of $S(f)$ and $N(f)$. To test whether there is still a signal present in the noise, calculate the *residuals* $\bar{X}_n - X_n$, and see whether their autocorrelation function in eqn (9.18) is significantly different from zero for $k \neq 0$.

10.3.2 State-space averaging

As with prediction, linear noise-reduction methods usually fail with chaotic data since the power spectrum of the signal often resembles the power spectrum of noise. A better approach is to use *state-space averaging.* A method that does not work is to replace each data point with a prediction from eqn (10.13). The reason is that the prediction is based on noisy data, and the sensitive dependence on initial conditions guarantees that on average it can be no better than the actual data. However, since you have access to *future* data, you can backwards predict ('postdict' or 'hindcast'). The best strategy is to do both and average segments of the data in which a window of $\pm m$ points on each side of X_n is close to it in state space

$$\bar{X}_n = \frac{\sum_{k=m}^{N-m} w_n(k) X_k}{\sum_{k=m}^{N-m} w_n(k)} \tag{10.17}$$

where

$$w_n(k) = \exp\left[-\frac{1}{\sigma_N^2} \sum_{j=-m}^{m} (X_{k-j} - X_{n-j})^2\right] \tag{10.18}$$

is a function that weights the nearer points more heavily. This method is a variant of one proposed by Schreiber (1993).

10.3.3 Noisy Hénon map example

A standard example is the Hénon map to which Gaussian white noise with a standard deviation of 0.1 is added, as in Fig. 10.5. With a time series of $N = 6000$, eqn (10.17) with $m = 3$ and $\sigma_N = 0.3$ gives the result at the right of Fig. 10.5. About half the noise is removed. Best results

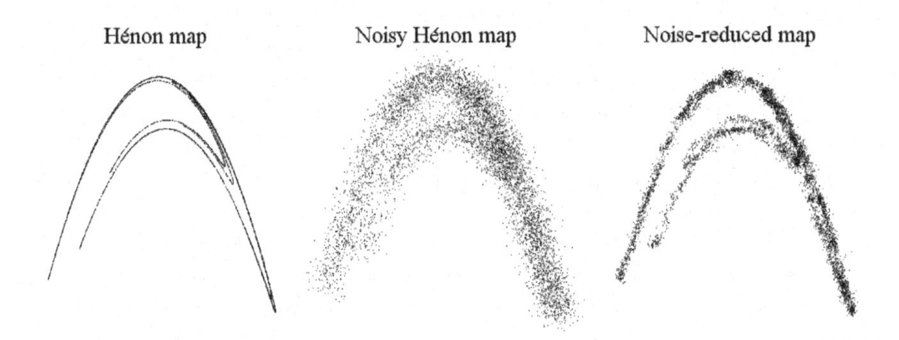

Fig. 10.5 State-space averaging applied to noisy Hénon map.

require m somewhat larger than the usual optimal embedding and σ_N several times the actual noise level. The procedure can be applied iteratively with successively smaller values of σ_N to remove additional noise. Noise reduction is never total and always comes at the expense of some distortion of the signal.

10.4 Lyapunov exponents from experimental data

Predictability of a chaotic system is ultimately limited by the sensitive dependence on initial conditions, and thus it is useful to calculate the largest Lyapunov exponent. If you can model the data with equations, you can use the methods in Chapter 5 to find the Lyapunov exponents (Gençay and Dechert 1992). Otherwise, you have a much harder task, especially if you want the whole spectrum of exponents.

One method (Wolf *et al.* 1985, Kruel *et al.* 1993) is a simple extension of the one described in §5.6, except instead of perturbing the orbit, you search the time series for nearby points in state space whose orbits you follow for several time steps or until they get too far apart, whereupon you choose other nearby points displaced in the same direction. The Wolf algorithm assumes but does not verify exponential divergence, and hence it cannot distinguish chaos from noise.

An alternate method is to advance through the time series looking for the nearest point X_l to each X_n in an m-dimensional time-delay embedding. Then average the logarithmic rate of separation of these two points for the next k time steps (Rosenstein *et al.* 1993)

$$L_k = \frac{1}{2(N-k-m+1)} \sum_{n=m}^{N-k} \log \sum_{j=0}^{m-1} (X_{l-j+k} - X_{n-j+k})^2 \qquad (10.19)$$

The largest Lyapunov exponent is given by $\lambda_1 = dL_k/dk$ at intermediate values of k. Small values of k should be ignored since the points have not yet aligned along the direction of maximum expansion and there may be

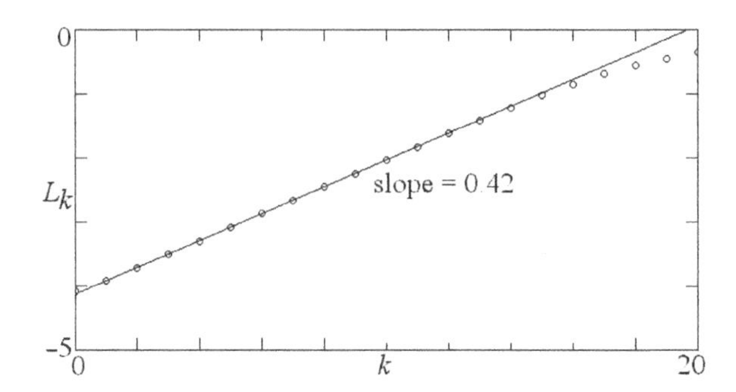

Fig. 10.6 Largest Lyapunov exponent from a time series of the Hénon map.

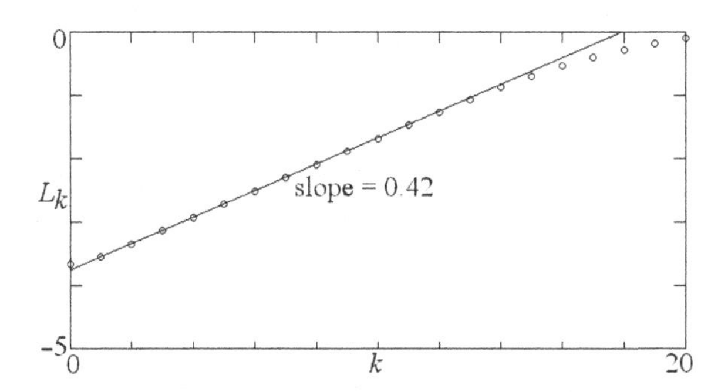

Fig. 10.7 Largest Lyapunov exponent from a time series of the Hénon map for an observer function $X + Y$ and embedding $m = 3$.

noise and round-off errors in calculating the separation. Large values of k should be ignored since the separation eventually approaches the size of the attractor and folding causes deviation from an exponential. In principle, you can distinguish chaos from $1/f^\alpha$ noise since the noise should have $e^{L_k} \propto k^H$, where $H = (\alpha - 1)/2$ is the Hurst exponent (see §9.4.6). Thus dL_k/dk is H/k, and the slope of the function $L(k)$ not constant. The Lyapunov exponent calculated this way in the limit $k \to 0$ is infinite for noise. A substantial region with $L_k \propto k$ (constant dL_k/dk) is evidence of chaos.

10.4.1 Hénon map example

The method is illustrated in Fig. 10.6 which shows L_k versus k for X_n data from the Hénon map with $N = 3.2 \times 10^4$ and $m = 2$. The slope of the curve ($\lambda_1 = 0.42$) agrees well with the largest Lyapunov exponent calculated from the equations (see §5.2.2).

Note that the Lyapunov exponents, like other invariant measures such as the fractal dimension, are independent of the observer function, assuming a smooth transformation between it and the dynamical variables. Furthermore, the embedding dimension should not matter if it suffices to remove overlaps. This property is illustrated in Fig. 10.7 where the observer function is taken as $X + Y$ with an embedding $m = 3$. The Lyapunov exponent is identical to that in Fig. 10.6.

10.4.2 Improvements

There are many ways to improve this simple algorithm. Kantz (1994) suggests using not just the nearest neighbor of each point but all points within some small neighborhood. Other methods that work even better estimate the local Jacobian matrix at each point along the orbit by least-squares fitting its elements to a collection of nearby points (Eckmann and Ruelle 1985, Sano and Sawada 1985). The Lyapunov exponents are the eigenvalues of the product of these matrices over many iterations. A technical difficulty is that some elements of the matrix grow exponentially, while others decay, requiring occasional reorthonormalization (Shimada and Nagashima 1979) using the Gram–Schmidt procedure (Press 1992). See Darbyshire and Broomhead (1996) for details.

This method works reasonably well for the positive Lyapunov exponents, but the negative exponents are hard to determine. If the orbit is on an attractor, there is no information from which to calculate the convergence to the attractor since the attractor will be locally very thin in the directions corresponding to strong contraction. Furthermore, the time-delay embedding dimension may be larger than the number of variables, giving spurious exponents. These exponents tend to be negative, but they can be more positive than even the most positive real exponent (Gençay and Dechert 1996, Sauer *et al.* 1998), especially in the presence of noise (Bryant *et al.* 1990). With extra dimensions, the dynamics are not unique, and any orbit, including one with strong stretching in the extra dimensions, that projects back onto the attractor is indistinguishable from the real one.

There are several methods to detect and avoid spurious exponents (Brown *et al.* 1991, Stoop and Parisi 1991, Bryant 1992, Sauer *et al.* 1998, Sauer and Yorke 1999). The real exponents should be independent of the choice of embedding dimension. Intentionally adding noise should affect the spurious exponents more than the real ones. The Kaplan–Yorke dimension in eqn (5.29) derived from the Lyapunov exponents can be compared with the dimension derived in other ways (see Chapter 12). You can also analyze the time series in reverse, in which case the real exponents should reverse their signs (Parlitz 1992). However, this method only works for data derived from flows and invertible maps, and even then it tends not to give accurate results, as Fig. 10.8 shows for the Hénon map. Because of the rapid contraction and absence of data off the attractor, the exponent derived in this way is $\lambda_2 \simeq -1.13$, whereas the correct value is $\lambda_2 \simeq -1.62$.

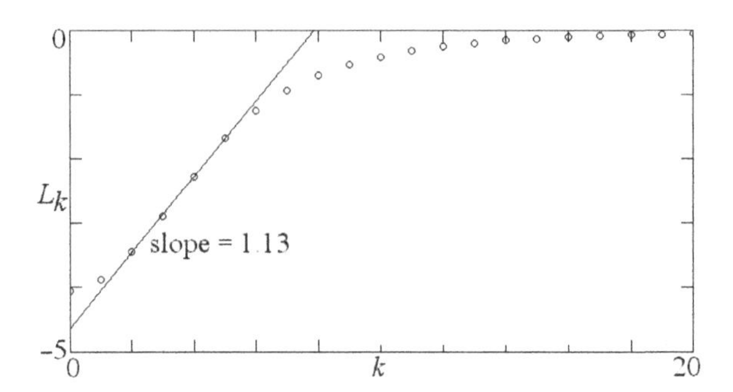

Fig. 10.8 Most negative Lyapunov exponent from a time series of the Hénon map.

10.5 False nearest neighbors

So far, we have not discussed how to choose the optimal embedding dimension but have used a value known to be appropriate. You might think that choosing a high embedding is advantageous, but there are severe penalties when it is unnecessarily high. The data become sparse in high embeddings, and each embedding dimension introduces additional noise. With experimental data, you need an objective criterion, and the optimal embedding is probably different for the various tests that use it. For example, prediction requires sufficient points in the neighborhood of the current point, and the number of such points decreases as the dimension increases. A test that guarantees an adequate embedding for most purposes and quantifies the penalty for choosing one too small is the method of *false nearest neighbors* (Kennel *et al.* 1992).

The method is conceptually simple. Find the nearest neighbor X_l for each point X_n in a time-delay embedding m, and call the separation between these points $R_n(m) = \sqrt{(X_l - X_n)^2 + (X_{l-1} - X_{n-1})^2 + \cdots}$. Then calculate the separation $R_n(m+1)$ in an embedding $m+1$. If $R_n(m+1)$ significantly exceeds $R_n(m)$, then the neighbors are close only because of overlap and are false. The criterion for falseness is thus

$$\frac{|X_{l-m} - X_{n-m}|}{R_n(m)} > R_T \tag{10.20}$$

where R_T is a threshold value. Abarbanel (1996) recommends $R_T = 15$, but the value is not critical.

With a small data set embedded in a high dimension, the nearest neighbors may not be very close and tend to lie at the edge of the space. In such a case, Kennel *et al.* (1992) recommend also considering a neighbor as false if

$$\frac{|X_{l-m} - X_{n-m}|}{\sigma} > A_T \tag{10.21}$$

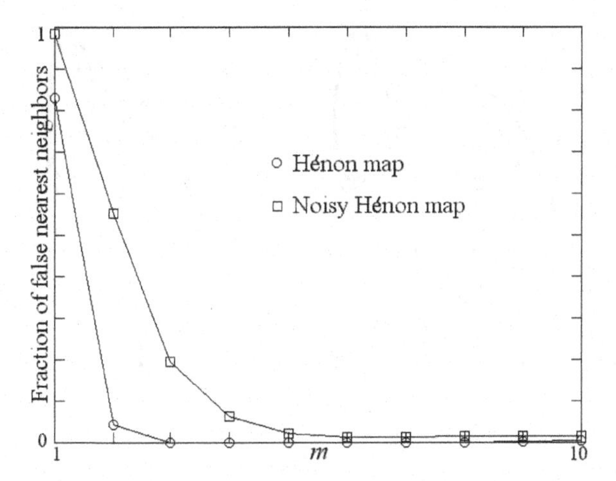

Fig. 10.9 Fraction of false nearest neighbors for the Hénon map with an observer function of $X + Y$.

where A_T is a threshold value and σ is the standard deviation of the data. Abarbanel (1996) recommends $A_T = 2$, but the value is not critical. A neighbor is considered false if either eqn (10.20) or (10.21) is satisfied.

10.5.1 Hénon map example

Combining the two criteria for the Hénon map with an observer function $X + Y$ (see the middle plot in Fig. 9.11) and $N = 6000$ gives the results in Fig. 10.9. As expected, there are essentially no false nearest neighbors for $m \geq 3$. Figure 10.9 also shows that the method degrades gracefully for added Gaussian white noise with a standard deviation of $\sigma_N = 0.1$. The cause of the false neighbors is evident in the stereo pair for the time-delay Hénon map with an observer function of $X+Y$ in Fig. 10.10, which requires viewing as described in §6.6.6.

10.5.2 Recurrence plots

A byproduct of the false nearest neighbor calculation is a set of indices n and l representing nearest neighbors in each embedding dimension. A plot of l versus n is called a *recurrence plot* and was introduced by Eckmann *et al.* (1987) and studied in detail by Casdagli (1997). Figure 10.11 shows such a plot for the Hénon map with an observer function of $X + Y$ and $m = 3$. Recurrence plots are useful for identifying structure such as intermittency or nonstationarity in the data. A stationary ergodic process[4] will eventually populate the plane uniformly. For a nonstationary process, the points will often cluster near the diagonal ($l \simeq n$). They will also cluster

[4]An *ergodic process* is one for which statistical measures such as mean and standard deviation are constant in time, and the statistics are the same for any time series chosen from an ensemble of such processes. Other definitions of 'ergodic' are used, and there is no universally accepted definition (Eubank and Farmer 1990).

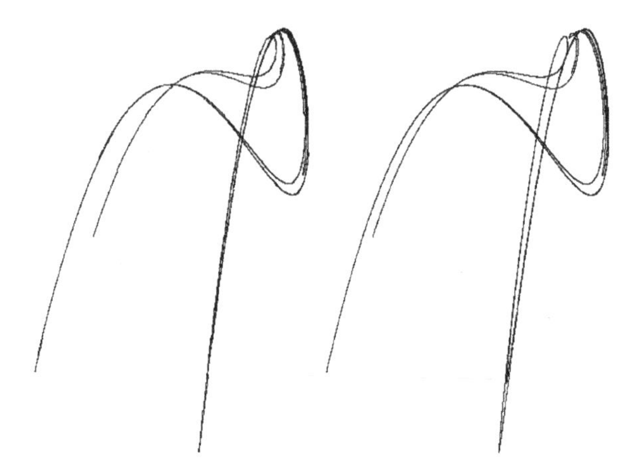

Fig. 10.10 Stereoscopic view of the Hénon map with an observer function of $X + Y$.

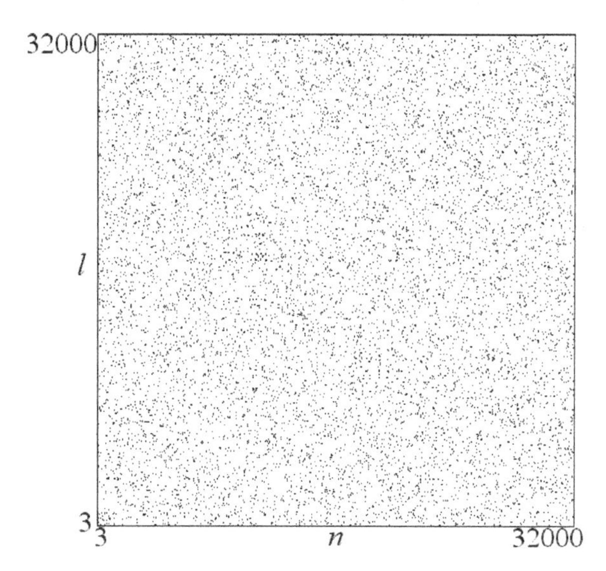

Fig. 10.11 Recurrence plot for the Hénon map with an observer function of $X + Y$ and $m = 3$.

near the diagonal for continuous-time data sampled too frequently, since temporal neighbors will then also be state-space neighbors. You can make the plot denser by including all neighbors within some small distance ϵ, but the results are similar. The method is not very sensitive to the choice of embedding.

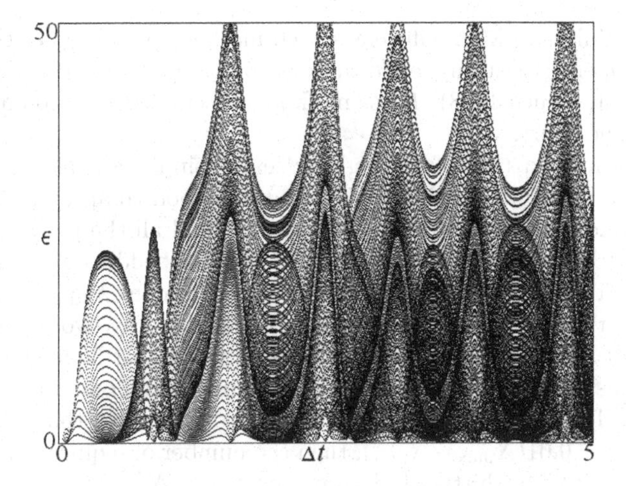

Fig. 10.12 Space–time separation plot for the Lorenz attractor with $m = 3$.

10.5.3 Space–time separation plots

In recurrence plots, all pairs of near neighbors are plotted, independent of the time at which they occur. A related method involving *space–time separation plots* was introduced by Smith (1992b, 1997) and Provenzale *et al.* (1992). This method integrates along parallels to the diagonal in the recurrence plots and thus only shows relative times. Usually one makes a scatter plot of the spatial separation ϵ versus the temporal separation Δt for every pair of points, or plots contours of constant probability per unit time for points to be within a distance ϵ in state space when their temporal separation is Δt. Such plots help identify temporal correlations, determine optimal time delays for the embedding, detect nonstationarity, and warn you when you have too few data points. They show how large Δt should be to ensure that the density of near neighbors is determined by the invariant measure rather than temporal correlations. To have a credible attractor, there must be points nearby in space that are not close in time, indicating *recurrence*.

Figure 10.12 shows a scatter plot for the Lorenz attractor with the standard parameters in a three-dimensional embedding. It is evident that for Δt less than about 0.1, all near neighbors are temporally correlated. Even values of Δt as large at 5 are noticeably correlated and do not sample the attractor uniformly.

10.5.4 Fast neighbor search

Identifying false nearest neighbors involves searching an array of length N for two points that lie within some distance ϵ of one another, a common operation in chaotic time-series analysis. The most straightforward implementation requires on the order of N^2 operations, since each of the N points must be compared with N others. For large data sets, the computation is prohibitive. More efficient search methods, requiring of the order of $N \ln N$ operations using *multi-dimensional trees* have been developed (Sproull 1991, Knuth 1998). These methods are cumbersome to implement but worth the effort.

You can instead use a fast neighbor search similar to the *box-assisted method* described by Schreiber (1995) that is a good compromise between simplicity and efficiency. Suppose you want to find all the points within a distance ϵ of X_n in an m-dimensional time-delay embedding of a time series with N points. Find the maximum X_{max} and minimum X_{min} of the scalar X_n, and partition that interval into bins of width ϵ. Any two points in the m-dimensional embedding separated by less than ϵ must lie in the same or adjacent bins since the additional $m-1$ dimensions can only increase their separation. Thus you need only search three bins rather than the entire data set. If $\epsilon = 0.01(X_{\text{max}} - X_{\text{min}})$, then the number of required operations is of order $0.03N^2$, a thirty-fold improvement over N^2.

To implement this procedure with minimal memory, you must establish a one-dimensional integer array I_k with N elements to hold the indices (values of n) of the data points X_n. Scan the data to determine how many points are in each bin. With 100 bins, you would need a second integer array with 100 elements to record the smallest value of k in each bin. Then populate the array I_k with the N indices of X_n by saving the value of each n in the next available I_k that lies in the proper bin. In essence, you are coarsely sorting the data by index without actually moving the points. To find the nearest neighbor to X_n, you need only search the portion of I_k with the bin that contains n and its two adjacent bins. If you have a fast sorting routine, it may be simpler and more expedient to sort I_k completely according to the values of X_n, in which case you can stop searching when the scalar separation exceeds ϵ.

10.6 Principal component analysis

When there is an attractor, the data points will reside in a small subspace of the time-delay embedding space, and the attractor may be very thin in certain directions. Furthermore, the coordinates of the embedding space contain redundant information, since in the long-time limit the projection of the orbit onto each axis is identical. Therefore, it is useful to construct a new set of basis vectors that are orthogonal and thus independent, and to choose them in directions in which the variance is largest. The first component is aligned in the direction of maximum variance. The second component is the direction perpendicular to the first in which the remaining variance is largest, and so forth (Broomhead and King 1986, Fraedrich 1986, Broomhead *et al.* 1996).

The method is variously known as *principal component analysis*, *principal value decomposition*, *singular value decomposition*, *singular system analysis*, *singular spectrum analysis*, *bi-orthogonal decomposition*, *proper orthogonal decomposition*, *empirical orthogonal functions*, and *Karhunen–Loéve decomposition* (Devijver and Kittler 1982, Jolliffe 1986), in what may be a record number of names for the same thing. The linear transformation amounts to a rotation of the attractor in the embedding space so as to expose its largest face and reduce folding, thus aiding visualization. It typically reduces the dimension of the system, but it may not completely eliminate overlaps (false nearest neighbors) unless the Takens' criterion (see §9.6.2) is satisfied. If the number of basis vectors is smaller than the original embedding, there is some noise reduction, since the noise is uniform in all directions whereas the signal is concentrated along the first few directions.

To implement the method, start with the $m \times m$ symmetric autocovariance matrix in eqn (10.9), which can be written in terms of the autocorrelation function $G(k)$ in eqn (9.18) as $C_{ij} = G(|i-j|)$ or

$$C = \begin{pmatrix} G(0) & G(1) & G(2) & \cdots \\ G(1) & G(0) & G(1) & G(2) \\ G(2) & G(1) & G(0) & G(1) \\ \cdots & G(2) & G(1) & G(0) \end{pmatrix} \tag{10.22}$$

whose eigenvalues are $\lambda = 1$ for $m = 1$, $\lambda = 1 \pm G(1)$ for $m = 2$, and so forth, assuming G is normalized such that $G(0) = 1$. The eigenvalues sum to m. Use a value of m such that $G(k) \simeq 0$ for $k \geq m$.

The eigenvectors corresponding to these eigenvalues (see §4.4) are the *principal components*, collectively called the *singular spectrum*, and they can be ordered by the size of their eigenvalues. Then the data points are projected onto these directions giving m new scalar functions of time $\psi_j(n)$, whose linear superposition is the original time series

$$X_n = \sum_{j=1}^{m} \psi_j(n) \tag{10.23}$$

For $m = 1$, the single function is $\psi_1(n) = X_n$. For $m = 2$, the two functions are $\psi_1(n) = (X_n + X_{n+1})/2$ and $\psi_2(n) = (X_n - X_{n+1})/2$, and so forth. Often the functions ψ_j are normalized so that

$$\sum_{n=1}^{N} \psi_j^2(n) = N\lambda_j \tag{10.24}$$

To reduce noise, replace eqn (10.23) with

$$\bar{X}_n \simeq \sum_{j=1}^{D} \psi_j(n) \tag{10.25}$$

where $D < m$. The identification of a few (D) dominant $\psi_j(n)$ functions is analogous to identifying the dominant frequencies in a discrete Fourier

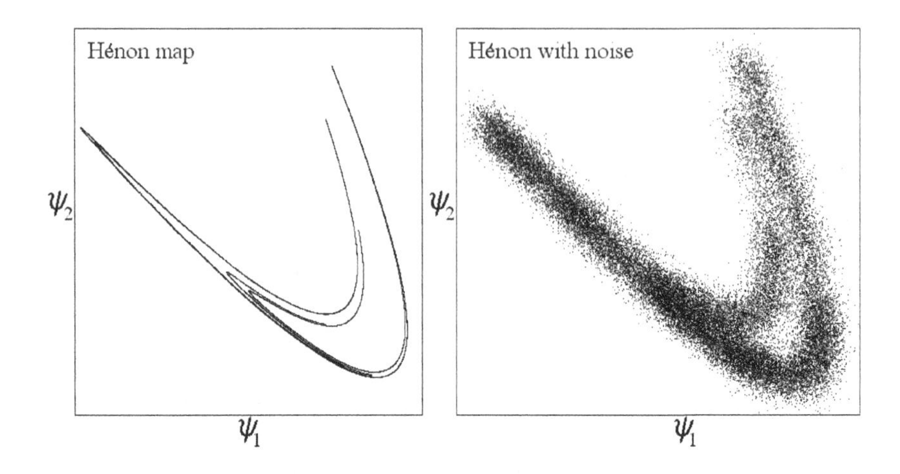

Fig. 10.13 Hénon map without and with noise projected onto the first two principal components with $m = 5$.

transform. In the limit of infinite m, principal component analysis reduces to discrete Fourier analysis (Vautard and Ghil 1989).

It is often incorrectly claimed that the number of eigenvalues that lie above the noise floor is a measure of the system dimension. In fact, a chaotic flow will have a single dominant eigenvalue if sampled at closely spaced intervals where successive points are highly correlated, and a large number of nearly equally eigenvalues if sampled infrequently so that there is little correlation between successive data points. With a judiciously chosen time lag, the number of significant eigenvalues may be an approximate upper limit for the attractor dimension.

The method is illustrated in Fig. 10.13 which shows $N = 3.2 \times 10^4$ data points for the X_n variable of the Hénon map decomposed into $m = 5$ components for which only the first two are plotted. They completely capture the structure of the map. The normalized eigenvalues are $\lambda \simeq 1, 0.45, 0.34, 0.34, 0.21$. Also shown is the result for the Hénon map with added Gaussian white noise with $\sigma_N = 0.1$. In this case, the eigenvalues are nearly unchanged, and the noise reduction is minimal.

You can use the method to construct a set of model equations

$$\psi_j(n+1) = F_j(\psi_1(n), \psi_2(n), \ldots) \tag{10.26}$$

such as polynomials with adjustable parameters $a_1, a_2, \ldots$ chosen by least-squares fit to the in-sample data (Rowlands and Sprott 1992). An example using a pair of quadratic polynomials for the noisy Hénon map data in Fig. 10.13 is shown in Fig. 10.14. The noise is completely removed, but the attractor is considerably distorted. The noise-free case (not shown) is fitted almost exactly since the Hénon map is quadratic. For more complicated chaotic systems, polynomial models that give good near-term fits seldom

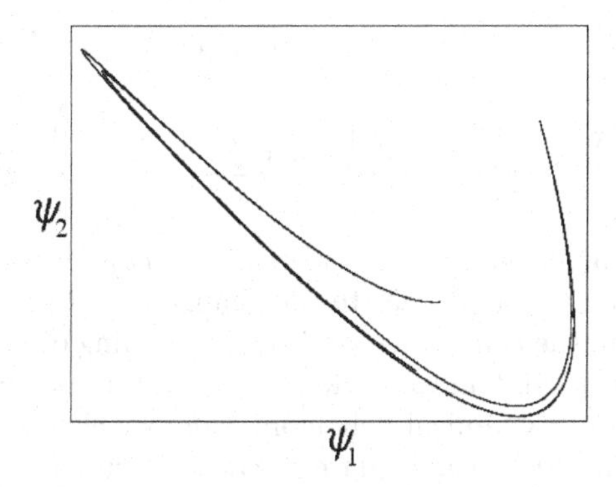

Fig. 10.14 Solution of quadratic polynomial model equations for the noisy Hénon map in Fig. 10.13.

have chaotic solutions. More typically, the solutions attract to a stable fixed point or periodic cycle, or are unbounded, unless they are somehow constrained to prevent this from happening.

10.7 Artificial neural network predictors

To overcome some of the problems with polynomial models, *artificial neural networks* (also called *connectionist networks* or *parallel distributed processors*) are used (Weigend *et al.* 1990). Such networks have a certain mystique since they try to imitate the brain,[5] but they are just another iterated map. One particular type, the *single-layer feed-forward network* with a hyperbolic tangent (*sigmoid*) squashing function (or *activation function*), has been mentioned in §6.4.2. As a general nonlinear predictor, it has several virtues:

1. It is a universal approximator that can represent any measurable function to arbitrary precision with sufficiently many neurons (Hornik 1989, Hornik *et al.* 1990).
2. The Jacobian matrix is easily calculated, simplifying calculation of eigenvalues and Lyapunov exponents.
3. Its solutions are always bounded since the squashing function has $-1 \leq \tanh^{-1}(X) \leq 1$ for all X.
4. Relatively small networks can be used to model highly nonlinear systems, and the adjustable parameters have a symmetry not shared by other models such as polynomials (Mitchison and Durbin 1989).
5. There is considerable experience and a large literature on the optimization and training of such networks.

[5]The human brain has about ten billion neurons, each connected to about a hundred others, giving a trillion (10^{12}) connections. The brain scientist, Sir John Eccles (1903–1997), Nobel laureate in Medicine (1963), described the brain as the most complex system in the universe.

Single-layer feed-forward networks are described by

$$\bar{X}_n = \sum_{i=1}^{N} b_i \tanh\left(a_{i0} + \sum_{j=1}^{D} a_{ij} X_{n-j}\right) \qquad (10.27)$$

where N is the number of *neurons* (or *cells*) and D is the dimension (number of time lags) as in Fig. 6.5. Note the change in notation; N is *not* the number of points in the time series, and the embedding dimension is *not* m, but D. In Chapter 6, random values were used for the coefficients a_i and b_{ij} to explore a range of dynamical behaviors. Here we choose the coefficients to minimize the one-step mean-square prediction error $E = \langle(\bar{X}_n - X_n)^2\rangle$. There are many ways to do this, but we mention two methods that are conceptually simple and easy to implement. Without the hidden layer of neurons, the method reduces to standard linear regression.

10.7.1 Multi-dimensional Newton–Raphson

Start with random values for all coefficients in the range 0 to 1, and form arrays of incremental values δa_{ij} and δb_i, initialized to -0.5. These values are not critical. Calculate the initial error E_0. Then add the increments (generically called δw) one-by-one to the corresponding coefficient w, each time calculating a new error E and updating the increment to $(E_0 - E)/\delta w$, from which the new coefficient is $w + \eta\delta w$, where $\eta \ll 1$ is the learning rate. Large values of η hasten the convergence but may lead to instability and failure to converge. Repeat until E stops changing or you lose patience. This training scheme is a multi-dimensional variant of the Newton–Raphson method (see §4.6.1) for finding the zero of dE/dw (the minimum E). Multi-dimensional root finding of nonlinear equations is an important but difficult subject (Acton 1970, Press *et al.* 1992).

10.7.2 Simulated annealing

Even easier is a simplified variant of *simulated annealing* (Press *et al.* 1992) in which you take the coefficients as $w = \bar{w} + r$, where $\bar{w}$ is the best estimate of w so far, initialized to zero, and r is a Gaussian random number with mean zero and standard deviation σ. Note that you can constrain the values of b_i to be positive, since the signs of a_{ij} are arbitrary. Whenever the error E is less than the smallest previous error, set each $\bar{w}$ to the corresponding w. Decrease σ slowly (the *annealing schedule*) until E stops changing or you lose patience.

With this method, you randomly explore shrinking Gaussian neighborhoods of the best solution. As with the previous method, convergence can be slow, and there is a chance of getting stuck in a local minimum that is much larger than the global minimum. Usually the 'fitness landscape' (contours of constant E) is *rugged* rather than *smooth* in a highly connected network. Using Gaussian random numbers helps avoid this problem because the long tail gives occasional points far from the local minimum. This

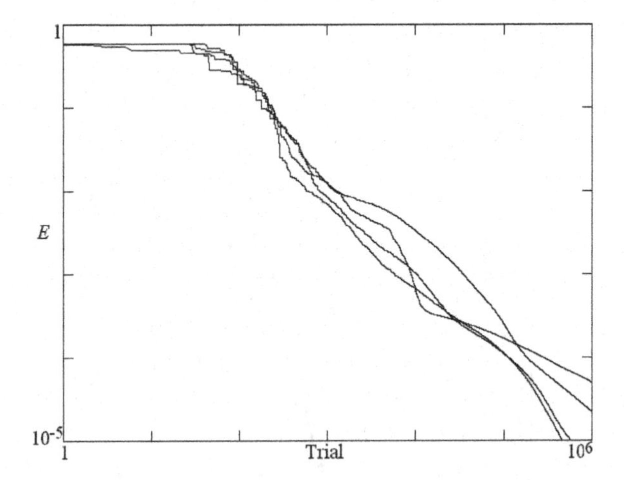

Fig. 10.15 Learning curve for four instances of an artificial neural network trained on Hénon map data.

scheme is slightly different from the classical simulated annealing method (Metropolis *et al.* 1953) in which worse solutions are occasionally accepted, but it is arguably as efficient. Methods for improving the convergence are described by Szu and Hartley (1987).

Simulated annealing is a very general method for solving a wide class of multivariate nonlinear optimization problems (Kirkpatrick *et al.* 1983). It takes its name from the annealing process in which a heated metal is slowly cooled, allowing it to form a crystalline state with minimum thermodynamic energy and thus greatest strength. Another analogy is shaking a canister of sugar at first vigorously and then more gently until the grains reach their lowest gravitational energy. With sufficiently slow cooling, the global optimum is always reached (Hajek 1988), but the required cooling rate may be unacceptably slow. Simulated annealing has been used to solve the 'traveling salesman problem' (finding the sequence of cities that minimizes the driving distance[6]) and arranging electrical components on printed-circuit boards and integrated circuits to minimize interconnection distances and crossovers. For more detail, see Press *et al.* (1992).

10.7.3 Hénon map example

As an example, Fig. 10.15 shows the learning rate (E versus the trial number) for a neural network with $N = 8$ and $D = 2$ trained on 6000 points from the X_n variable of the Hénon map using simulated annealing. The

[6]The traveling salesman problem is a classic example of an *NP* (nondeterministic polynomial) *problem*, one whose difficulty, as measured by the required number of mathematical operations, increases faster than a polynomial function of size, in this case the number of cities visited. It is an *NP-complete* problem in that any solution in polynomial time would imply a solution in polynomial time for all NP problems, thus eliminating their existence.

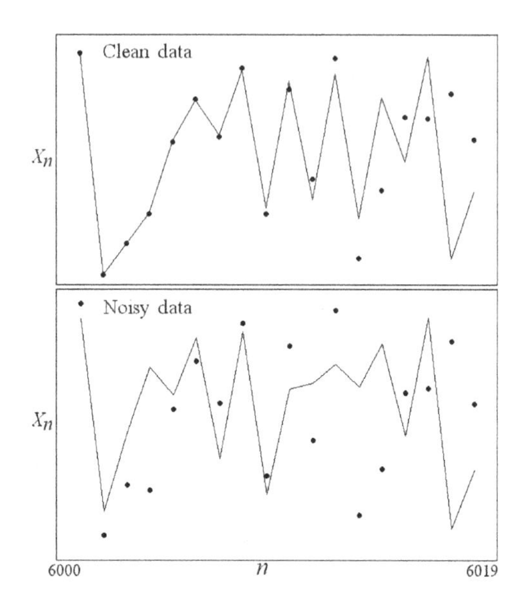

Fig. 10.16 Out-of-sample fit for a neural network trained on Hénon map data.

initial σ was 0.1, and it was reduced by a factor 0.9 each time a larger E resulted and increased by a factor of two each time a smaller E resulted. This schedule keeps σ large whenever progress is being made, but slowly shrinks the neighborhood when a minimum is found. The four cases differ only in the seed of the random number generator, illustrating that some paths toward the solution are more efficient than others but that all cases continue to improve even after a million trials.

The mean-square in-sample next-step error for the Hénon map typically reaches $E \simeq 1 \times 10^{-5}$ after about 10^6 trials and is still slowly decreasing. A typical out-of-sample fit for 18 time steps is in Fig. 10.16. The fit is good for a few steps but fails rapidly. The error after k time steps is consistent with $\sqrt{E}\mathrm{e}^{\lambda_1(k-1)}$, where E is the mean-square in-sample error for one time step. For Hénon map data with noise having a standard deviation of $\sigma_N = 0.1$, the fit is significantly worse as expected, since the initial next-step error is $E \simeq 4 \times 10^{-2}$.

It is generally difficult to replicate the attractor, although the Hénon map is simple enough that a neural network gives an attractor very similar in appearance to the one obtained from the data, as in Fig. 10.17, even with added noise having $\sigma_N = 0.1$, although the case with noise is noticeably different. Thus the method can be used for noise reduction at the expense of significantly distorting the attractor and the corresponding dynamics.

A global model such as a neural network optimized for short-term prediction usually fails to reproduce the topology of the data, and thus it is not surprising that it can generate spurious positive Lyapunov exponents.

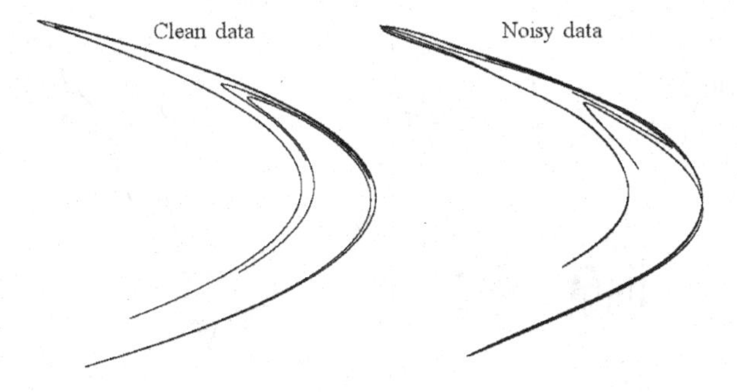

Fig. 10.17 Attractor from a neural network trained on Hénon map data.

Minimizing the sum of the errors in the prediction over K time steps

$$E = \sum_{k=1}^{K} \langle (\bar{X}_{n+k} - X_{n+k})^2 \rangle \tag{10.28}$$

is often helpful, but it is difficult to get good fits for large K with chaotic data.

For most cases, the long-term solution of a simple (small N and D) neural network fit to chaotic data is a periodic cycle. With polynomial models, the solutions are often unbounded. More complicated neural networks are more often chaotic (see §6.4) but with an attractor that does not resemble the one being fit. The problem of finding a dynamical model that accurately replicates the structure of a given attractor is much harder and has no known general solution.

10.7.4 Deterministic $1/f$ noise

Neural networks can be used to find a deterministic system whose time series has some desired property. For example, suppose you want a simple map whose iterates have a $1/f$ power spectrum. Equivalently, minimize the mean-square deviation E between the autocorrelation function of the model $\bar{G}(k)$ and the desired autocorrelation function $G(k)$

$$E = \sum_{k=1}^{m} \frac{1}{k} (\bar{G}(k) - G(k))^2 \tag{10.29}$$

which fits frequencies from f_c to f_c/m, where f_c is the Nyquist frequency (see §9.4.4). The factor $1/k$ weights the frequencies evenly on a logarithmic scale.

The results for a neural network with $N = 8, D = 2, m = 100$, and $a_{i0} = 0$, trained on a time series with 2000 points of $1/f$ noise is shown in Fig. 10.18. The conclusion is that even relatively simple neural networks can produce a time series with some desired property. Table 10.2 lists the coefficients of eqn (10.27) that produce a $1/f$ power spectrum.

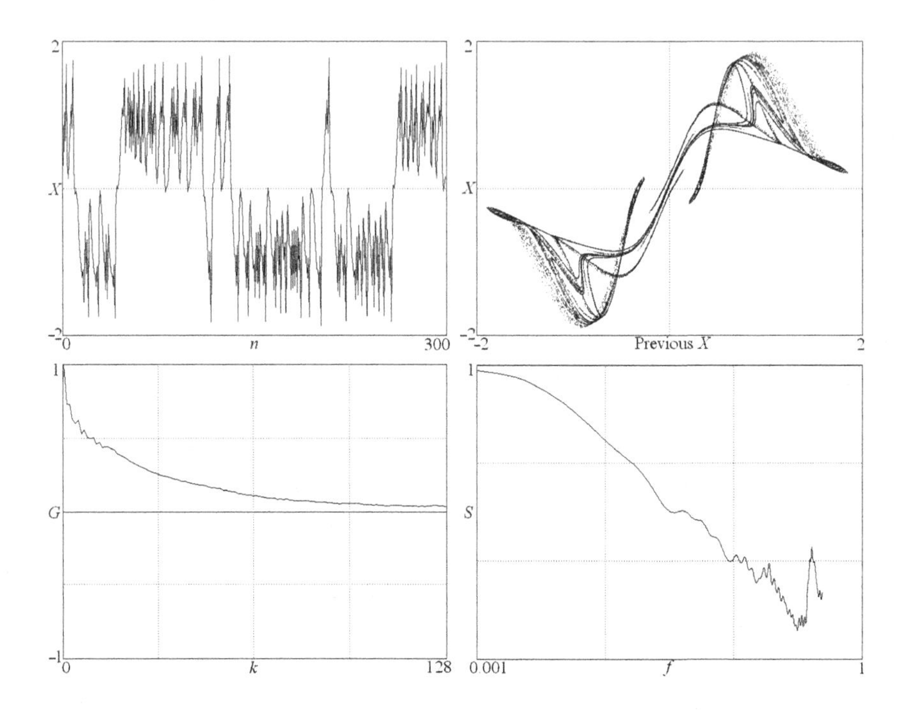

Fig. 10.18 Behavior of a two-dimensional neural network trained to give a $1/f$ power spectrum.

Table 10.2 Neural network coefficients of eqn (10.27) that give a $1/f$ power spectrum.

i	a_{i1}	a_{i2}	b_i
1	0.74073	−0.19776	1.98331
2	0.78255	0.82671	1.95970
3	3.01602	−1.00483	1.66248
4	−1.36676	1.48674	1.07933
5	−1.88023	2.50657	0.70363
6	1.68119	−2.37619	0.59973
7	0.57455	−1.51264	0.54699
8	−2.01825	−0.53433	0.31743

There is nothing special about neural networks for producing deterministic $1/f$ signals. A two-dimensional quadratic map optimized in a similar way to give a time series with a $1/f$ power spectrum is

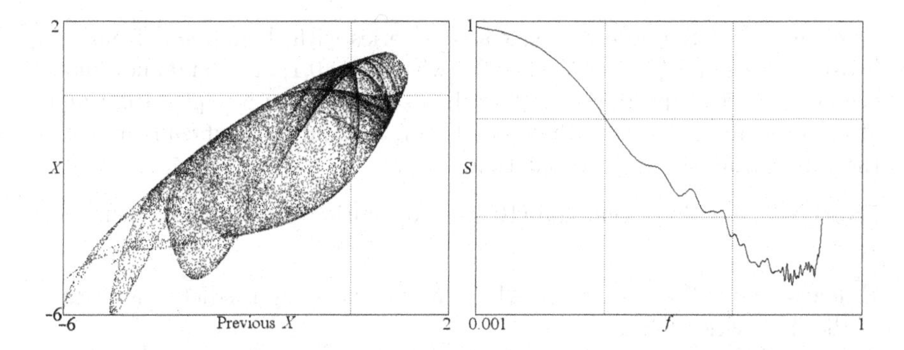

Fig. 10.19 Behavior of a two-dimensional quadratic map trained to give a $1/f$ power spectrum.

$$X_{n+1} = 0.671 - 0.416X_n - 1.014X_n^2 + 1.738X_nX_{n-1} + 0.835X_{n-1} - 0.814X_{n-1}^2 \qquad (10.30)$$

The attractor for this map and the corresponding power spectrum are in Fig. 10.19. Why the attractors for the two cases look like animals is a deep mystery!

10.8 Exercises

Exercise 10.1 Show that the linear extrapolation in eqn (10.1) can only lead to linear growth, but that $X_{n+1} = a_1X_n + a_2X_{n-1}$ has exponentially growing and damped solutions that oscillate for appropriate values of a_1 and a_2.

Exercise 10.2 Show that the quadratic extrapolation in eqn (10.2) is consistent with fitting a parabola to three points of the time series, and evaluate its value for the next point.

Exercise 10.3 Show that the cubic extrapolation in eqn (10.3) is consistent with fitting a cubic polynomial to four points of the time series, and evaluate its value for the next point.

Exercise 10.4 Show that the quartic extrapolation in eqn (10.4) is consistent with fitting a quartic polynomial to five points of the time series, and evaluate its value for the next point.

Exercise 10.5 Show that the quintic extrapolation in eqn (10.5) is consistent with fitting a quintic polynomial to six points of the time series, and evaluate its value for the next point.

Exercise 10.6 Suppose you have a time series with ten points from $X_n = \sin n$ for $1 \leq n \leq 10$ (n is in radians), but you do not know the equation that produced it. Use the extrapolations in eqns (10.1)–(10.5) to estimate the next three points X_{11}, X_{12}, and X_{13}, and compare with the values from $\sin n$.

Exercise 10.7 Suppose you have a time series with three points from the logistic equation $X_{n+1} = 4X_n(1-X_n)$ with $X_1 = 0.1$, but you do not know the equation that produced it. Use the extrapolations in eqns (10.1) and (10.2) to estimate the next three points X_4, X_5, and X_6, and compare with the values from the logistic equation.

Exercise 10.8 Show that eqn (10.10) is a solution of eqn (10.8) with $m = 1$.

Exercise 10.9 Show that the AR(1) model correctly predicts the future of the time series $1, 2, 4, 8, \ldots$.

Exercise 10.10 Use the AR(1) model to estimate the next three points in the time series $1, 5, \ldots$.

Exercise 10.11 Use the AR(2) model to estimate the next three points in the time series $1, 0, 2, \ldots$.

Exercise 10.12 Show that the AR(2) model correctly predicts the next three Fibonacci[7] numbers $1, 1, 2, \ldots$, in which each term is the sum of the two previous ones.

Exercise 10.13 Show that the iterates of the map $X_{n+1} = 1.080605X_n - X_{n-1}$ are given by $X_n \simeq \sin n$.

Exercise 10.14 Calculate the period and damping rate for the AR(2) model $X_{n+1} = 1.03671X_n - 0.95714X_{n-1}$.

Exercise 10.15 Show that the AR(2) model $X_{n+1} = a_0 + a_2X_n + a_1X_{n-1}$ produces an undamped oscillation if $a_1 = -1$, and calculate the period of the oscillation.

Exercise 10.16 Calculate the fixed point for the AR(2) model $X_{n+1} = 0.25692 - 0.26173X_n + 0.16583X_{n-1}$ to the Hénon map, and show that it is attracting.

Exercise 10.17 Evaluate $(m + l)!/m!l!$ to determine the number of coefficients for a polynomial with m variables and order l for $1 \leq l \leq 4$ and $1 \leq m \leq 4$.

Exercise 10.18 Derive a formula for the amount by which the amplitude of a sine wave of period T sampled at intervals of Δt with $\Delta t \ll T$ is diminished by the smoothing formula in eqn (10.15) as a function of m.

Exercise 10.19 Show that the eigenvalues λ of the autocovariance matrix in eqn (10.22) with $m = 3$ are given by $(1-\lambda)^3 - (1-\lambda)[2G(1)^2 + G(2)^2] + G(1)G(2)[G(1) + G(2)] = 0$.

[7] Leonardo of Pisa (*c.* 1175–1250), better known by his nickname 'Fibonacci,' was an Italian merchant and the greatest European mathematician of the Middle Ages, whose work in number theory was unknown until modern times.

Exercise 10.20 Design a single-layer feed-forward neural network that will produce a period-2 output given by $X_n = (-1)^n$.

Exercise 10.21 Design a single-layer feed-forward neural network that will produce a period-3 output given by $0, 1, 2, 0, 1, 2, \ldots$.

Exercise 10.22 Design a single-layer feed-forward neural network that will map the unit interval $0 \leq X \leq 1$ back onto itself twice with $f(0) = f(1) = 0$ and $f(0.5) = 1$ as with the logistic map so as to produce a chaotic time series.

Exercise 10.23 Show with the variant of simulated annealing in §10.7 in which you decrease σ by a factor of 0.9 wherever E increases and increase it by a factor of two whenever E decreases, that you expect one improvement for about every seven trials.

10.9 Computer project: Nonlinear prediction

This project is a continuation of the development of tools for the analysis of a chaotic time series. This time you will develop a nonlinear predictor. Prediction of the future of a time series can be an end in itself, but it can also be a noise-reduction method, a way of estimating the sensitivity to initial conditions, and a way to generate model equations whose solution can provide an infinite amount of data and whose form might provide physical insight into the nature of the underlying dynamical system.

1. The first step is to generate a time series for analysis. For this purpose, use the Hénon map with parameters chosen to give chaos in eqns (5.12) and (5.14). Perform a few iterations to ensure that the orbit is on the attractor, and then generate a time series X_n for $n = 1$ to 1000 from the variable X. Also generate the next 10 points ($n = 1001$ to 1010) to use as a test of the quality of your prediction. Make a plot of X_n versus X_{n-1} for $n = 2$ to 1000, and convince yourself that this plot replicates the topology of the original Hénon map in the XY plane.
2. Now using your 1000 points as input data, search through the data record for the k points in an m-dimensional embedding that are closest to the last point in the time series ($n = 1000$). Choose $k = 8$ and $m = 2$, but write your code generally so that you can adjust these parameters. Now make a list of how much these eight points change (ΔX, a scalar) on the next iteration. Take the average of these eight changes and add it to X_{1000} to form your prediction of X_{1001} and compare it with the actual value X_{1001} that you saved from part 1 above.
3. Continue this process for another nine time steps, using the k nearest neighbors to your predicted value of X_{1001} in an m-dimensional embedding to get a prediction of X_{1002}, and so forth up to X_{1010}. Make a table or graph of your predictions for $n = 1001$ to 1010 along with the actual values of X_n that you saved from part 1 above. Is the

growth rate in the error of your prediction consistent with the largest Lyapunov exponent of the Hénon map ($\lambda_1 \simeq 0.419$)?

4. (optional) Make a new time series that is the same as your time series above but that has a small amount of noise added (say 10%). See how the quality of your prediction is degraded by this noise, and see if there are optimal values of k and m that give the best prediction.

11 Fractals

Identifying chaos in experimental data usually includes searching for a strange attractor in the state-space dynamics, identified by its fractal structure. Having found such an attractor, you can try to estimate its dimension, which is a measure of the number of active variables and hence the complexity of the equations required to model the dynamics. Thus it is useful to digress from the discussion of time-series analysis to show examples of fractals and discuss their properties so you will recognize them when they occur in an experiment. Fractals are to chaos what geometry is to algebra. They are the usual geometric manifestation of the chaotic dynamics. They have been called 'the fingerprints of chaos' (Richards 1999), but they are interesting and important in their own right, independent of their relation to chaos. Many natural objects in real three-dimensional space have fractal qualities.

It is difficult to define a fractal precisely, but instead we list some properties shared by most fractals (Falconer 1990):

1. They have fine structure (detail on arbitrarily small scales) and no characteristic scale length.
2. They are too irregular to be described by ordinary geometry, both locally and globally.
3. They have some degree of self-similarity (perhaps approximate or statistical), meaning that small pieces of the object 'resemble' the whole in some way.
4. Their fractal dimension (suitably defined) is greater than their topological dimension.
5. They often have unusual statistical properties such as zero or infinite average and variance.
6. They are defined in a simple way, perhaps recursively.

Dynamical systems are only one way to produce fractals. In this chapter we will describe fractals produced in various ways. Some of these methods are sufficiently important to warrant further discussion in subsequent chapters. Fractals can be either *deterministic*, where they are *exactly* (or *geometrically*) self-similar, at least in the limit of small scale, or *random*, where they are only *statistically* self-similar, although random rules can produce deterministic objects (see Chapter 14) and deterministic rules can produce random objects (see the figure on page **xx**). We will illustrate both types of fractals as well as hybrid cases involving both determinism and randomness and cases where the self-similarity is not obvious. The self-similarity implies that the object or the dynamical process that produced it is *scale invariant* (there is no characteristic scale length).

The term 'fractal' (both a noun and an adjective) was coined in 1975 by Benoit B. Mandelbrot (1977). It comes from the Latin adjective *fractus*, meaning 'irregular.' The corresponding Latin verb *frangere* means 'to break into irregular fragments.' It also suggests an object with fractional dimension, which is usually the case. Many of the examples in this chapter were known and studied over a hundred years ago, but they were considered mathematical curiosities of no practical interest. We now recognize that they are good models for many natural processes and forms. In his book *Fractals everywhere*, Michael F. Barnsley (1988) begins with the admonition, also appropriate here:

> Fractal geometry will make you see everything differently. There is danger in reading further. You risk the loss of your childhood vision of clouds, forests, galaxies, leaves, feathers, flowers, rocks, mountains, torrents of water, carpets, bricks and much else besides. Never again will your interpretation of these things be quite the same.

11.1 Cantor sets

Perhaps the simplest, and in some way the prototypical fractal, is the Cantor set[1] (Cantor 1883), already mentioned in §2.3, §5.9, and §7.5.3. Imagine taking a string of unit length, say one meter, and removing the middle third.[2] Then take the remaining pieces and remove their middle thirds, and so forth, as in Fig. 11.1. After infinitely many such operations, the resulting object is a *triadic* (or *middle-third*) *Cantor set* (also called a *Cantor ternary set*). The initial object (the one-meter-long string) is called the *initiator*, and the object after one step (the two line segments) is called the *generator* or *motif* of the set. The Cantor set is produced by infinitely many applications of the generator. The intermediate objects formed by finitely many applications are called *prefractals*.

The Cantor set is a fractal in the sense that at any step, each remaining third is identical to the entire object in the previous step except three times smaller. Thus it is *self-similar*. As with similar triangles, which are identical except for size, a self-similar object contains infinitely many copies of itself with ever smaller sizes. The final object will contain infinitely (in fact,

[1]Details of the set were published by the Russian-born, German-immigrant mathematician Georg Ferdinand Ludwig Philipp Cantor (1845–1918) in 1883, but it was invented eight years earlier by the British mathematician Henry John Stephen Smith (1826–1883).

[2]Technically, you remove the *open* middle third, meaning the interval from 1/3 to 2/3, excluding the endpoints.

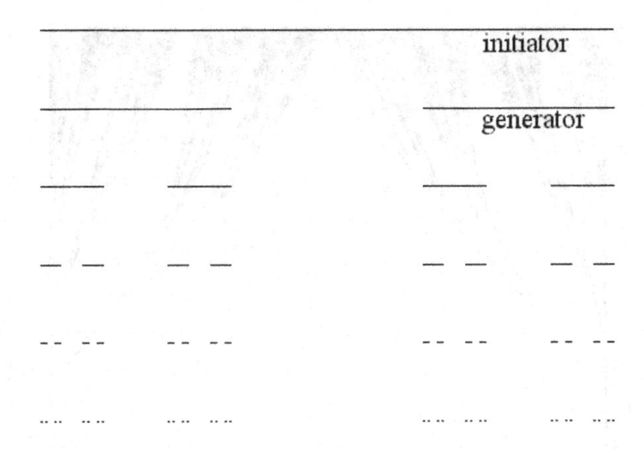

Fig. 11.1 First six steps in the construction of a triadic Cantor set.

uncountably[3]) many pieces, each of which has infinitely many neighbors within any nonzero neighborhood (a *Cantor dust* or *Cantor discontinuum*). However, the set is *totally disconnected*, meaning that each piece of the set is separated from its neighbors by a gap. For more information about the orders of infinity, see Rucker (1982).

Surprisingly, the total length of the pieces is zero, as you can confirm by considering that each step reduces the length to two-thirds of the length at the previous step. Thus after k steps, the total length is $(2/3)^k$, which approaches zero as k goes to infinity. Thus the triadic Cantor set is more than a finite collection of points but less than a collection of line segments. It has a dimension greater than zero (a collection of points) but less than one (a line). In the next chapter we will show that its similarity dimension is $\log 2/\log 3 = 0.63092975\ldots$. It consists of all those points on the unit interval whose base-3 (*ternary*) representation[4] contains only the symbols 0 and 2. It has other surprising properties, such as the fact that *any* number in the interval $0 < X < 2$ can be represented exactly as a sum of two Cantor numbers.

There is nothing special about removing the middle third at each step. The portion removed need not be in the middle, and the fraction can be anything between zero and one. A stack of Cantor sets with different middle fractions removed produces the *Cantor curtains* in Fig. 11.2 (Mandelbrot 1983). The dimension of this object varies linearly from 1.0 at the bottom

[3]A set is *countable* if its elements can be listed as a sequence of numbers with every element in the set appearing at a specific place in the list. The rational numbers are countable, but the irrationals are uncountable (Kline, 1980).

[4]Ternary numbers are less common than binary and decimal numbers, but they have advantages such as minimizing the product of the *radix* (3) and the number of digits required to represent a number (Hayes 2001b) and are thus in a sense the most compact representation of a number. Computers using *trits* rather than bits have been proposed and built, but they never caught on.

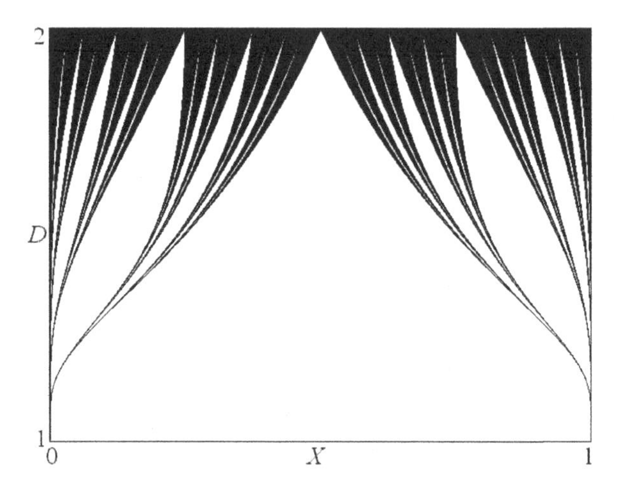

Fig. 11.2 Cantor curtains.

where it looks like a pair of lines to 2.0 at the top where it looks like a surface. Thus a cross-section through it at the place where $D = 1.6309\ldots$ would give the triadic Cantor set with $D = 0.6309\ldots$. This example illustrates that the fractal dimension of an object may not be unique, but can change with location.

11.2 Fractal curves

Another way to produce fractals is to start with a line, but rather than remove portions of it, you bend it in a self-similar manner, extending its length as necessary. There are countless variants on this theme, a few examples of which follow.

11.2.1 Devil's staircase

The *Devil's staircase* in Fig. 11.3 is obtained by integrating the triadic Cantor set along its extent (Hille and Tamarkin 1929). It has infinitely many steps, but only rises to a finite height. It has obvious self-similarity, but a well-defined length of $x + y$, where x is the horizontal span, and y is the vertical rise. More precisely, the Devil's staircase is *self-affine* since the horizontal scaling factor is three and the vertical scaling factor is two. Since the length is neither zero nor infinite, the dimension of the Devil's staircase is exactly 1.0. For this reason it is not universally considered a fractal. A line with a uniquely defined length is called *rectifiable*. This example contradicts the common presumption that the dimension of a fractal cannot be an integer.

The Devil's staircase occurs in many physical systems (Bak 1986), such as a plot of the frequency of a driven van der Pol oscillator as a function of the drive frequency, where mode-locking occurs over a Cantor set of fre-

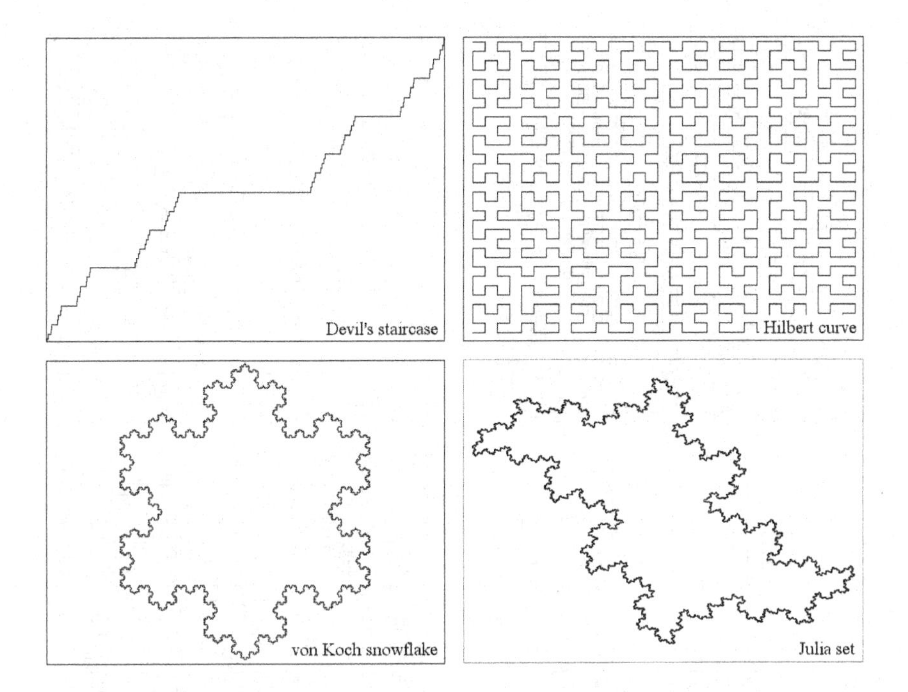

Fig. 11.3 Fractal curves.

quency ratios (see §4.6.3). You might note a connection to *Zeno's paradox*,[5] which asserts that no motion is possible, because to move from one place to another requires that you first move half way, and before that one-quarter of the way, and so forth, requiring infinitely many steps in finite time.

11.2.2 Hilbert curve

The *Hilbert curve* is an example of a nonintersecting, *space-filling* curve (Sagan 1994), shown after the first few steps in Fig. 11.3. At each step, four segments with the shape ⊏ (the initiator) are replaced by pieces with shape ⊏ (the generator) having a proper orientation. As with the Cantor set, it would be hard to view the infinite limit where the line comes arbitrarily close to every point in the plane. An infinitely long line is required to cover the plane, and the dimension of the final object is 2.0 despite its construction from line segments of finite length. Adjacent points along the curve are adjacent in the plane, but not vice versa. The Hilbert curve

[5]Zeno of Elea (*c.* 535–495 BC) was a Greek philosopher and logician who was famous for his paradoxes that puzzled mathematicians for millennia. He wrote only one book of which only 200 words survive, but it was quoted by Aristotle and others, and he was tortured to death for plotting against Elea's tyrant, Nearchus.

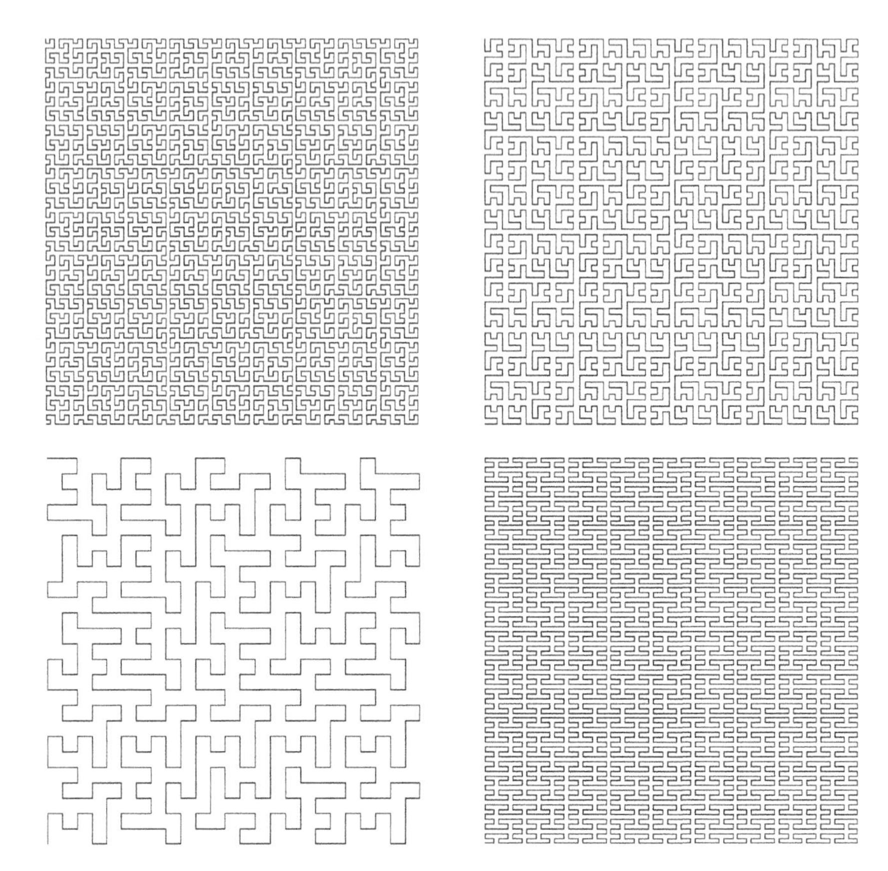

Fig. 11.4 Peano curves.

is one example of a large class of *Peano curves*[6] (Peano 1890), some other examples of which are in Fig. 11.4.

11.2.3 Von Koch snowflake

The *von Koch snowflake*[7] (von Koch 1904) in Fig. 11.3 is formed by starting with an equilateral triangle and then replacing each side of the triangle (the initiator) by the pattern __/__ (the generator) with a total length 4/3 times the length of the side that is replaced. The process is repeated, giving a total length of $3(4/3)^k$ after k steps. The resulting line as k approaches infinity is infinitely long, and the dimension is $\log 4/\log 3 = 1.26185950\ldots$. It is a continuous curve that nowhere has a tangent or derivative, and is

[6]Guiseppe Peano (1858–1932) was an Italian mathematician who invented space-filling curves in 1890 and laid the foundations of set theory, but was known for finding flaws in others' works and for developing an international language 'Latino sine flexione' based on Latin but without grammar.

[7]Niels Fabian Helge von Koch (1870–1924) was a Swedish mathematician, known more for his lengthy calculations than for his originality.

Fig. 11.5 Von Koch curves.

one example of a large class of *von Koch curves*, some other examples of which are in Fig. 11.5. All these curves are infinitely long but enclose a finite area. In fact any finite portion of these curves is also infinitely long, no matter how small the portion.

The von Koch snowflake also resembles the shoreline of a rather too symmetrical island or lake. Indeed, a shoreline is not a line at all, but a fractal whose length depends on the length of the ruler used to measure it, and approaches infinity at the highest resolution, as was pointed out by Richardson[8] (1961). With a shorter ruler, you can resolve more detail

[8]Lewis Fry Richardson (1881–1953) was an English mapmaker, physicist, and meteorologist who after serving as an ambulance driver during World War I researched the causes of war, discovering that nations sharing a common border are more likely to wage war, but that the length of their common border is a bad metric because its value depends on the resolution of the measurement (Richardson 1961). He is also famous for the poem 'Big whorls have little whorls that feed on their velocity, and little whorls have smaller whorls and so on to viscosity,' which summarizes his paper on atmospheric eddies (Richardson 1920).

in the shoreline. Most real shorelines, such as the coast of Britain, have a fractal dimension (see Chapter 12) of about 1.2, although the ragged coast of Norway has a dimension closer to 1.5.

11.2.4 Julia set

Another fractal curve is the basin boundary for certain attractors such as the *Julia sets* mentioned in §6.8.4 and described more fully in §14.4. Figure 11.3 shows the boundary for the complex map $Z_{n+1} = Z_n^2 + C$ with $C = 0.62i$, inside of which the orbits are bounded and outside of which they escape to infinity. The map can be written in terms of real variables as

$$X_{n+1} = X_n^2 - Y_n^2 \tag{11.1}$$

$$Y_{n+1} = 2X_nY_n + 0.62 \tag{11.2}$$

The basin boundary is a repellor (a *strange repellor* if it is a fractal) and can be found from almost any initial condition (except a periodic point) by running time backwards to make it an attractor $Z_n = \pm\sqrt{Z_{n+1} - C}$, which is equivalent to the map

$$X_n = \pm\sqrt{\frac{X_{n+1} - A}{2} + \frac{1}{2}\sqrt{(X_{n+1} - A)^2 + (Y_{n+1} - B)^2}} \tag{11.3}$$

$$Y_n = \frac{Y_{n+1} - B}{2X_n} \tag{11.4}$$

where $C = A + iB$. To produce the curve, choose the positive or negative root randomly at each iteration. Fractal curves of this type are called *dragons* because of their appearance, although many of them do not look much like dragons. Sometimes the Julia set is taken as the entire basin rather than its boundary, in which case it is more properly called the *filled-in Julia set* or *prisoner set* (orbits that cannot escape). Note that although the roots of eqn (11.3) are chosen randomly, the resulting pattern is completely deterministic after many iterations.

11.2.5 Weierstrass function

Another class of fractal curve is produced by the graph of a mathematical function such as the *Weierstrass function*[9] (Weierstrass 1895)

$$W(t) = \sum_{n=0}^{\infty} \frac{\cos 2^n t}{2^{(2-D)n}} \tag{11.5}$$

[9]The German mathematician Karl Theodor Wilhelm Weierstrass (1815–1897) had tried unsuccessfully to prove the stability of the Solar System, inspiring two of his former students who were professors at Stockholm to persuade King Oscar to offer a prize for finding the solution to be awarded on the King's sixtieth birthday. One of these students, Magnus Gsta Mittag-Leffler (1846–1927), along with Weierstrass and Charles Hermite (1822–1901), a professor at Paris, were judges, awarding the prize to Poincaré, who was Hermite's former student (Barrow-Green 1997).

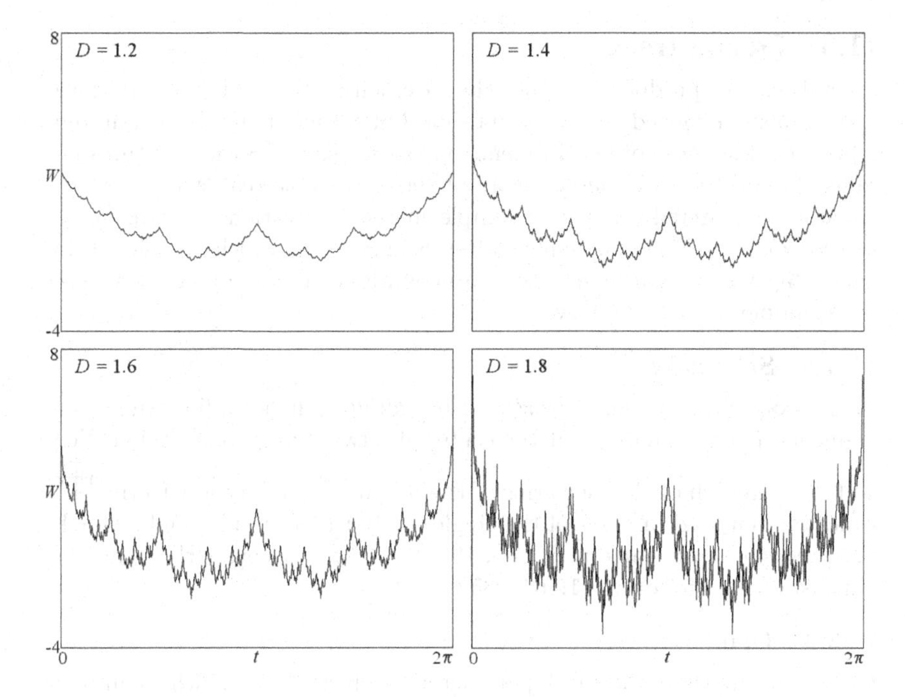

Fig. 11.6 Weierstrass function.

for $0 < t < 2\pi$ in which the parameter D is the dimension of the resulting curve for $1 < D < 2$. You will recognize this formula as a variant of a Fourier series (see §9.4.4) with frequencies extending to infinity. It is a continuous but nowhere differentiable function, shown in Fig. 11.6 for various values of D. The line is infinitely long for any $D > 1$. You should calculate it in at least double precision and truncate the summation when 2^n exceeds the number of horizontal pixels in your graphic display. Fractals of this sort violate the principles upon which differential calculus is based (that a curve is sufficiently smooth upon magnification that a derivative can be defined).

A graph of the Weierstrass function resembles the surface of a fractured metal or the horizon in the mountains. The case with $D = 1.2$ most nearly resembles terrestrial mountains, which have a similar fractal dimension (Turcotte 1992). Extraterrestrial mountains such as on the Moon or Mars have a higher dimension because of less erosion from wind and water and perhaps a wider range of scales over which they are self-similar. The fractal quality explains the difficulty of judging the distance to a mountain. Lacking other visual cues, a 10 000-foot mountain 100 miles away looks similar to a 1000-foot mountain 10 miles away.

11.3 Fractal trees

Fractals can be produced by lines that are neither infinitely long nor have had segments removed from their interior, but rather are made by infinitely replicating a pattern of line segments on various scales. Such structures are called *fractal trees*, although not all of them resemble real trees. We have already encountered one such example in the bifurcation diagram of the logistic map in Fig. 2.4, named after Feigenbaum, which coincidentally means 'fig tree' in German. There are countless variants on this theme, a few examples of which follow.

11.3.1 Snowflake

As an example of a fractal produced by infinitely many infinitesimal line segments, place a small '+' at the center of a two-dimensional region. Now add a '+' to each of the four arms of the original '+' to get a pattern which is then used in place of the single '+' in the next step, and so forth, until the *snowflake* pattern (Vicsek 1983) in Fig. 11.7 covers the plane. Its dimension is $\log 5/\log 3 = 1.46497352\ldots$.

11.3.2 Dendrite

For an appropriate choice of parameters such as $C = i$, which implies $A = 0$ and $B = 1$, the time-reversed Julia set in eqns (11.3) and (11.4)

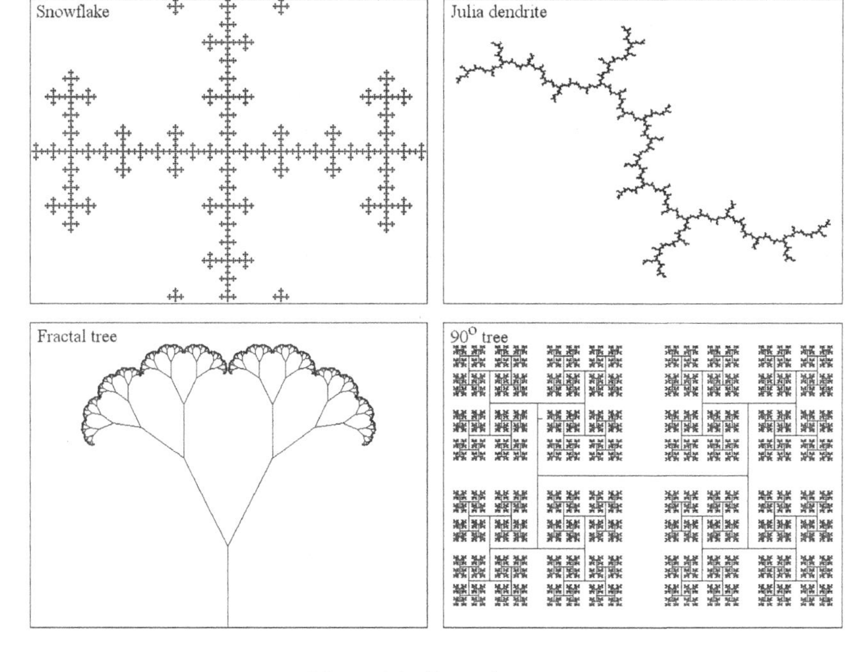

Fig. 11.7 Fractal trees.

has an attractor called a *dendrite*, as in Fig. 11.7. It is suggestive of a river basin, albeit with no outlet. Its dimension is about 1.3, which is only slightly higher than the dimension of about 1.2 for real river systems on Earth (Hack 1957).

11.3.3 Stick trees

The highly idealized *stick fractal tree* in Fig. 11.7 is produced starting with a vertical line segment (the trunk), at the top of which are two branches that make angles $\pm 27.5°$ from the trunk and whose length is reduced by a factor 0.544. The process is repeated at the end of each branch until the line segments become too short to display. The constants are chosen so the canopy is essentially two-dimensional, with the smallest twigs just touching, but not overlapping. Nature presumably grows trees this way to maximize the captured sunlight. Like the Cantor curtains in Fig. 11.2, the tree varies from one-dimensional at the bottom to two-dimensional at the top. Also shown in Fig. 11.7 is a variant of the fractal tree in which the angle is 90° at each branch and the length reduction per step is 2/3. Such a model might also describe the human circulatory system, which brings blood close to every point in the body, while occupying a relatively small volume.

11.3.4 Lindenmayer systems

Lindenmayer[10] (1968) proposed an axiomatic theory of biological development subsequently called *Lindenmayer systems* (or *L-systems*). These systems consist of commands applied *recursively* (the command can operate on itself) through string rewriting operations. The commands are relatively simple and resemble *turtle graphics*[11] used in the *LOGO* computer language:

F	Move forward one step while drawing a line
G	Move forward one step without drawing a line
+	Turn left through a specified angle
−	Turn right through a specified angle
[	Store the current position and angle (PUSH)
]	Return to the location and angle last stored (POP)

Different implementations use different commands and conventions, and some implementations allow additional commands such as to change the angle, step size, color, and line thickness.

You begin by specifying an axiom, an angle (an integer n such that the angle is $360°/n$), and the transformation rule(s). The initial step size is usually the size of the screen. For example, the system in the upper left of

[10] Aristid Lindenmayer (1925–1989) was a Dutch theoretical biologist who worked with yeast and filamentous fungi and developed a mathematical description of cell division.

[11] Turtle graphics is so named because you can imagine having a turtle that moves around the screen in response to your commands.

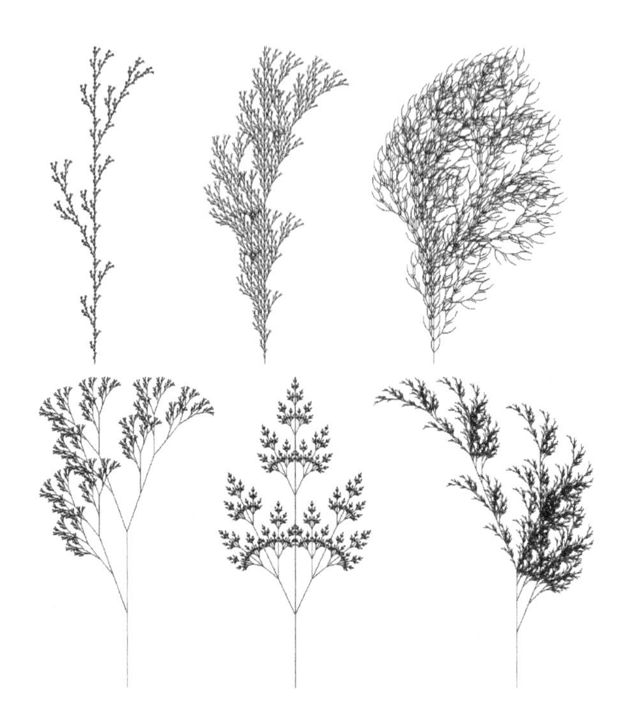

Fig. 11.8 Lindenmayer systems.

Fig. 11.8 has axiom = F, which would simply draw a line from the bottom to the top of the screen, an angle = 14 (25.7142857...°), and a transformation F = F[+F]F[−F]F. With each recursion,[12] every occurrence of the axiom F is replaced with the transformation, and the step size is reduced by the appropriate factor (3.0 in this case) until the resulting features are below the resolution of the screen. Sometimes the results are more realistic if the recursion is truncated after just a few levels, but then the object is a prefractal and not a true fractal. The resulting images in Fig. 11.8 resemble trees, but L-systems can generate many other fractal types (Prusinkiewicz and Hanan 1989).

11.4 Fractal gaskets

Fractals can also be produced from surfaces with infinitely many holes of various sizes, like Swiss cheese.[13] These *fractal gaskets* come in infinite variety, a few examples of which are given below. These examples were

[12]For an entertaining and provocative discussion of recursion in a wide variety of contexts, see *Gödel, Escher, Bach: an eternal golden braid* by Hofstadter (1980).

[13]Swiss cheese is a poor example of a fractal, since the U.S. Department of Agriculture has a standard that most of the holes, which are created by carbon-dioxide-emitting bacteria, should have diameters in the range of 3/8 to 13/16 inch.

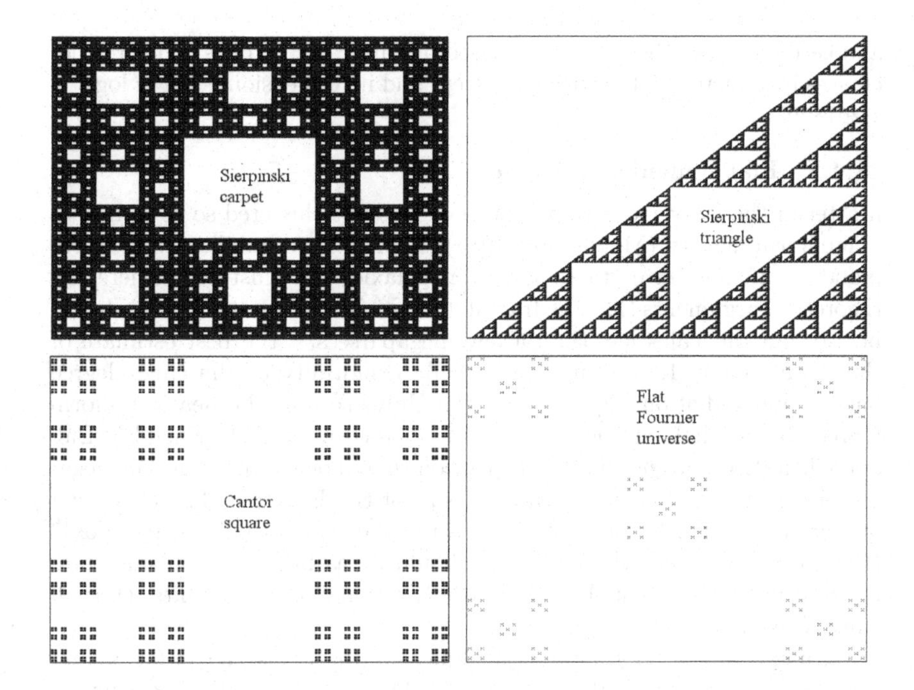

Fig. 11.9 Fractal gaskets.

produced using iterated function systems, described in §14.3, which is a generalization of the method used to find the Julia set boundaries in §11.2.4.

11.4.1 Sierpinski carpet

The *Sierpinski carpet*[14] (Sierpinski 1916) in Fig. 11.9 has nine regions, the middle of which is empty, and the other eight of which are reduced versions of the whole. Its area decreases by a factor of 8/9 with each step, and hence it approaches zero after infinitely many steps. Its dimension is $\log 8/\log 3 = 1.89278926\ldots$.

11.4.2 Sierpinski triangle

A related figure is the *Sierpinski triangle* in Fig. 11.9. The large triangle contains four smaller triangles, the middle of which is empty, and the other three of which are reduced versions of the whole. Its area decreases by a factor of 3/4 with each step, and its dimension is $\log 3/\log 2 = 1.58496250\ldots$.

11.4.3 Cantor square

The two-dimensional generalization of the Cantor set is the *Cantor square* (also called a *triadic Sierpinski carpet*) in Fig. 11.9. It consists of a square (or rectangle) from which the middle third is removed both horizontally

[14]Waclaw Sierpinski (1882–1969) was a prolific Polish mathematician and physicist who devoted most of his career to set theory.

and vertically, and the process repeated infinitely many times. The area decreases by a factor of $4/9$ with each step, and its dimension is $\log 4/\log 3 = 1.26185950\ldots$.

11.4.4 Flat Fournier universe

Matter in the Universe may be homogeneous but clustered so as to have a nonuniform density (Mandelbrot 1983). Planets have satellites, stars have planets, galaxies have stars, and even galaxies are clustered. There are theoretical arguments that the fractal dimension of the real Universe should be 1.0, but on scales less than a few megaparsecs,[15] the best estimate of the dimension is 1.23 (Wu *et al.* 1999). Uniformity of the three-degree background radiation requires that the Universe must be nearly uniform (three-dimensional) at the largest scale (the cosmological principle). Bak and Chen (2001) argue that the transition should occur at about 300 megaparsecs, or about 10% the estimated size of the Universe. The clustering of stars provides one (but not the best) explanation of Olbers' paradox[16] (Harrison 1987), that the night sky should be as bright as the surface of the Sun if stars are distributed uniformly throughout an infinite, eternal, static Universe.

A highly idealized fractal with dimension of 1.0 is the *flat Fournier universe*[17] (Fournier d'Albe 1907) in Fig. 11.9. It consists of a 5×5 matrix of squares, with the center and each corner square filled with a miniature replica of the whole. In this case the area decreases by a factor of $1/5$ with each step, and its dimension is $\log 5/\log 5 = 1.0$. The flat Fournier universe illustrates that fractals can have integer dimension and that everything with a dimension of 1.0 does not resemble a line. Instead, it fills space with the same density as would a line, which accounts for the difficulty of seeing it. Figure 11.9 illustrates how the dimension of these fractals affects their space-filling property.

The flat Fournier universe illustrates that dimension is not a sufficient criterion for characterizing a fractal since fractals of the same dimension can have very different textures. This quality is called *lacunarity* (Mandelbrot 1983). A fractal with high lacunarity is heterogeneous and consists of a wide range of gap sizes, whereas one with a low lacunarity is more homogeneous and translationally invariant with similar gap sizes, but lacunarity requires more than one numerical value to quantify it.

[15]A *megaparsec* is a million parsecs, where a *parsec* is defined as the distance at which parallax causes a nearby astronomical object to move by one second (1/3600 degree) of arc relative to the distant stars over the six months during which the Earth is at the extremes of its orbit around the Sun (1 parsec $\simeq 3.08 \times 10^{16}$ meters $\simeq$ 3.26 light-years).

[16]Olbers' paradox is named after the German astronomer Heinrich Olbers (1758–1840), although it was discussed much earlier by others, including Johannes Kepler (1571–1630) and Edmond Halley (1656–1742), after whom the famous comet was named.

[17]Fournier d'Albe (1868–1933) was an Irish patriot, science journalist, and inventor who constructed a device, called an *optophone*, that enabled the blind to 'hear' written text and who transmitted the first television signal from London.

Note that all these fractals with dimension less than 2.0 are a set of measure zero in the plane. Thus you would not be able to see them if they were plotted correctly, just as you would not be able to see a line if it were plotted with zero width. You could throw darts at these objects, even the Sierpinski carpet with dimension nearly 1.9 that looks mostly filled in, and you would never hit them.

11.5 Fractal sponges

Any of the preceding examples can be generalized to three-dimensional embeddings. Such objects are variously called *sponges*, *foams*, or *webs*, depending on their shape and dimension. It is difficult to display them in two dimensions, but Fig. 11.10 shows a stereoscopic view of a three-dimensional generalization of the Sierpinski triangle, which requires viewing as described in §6.6.6. It consists of four self-similar tetrahedrons with a fifth in the center empty, and thus its volume decreases by a factor of 4/5 at each step. It has a fractal dimension of $\log 4/\log 2 = 2.0$, and thus the projection of any view onto the plane has finite area. However, the object does not look like a surface, just as the flat Fournier universe with dimension 1.0 does not resemble a line. Mandelbrot (1983) calls this case a *fractal skewed web*, but a better name might be a *Sierpinski tetrahedron*, although there is no evidence Sierpinski ever considered it. It resembles an upward view of the interior of the Eiffel Tower, which is constructed from trusses whose members are smaller trusses.

The three-dimensional generalization of the Sierpinski carpet in Fig. 11.9 is called a *Menger sponge*[18] (Blumenthal and Menger 1970), sometimes

Fig. 11.10 Stereoscopic view of the Sierpinski tetrahedron.

[18]Karl Menger (1902–1985) was an Austrian mathematician who was also deeply interested in philosophy, economics, and ethics. He constructed the sponge that now bears his name in 1926 as a 'universal one-dimensional set.'

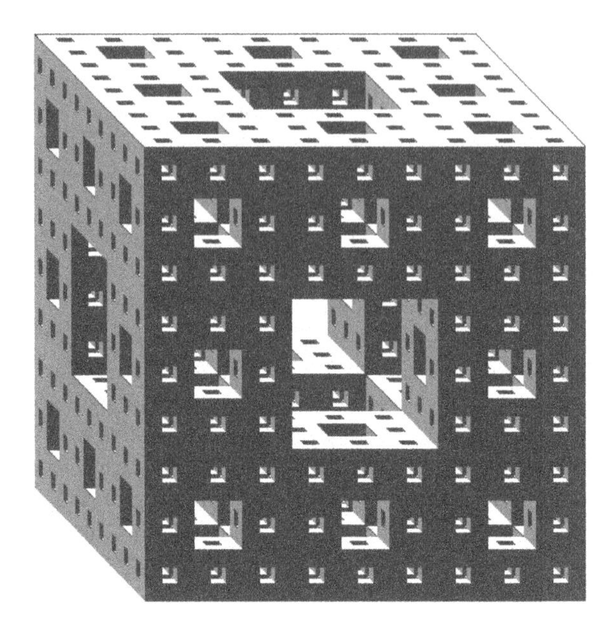

Fig. 11.11 First few generations of the Menger sponge.

wrongly called a *Sierpinski sponge*, and has a dimension of $\log 20/\log 3 = 2.72683302\ldots$. Figure 11.11 shows the first few generations of the Menger sponge. Since the sponge has no volume, it has no weight and could soak up a volume of water equal to its size.

Objects with fractal dimension $2 < D < 3$ have biological analogs as with the lungs, whose area increases with resolution, implying a fractal dimension of about 2.17 (Weibel 1979). The branching of blood vessels into ever smaller capillaries is another example of a biological fractal. Nature exploits fractal geometry to increase fluid exchange in a small volume by making structures with essentially infinite area.

Fractals can be embedded in dimensions higher than three and their properties calculated by the above methods, but it is increasingly hard to visualize them as the dimension increases. The visualization methods for strange attractors in §6.6 can be applied to other types of fractals.

11.6 Random fractals

The fractals illustrated above are produced by simple deterministic rules, and thus have exact self-similarity. With more complicated nonlinear rules giving chaotic orbits, the resulting strange attractors are usually self-similar only in the limit of small scales. The strange attractors in Fig. 6.1 seem to have structure on all scales, but in most cases the self-similarity is not apparent, and you would not expect to find miniature versions of the attractors embedded within them. However, if you magnify some appropriate detail, as shown with the Hénon map in Fig. 5.6, you will see the self-similarity.

Fractals produced by random rules, such as the randomly chosen positive or negative square roots in the Julia set examples in Figs 11.3 and 11.7, can have exact self-similarity. However, random rules do not usually give exact self-similarity even in the limit of infinite resolution, but only *statistical self-similarity*. That is to say, a small portion of the object is not an exact replica of the whole, but it should have the same distribution. Most natural fractals, such as trees, rivers, mountains, and coastlines have this quality. The following examples show some fractals produced by simple random rules. They are representative of an infinite variety of such objects, not only because there are infinitely many possible rules, but because a given rule will produce a different image depending on the sequence of random numbers used to generate it. Furthermore, not every shape produced by a random process is a fractal, since self-similarity, even over a limited range, is not guaranteed.

11.6.1 Random Cantor sets

Perhaps the simplest example of a random fractal is the Cantor set in which one-third of each line segment is removed at each step, but the portion removed is chosen randomly rather than from the middle. The resulting set has the same statistical properties as the triadic Cantor set in §11.1, but it is not exactly self-similar. Another way to produce a random Cantor set is to replace each line segment at each step by two lines with lengths chosen randomly but whose sum is less than the length of the line segment being replaced. In such a case the self-similarity is even less exact.

11.6.2 Random fractal curves

The fractal curves in Figs 11.3–11.5 can be made random by choosing more than one generator to apply randomly to the initiator. For example, the von Koch snowflake in §11.2.3 has a generator _/_ that can be applied either as shown or upside down. If the two variants are chosen randomly, then the resulting von Koch snowflake in Fig. 11.12 is only statistically self-similar but better resembles a natural coastline.

11.6.3 Random fractal gaskets

The methods used to produce the fractal gaskets in §11.4 can be modified to produce the *random fractal gaskets* in Fig. 11.13. At each step the image is divided into nine regions, as with the Sierpinski carpet and Cantor square in Fig. 11.9, but rather than replicating the whole pattern in a fixed subset of the regions, it is replicated in a random subset whose number is constant only on average. This process is called *curdling*. The cases in Fig. 11.13 correspond to average area contractions of 5/9, 6/9, 7/9, and 8/9, giving the fractal dimensions shown in the figure. Such images might model the distribution of vegetation (or people) over the surface of the Earth.

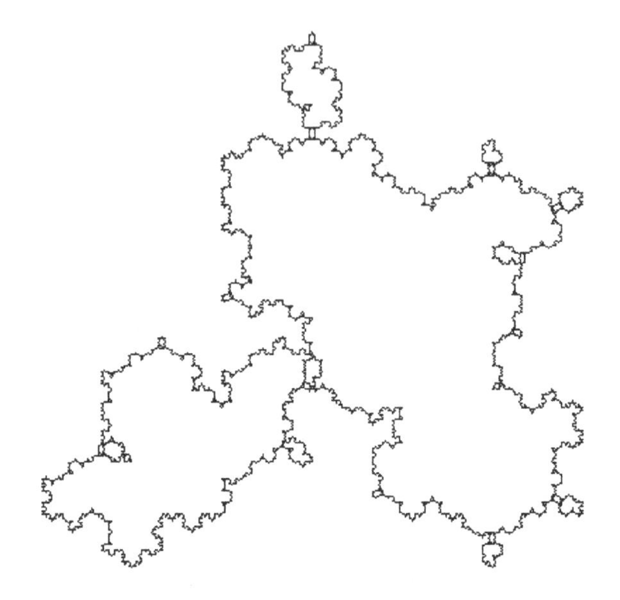

Fig. 11.12 Random variant of the von Koch snowflake.

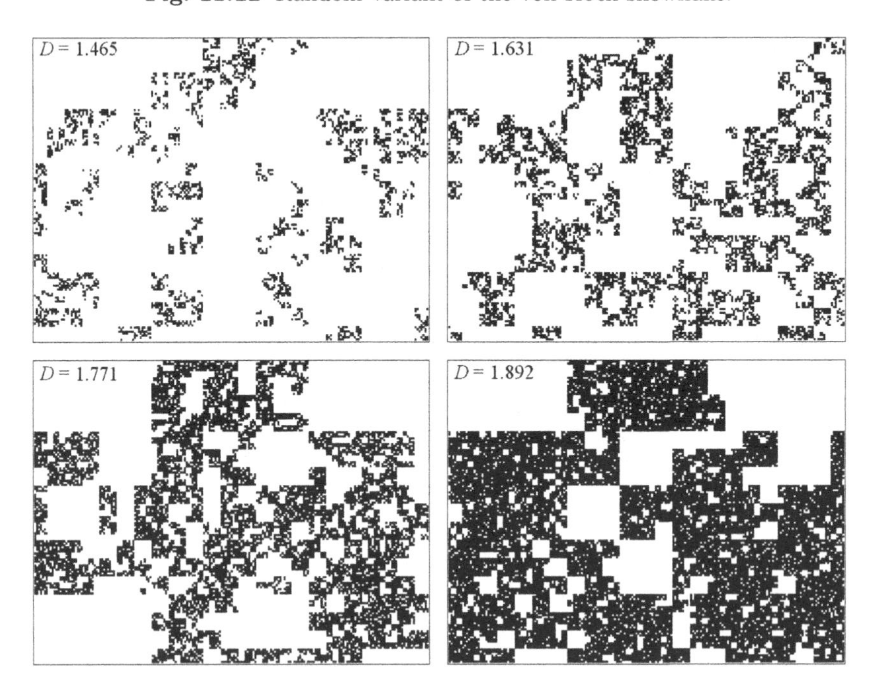

Fig. 11.13 Random fractal gaskets of various dimensions.

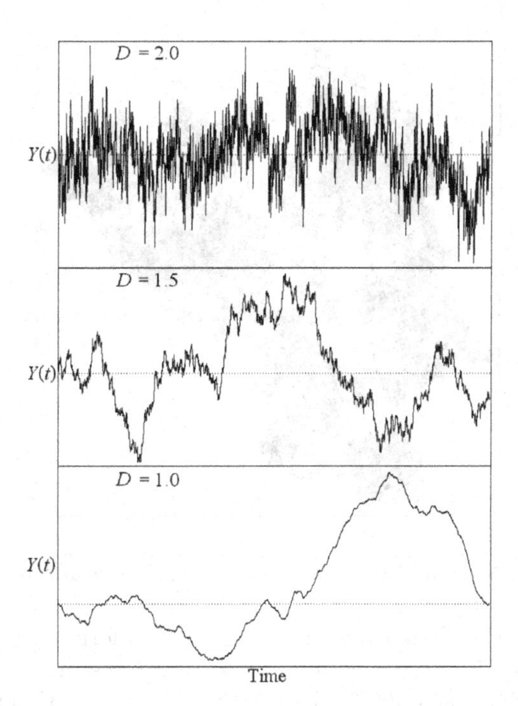

Fig. 11.14 Time series for fractional Brownian motion with various power laws.

11.6.4 Fractional Brownian motion

We already mentioned in connection with the Hurst exponent in §9.4.6 that *fractional Brownian motion* (Mandelbrot and van Ness 1968) can produce random fractals, both in the trace of $X(t)$ and in the resulting trail (or trajectory) in some embedding space. Figure 11.14 shows $Y(t)$ for $1/f^\alpha$ noise with $\alpha = 1, 2$, and 3, respectively, whose dimension is $D = (5 - \alpha)/2$. The middle case is ordinary Brownian motion identical to Fig. 9.6. The plots resemble the Weierstrass function in Fig. 11.7 except they are only statistically self-similar. Thus it is a much more realistic representation of the horizon in the mountains, whose dimension is about 1.2. Fractals produced by the trace of fBm are actually *self-affine*, since scaling the time by a factor of A scales the amplitude by a factor of A^H, where H is the Hurst exponent.

Fractional Brownian motion gives a trail in space as in Fig. 11.15 for the case of $\alpha = 2$ (ordinary Brownian motion). Although this figure was produced by having the computer move one step randomly either up, down, left or right at each iteration, the same plot is produced by two independent Brownian functions as in Fig. 11.14, one for the horizontal position and the other for the vertical position, or by choosing the step size from a Gaussian distribution and the angle from a distribution uniform over the range zero to 2π. The resulting plot has a fractal dimension of 2.0. The zero crossings

Fig. 11.15 Trail of Brownian motion (random walk) in a plane.

(called the *zeroset*) of the Brownian function $Y(t)$ form a fractal *Lévy dust* with dimension 0.5. More generally, the *isosets* (crossings of any Y value) for fBm have dimension $D = (3 - \alpha)/2$ for $1 \leq \alpha \leq 3$, and the time t_c between crossings obeys a *Lévy distribution* $P(t_c) \propto t_c^{-D}$.

The Brownian trajectory will come arbitrarily close to every point in the plane infinitely many times, including the starting point, but the probability that it will hit a given point exactly is zero. Curiously, for Brownian motion in three or more dimensions, the fractal dimension is still 2.0, and the trajectory does *not* return arbitrarily close to its starting point or intersect itself. Even more curiously, a Brownian trail of *finite* length (obtained by terminating the motion after a finite time) also has dimension 2.0 but has zero area, since a line of finite length has no area. Furthermore, the self-intersections of the trail form a set of dimension 2.0. The trail resulting from fBm has a fractal dimension of $2/(\alpha - 1)$, provided the dimension is not less than 1.0 or greater than the dimension of the space.

11.6.5 Diffusion-limited aggregation

Random fractals can be produced by a process called *diffusion-limited aggregation* or *DLA* (Witten and Sander 1983). Start with a single point at the center of a plane. A particle (called a *monomer*) begins a random walk from a random point on a circle surrounding this point. If it contacts the point (the point becomes one of its eight nearest neighbors on a two-dimensional square grid), it sticks, and it too becomes sticky. If it intersects the circle, it is reflected so as to remain always inside the circle. The circle should be very large, but to speed the development, it is made just slightly larger than the maximum radius of the pattern and grows as necessary.

The resulting pattern after many steps is on the left of Fig. 11.16. The

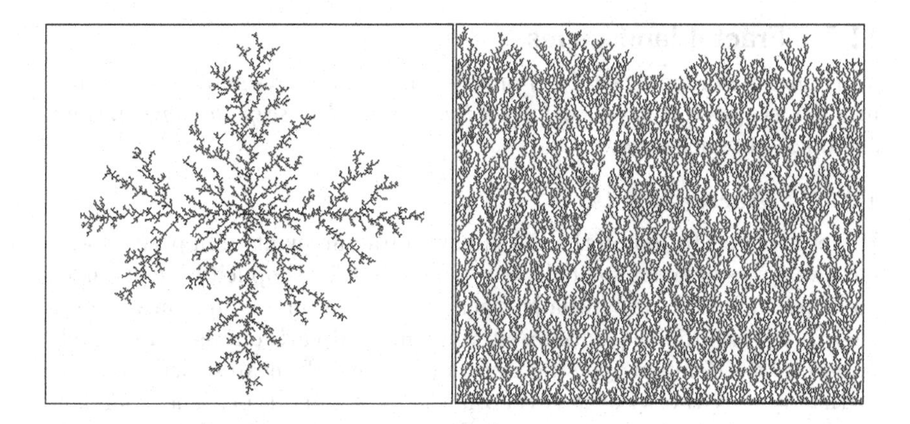

Fig. 11.16 Examples of diffusion-limited aggregation.

pattern is a type of dendrite similar to what you might see in a lightning discharge or the patterns of ice crystals that form on your windows in the winter. Similar patterns are produced by *electrodeposition*, *corrosion*, and *viscous fingering*, such as the interface of water pumped through a porous medium like oil shale in the standard method of squeezing oil from rocks (Stokes *et al.* 1986). The average number of particles contained in the pattern increases as a power of its radius, $N \propto R^D$. The fractal dimension in a plane is about $D \simeq 1.71$. Such patterns can also be produced in higher dimensions (Meakin 1983a). In three dimensions, the fractal dimension is about 2.50. The tendency of the pattern to have a four-fold symmetry is a consequence of growing it on a rectangular grid. The exact value of the dimension of a natural object of this sort depends on the physical and chemical processes that produced it. Measuring the fractal dimension is thus a useful way to identify these mechanisms.

On the right of Fig. 11.16 is a similar pattern in which the particles rain down on the object and stick to the first surface they contact. A less dense pattern is produced if the particles random walk horizontally as they move downward, reflecting off the walls or wrapping to the opposite wall as desired.

Related to diffusion-limited aggregation is *cluster agglomeration* where a collection of particles random walk until one touches another, whereupon they stick and thereafter move together as a small cluster (Meakin 1983b). The clusters in turn form larger clusters, eventually producing a fully connected self-similar object with a fractal dimension of about 1.4 in two dimensions and 1.8 in three dimensions (Kolb *et al.* 1983).

Cluster agglomeration also occurs with urbanization, epidemics, and information flow in society. By analogy with *genes*, information replicators that spread though communication have been called '*memes*' (Dawkins 1976) and include ideas, beliefs, habits, morals, fashions, and techniques. They inflict not only human societies, but increasingly computer networks.

11.7 Fractal landscapes

You can use random fractals to produce images that resemble various natural landscapes (often called *fractal forgeries* or *landscape forgeries*), some examples of which follow.

11.7.1 Fractal forests

Patterns of vegetation on the Earth have a fractal quality. Suppose a forest has two types of trees, say oaks and pines, that for simplicity are arranged on a rectangular grid. Initially, each point contains a randomly chosen tree. The trees die sequentially at random, and each one is immediately replaced by a different type if six or more of its eight nearest neighbors are of that type. Otherwise, it is reborn with the same type. The boundary conditions are periodic, meaning that a point on the right edge is assumed to be adjacent to the corresponding point on the left edge, and similarly for the top and bottom. You can think of the forest as existing on a toroidal surface. After many generations, the forest pattern in Fig. 11.17 emerges. There are regions of oaks and pines intertwined in a complicated way. The pattern is not strictly self-similar, and the boundary structure is a remnant of the random initial condition.

This example is a *cellular automaton*, described in §15.1. You could extend the method to forests with more than two tree types, different rules,

Fig. 11.17 Examples of fractal landscapes.

different neighborhood sizes, and different boundary conditions such as might exist on an island surrounded by an ocean (Sprott *et al.* 2002). The same model could be used for other systems such as the distribution of Democrats and Republicans in a United States voting population, the spread of epidemics, or the elevation contours on a topographical map.

11.7.2 Fractal craters

Craters on the Earth are mostly volcanoes, but those on the Moon, Mars, and other satellites in the Solar System are usually formed by meteorites. Observations (Arthur 1954) suggest that lunar craters occur with a probability inversely proportional to their area (or square of their diameter since most are nearly circular). Distributing circles randomly in the plane with such a self-similar distribution gives the fractal craters in Fig. 11.17. After an infinite time, the entire area of the body would be covered by craters, unless some process occasionally removes them or unless they are a remnant of an earlier phase of the Solar System in which most of the bombardment occurred. The craters on some satellites of the outer planets have a different scaling (Soderblom 1980). If you were shown a photograph of the surface of a planet with fractal craters, you would be unable to determine its scale.

11.7.3 Fractal terrain

We already noted that the time series for fractional Brownian motion (Fig. 11.14) resembles the silhouette of mountains. Thus fBm can be used to produce realistic landscapes. To do this, you need to generalize the method to two dimensions. Equation (9.35) gives a formula for producing one-dimensional fBm with a given power spectrum. It involves the sine of a random variable. The cosine of the same random variable gives an orthogonal time series

$$X_n = \sum_{m=1}^{N/2} m^{-\alpha/2} \cos 2\pi(mn/N + r_m) \tag{11.6}$$

$$Y_n = \sum_{m=1}^{N/2} m^{-\alpha/2} \sin 2\pi(mn/N + r_m) \tag{11.7}$$

where the constant $a_0/2$ has been omitted and the power spectrum taken as $S_m = 1/m^\alpha$. From these formulas you can construct a series of fBm curves from

$$Z_n = X_n \cos\phi + Y_n \sin\phi \tag{11.8}$$

where ϕ is a phase chosen randomly for each curve but kept constant for a given curve.

The terrain in Fig. 11.17 shows a sequence of 32 such curves with $\alpha = 2.6$, giving a fractal dimension of $D = (5 - \alpha)/2 = 1.2$ as appropriate for natural terrain. Each curve is displaced vertically from its neighbor, and only the portion of the curve that lies above the one below it is plotted, since the remainder would be occluded by the nearer terrain. Finally the sky is filled with gray, and an image of the Moon is added for effect. This simple method gives surprisingly realistic terrain and could be greatly improved with color. The self-similarity explains the difficulty estimating the distance to a mountain on the horizon at dusk.

11.7.4 Fractal oceans

If you can make artificial terrain, you can also make artificial oceans and continents by dividing the terrain into two regions according to its elevation. You can choose an appropriate elevation to mimic the fact that the Earth's surface is 71% water. The oceans in Fig. 11.17 were produced this way, except that the elevation was taken as $Z_{mn} = |X_m + Y_n|$, where m is the longitude (measured from east to west) and n is the latitude (measured from north to south) on a sphere.[19] To display the sphere, project it onto a plane using

$$x = R \cos \frac{\pi m}{N} \sin \frac{\pi n}{N} \tag{11.9}$$

$$y = R \cos \frac{\pi n}{N} \tag{11.10}$$

where R is the radius of the sphere and N is the number of values of n and m in the time series. This transformation gives a circle of radius R centered on $x = y = 0$. The resulting coastlines have a fractal dimension of 1.2 and are representative of continents on the Earth.

11.7.5 Plasma clouds

Some of the nicest natural fractals are clouds, whose billows have billows. Satellite photographs show structure from less than a kilometer to over a thousand kilometers. The perimeter of clouds has a fractal dimension 1.35 ± 0.05 over the entire range (Lovejoy 1982). Furthermore, the regions of precipitation on radar maps have the same dimension. The self-similarity explains the difficulty estimating the height of clouds. A cloud with a diameter of 1 kilometer at a height of 10 kilometers looks the same as one with a diameter of 100 meters at a height of 1 kilometer. The self-similarity is remarkable since the length scales encompass the atmospheric thickness (about 10 km). Theoretical estimates (Hentschel and Procaccia 1984) based on atmospheric turbulence predict a dimension $1.37 < D < 1.4$.

The algorithm that produced the fractal terrain in Fig. 11.17 can be modified to produce clouds in a manner similar to the way it was used to create oceans and continents. Regions are black where the altitude is less than some fixed value, above which it is given a shade of gray proportional to the logarithm of the altitude (Voss 1985).

[19]Mathematicians always take a sphere to be a two-dimensional object defined as the locus of points equidistant from its center, whereas physicists (and most other normal people) use the term to mean the three-dimensional space enclosed by it. To avoid confusion, you can use the term 'spherical shell' for the former and 'ball' for the latter.

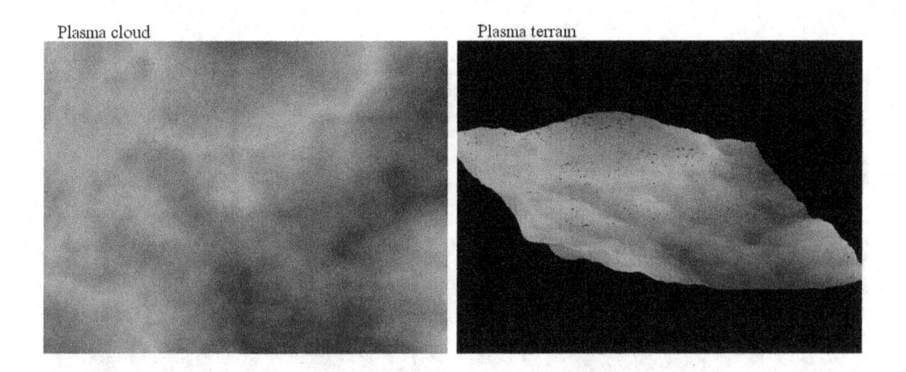

Fig. 11.18 Plasma clouds and corresponding terrain.

Another variant is called a *plasma cloud*, an unfortunate terminology since the word 'plasma' has three other unrelated meanings (the liquid component of blood and other fluids, an ionized gas, and a green quartz crystal). Choose random numbers for the corners of a rectangle. Then divide the rectangle into four equal smaller rectangles and linearly interpolate the original four values to get values for the five new corners, but add a uniform random value with mean zero and a maximum equal to their difference to each one. Repeat the process until you reach the resolution of the screen. The size of the random numbers determines the graininess of the image. Convert the resulting values to a gray scale and plot the result[20] as on the left of Fig. 11.18. The values so obtained can also be used to create a fractal terrain as on the right of Fig. 11.18. These images, in contrast to the previous ones in this chapter, have 256 gray levels. They are especially stunning when viewed in color.

11.8 Natural fractals

When you walk in the woods, you rarely see forms that resemble classical Euclidean shapes, such as lines, planes, circles, and spheres. More often, the objects are better described as fractals (Porter and Gleick 1990, McGuire 1991, Guyon and Stanley 1991, Briggs 1992, Hirst and Mandelbrot 1994, Ruderman and Bialek 1994). As Mandelbrot (1983) puts it,

> Clouds are not spheres, mountains are not cones, coastlines are not circles, and bark is not smooth, nor does lightning travel in a straight line.

We thus conclude this chapter with a collection of photographs in Fig. 11.19 showing familiar objects in nature that have a fractal quality, although none of them is precisely self-similar, and what self-similarity that does exist is usually confined to a small range of scales. The ubiquity of fractal forms in nature is probably a consequence of the ubiquity of chaos in nature.

[20]This figure was produced with the freeware program FRACTINT.

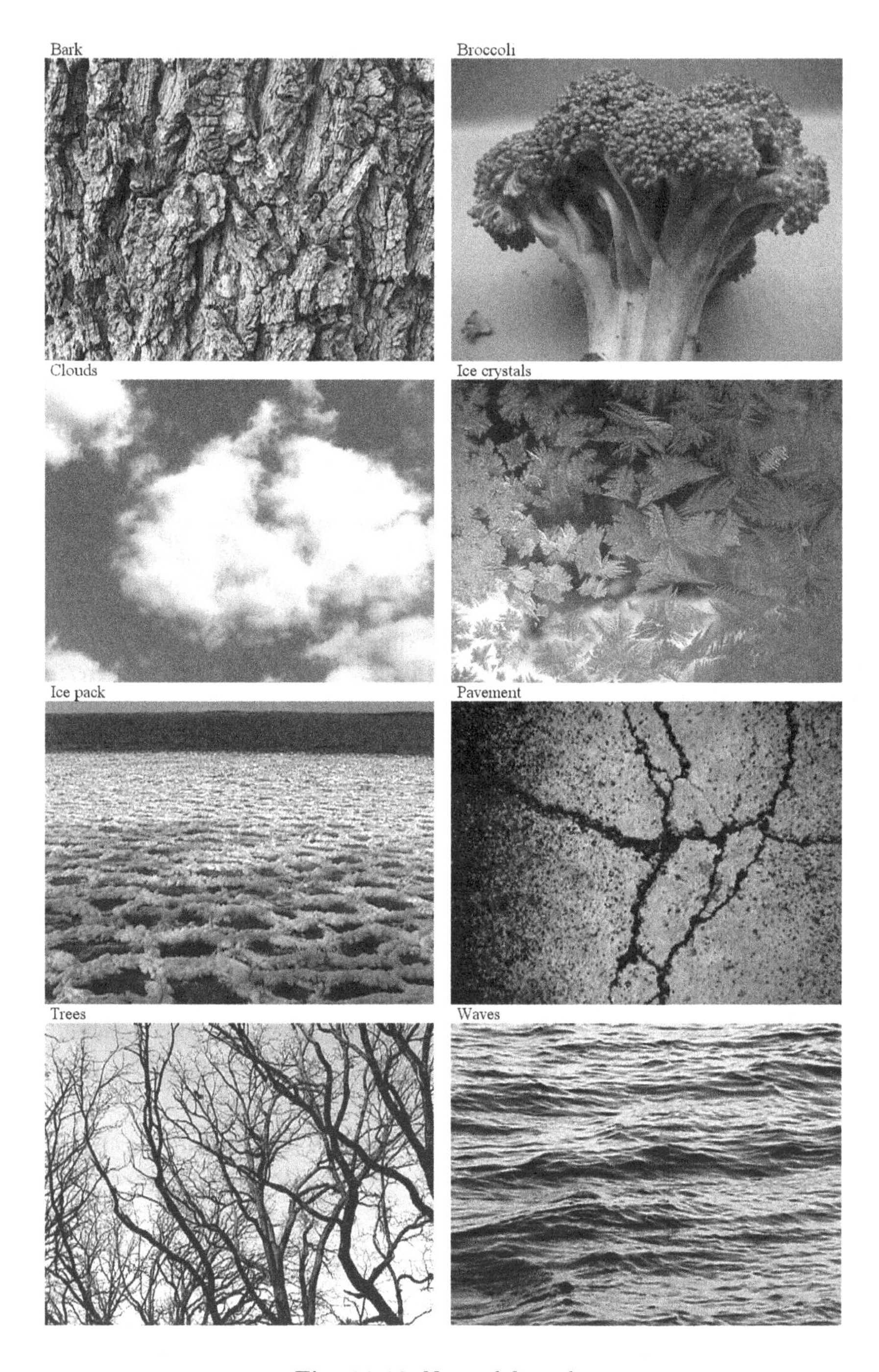

Fig. 11.19 Natural fractals.

11.9 Exercises

Exercise 11.1 Calculate the number of line segments after k steps in the construction of the Cantor set, and show that this number goes to infinity for $k \to \infty$.

Exercise 11.2 Show that the triadic Cantor set in the interval $0 < X < 1$ consists of those points whose base-3 representation contains only the symbols 0 and 2.

Exercise 11.3 Show that any number in the interval $0 < X < 2$ can be represented exactly as a sum of two Cantor numbers.

Exercise 11.4 Show that the length of the Devil's staircase is $x+y$, where x is the horizontal distance and y is the vertical rise.

Exercise 11.5 Calculate the factor by which the length of the Hilbert curve increases with each step in its generation.

Exercise 11.6 Show that the area bounded by the von Koch snowflake in Fig. 11.3 is $\frac{2}{5}L^2\sqrt{3}$, where L is the length of each side of the original equilateral triangle from which it is formed.

Exercise 11.7 Show that eqns (11.1) and (11.2) are equivalent to the complex map $Z_{n+1} = Z_n^2 + 0.62i$.

Exercise 11.8 Show that eqns (11.3) and (11.4) are the time-reversed iteration of $Z_{n+1} = Z_n^2 + C$.

Exercise 11.9 Show that the Weierstrass function in eqn (11.5) is self-similar by finding a relation between $W(t/2^m)$ and $W(t/2^{m+1})$, where m is an arbitrary integer.

Exercise 11.10 Derive an expression for the derivative of the Weierstrass function as a function of the number of terms N in the summation, and show that the derivative is everywhere infinite for $N \to \infty$.

Exercise 11.11 Calculate the power spectrum $S(f)$ for the Weierstrass function in eqn (11.5).

Exercise 11.12 Write an L-system description of the triadic Cantor set in Fig. 11.1.

Exercise 11.13 Write an L-system description of the Devil's staircase in Fig. 11.3.

Exercise 11.14 Write an L-system description of the Hilbert curve in Fig. 11.3.

Exercise 11.15 Write an L-system description of the von Koch snowflake in Fig. 11.3.

Exercise 11.16 Write an L-system description of the fractal tree in Fig. 11.7.

Exercise 11.17 From the area reduction factors for the fractal gaskets in §11.4 and the quoted fractal dimensions, see if you can deduce the rule by which the dimension of these objects is calculated.

Exercise 11.18 Generalize the Cantor square to three dimensions, and calculate the factor by which its volume changes at each step in the construction.

Exercise 11.19 Generalize the Cantor square to four dimensions, and calculate the factor by which its hypervolume changes at each step in the construction.

Exercise 11.20 Generalize the flat Fournier universe to three dimensions, and calculate the factor by which its volume changes at each step in the construction.

Exercise 11.21 Generalize the flat Fournier universe to four dimensions, and calculate the factor by which its hypervolume changes at each step in the construction.

Exercise 11.22 For two adjacent isolated particles subject to diffusion-limited aggregation, show that the cluster is more likely to grow in length than breadth, and calculate the relative probabilities for a new particle to attach at various sites.

Exercise 11.23 Suppose the largest crater on the Moon has area A_0, the radius is of the Moon is R, and the probability distribution of craters is inversely proportional to their area. Calculate the fraction of the surface of the Moon that is covered with craters, considering that some of the smaller craters are within larger craters and thus do not contribute to the area covered by craters.

11.10 Computer project: State-space reconstruction

This project is intended to convince you that you can take a single variable of a chaotic system (in this case the Lorenz attractor) sampled at discrete times and reconstruct an attractor in arbitrary dimension that has the same topological properties as the original attractor. This procedure is a necessary step in the analysis of chaotic experimental data. Examination of such plots will often reveal periodicities that are not apparent in a graph of the original time series. This project is an extension of the one in §9.9. In fact, you can use the same data set you generated there.

1. Using a time series of 1000 points from the x variable of the Lorenz attractor with the usual parameters of $\sigma = 10, r = 28$, and $b = 8/3$, sampled at equally spaced time intervals of $\Delta t = 0.05$, after discarding

the initial transient, make a plot of X_n versus X_{n-k} for $k = 1, 2, 4$, and 8. Convince yourself that the resulting plot resembles the plot of the original Lorenz attractor. Is there an optimal value for k, and how does it relate to the correlation time you calculated in §9.9?

2. Make a phase-space plot of your time series (plot dx/dt versus x). For this purpose, it suffices to approximate dx/dt by $(X_{n+1} - X_{n-1})/2\Delta t$. Convince yourself that this plot also resembles the original Lorenz attractor. This method tends to be inferior to the method of time delays above whenever the data record contains significant noise.
3. Make a return-map plot of your data, for example by plotting the local maximum of x versus the previous local maximum, and show that the plot appears to be one-dimensional. (It is actually a fractal with small-scale structure that would appear under high magnification if you had a sufficient number of data points.) To find the local maximum, it will suffice to plot those points whose two nearest neighbors are smaller. However, to see the fractal structure, you would need to do slightly better by fitting these three points to a parabola whose maximum you calculate analytically and/or by using a smaller value of Δt in the original time series.
4. (optional) Repeat parts 1 and 2 above, except using three-dimensional plots, in the first case with axes of X_n, X_{n-k}, and X_{n-2k}, and in the second case with axes of $x, dx/dt$, and d^2x/dt^2. You may approximate d^2x/dt^2 by $(X_{n+1} + X_{n-1} - 2X_n)/\Delta t$. Use your ingenuity and some of the ideas discussed in Chapter 6 to display the attractors in spaces with dimension higher than two.

12

Calculation of the fractal dimension

Much of the interest in chaos is motivated by the hope that an apparently random time series from some experiment arises from a low-dimensional dynamic, in which case it is amenable to simple modeling and short-term prediction. Chapters 9 and 10 described a few such models and discussed how you might fit them to data and test for predictability. Another important indicator of low-dimensional chaos is the existence of a strange attractor. Detecting chaos in experimental data involves estimating the dimension of this attractor embedded in some space, usually with time delays. The dimension measures the minimum number of variables required to model the process.

The Kaplan–Yorke dimension discussed in §5.9 is a dynamical measure of the attractor dimension. Unfortunately, its calculation requires that you determine the spectrum of Lyapunov exponents, which is difficult if you only have a time series rather than the equations that produced it. If the dimension is sufficiently small, you need only plot each data point versus its predecessor or perhaps take a Poincaré section to reveal the fractal structure. If the dimension is too high for this method to work but low enough to be interesting (say less than about five or six), better methods are available. In higher dimensions where the governing equations are known, the Kaplan–Yorke conjecture provides the best estimate of the attractor dimension. The difficulty of calculating it scales linearly with the attractor dimension times the embedding dimension, whereas the other methods become exponentially more difficult.

There are many (in fact infinitely many) ways to define and calculate an attractor dimension, none of which can claim to be the 'true' fractal dimension. If the orbit visits the regions of the attractor uniformly, the various dimensions should be identical, and the convergence is rapid, but this case is rare. However, you are probably interested only in an approximate value of the dimension, in which case the methods in this chapter will suffice since the various dimensions are usually not very different. More refined dimension estimates for attractors with nonuniform measures are described in Chapter 13.

When you calculate the dimension, with very little extra effort you can also calculate the *entropy*, which is the sum of the positive Lyapunov exponents and measures the average rate at which predictability is lost, and the *BDS statistic*, which measures the departure of the data from pure randomness. We will show examples of these calculations and discuss practical considerations in calculating these quantities from a limited amount of perhaps noisy or otherwise corrupted data.

12.1 Similarity dimension

Fractals with exact and obvious self-similarity can be characterized by their *similarity* (or *self-similarity*) *dimension*), best explained by example. The triadic Cantor set in Fig. 11.1 has the property that each step in its construction produces two pieces, each a factor of three shorter than the corresponding piece in the previous step. If you imagine forming the set by successively cutting a uniform string, the number of pieces N at each step is related to the length reduction R at that step by $N \propto R^{D_s}$ or

$$D_s = \frac{\log N}{\log R} \tag{12.1}$$

where D_s is the similarity dimension. For the triadic Cantor set with $N = 2$ and $R = 3$, the dimension is $D_s = \log 2/\log 3 = 0.63092975\ldots$. Note that the dimension is a ratio of two logarithms, and thus it does not matter what base you use for the logarithms since $\log_a x = \log_b x/\log_b a$. If you were to cut the string successively into two equal halves without removing any string at each step, the dimension would be $\log 2/\log 2 = 1.0$, as you would expect for a line. If you were to cut it into two pieces of negligible length and discard the middle part, the dimension would be $\log 2/\log \infty = 0$, as you would expect for a pair of points.

As another example, consider the Sierpinski carpet in Fig. 11.9. It contains nine regions, each smaller than the whole by a factor of three in linear dimension, but the center piece is missing, leaving only eight. Thus the similarity dimension is $D_s = \log 8/\log 3 = 1.89278926\ldots$. Remember that the numerator is always the logarithm of the number of self-similar pieces, and the denominator is always the factor by which the *linear* dimension (length) changes at each step. This simple rule allows you to verify the dimension of those fractals in the previous chapter that are exactly self-similar.

Imagine that the Cantor set is formed with a rubber band rather than a string or that the Sierpinski carpet is made of rubber that you stretch nonuniformly to destroy the self-similarity. The fractal dimension is not changed by such a stretching process since it preserves the topology (the connectedness of the set). The topological properties, including fractal dimension, are more robust than the geometrical properties such as self-similarity. Thus you can often determine the fractal dimension of an object by stretching it to make it self-similar, and then calculating the similarity dimension.

12.2 Capacity dimension

We need a general method to calculate the dimension of objects that are not self-similar such as the random fractal gaskets in Fig. 11.13. The method should give a value of 1.0 for a line with a well-defined length, even one like the Devil's staircase in Fig. 11.3, and it should give a value of 2.0 for a filled-in, nonzero area, even with a very complicated perimeter and many holes like the trail of Brownian motion in Fig. 11.15.

Suppose you had a large number of square tiles of various sizes $L_1, L_2, L_3, \ldots$, where L is the length of a side. Count the *minimum* number of tiles N_1 of size L_1 required to cover the object whose dimension you want to calculate, as in Fig. 12.1, for the *von Koch curve* (one-third of the von Koch snowflake). Repeat with tiles of different sizes, and plot N versus L on a log–log scale. If the object being covered is a line, you would expect $N \propto 1/L$, and if it is a surface, you would expect $N \propto 1/L^2$ since the area each tile covers is L^2. More generally, you expect $N \propto 1/L^{D_0}$ or

$$D_0 = \lim_{L \to 0} \frac{\log N}{\log(1/L)} \tag{12.2}$$

where D_0 is called the *capacity dimension*, *cover dimension*, *grid dimension*, or *box-counting* (or simply *box*) *dimension* (Russell *et al.* 1980) since the generalization to three dimensions would cover the object with cubic boxes of size L.

Another way of thinking about the definition is that D_0 is the dimension such that NL^{D_0} is neither zero nor infinite in the limit $L \to 0$. If D_0 is altered even by an infinitesimal amount, then the product will diverge either

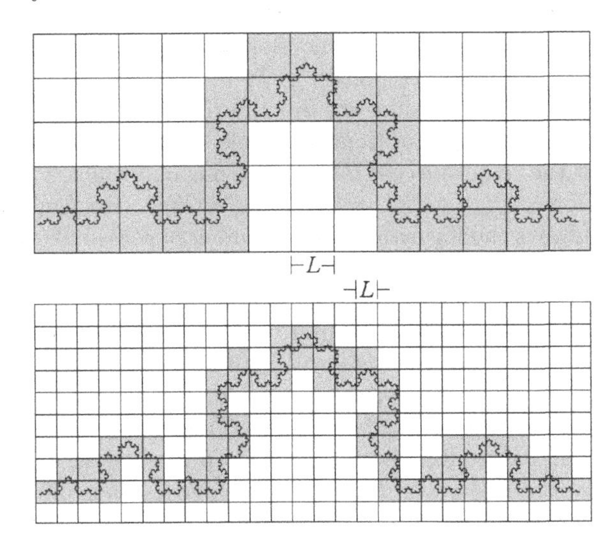

Fig. 12.1 Calculation of the capacity dimension for the von Koch curve using box counting.

to zero or infinity. This definition is the basis for the *Hausdorff–Besicovitch dimension*, often simply called the *Hausdorff*[1] *dimension* (Hausdorff 1919). The Hausdorff dimension uses boxes of varying sizes and has nice mathematical properties but is hard to calculate.

The capacity dimension is easier to calculate and is an upper bound on (greater than or equal to) the Hausdorff dimension, and sets can be constructed in which they are very different (Essex and Nerenberg 1990, Mainieri 1993). In many cases, especially for strange attractors, they are apparently equal (Ott 1993). For the von Koch curve, the boxes in the lower half of Fig. 12.1 with $N = 64$ and $L = 1/26$ lead to a predicted dimension of $D_0 \simeq \log 64/\log 26 \simeq 1.276476\ldots$, whereas the similarity dimension is $D_s = \log 4/\log 3 = 1.261859\ldots$. Mandelbrot's early definition of a fractal was a set whose Hausdorff dimension strictly exceeds its topological dimension, but that definition excludes sets that are generally considered fractal, such as the Devil's staircase (see §11.2.1).

You can calculate the capacity dimension for any embedding dimension using hypercubes.[2] Note that the units in which you measure L do not matter as long as the unit is large enough that the limit above can be calculated for $L \ll 1$. For best convergence of the limit, measure L in units comparable to the size of the object. It is usually easier to determine the slope of the $\log N$ versus $\log 1/L$ curve than to calculate the $L \to 0$ limit of eqn (12.2). Sets can be constructed for which the limit in eqn (12.2) does not exist, in which case the capacity dimension is not defined.

In practice, fill the plane completely with tiles of size L, and count the number of these tiles that cover some portion of the object for various L. The result will depend slightly on where the tiles are located, and for careful work you might want to use various placements and take the minimum (Appleby 1996). In the limit of infinitesimal boxes, the placement does not matter, and the result is identical to that for optimally placed boxes. The same is not true for the Hausdorff dimension, however.

Results for a number of ordinary and fractal objects are shown in Fig. 12.2. These examples were produced by plotting the objects at a resolution of 640×480 and counting the occupied pixels for tiles of size $1, 2, 4, \ldots$ pixels on a side, and measuring L in units of the width (640). The convergence is relatively slow, and even these simple low-dimensional objects give dimensions that differ from expectation (the slope of the straight line in the plots on the right) by typically 10%. To do better requires much higher resolution (smaller L), but that taxes the computer memory, especially in higher embedding dimensions where most of the boxes are empty

[1]Felix Hausdorff (1868–1942) was a German mathematician, philosopher, poet, and playwright who, along with his wife and sister, committed suicide when sentenced to an internment camp by the Nazis during World War II.

[2]A four-dimensional hypercube is called a *tesseract* and contains 16 vertices (corners), 32 edges, 24 squares, and 8 cubes. More generally, an n-dimensional hypercube has 2^n vertices. Do not try to visualize it!

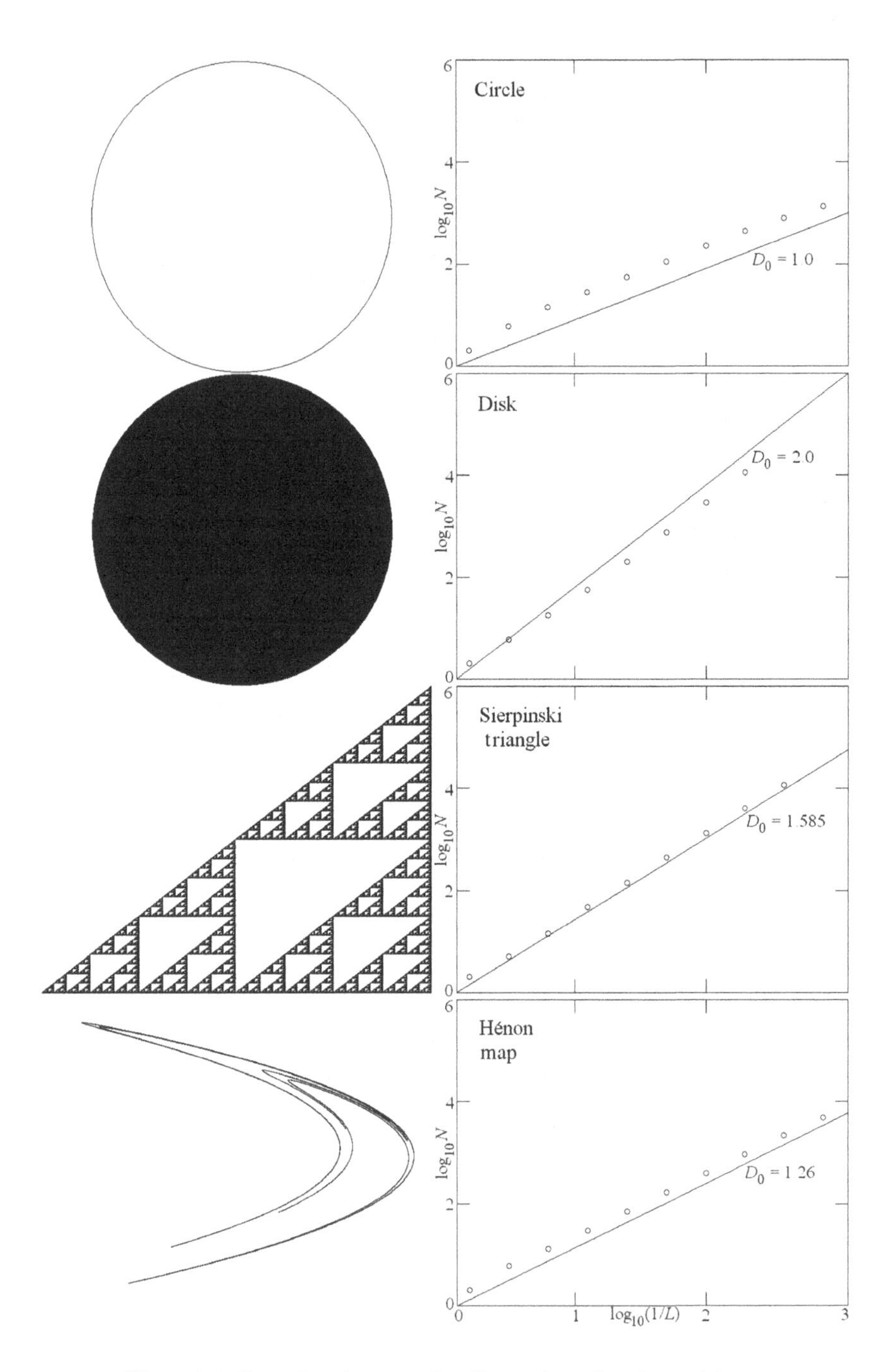

Fig. 12.2 Data for the capacity dimension of various objects.

(Grassberger 1983a), and the method becomes impractical in dimensions higher than about three (Greenside *et al.* 1982). However, see Liebovitch and Toth (1989) for a fast box-counting algorithm. Note that the limit $L \to 0$ corresponds to the right side of the plot where $\log(1/L) =\!\!\longrightarrow \infty$.

12.3 Correlation dimension

The difficulties in calculating the capacity dimension are largely circumvented by instead calculating the *correlation dimension.* Suppose the object is formed from N points in some embedding space, taken as two-dimensional for ease of visualization. This case would correspond to a time series in which X_n is plotted versus X_{n-1}. Now pick one of the points and draw a circle of radius r around it, and count the number of other points within that circle, as in Fig. 12.3. Repeat for circles of various radii and for all the points. If the data come from a uniformly dense line, you would expect the number of such points to increase linearly with r for small r, and if from a uniformly dense surface, to increase with the square of r, since the area of a circle is πr^2. In general, the power of r is not an integer and is a measure of the fractal dimension.

Let $C(r)$ be the number of points within all the circles of radius r, normalized so that $C(r)$ is 1.0 when r is sufficiently large that it includes all the points without double counting

$$C(r) = \frac{2}{N(N-1)} \sum_{j=1}^{N} \sum_{i=j+1}^{N} \Theta(r - r_{ij}) \tag{12.3}$$

where Θ is the *Heaviside function*[3]

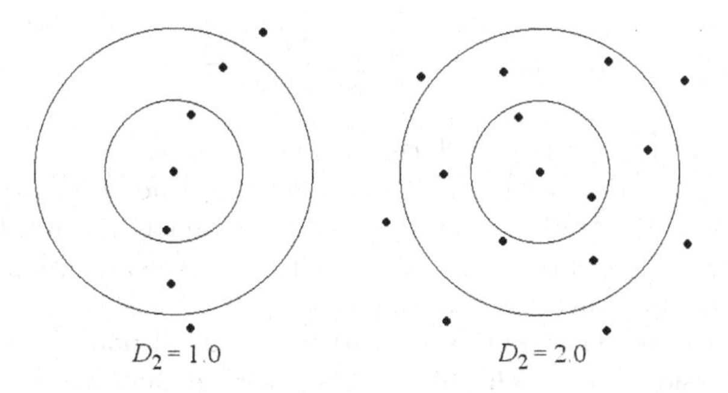

Fig. 12.3 Calculation of the correlation dimension for a line and a surface represented by a finite number of data points.

[3]The Heaviside function, also called the *Heaviside step function*, was named after Oliver Heaviside (1850–1925), an English mathematician, who formulated Maxwell's equations of electromagnetism and predicted the existence of the Earth's ionosphere, also called the *Heaviside layer*.

$$\Theta(x) = \begin{cases} 0 & \text{for } x < 0 \\ 1 & \text{for } x \geq 0 \end{cases} \tag{12.4}$$

and r_{ij} is the spatial separation between two points labeled i and j, usually given in an m-dimensional time-delay embedding by the *Euclidean norm*[4] (*Pythagorean theorem*)

$$r_{ij} = \sqrt{\sum_{k=0}^{m-1} (X_{i-k} - X_{j-k})^2} \tag{12.5}$$

$C(r)$ is called the *correlation sum*, and it converges to the *correlation integral* for $N \to \infty$. Note that $m-1$ data points are lost from the sum in eqn (12.3) to allow for the time-delay embedding. In dimensions higher than two, the circles become (hyper)spheres. This procedure, due to Grassberger and Procaccia (1983a,b), is called the *Grassberger–Procaccia algorithm* and has been very widely used (Grassberger *et al.* 1991, Berliner 1992). The importance of excluding $i = j$ is discussed by Grassberger (1988) but is often overlooked. Calculating it both ways can be enlightening.

Equivalently, think of $C(r)$ as the probability that two different randomly chosen points will be closer than r. In fact, choosing points randomly from the time series will often lead to faster convergence when you have more data than you can reasonably analyze sequentially. You would expect $C(0) = 0$ for a chaotic system since the points never repeat in a nonperiodic system embedded without false nearest neighbors. A plot of $\log C(r)$ versus $\log r$ should give an approximately straight line whose slope in the limit of small r and large N is the correlation dimension

$$D_2 = \lim_{r \to 0} \lim_{N \to \infty} \frac{d \log C(r)}{d \log r} \tag{12.6}$$

Lacunarity in the set often produces oscillations about the line and makes $D_2(r)$ fluctuate somewhat (Smith *et al.* 1986, Arnéodo *et al.* 1987, Theiler 1988b). In fact the trace of the $D_2(r)$ curve is apparently itself a fractal for most chaotic systems, even at large values of r where statistical fluctuations are unimportant. The subscript on D_2 distinguishes the correlation dimension, which considers the separation of pairs of points, from the capacity dimension D_0, which only counts boxes. It anticipates an infinite sequence of *generalized dimensions* D_q for $-\infty < q < \infty$, described in the next chapter.

[4]Other norms such as the L_s norm $r_{ij} = [\sum_{k=0}^{m-1} |X_{i-k} - X_{j-k}|^s]^{1/s}$ can be used and in principle give identical dimensions. L_1 is the *absolute norm* (also called the *taxicab norm* or *Manhattan norm* for reasons left to your imagination), and L_∞ is the *maximum norm* or *supremum norm*, which is easy to compute, but only L_2 (the Euclidean norm) is *rotationally invariant* (independent of the orientation of the coordinate systems in which it is measured).

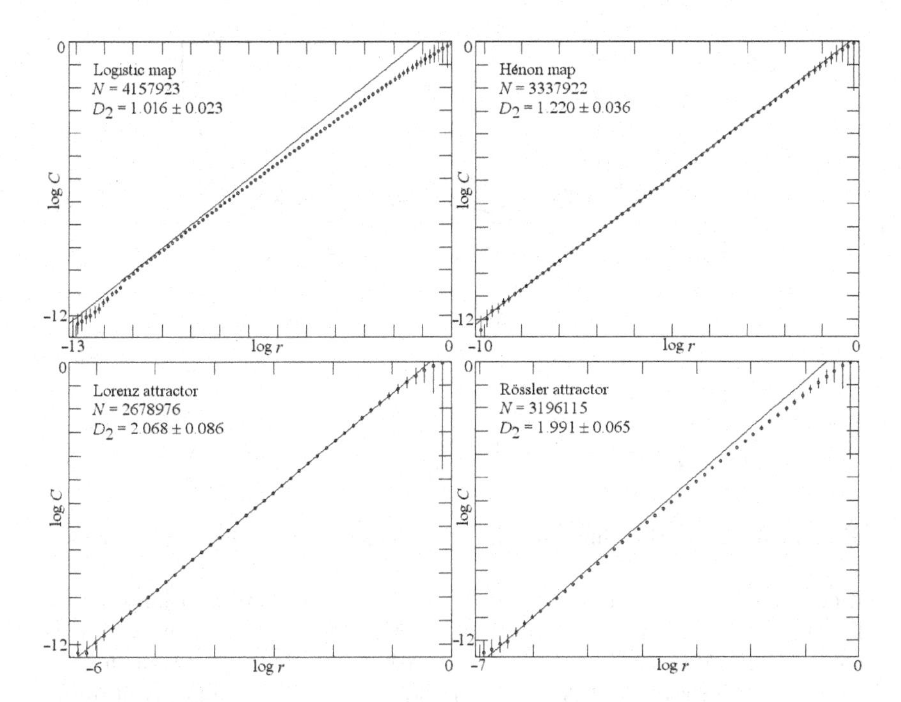

Fig. 12.4 Correlation dimension for various chaotic systems (base-10 logarithms).

The Grassberger–Procaccia algorithm is applied to various chaotic systems in Fig. 12.4. These systems use on the order of 3×10^6 data points, requiring about 10^{13} distance calculations. Data are binned according to their value of $\log_{10} r$ with $2 \log_2 10 = 6.643856\ldots$ bins per decade of r. For each case, r is normalized to the largest value of r_{ij} (the attractor size). The logistic map is embedded in one dimension with $r_{ij} = |X_i - X_j|$, the Hénon map is embedded in two dimensions with $r_{ij} = \sqrt{(X_i - X_j)^2 + (Y_i - Y_j)^2}$, and the Lorenz and Rössler attractors are embedded in three dimensions with $r_{ij} = \sqrt{(X_i - X_j)^2 + (Y_i - Y_j)^2 + (Z_i - Z_j)^2}$, sampled at time intervals of 0.001 and 0.01, respectively.

In these cases, we have the luxury of knowing the relevant variables, and we do not have to resort to time-delay embedding. When analyzing a scalar time series in which you do not know the optimal embedding, you should calculate D_2 in increasing embeddings until it ceases to change, as in Fig. 12.5 for the Hénon map. A clear saturation is evident for $m \geq 2$. For attractors with less uniform measures, the saturation is less pronounced, and there is some benefit in using an embedding higher than theoretically necessary (Grassberger and Procaccia 1983b).

Figure 12.5 also shows the typical oscillations caused by the lacunarity of the set, about which more will be said in the next chapter. These oscil-

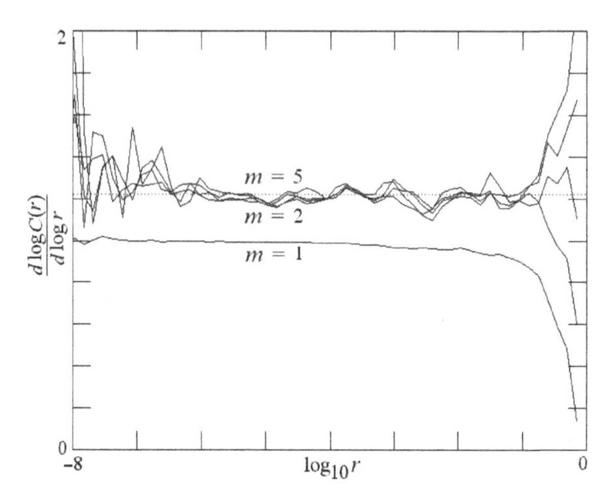

Fig. 12.5 Correlation dimension for the Hénon map in various embeddings.

lations, as well as the departure from constancy at small and large r, are all but invisible in the undifferentiated plot in Fig. 12.4, showing why it is usually more informative to plot $D_2(r)$ rather than $\log C(r)$ versus $\log r$.

Rather than retain all $N \sim 3 \times 10^6$ points, a circular buffer containing the most recent 3.2×10^4 points is used in these examples to conserve computer memory. This shortcut suffices because the transitivity of chaotic systems ensures that any portion of the orbit samples the attractor with the same probability as any other. Each time a new point is generated in the time series, it is correlated with all the points in the buffer and it then replaces the least recent point. Thus N is an equivalent value, much smaller than the actual number of points generated. It is often more computationally expedient to generate additional data as needed than to store a large array.

You will notice in Fig. 12.5 that the calculated dimension has large statistical errors for small r where the bins contain relatively few data points, and that it departs from constancy when r becomes comparable to the size of the attractor. We expect a *scaling region* where the value is approximately constant, aside from small oscillations, at intermediate values of r. This region is always sandwiched between a region of saturation and a region of poor statistics, noise, or round-off errors. Note from Fig. 12.4 that the scaling is good for the Hénon map and Lorenz attractor, but that the convergence for $r \to 0$ is slow for the logistic map and the Rössler attractor. The cause of this problem and its cure will be discussed in the next chapter as well as the method for estimating the error in the calculated dimension.

The problem of choosing a scaling region is partially resolved by using the *Takens' estimator* (Takens 1985, Theiler 1988b), in which the correlation dimension T_2 for scale lengths smaller than r is

$$T_2(r) = \frac{C(r)}{\int_0^r C(r)/r dr} = \frac{1}{\langle \ln(r/r_{ij}) \rangle} = \frac{\sum \Theta(r - r_{ij})}{\sum \ln(r/r_{ij})\Theta(r - r_{ij})} \tag{12.7}$$

where the sums are over all pairs of points with separation $r_{ij} < r$. This value is simply an average dimension for spatial scales smaller than r, with higher weights for the more nearby points. Kantz and Schreiber (1997) recommend taking r about half the rms amplitude of the data, but you can also use the largest r for which the calculated error is consistent with the prediction of $T(r)$ for all smaller values of r. The Takens' estimator is optimal (Theiler 1993) since it has a Gaussian distribution of values and an unbiased mean. If offers little advantage over eqn (12.6) for clean, abundant, low-dimensional data such as from the Hénon map in Fig. 12.5, but it is better for sparse, high-dimensional, noisy data from real experiments. See also Judd (1992) and Borovkova *et al.* (1999) for further discussion and critique.

The values of D_2 (or T_2) are similar to the Kaplan–Yorke dimension D_{KY} (see Appendix A). They are also similar to the capacity dimension D_0, but are easier to calculate and more accurate. The correlation dimension is more efficient at quantifying the fractal nature of an attractor because it emphasizes regions of state space that contain the most data. For an attractor with uniform measure, we expect $D_2 = D_0$; otherwise $D_2 < D_0$ (Grassberger and Procaccia 1983b). The correlation dimension may be more physically meaningful since it emphasizes regions of the attractor visited most frequently by the orbit, rather than being a purely geometric quantity like the capacity dimension, although time-delay reconstruction and prediction depend more on D_0.

12.4 Entropy

The Kaplan–Yorke conjecture in eqn (5.29) provides an estimate of the attractor dimension from the Lyapunov exponents. Conversely, the dimension is useful for determining the Lyapunov exponents. According to *Pesin's identity* (Pesin 1977), the sum of the positive Lyapunov exponents is the *Kolmogorov–Sinai* (or *K–S* or *metric*) *entropy* (or *invariant*). Kolmogorov (1958) proposed applying the concept of entropy to dynamical systems, and Sinai (1959) gave a refined definition and proof. More precisely, according to *Ruelle's inequality* (Ruelle 1978), the K–S entropy is a lower bound on the sum of the Lyapunov exponents, but they are equal when the natural measure is continuous along the unstable directions, as is usually the case for chaotic flows. It is related to the usual thermodynamic entropy, which is a measure of the disorder[5] of a system, since it measures the expansion of nearby trajectories into new regions of state space. However,

[5]The concept of thermodynamic entropy as a measure of disorder traces back to Ludwig Boltzmann (1844–1906) and Josiah Willard Gibbs (1839–1903), the latter of whom called it 'mixedupness,' but the term 'entropy' was coined by Rudolph Julius Emmanuel Clausius (1822–1888), a German physicist and mathematician who wanted to stress its similarity to 'energy.' Entropy means evolution in Greek.

unlike the thermodynamic entropy, the K–S entropy has units of *inverse time* (or *inverse iterations* for maps) and is a measure of the average rate at which predictability is lost. Its inverse is a rough estimate of the time for which reasonable prediction is expected. A purely random system has infinite entropy, and a periodic system has zero entropy.

The entropy can also be considered as the rate of creation of information as a chaotic system evolves, called *Shannon's entropy*.[6] The sensitive dependence on initial conditions means that two indistinguishable nearby points in state space separate in time. As time advances, we learn more and more about the initial condition as initially insignificant digits in its specification make themselves felt. A chaotic system is an endless source of new information (Shaw 1981). It has been said that information is a measure of the *surprise* of an occurrence or the *distinguishability* of a configuration, and that entropy is the information we do not yet have. The entropy is always positive and can be large for a chaotic system. Sometimes this process is viewed as a *loss* of information since predictions from a given initial state become *less* accurate as time advances. Which viewpoint you adopt is a matter of taste.

Unfortunately, the K–S entropy is difficult to calculate for the same reasons the capacity dimension is difficult to calculate. Instead, we calculate the *correlation entropy*, analogous to the correlation dimension, which is a close lower bound on the K–S entropy in the same way the correlation dimension is a close lower bound on the capacity dimension (Grassberger and Procaccia 1983b,c, Cohen and Procaccia 1985). It is given by

$$K_2 = \lim_{m \to \infty} \lim_{r \to 0} \lim_{N \to \infty} \log \frac{C(m,r)}{C(m+1,r)} \tag{12.8}$$

where $C(m,r)$ is the correlation sum in eqn (12.3) for an embedding dimension m. Thus you get the correlation entropy for free whenever you calculate the correlation dimension in various embeddings. You must be careful to use the same normalization of $C(m,r)$ for the different embeddings. There is a spectrum of *metric entropies* K_q for $-\infty < q < \infty$, called *Rényi's*[7] *q-order entropies* (Rényi 1961), but they are identical for attractors with uniform measure (see §13.6).

[6]The connection between information and entropy was developed by Claude Elwood Shannon (1916–2001), an eccentric American mathematician who founded information theory (Shannon 1948) but who also invented curious machines such as a gasoline-powered pogo stick, a rocket-powered Frisbee, and a computer called THROBAC (THrifty ROman-numeral BAckward-looking Computer) based on Roman numerals and who enjoyed diversions such as riding a unicycle while juggling down the halls at Bell Labs and MIT. He credits the great mathematician John von Neumann with recommending the term 'entropy,' since no one really knows what it means and thus he would have an advantage in debates about his theory.

[7]Alfréd Rényi (1921–1970) was a Hungarian mathematician, known as 'Buba,' who is remembered for saying that 'a mathematician is a machine for converting coffee into theorems.'

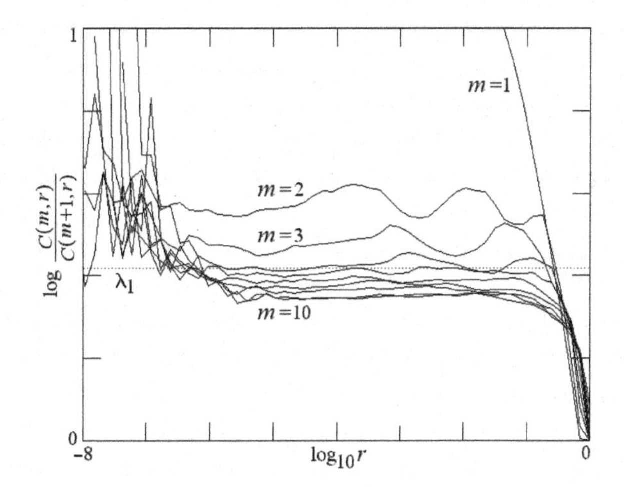

Fig. 12.6 Correlation entropy for the Hénon map in various embeddings.

The limits in eqn (12.8) are somewhat incompatible, and so you should plot $\log[C(m,r)/C(m+1,r)]$ versus $\log r$ for various m as in Fig. 12.6 for the Hénon map and look for a plateau. Since $C(m,r)$ increases with $\sqrt{m}$ for a time-delay embedding with the Euclidean metric, convergence can be improved by rescaling r (Frank *et al.* 1993). The Hénon map has a single positive exponent whose value should equal the entropy. The plateau occurs close to the expected value of $K_2 \simeq \lambda_1 \simeq 0.41922$, and for a value of m not much greater than required to embed the attractor ($m = 2$). In fact, taking m too large degrades the result because of the sparseness of data. This method is not an accurate way to calculate the sum of the positive exponents, but it can provide a consistency check with other methods.

12.5 BDS statistic

The correlation sum in various embeddings can also be used as a measure of determinism in a time series. Brock *et al.* (1996) showed that for a purely random time series with no memory (*independent and identically distributed*, or *IID*, random values, signifying white noise), the correlation sum satisfies $C(m,r) = C(1,r)^m$. The reason is simply that random data should fill any chosen embedding, and thus the density of points in a hypersphere of radius r centered on any point is inversely proportional to its hypervolume, which scales as r^m. Departure from equality is a signature of underlying determinism, whether linear or nonlinear. The corresponding *BDS* (Brock, Dechert, Scheinkman) *statistic* is

$$W(m,r) = \lim_{N\to\infty} \sqrt{N}[C(m,r) - C(1,r)^m]/\sigma(m,r) \tag{12.9}$$

where $\sigma(m,r)/\sqrt{N}$ is a complicated normalization that you can ignore when comparing a time series with appropriate surrogate data (Brock *et*

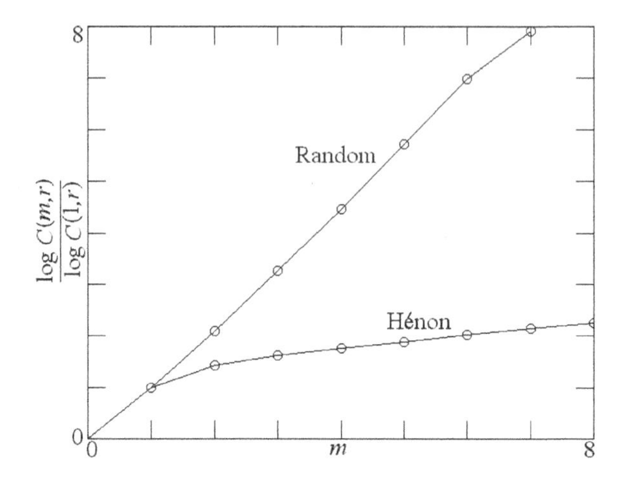

Fig. 12.7 Ratio used to calculate the BDS statistic for random and Hénon map data.

al. 1991). A statistically significant nonzero value of $W(m, r)$ is evidence for determinism in the time series.

The method is illustrated in Fig. 12.7 in which $\log C(m, r)/\log C(1, r)$ is plotted versus m for the Hénon map and for a sequence of random numbers distributed uniformly over the range 0 to 1. The Hénon case saturates for $m \geq 2$, whereas the random case increases linearly up to at least $m = 7$. These results were obtained with $r = 1 \times 10^{-4}$.

12.6 Minimum mutual information

When calculating a fractal dimension from a time-delay embedding, you must choose an appropriate embedding dimension using false nearest neighbors (see §10.5) or another method such as looking for saturation of the measured dimension as m increases. If the time series comes from a flow sampled at discrete times, you must also choose a time step Δt. In principle, this choice should not matter. In practice, if the time step is too small, each data point is very close to its predecessors, and the attractor is stretched out along the diagonal in the embedding space, requiring a very small r to see a dimension different from one. At the other extreme, taking the time step too large leads to excessive folding of the attractor, since temporally distant points are completely uncorrelated in a chaotic system.

Figure 12.8 shows the Lorenz attractor embedded in $m = 2$ for four different time lags, indicating how the attractor lies along the diagonal for small Δt but is excessively folded for large Δt. Although the correlation dimension is about 2.07, it appears lower for small time lags and higher for large time lags unless you consider very small values of r.

One rule of thumb is to choose the time delay less than the time at which the autocorrelation function in eqn (9.18) decays to $1/\mathrm{e}$, but greater

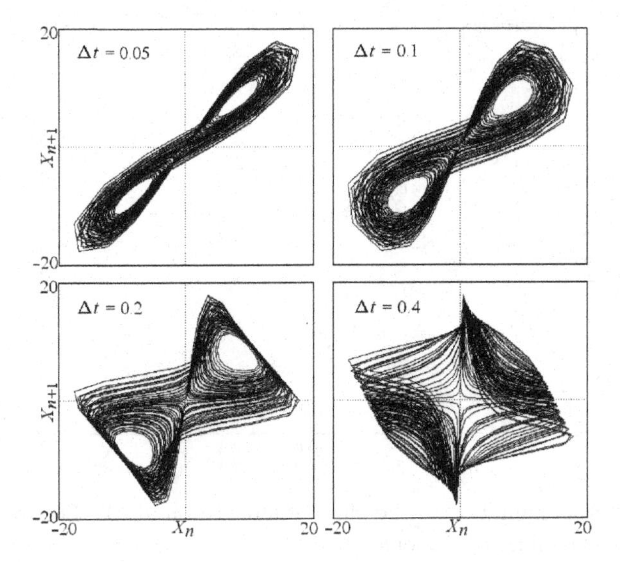

Fig. 12.8 Lorenz attractor embedded in two dimensions for various time lags.

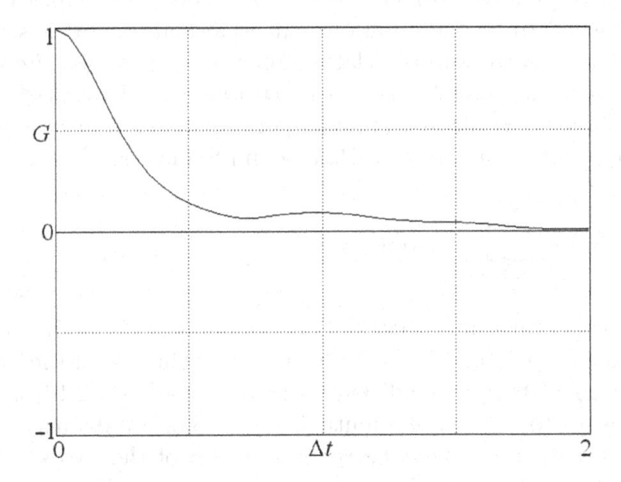

Fig. 12.9 Autocorrelation function for the Lorenz attractor, showing a correlation time of about 0.3.

than this time divided by $m - 1$. Figure 12.9 shows the autocorrelation function for the Lorenz attractor with a correlation time of about 0.3. When embedded in a space with $m = 3$, the predicted optimal time delay is in the range 0.13 to 0.3, which agrees with Fig. 12.8. More simply, choose a delay large enough that the attractor begins to fill the embedding space, but no larger (Buzug and Pfister 1992, Rosenstein *et al.* 1994). In practice, experimental data often have more than one correlation time-scale, in which case the calculated dimension and other quantities may change with Δt,

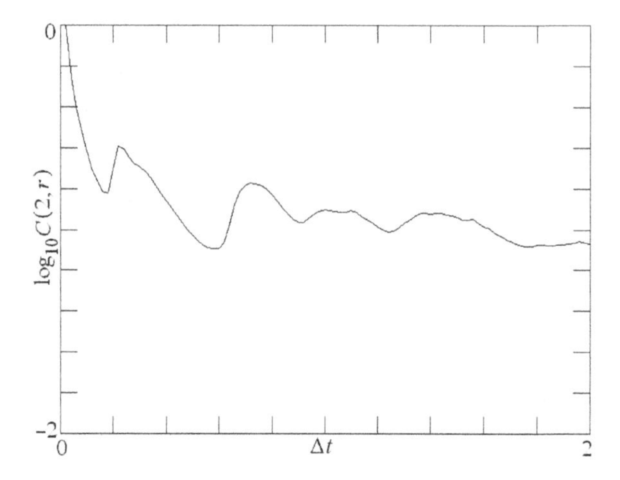

Fig. 12.10 Approximation to the mutual information for the Lorenz attractor, showing a first minimum at about 0.15.

and there is no unique 'correct' value. For a noise-free map, the optimal time delay seems to be one iteration, but no formal proof exists.

A problem with the autocorrelation function is that it is a *linear* statistic and does not account for *nonlinear* correlations. To capture the nonlinear correlation, use the first minimum of the *mutual information* (Fraser and Swinney 1986, Fraser 1989a, Pineda and Sommerer 1993)

$$I(\Delta t) = \sum_{i=1}^{N} \sum_{j=1}^{N} P_{ij}(\Delta t) \ln P_{ij}(\Delta t) - 2 \sum_{i=1}^{N} P_i \ln P_i \tag{12.10}$$

where P_i is the probability that $X(t)$ is in bin i and P_{ij} is the probability that $X(t)$ is in bin i and $X(t + \Delta t)$ is in bin j. The size of the bins is not critical as long as they are sufficiently small. Equation (12.10) is only one of several ways to express the mutual information (Williams 1997). The mutual information (or *redundancy*) is the sum of the two self-entropies minus the joint entropy, and is a measure of how much you know about $X(t+\Delta t)$ if you know $X(t)$. Mutual information is also useful for estimating the optimum embedding dimension, determining the accuracy of the data, estimating the K–S entropy (Fraser 1989b), assessing predictability, and testing for nonlinearity (Paluŝ 1993).

In practice, the quantity $\log C(m, r)$ for $m = 2$ and small fixed r has a shape similar to the mutual information and can be used in its place to determine the optimal time delay (Prichard and Theiler 1995). For example, Fig. 12.10 shows this quantity for the Lorenz attractor with $r = 0.625$, which is about 10^{-3} times the size of the attractor, with a first minimum near 0.15. There is a deeper minimum near 0.6, and so this estimate is consistent with other methods. There is no theoretical reason that the mutual information must have a minimum for finite Δt, but the results are eventually dominated by statistical fluctuations, at which point a further increase in delay time is counterproductive.

12.7 Practical considerations

Calculation of the correlation dimension is a very powerful method for detecting low-dimensional chaos in experimental data, but it is surprisingly easy to get spurious results without appropriate precautions (Ruelle 1990, Hamburger *et al.* 1996). Thus we discuss some practical considerations that commonly arise with real data. Many of these considerations apply also to other types of dimension, entropy, and related quantities. A good summary of these issues is provided by Theiler (1990).

12.7.1 Calculation speed

The most straightforward implementation of the Grassberger–Procaccia algorithm requires of the order of N^2 calculations, where N is the number of points in the time series. Since the dimension is defined in the limit $N \to \infty$, calculation speed is important. Here are some suggestions for speeding the calculation, ordered roughly by increasing difficulty:

1. Avoid double counting by summing $\Theta(r - r_{ij})$ from $j = i + 1$ to N as indicated in eqn (12.3), rather than 1 to N.
2. Discard a few temporally adjacent points (sum from $j = i + w$ to N, where w is an appropriate minimum time separation), which also avoids spurious correlations when the sample time of a flow is small (Theiler 1986, Provenzale *et al.* 1992). You might just set w to a few percent of N.
3. Use a norm that is faster to calculate instead of the Euclidean norm.
4. Collect all the r values at once by binning the values of r_{ij}. That is, test each separation r_{ij}, and increment a counter depending on the (narrow) range in which its value occurs.
5. Avoid taking square roots by binning r_{ij}^2 rather than r_{ij}, using the fact that $\log r^2 = 2 \log r$.
6. Avoid calculating a logarithm (since you will probably want the bin widths to be a logarithmic function of r so that $\Delta r/r$ is a constant) by using the exponent (base-2) of the computer's floating-point number representation as a bin index. When combined with 5 above, this trick gives $\log_2 10^2 \simeq 6.64$ bins per decade of r, which is reasonable.
7. Use the fast neighbor search described in §10.5.4.

12.7.2 Required number of data points

Any finite data set will have a dimension of zero since it consists of a finite number of points in any embedding space. You want to infer from this finite record the dimension of an infinite time series obeying the same dynamics. You can never be completely confident of your inference since

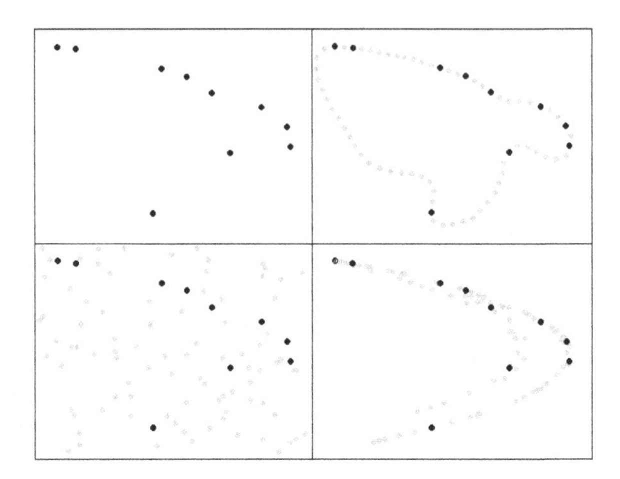

Fig. 12.11 Ten points from a time series embedded in two dimensions could be a cyclic orbit (upper left), limit cycle (upper right), torus (lower left), or strange attractor (lower right).

a high-dimensional attractor is indistinguishable from a periodic point attractor with a very long period. Thus there is no definitive criterion for the number of data points required to determine a dimension. However, there are general rules that give you some confidence.

Suppose you have a time series with ten points embedded in a plane ($m = 2$). The points could be part of a cyclic orbit with a period of ten or more (dimension zero), or part of a limit cycle (dimension one), or a torus (dimension two, three, ...), or a strange attractor (noninteger dimension) as in Fig. 12.11. Generally speaking, as the attractor dimension gets larger, more points are required to determine the dimension with a given confidence. If you assume ten points are required to distinguish a cyclic orbit from a limit cycle (dimension one), then 10^2 points are required for an attractor with dimension two, 10^3 points for dimension three, and so forth. Generally, the number of points required is of the order of $N \simeq 10^D$, where D is the dimension of the attractor (the 'curse of dimensionality'). Smith (1988) argues for an even more conservative value of $N = 42^m$, where m is the embedding (not attractor) dimension. At the other extreme, Ding *et al.* (1993) claim that only $N = 10^{D/2}$ is required. A compromise suggested by Tsonis (1992) is

$$N = 10^{2+0.4D} \tag{12.11}$$

This criterion implies that the highest dimension attractor that can be discerned in a time series with N points is

$$D = 2.5 \log_{10} N - 5 \tag{12.12}$$

Thus a million points are required for a correlation dimension of $D = 10$. You can also calculate the dimension for various N to judge whether

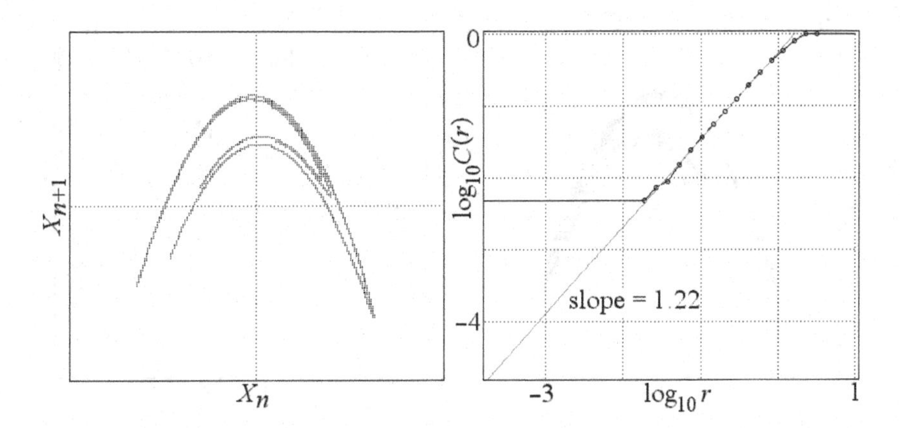

Fig. 12.12 Correlation dimension for the Hénon map with data rounded to 0.02.

it has converged. These estimates reflect the *necessary* number of data points but do not guarantee that the number is *sufficient*. Examples can be constructed for which the required number of points vastly exceeds even the most conservative of the estimates.

12.7.3 Required precision of the data

Often in collecting experimental data the values are discretized, or quantized in some way, usually by rounding or truncating. For example, an analog-to-digital converter has a certain number of bits of resolution. Digital values are rounded to a certain number of digits, and so forth. In such cases, the points in an embedding space lie on a rectangular lattice with a smallest nonzero separation ϵ equal to the lattice spacing and some points coincident with others (zero separation). It should be evident that self-similarity fails at sizes comparable to ϵ, and the dimension approaches zero at smaller scales. Thus the scaling region over which $\log C(r)$ is proportional to $\log r$ is correspondingly reduced. To leading order in ϵ, we expect $C(r) \sim (r + \epsilon/2)^{D_2}$ (Theiler 1988a), and so it is better to plot $\log C(r)$ versus $\log(r + \epsilon/2)$.

As an example, Fig. 12.12 shows 1.6×10^4 points from the Hénon map rounded to 0.02 and embedded in $m = 2$. The correlation sum $C(r)$ is constant for $r < 0.02$ as expected, and the narrow scaling region gives a slope consistent with the expected correlation dimension of $D_2 \simeq 1.22$. Sometimes adding noise to a time series (called *dithering*) can reduce the discretization error (Möller *et al.* 1989), but it leads to other concerns as we now show.

12.7.4 Noisy data

Rather than rounding the data, suppose there is low-level superimposed noise (Moss and McClintock 1989, Casdagli *et al.* 1991). On a scale much larger than the noise, the fractal structure should be evident, but on smaller

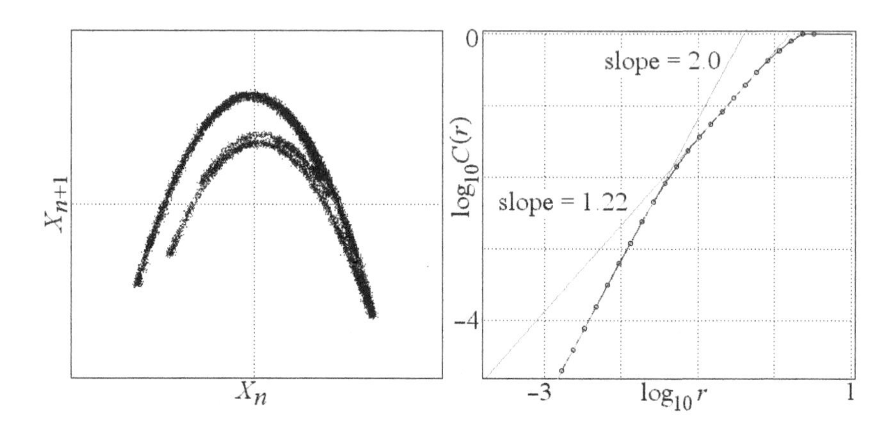

Fig. 12.13 Correlation dimension for the Hénon map with Gaussian white noise having a standard deviation of 0.02.

scales the data should fill the embedding space. The scaling region over which $\log C(r)$ is proportional to $\log r$ is reduced, and there should be a change in slope at the value of r where the noise begins to dominate. For example, Fig. 12.13 shows 1.6×10^4 points from the Hénon map with Gaussian white noise having a standard deviation of 0.02 embedded in $m = 2$. The slope of $\log C(r)$ versus $\log r$ is near 2.0 for $r < 0.02$ as expected, and the narrow scaling region gives a slope consistent with the expected correlation dimension of $D_2 \simeq 1.22$. The knee in the curve indicates the noise level.

Even worse is that colored noise can masquerade for low-dimensional chaos in correlation dimension calculations. For example, Fig. 12.14 shows 1.6×10^4 points from *brown noise* (the integral of Gaussian white noise) embedded in $m = 2$. The plot of X_{n+1} versus X_n shows the data spread along the diagonal, since nearby points in time are also nearby in space for a random walk. Such a plot usually suggests that the sample time is too small, but in this case it indicates a lack of stationarity as is typical of strongly correlated noise.

The plot of $\log C(r)$ versus $\log r$ embedded in $m = 2$ in Fig. 12.14 appears to have a narrow scaling region with a dimension of $D_2 \simeq 1.0$, but that is an illusion. The curve resembles the one for the Hénon map with added noise in Fig. 12.13. This example should warn you not to trust results from a narrow scaling region. Osborne and Provenzale (1989) showed that noise with a $1/f^\alpha$ power spectrum resembles a chaotic system with $D_2 \simeq 2/(\alpha - 1)$ for $1 < \alpha < 3$. This result is not an artifact of the finite number of data points but is a fundamental property of the power-law correlation, which is necessarily nonstationary. Such noise can be detected by testing whether the calculated correlation dimension is independent of the length of your data set. Since colored noise can mimic chaos, you should always test your conclusions using surrogate data (see §9.5.5). Space–time

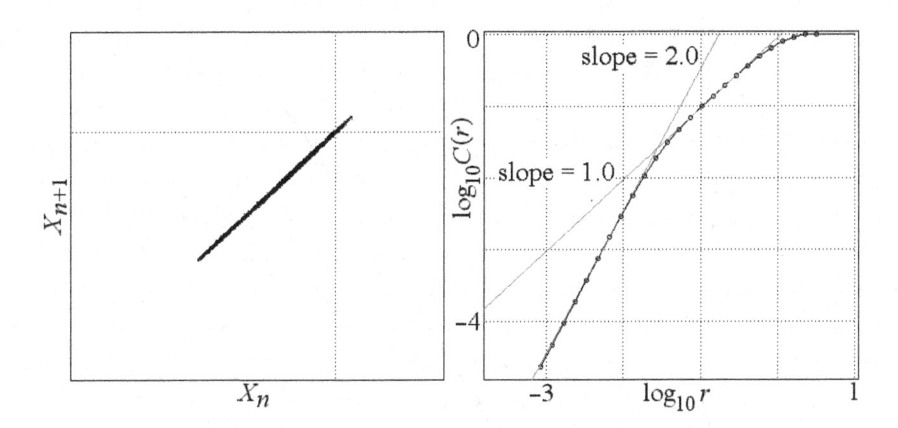

Fig. 12.14 Correlation dimension for brown noise showing a spurious scaling region.

separation plots (see §10.5.3) or plots of X_n versus n are also helpful in avoiding this pitfall.

12.7.5 Multivariate data

Although the dynamical and geometrical quantities such as Lyapunov exponent and dimension are contained in the time series of (almost) any scalar dynamical variable or observable in a system, it is always better to measure simultaneous quantities, which form a time-dependent *vector* (Guckenheimer and Buzyna 1983, Buzug *et al.* 1994). If the number of quantities does not provide a sufficient embedding, you can combine them with their time delays as needed. You should scale the quantities so they have roughly the same mean and range so that the attractor is not overly compressed or expanded in any direction.

Conversely, if you have a program for calculating the correlation dimension of a *scalar* time series, you can analyze *multivariate* data without modifying your program. Construct a single time series with the variables interweaved such as $X_n, Y_n, Z_n, X_{n+1}, Y_{n+1}, Z_{n+1}, X_{n+2}, \ldots$ after adjusting the variables to have similar mean and range. Then analyze the time series as if it were a single scalar variable. Figure 12.15 shows the result of this method for the Lorenz attractor with $N = 3.2 \times 10^4$, $\Delta t = 0.1$, and $m = 3$. In this case, no normalization of the variables was performed, and yet the resulting dimension is close to the expected value of $D_2 \simeq 2.07$.

12.7.6 Filtered data

Seldom do you have the luxury of collecting data directly from one of the fundamental dynamical variables, even if you knew what they were. More often, the observed quantity is some combined, filtered, or transformed version of the variables. Your detector might be sensitive to the power (usually the square of the variables), or it might have a limited bandwidth, attenu-

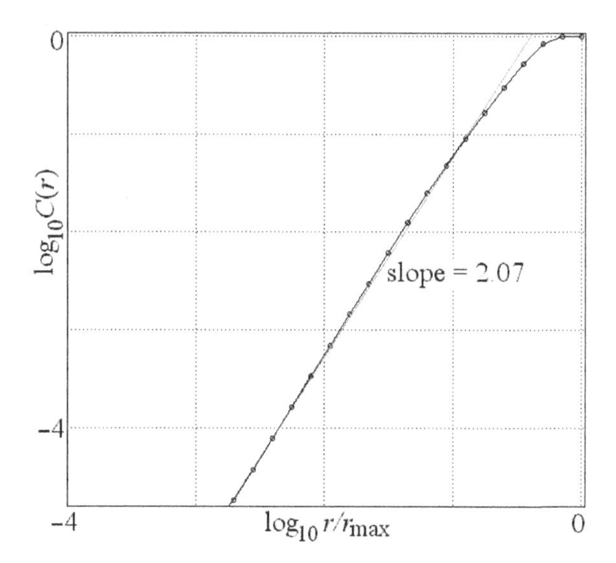

Fig. 12.15 Correlation dimension for the Lorenz attractor with interweaved X, Y, and Z values.

ating either the high or low frequencies or both. In some cases, you might be measuring the time derivative or integral of the variable. In many cases, such filtering has little or no effect except perhaps to degrade the signal-to-noise ratio, limit the scaling range, or require a higher embedding. However, under certain conditions, the result raises the dimension by one (Badii *et al.* 1988, Mitschke *et al.* 1988). For example, differentiating the data adds another differential equation and variable to the system, and integrating the data can destroy the stationarity if the quantity being integrated has a nonzero mean. Since noise is often broadband, it is tempting to filter the data to reduce its effect, but this is not usually recommended since it can also remove important dynamics. *Nonlinear* filters can sometimes be successfully employed, however (Kostelich and Yorke 1988).

12.7.7 Missing data

Often a time series has missing data or points flawed in some way. The proper and honest way to deal with this problem is to eliminate from the correlation sum any points whose time-delay reconstruction involves these points, but this correction requires additional computational bookkeeping and reduces the number of points. It is a bad idea to ignore the gap or to set the missing or corrupted data to zero or to the average of the data. A better approach for data from a smooth flow is to estimate the missing values by interpolation of the temporally adjacent values or use the nonlinear prediction methods in §10.2 to estimate the missing values.

12.7.8 Nonuniformly sampled data

A time series is usually constructed by measuring some quantity at equal time intervals. For example, you might take a temperature reading at noon every day. Suppose one day you took the reading an hour later than usual or that you did not have a clock and only took the data near midday. Generally speaking, if the sampling is *deterministically nonuniform* (you are exactly 1 hour late every Tuesday, for example), the effect is only a distortion of the attractor that does not affect its dimension. On the other hand, if the sampling is *randomly nonuniform*, noise is added to the data, raising the calculated dimension. If the sample times are known, you can construct a new uniformly sampled data set by interpolating the measured data (linearly or nonlinearly), perhaps using *splines* (Press *et al.* 1992). Another method, called *fuzzy delay coordinates*, uses a large fixed time delay and considers only those points closest to the desired sample time (Breeden and Packard 1992).

12.7.9 Nonstationary data

Another common problem is lack of stationarity, usually evidenced by a low-frequency component or slow drift in the mean or variance of the data (see §9.4.1). Stock market averages during a bull market and weather data during a changing season are familiar examples. The effect is to narrow the scaling region and perhaps raise the measured dimension. For example, suppose you had a dynamical system given by $d\mathbf{x}/dt = f(\mathbf{x}, t)$, where $\mathbf{x}$ is a vector of dynamical variables and there is an explicit time dependence in f that might represent a slow change in some system parameter. The time dependence may be monotonic, periodic, or something more complicated, but it can be removed by adding a variable z and an equation $dz/dt = 1$. Hence the space of dynamical variables is increased by one, and the attractor dimension is increased by ≤ 1.

One test for stationarity is to compare the correlation dimension calculated from the first half of the data with that calculated from the second half. Alternately, use a recurrence plot (see §10.5.2) or a space–time separation plot (see §10.5.3) to determine the minimum temporal separation w. Only points with w sufficiently large that the contours are independent of w should be included in the calculation, and if no such points exist, then a meaningful dimension cannot be determined. The methods described in Chapter 9 can be used to detrend the data, such as using log first differences, but this procedure is risky (Theiler and Eubank 1993). Sometimes, the nonstationarity is the interesting feature (Schreiber 1997), such as when using an electroencephalogram (EEG) to identify sleep stages or seizures.

12.8 Fractal dimension of graphic images

Often the data for which you want to calculate a fractal dimension consist of a graphic image rather than a time-series or mathematical formula.

Fig. 12.16 Calculation of the fractal dimension of graphic images.

The foregoing discussion suggests that you should not expect a highly accurate result because many objects are only approximately fractal or are self-similar over only a small scale range. Furthermore, the dimension depends on the density of points which may be unobservable with the finite number of pixels in the image. Many points may fall on top of others at the resolution of the image. If the image is in color or in a gray scale, you must decide what you mean by the fractal dimension, in particular, whether the image is assumed to be embedded in two dimensions or whether the additional color and intensity information is sufficient to infer a dimension greater than 2.0. Probably the best use of such methods is to compare images rather than deduce a useful numerical value and to test their self-similarity. We will limit the discussion to images having only two colors, assumed to be black on a white background.

Suppose your image contained lines, a portion of which is shown at the pixel level in Fig. 12.16. One method for estimating the fractal dimension would be to look at each black pixel such as the one at the center of the figure and construct boxes of various sizes centered on that pixel. You would expect the number of black pixels within a box of linear size d to be proportional to d for a line and proportional to d^2 for a surface. If you choose $d = 3$ and $d = 5$, the dimension is given by

$$D = \log(N_5/N_3)/\log(5/3) \tag{12.13}$$

where N_3 is the number of black pixels within the region with $d = 3$ (maximum of 9), and N_5 is the number within the region with $d = 5$ (maximum of 25) summed over all the black pixels. The numbers three and five are the two smallest odd integers greater than one, chosen to avoid round-off at the smallest scale ($d = 1$), while maintaining symmetry about each point. The choice is equivalent to calculating the fractal dimension at a size scale of about four pixels.

Table 12.1 shows the dimension calculated by this method on a 475×475 grid for a number of fractal objects. The first twelve cases are chaotic systems from Appendix A. The expected values for these cases are assumed to be the calculated correlation dimension (see §12.3). The next nine cases are

Table 12.1 Calculated fractal dimension of graphic images.

Case	From	D (calc)	D (actual)	Error
Logistic map	§A.1.1	1.036	1.000	4%
Tent map	§A.1.3	1.118	1.000	12%
Gauss map	§A.1.7	1.603	1.000	60%
Hénon map	§A.2.1	1.348	1.220	10%
Lozi map	§A.2.2	1.618	1.384	17%
Tinkerbell map	§A.2.4	1.517	1.329	14%
Burger's map	§A.2.5	1.739	1.462	19%
Kaplan–Yorke map	§A.2.7	1.522	1.432	6%
Dissipative standard map	§A.2.8	1.713	1.356	26%
Ikeda map	§A.2.9	1.735	1.690	3%
Chirikov map	§A.3.1	1.900	2.000	–5%
Arnold's cat map	§A.3.3	1.784	2.000	–11%
Triadic Cantor set	Fig. 11.1	0.713	0.631	13%
Weierstrass function	Fig. 11.6	1.486	1.500	–1%
Snowflake	Fig. 11.7	1.637	1.465	12%
Sierpinski carpet	Fig. 11.9	1.930	1.893	2%
Sierpinski triangle	Fig. 11.9	1.672	1.585	5%
Cantor square	Fig. 11.9	1.425	1.262	13%
Flat Fournier universe	Fig. 11.9	1.162	1.000	16%
Brownian $Y(t)$ function	Fig. 11.14	1.633	1.500	9%
Brownian 2-D trail	Fig. 11.15	1.873	2.000	–6%
Trees photograph	Fig. 11.19	1.788	?	–

self-similar fractals from Chapter 11. Their expected values are assumed to be their similarity dimension (see §12.1). The last case is from the photograph of trees in Fig. 11.19 converted to black and white. The calculated dimensions are within 20% of the expected values except for the Gauss map (60%) and the dissipative standard map (26%) which are problematic because their fractal structure is only evident on very small scales.

If you want to calculate a fractal dimension in real time as you generate the data, it is probably more convenient to calculate a running sum of the points in the neighborhood of each point as it is plotted. This method naturally gives more weight to regions with a high density of points and is thus more nearly akin to the correlation dimension, whereas the method above approximates the capacity dimension. When calculating in real time, the dimension will typically start from a value near zero when the points are sparse and increase asymptotically to the final value. One criterion for terminating the calculation is when the dimension first decreases due to the inevitable fluctuations about the asymptotic value.

12.9 Exercises

Exercise 12.1 Show that the similarity dimension for the *even-fifths Cantor set* below is 0.68260619....

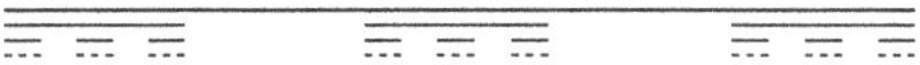

Exercise 12.2 Show that the similarity dimension for the *odd-sevenths Cantor set* below is 0.56457503....

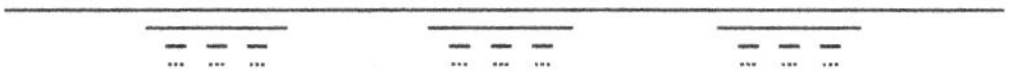

Exercise 12.3 Show that the similarity dimension for the von Koch snowflake in Fig. 11.3 is 1.26185950....

Exercise 12.4 Calculate the similarity dimension for the fractal gaskets in Fig. 11.9.

Exercise 12.5 Show that the similarity dimension for the Sierpinski tetrahedron in Fig. 11.10 is 2.0.

Exercise 12.6 Show that the capacity dimension for the triadic Cantor set in Fig. 11.1 is 0.63092975....

Exercise 12.7 Show that the capacity dimension in eqn (12.2) is independent of the units used for L in the limit $L \to 0$.

Exercise 12.8 Show that the set $0, 1, 1/2, 1/3, 1/4, \ldots$ has a capacity dimension of 0.5. The Hausdorff dimension of this set is zero because the set is countable.

Exercise 12.9 Show that the set $0, 1, 1/2^\alpha, 1/3^\alpha, 1/4^\alpha, \ldots$ has a capacity dimension of $1/(1+\alpha)$. The Hausdorff dimension of this set is zero because the set is countable.

Exercise 12.10 Show that the correlation dimension is independent of the L_s norm used to calculate it.

Exercise 12.11 Show from eqn (12.3) that the correlation sum $C(r)$ has $C(0) = 0$ and $C(\infty) = 1$.

Exercise 12.12 Modify the definition of the correlation sum in eqn (12.3) for the case of a continuum of points by replacing the sums over points by integrals over the distribution of points.

Exercise 12.13 Show that the correlation dimension of a plane filled uniformly and densely with points is 2.0.

Exercise 12.14 Show that the correlation dimension of the tent map in eqn (2.15) is 1.0.

Exercise 12.15 Show that $C(m, r) = C(1, r)^m$ for an IID random time series.

Exercise 12.16 Show that for $\Delta t = 0$ the mutual information in eqn (12.10) is equal to the entropy ($\sum P \ln P$) and for $\Delta t = \infty$ the mutual information is zero for a chaotic system.

Exercise 12.17 Calculate the number of data points required to detect an attractor with dimension $1, 2, \ldots, 10$ using the Tsonis criterion in eqn (12.11).

Exercise 12.18 Show that the correlation sum for a time series discretized in increments of ϵ is of the form $C(r) \sim (r + \epsilon/2)^m$ for small ϵ.

Exercise 12.19 Show that the minimum number of data points that will give a scaling region for the correlation sum from r_{min} to r_{max} is $N = \sqrt{2(r_{\text{max}}/r_{\text{min}})^D}$.

Exercise 12.20 Calculate the fractal dimension given by eqn (12.13) for a straight line 100 pixels long.

Exercise 12.21 Calculate the fractal dimension given by eqn (12.13) for a 100×100 square of black pixels.

Exercise 12.22 Derive an expression for the fractal dimension given by eqn (12.13) for a black rectangle $a \times b$ pixels in size.

12.10 Computer project: Correlation dimension

This project concludes the development of tools for the analysis of chaotic data with an implementation of the Grassberger and Procaccia (1983a,b) algorithm for determining the correlation dimension. This is the algorithm that opened the floodgates for seeking chaos in experimental data. It is still the best method for distinguishing chaos from colored noise. When chaos is found, it provides a measure of the attractor dimension (actually a close lower bound on the dimension), which gives an indication of the minimum number of dynamical variables required to model the system. The method is described in §12.3 and in the original references. This is perhaps the most useful analysis program you will develop, and so you are advised to give it special effort. It is also one of the harder projects.

1. The first step is to generate a time series for analysis. For this purpose, use the Hénon map (see §A.2.1) with parameters ($a = 1.4$ and $b = 0.3$) chosen to give chaotic behavior. Perform a few iterations to ensure that the orbit is on the attractor, and then generate a time series X_i for $i = 1$ to 1000 from the variable X. You may use the data you generated in §10.9.
2. Develop a program to calculate the correlation sum $C(r)$ in eqn (12.3) for your data in an m-dimensional time-delay embedding. Note that for $m > 1$, the data points are vectors whose components are X_{i-k}

and X_{j-k}, and thus you need to use eqn (12.5) to determine r_{ij}. Make a plot of $\log C(r)$ versus $\log r$ for the Hénon map similar to Fig. 12.4. Write your code generally so you can apply it to any data set with arbitrary embedding dimension m.

3. Plot the slope $d \log C(r)/d \log r$ versus $\log r$, and show that there is a plateau whose value is close to the expected correlation dimension of $D_2 = 1.22$ for the Hénon map.
4. (optional) Plot a family of curves $\log C(r)$ versus $\log r$ for values of m from 1 to 8, and estimate the K–S entropy. How does the slope of these curves vary with m? With experimental data it is necessary to make such a plot to ensure that your chosen embedding is sufficiently (but not unnecessarily) large.

13 Fractal measure and multifractals

The foregoing discussion of fractals largely ignored the fact that most real fractal objects are not precisely self-similar and thus may have different dimensions on different size scales and on different parts of the object. When characterizing such an object by a single dimension, you must decide on a range of scale lengths and on how to weight the contributions to the dimension from its various regions. In the former case you normally calculate the dimension at a convenient scale for which sufficient data exist and extrapolate to the limit of infinitesimal scale. In the latter case you must specify a weighting or calculate a spectrum of dimensions for different choices. The previous chapter gave examples of fractals for which the capacity dimension and the correlation dimension differ.

Both problems involve determining the *fractal measure*, which quantifies how much weight to assign to each part of the fractal. It can be considered a literal weight or mass distribution for a fractal composed of real matter such as a string from which a Cantor set might be formed or a sheet of paper from which a Sierpinski gasket is cut or a brick of cheese from which a Menger sponge might be molded. For fractals produced by dynamical systems such as strange attractors, the *natural measure* is the frequency with which the orbit visits the various parts of the attractor and thus defines a natural weighting. A characteristic of strange attractors is that in the limit of infinite time, this quantity is the same for almost all initial conditions. The term 'almost all' means that the set of initial conditions that do not satisfy the condition has zero measure, in this case consisting of the infinitely many unstable periodic orbits embedded in the attractor.

Fractals that can be fully characterized only by specifying a *spectrum* of dimensions are called *multifractals*. Roughly speaking, a multifractal can be considered as an interwoven set of fractals of different dimensions, each having a different weight. We will return to their analysis later in the chapter, but we begin with a discussion of the measure of some simple dynamical systems and the effect that nonuniformities in this measure have on the correlation dimension, which is a particular but important weighting.

The results that follow are described more fully by Sprott and Rowlands (2001). This chapter is slightly more technical and mathematical than the others.

13.1 Convergence of the correlation dimension

When the probability distribution $P(X)$ is known for a one-dimensional map, the correlation sum in eqn (12.3) can be replaced with a correlation integral

$$C(r) = \int\int P(X)P(X')\Theta(r - |X - X'|)dXdX' \qquad (13.1)$$

where $\Theta(x)$ is the Heaviside function defined in eqn (12.4) and the integrals are over the range of X for which $P(X) > 0$. If we define

$$\nu(r) = \frac{d\log C(r)}{d\log r} = \frac{r}{C}\frac{dC}{dr} \qquad (13.2)$$

then the correlation dimension is given by

$$D_2 = \lim_{r\to 0} \nu(r) \qquad (13.3)$$

13.1.1 Logistic map

For the logistic map with $A = 4$, the probability distribution function is given by eqn (2.11), and the integral in eqn (13.1) gives the result

$$\nu(r) \simeq 1 + \frac{1}{\ln r - 2.386} \simeq 1 + \frac{1}{q} \qquad (13.4)$$

where $q = \ln r$ if terms proportional to powers of r are neglected since they rapidly go to zero as r approaches zero. The result is $D_2 = 1.0$, but the convergence of $\nu(r)$ to D_2 is very slow as r (or $1/q$) approaches zero. The slow convergence is a consequence of the singular $(1/\sqrt{X})$ nature of the measure $P(X)$ at $X = 0$ (and similarly at $X = 1$), which in turn results from the quadratic maximum of the logistic equation that maps many values near $X = 0.5$ into $X \simeq 1$ and thereafter into $X \simeq 0$.

Figure 13.1 shows data from the logistic map with $A = 4$, confirming convergence in accordance with eqn (13.4). A least-squares fit gives $D_2 = 1.016 \pm 0.023$ and a coefficient of $1/q$ given by $B = 0.990 \pm 0.346$. Even with the equivalent of over four million data points, the measured dimension would be about 10% low without the extrapolation, but it is within 2% when fit by eqn (13.4).

The behavior described above is not unique to the case of $A = 4$. For arbitrary A, the Frobenius–Perron relation in eqn (2.12) predicts the maximum at $X = 0.5$ maps to a square root singularity at $X = A/4$ with measure

$$P(X) \simeq P(0.5)/\sqrt{1 - 4X/A} \qquad (13.5)$$

which in turn maps to other values of X with the same type of singularity. An example of $P(X)$ for the logistic map with $A = 3.8$ is in Fig. 13.2. It

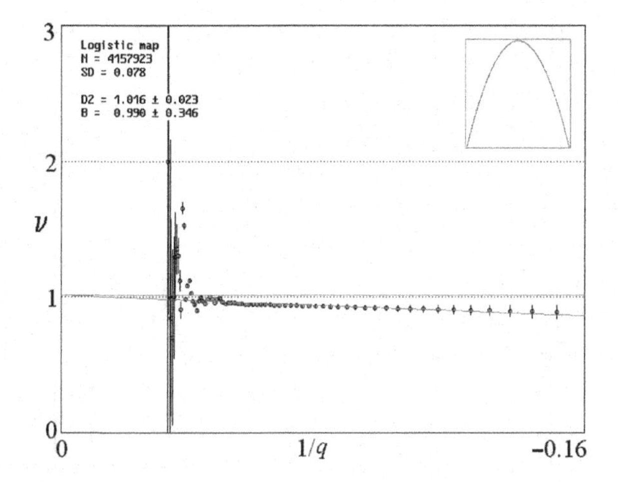

Fig. 13.1 Correlation dimension of the logistic map with $A = 4$ extrapolates to 1.0 for r (or $1/q$) $\to 0$.

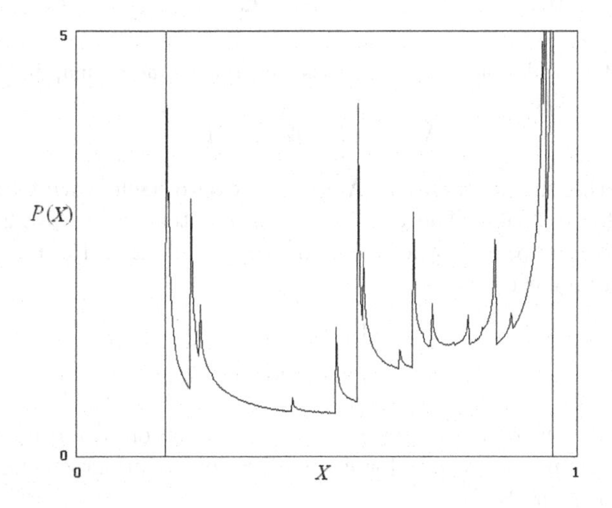

Fig. 13.2 Probability distribution function for the logistic map with $A = 3.8$.

contains infinitely many such singularities, and the convergence of $\nu(r)$ to D_2 is of the same form as eqn (13.4), provided the peaks are sufficiently separated. We expect any one-dimensional chaotic map with a quadratic maximum to have $D_2 = 1.0$ and a similarly slow convergence.

13.1.2 General one-dimensional maps

The convergence of the correlation dimension depends on the shape of the map near its maximum. Therefore, it is useful to consider the general symmetric map in §2.5.3, which for $A = 1$ is

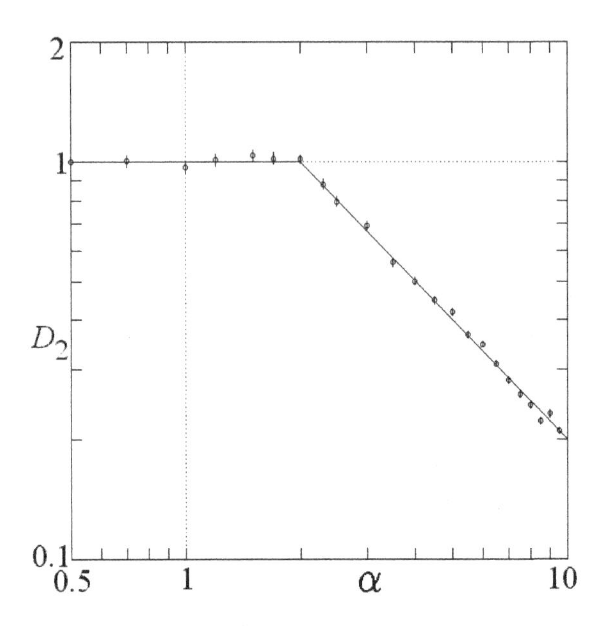

Fig. 13.3 Correlation dimension of the general symmetric map in Fig. 2.15.

$$X_{n+1} = 1 - |2X_n - 1|^\alpha \tag{13.6}$$

and maps the unit interval $0 < X < 1$ back onto itself twice with chaotic solutions for $\alpha \geq 0.5$. The case $\alpha = 1$ is the tent map (§2.5.2), $\alpha = 2$ is the logistic map, and other cases are shown in Fig. 2.15. The resulting correlation dimension is

$$D_2 = \begin{cases} 1 & \text{for } \alpha \leq 2 \\ 2/\alpha & \text{for } \alpha > 2 \end{cases} \tag{13.7}$$

with a rapid convergence (a power of r that depends on α) for $\alpha \neq 2$. Figure 13.3 shows numerical results for maps of various α in agreement with the prediction of eqn (13.7).

The correlation dimension of a one-dimensional chaotic map is determined solely by the shape of the map near its extrema according to eqn (13.7), which for $\alpha > 1$ produces singularities of the form $P(X) \sim X^{1/\alpha-1}$. If df/dX is discontinuous at all such extrema, then the correlation dimension will be 1.0. Otherwise, focus on the extremum X_m with the smallest $|d^2f/dX^2|$ (the flattest region). Take a few points in the vicinity of X_m, fit the results to a function of the form $f = f_m|X - X_m|^\alpha$, and perform a least-squares fit to determine f_m and α. For $\alpha \leq 2$, the dimension is 1.0, and for $\alpha > 2$, the dimension is $2/\alpha$.

A general strategy for determining the correlation dimension of a time series is to make the usual plot of $p = \log C(r)$ versus $q = \log r$ and fit the data to a function of the form

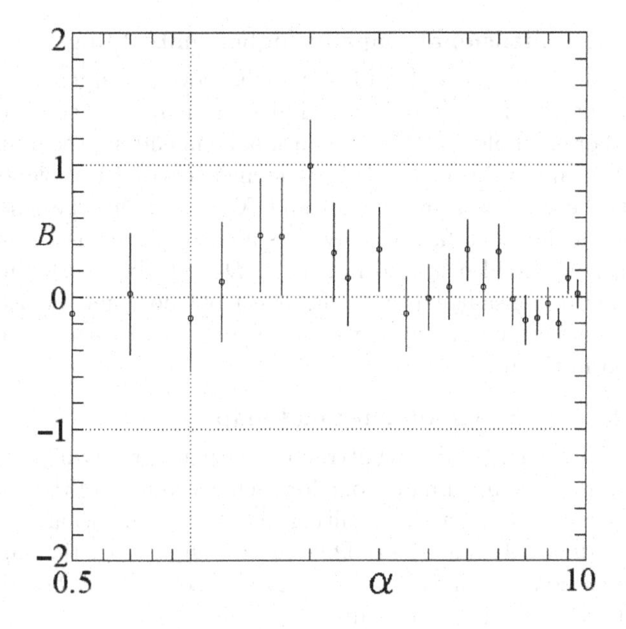

Fig. 13.4 The general symmetric map in Fig. 2.15 has a slow convergence of the correlation dimension only for $\alpha \simeq 2$ (a quadratic maximum).

$$p \simeq A + D_2 q + B \ln(-q) \tag{13.8}$$

using a least-squares fit to determine A, B, and D_2. The quantity B measures the slowness of convergence and should be small except for the common case of a map with a quadratic extremum. Figure 13.4 shows such behavior for the general symmetric map in eqn (13.6), where $B \simeq 1$ for $\alpha = 2$ and $B \simeq 0$ otherwise.

The least-squares fit (Press *et al.* 1992) also gives estimates of the uncertainty in the parameters A, B, and D_2 if the errors in the data are specified. For this purpose, p and q are normalized so that $p = 0$ for $q = 0$ (r is normalized to have a maximum of 1.0, and $C(r)$ is normalized to be 1.0 at $r = 1$), and the errors in p are assumed to be $\sqrt{1/p^2(r) + 4/N_c(r)}$, where N_c is the number of pairs of points included in the correlation sum. The first term approximates the systematic error in the data at large r from edge effects, and the second term approximates the statistical error in the data at small r from the finite number of correlations in each bin, and the errors are assumed to add in *quadrature* (by the Pythagorean theorem). These choices were made with considerable hindsight after examining dozens of low-dimensional data sets, some of which are given in Appendix A. The errors in the correlation dimension are calculated for a 95% confidence level (two standard deviations) but are evidently still too optimistic in some cases and too pessimistic in others.

13.1.3 One-dimensional maps in higher embeddings

Grassberger and Procaccia (1983b) noted the slow convergence of $\nu(r)$ for the logistic map and suggested embedding the map in a space of higher dimension. For example, in a two-dimensional embedding, the unit interval ($0 < X < 1$) along the real axis is stretched into a parabola such that for $A = 4$, the regions near $X = 0$ and $X = 1$ where the singularities occur are expanded by a factor of five. As a consequence, the convergence is faster, although still of the form $\nu(r) \simeq D_2 + 1/(\ln r - 4)$. Embedding in even higher dimensions has a similar effect but does not remove the $\ln r$ dependence. A better strategy is to convolve the map with the tent map, as suggested in the next section.

13.1.4 Decoupled one-dimensional maps

A useful technique for producing attractors with arbitrarily high dimension is to use several decoupled maps, one for each coordinate of the embedding space. It is easy to see that the resulting map has a dimension equal to the sum of the individual dimensions. Thus two decoupled logistic maps should have a correlation dimension of $D_2 = 2$, and so forth. A few examples, showing the rate of convergence are

1. Two logistic maps: $\nu(r) \simeq 2 + \frac{2}{q} \to 2$
2. Logistic map + tent map: $\nu(r) \simeq 2 + \frac{1}{q} \to 2$
3. Logistic map + general symmetric map ($\alpha > 2$): $\nu(r) \simeq 1 + \frac{2}{\alpha} + \frac{1}{q} \to 1 + \frac{2}{\alpha}$

Going to two dimensions, even when the maps are decoupled, can change the convergence of $\nu(r)$. In particular, convolving the logistic map with a tent map gives a two-fold improvement in the rate of convergence. This procedure could also improve the convergence of an experimental time series.

13.1.5 Two-dimensional maps

Whereas noninvertible maps such as the logistic map usually have chaotic solutions that fill intervals of the line with iterates of nonuniform measure, invertible, two-dimensional chaotic maps usually produce a strange attractor with fractal structure. The Hénon map in Fig. 5.6 is one such example. It consists of a direction in which the structure is one-dimensional and a perpendicular direction in which the structure resembles a Cantor set (see §11.1).

Broomhead and Rowlands (1984) argue that the resulting fractal dimension should be $D \simeq 1 + \log 2/\log(\sqrt{a}/b)$. For the usual values of $a = 1.4$ and $b = 0.3$, the prediction of $D = 1.505132\ldots$ differs considerably from the correlation dimension of 1.220 ± 0.036 because it assumes the fractal structure is precisely self-similar. The structure (or 'lacuna' from which the term *lacunarity* derives) produces oscillations (Badii and Politi 1984) with a period of $\Delta q \simeq \ln(\sqrt{a}/b) = 1.372208\ldots$ in $\nu(r)$ as is evident in

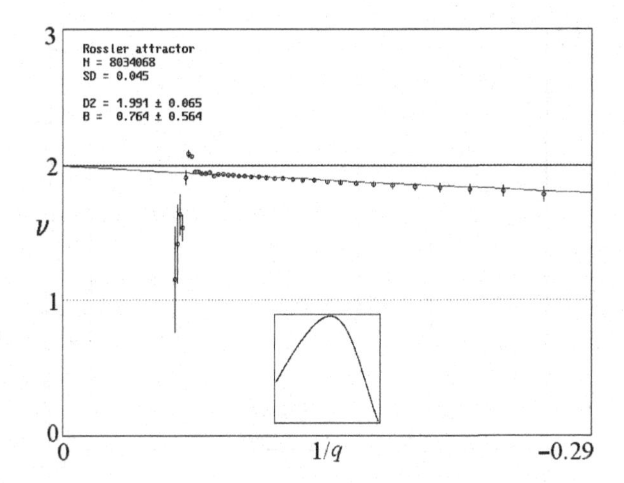

Fig. 13.5 Correlation dimension of the Rössler attractor converges slowly like the logistic map.

Fig. 12.5. Otherwise, the convergence of $\nu(r)$ is rapid for the Hénon map ($B = 0.077 \pm 0.427$) because the singularities that occur in the measure at each branch of the fractal crowd together on the smallest scales. Similar behavior occurs for other two-dimensional chaotic maps.

13.1.6 Chaotic flows

All chaotic three-dimensional flows have a direction parallel to the flow along which the measure should vary continuously, an expanding direction along which, by analogy with the logistic equation, we expect singularities in the measure, and a contracting direction along which, by analogy with the Hénon map, we expect Cantor-like structure. The parallel direction contributes 1.0 to the correlation dimension, as does the expanding direction unless the return map is flatter than quadratic near its extremum. The contracting direction contributes a fraction to the correlation dimension, so that the dimension is generally $2 < D_2 < 3$. The convergence of the correlation dimension is dominated by the shape of the return map near its flattest extremum. The return map of a chaotic flow cannot be precisely one-dimensional, however, since that would imply noninvertible dynamics, which contradicts the time-reversibility of the flow.

As an example, the Lorenz attractor with $\sigma = 10, r = 28$, and $b = 8/3$ has a map of successive maxima of z with a dimension of about 1.07 that resembles the $\alpha = 0.5$ case of Fig. 2.15 (Lorenz 1963). Thus we expect rapid convergence as evidenced by the computed value of $B = 0.120 \pm 0.728$ and a correlation dimension of $D_2 = 2.068 \pm 0.086$. On the other hand, the successive maxima of x for the Rössler attractor with $a = b = 0.2$ and $c = 5.7$ more resemble the case of $\alpha \simeq 1.8$, and so its convergence as in Fig. 13.5 is slower, as evidenced by the value of $B = 0.764 \pm 0.564$ and a

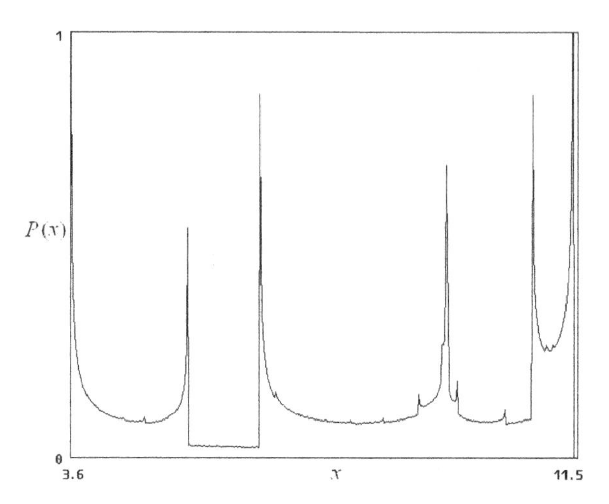

Fig. 13.6 Return map for the maximum x of the Rössler attractor gives singularities in the measure similar to those in the logistic map in Fig. 13.2.

correlation dimension of $D_2 = 1.991 \pm 0.065$. The corresponding $P(x)$ for this case, as in Fig. 13.6, resembles the logistic map for $A = 3.8$ in Fig. 13.2.

Other maps and three-dimensional flows have similar behavior as summarized in Appendix A, all of which have over 10^{12} correlations, corresponding to data sets of over a million points. Cases with $|B| > 0.5$ are labeled 'converges slowly,' showing that logarithmic convergence is not unusual. We expect the methods also to apply to systems higher than three-dimensional and to experimental data. In particular, eqn (13.8) seems to provide a good fit to most of the cases examined and allows extrapolation to the $r \to 0$ limit from a limited data set.

13.2 Multifractals

The difference between the capacity dimension and the correlation dimension in the previous chapter is a consequence of the nonuniform measure that weights different regions of the attractor differently. Highly nonuniform, even singular, measures are common, as evidenced by the examples in Figs 13.2 and 13.6. Thus it is not surprising that the capacity dimension, which simply counts boxes occupied by the orbit, and the correlation dimension, which involves the probability that two points are close together, have different values. The former weights all regions of the attractor equally, while the latter emphasizes regions of the attractor frequently visited by the orbit. Thus it is too simplistic to represent an attractor by a single dimension, and it is more informative instead to define a *multifractal spectrum* of dimensions. Such spectra have proved useful in the analysis of heartbeat dynamics (Amaral *et al.* 2001) and tropical rainforests (Solé and Manrubia 1995), among others.

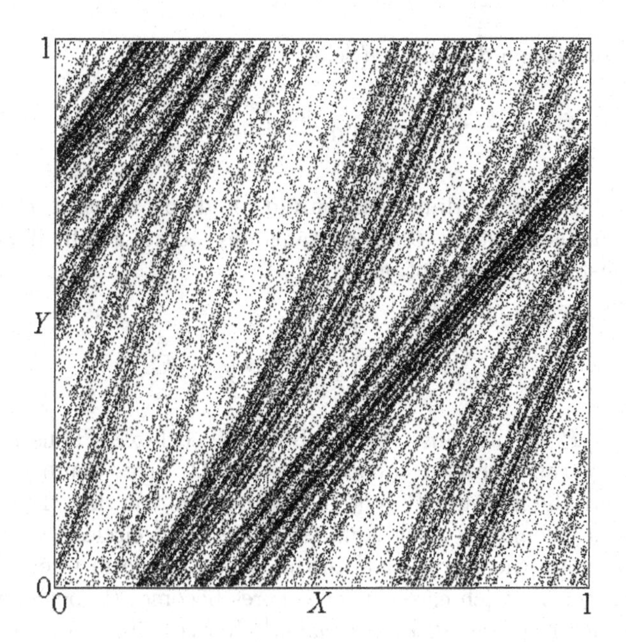

Fig. 13.7 Attractor for the Sinai map with $\delta = 0.1$ showing a highly nonuniform measure.

13.2.1 Sinai map

An example of an attractor in two dimensions with a very nonuniform measure is the *Sinai map* (Sinai 1972)

$$X_{n+1} = X_n + Y_n + \delta \cos 2\pi Y_n \pmod 1 \tag{13.9}$$

$$Y_{n+1} = X_n + 2Y_n \pmod 1 \tag{13.10}$$

For small δ, the attractor is the unit square ($0 < X < 1, 0 < Y < 1$), and typical orbits come arbitrarily close to any point in the square if you wait long enough. Hence, we expect the capacity dimension to be $D_0 = 2.0$. However, for $\delta > 0$, the measure is highly nonuniform, as indicated in Fig. 13.7. We expect the correlation dimension to be lower than 2.0 since it emphasizes the diagonal stripes. Numerical estimates give $D_2 = 1.779 \pm 0.063$. The Kaplan–Yorke dimension is $D_{\mathrm{KY}} \simeq 1.89075$. It is typical for chaotic attractors to have $D_2 \leq D_{\mathrm{KY}} \leq D_0$ and the values similar.

13.2.2 Generalized dimensions

To be more specific, suppose we cover the attractor with squares of linear size r as we did for the capacity dimension calculation in §12.2. Let P_i be the probability that a point is in square i. If there are N points on the attractor with N_i of them in square i, then $P_i = N_i/N$ and the capacity dimension is

$$D_0 = \lim_{r \to 0} -\frac{\log \sum_i P_i^0}{\log r} \tag{13.11}$$

while the correlation dimension (Schuster 1995) is

$$D_2 = \lim_{r \to 0} \frac{\log \sum_i P_i^2}{\log r} \tag{13.12}$$

Note that the capacity dimension only depends on whether P_i is zero or positive (that is, whether a box contains any points at all) if we use the slightly nonstandard convention

$$P^0 = \begin{cases} 0 & \text{for } P = 0 \\ 1 & \text{for } P \neq 0 \end{cases} \tag{13.13}$$

The sum $\sum_i P_i^0$ is then just the number of filled boxes. Thus D_0 weights all regions of the attractor equally, whereas D_2 gives emphasis to those regions where the measure P_i is large. If it bothers you that the correlation dimension is calculated using squares rather than circles, use the maximum (supremum) norm rather than the Euclidean norm in eqn (12.5) for calculating r_{ij}, in which case (hyper)spheres becomes (hyper)cubes.

We can define a *generalized dimension*[1] (Rényi 1970, Grassberger 1983b, Hentschel and Procaccia 1983, Paladin and Vulpiani 1987)

$$D_q = \lim_{r \to 0} \frac{1}{q-1} \frac{\log \sum_i P_i^q}{\log r} \tag{13.14}$$

The limit $q \to \infty$ gives the local dimension in the most densely populated region of the attractor, and the limit $q \to -\infty$ gives the local dimension in the least densely populated region. The former can be calculated more accurately than the latter because there are usually many more points in the dense region (which is how it got to be dense!).

13.2.3 Information dimension

The case $q = 1$ is special because the formal definition in eqn (13.14) is zero divided by zero. The numerator has $\sum_i P_i = 1$, whose logarithm is zero, and the denominator is $q - 1$. However, we can use *l'Hôpital's*[2] *rule*

$$\lim_{q \to 1} \frac{f(q)}{g(q)} = \lim_{q \to 1} \frac{f'(q)}{g'(q)} \tag{13.15}$$

where the primes denote derivatives with respect to q, to obtain

[1]Note that the q used here is an index unrelated to the $q = \ln r$ used earlier in the chapter.

[2]Guillaume de l'Hôpital (also spelled l'Hospital) (1661–1704) was a French mathematician who in 1696 published the first book on differential calculus. The rule that bears his name simply says that if two smooth functions of a variable simultaneously go to zero, change the variable by an infinitesimal amount to calculate the ratio, which is otherwise undefined. If the result is still zero, keep differentiating until it is not.

$$D_1 = \lim_{r \to 0} \frac{\sum_i P_i \log P_i}{\log r} \tag{13.16}$$

The dimension D_1 is called the *information dimension*. The quantity in the numerator of eqn (13.16) is *Shannon's entropy*, the amount of information required to specify the state of the system to within an accuracy r. If the probabilities of all states are unity ($P_i = 1$), then $\log P_i$ is zero, and no new information is required to specify the system. Conversely, if the probabilities of all states are zero ($P_i = 0$), no new information is required either. Only when the states have unknown probabilities $0 < P_i < 1$ is information required to specify them.

The information dimension describes how fast the information needed to specify a point on the attractor increases as r decreases. Other generalized dimensions besides capacity (D_0), information (D_1), and correlation (D_2) have no specific names (yet). Numerical calculation of the information dimension from a time series is difficult, but satisfactory results can sometimes be obtained using a 'fixed mass' algorithm (Kantz and Schreiber 1997) in which r is adjusted to include a given number of data points, and the limit taken as this number approaches zero.

13.2.4 Multifractal properties

We can now define a *monofractal* (or *homogeneous fractal*) as one in which D_q is independent of q and a *multifractal* (or *inhomogeneous fractal*) as one in which D_q depends on q. With this definition, almost all fractals are multifractals, either because the measure is nonuniform or because the local fractal dimension is different in different regions. As with the term 'fractal,' there is no universally accepted definition of a multifractal.

A characteristic of the generalized dimensions is that they generally decrease with increasing q so that

$$D_{q_1} \geq D_{q_2} \quad \text{if } q_2 > q_1 \tag{13.17}$$

In particular, the correlation dimension is a lower bound on the capacity dimension since $D_2 \leq D_0$. More generally

$$D_E \geq D_0 \geq D_H \geq D_1 \geq D_2 \geq \cdots \geq D_T \tag{13.18}$$

where D_E is the *Euclidean dimension* (the integer dimension of the space in which the set is embedded), D_H is the Hausdorff dimension (see §12.2), and D_T is the *topological dimension* (the largest integer dimension of the pieces that compose the set[3]). For example, a 2-torus has a topological dimension of two and a Euclidean dimension of three. Grassberger and Procaccia (1984) give for the Hénon map $D_0 = 1.272 \pm 0.006$, $D_H = 1.28 \pm 0.01$, $D_1 = 1.25826 \pm 0.00006$, and $D_2 = 1.224 \pm 0.006$. For a monofractal, all the dimensions are the same. Note also that q need not be an integer or positive.

[3] A more rigorous definition is that a set S has topological dimension D_T if each point in S has arbitrarily small neighborhoods whose boundaries meet S in a set of dimension $D_T - 1$ and D_T is the least non-negative integer for which this holds.

13.3 Examples of generalized dimensions

There are a number of cases for which the generalized dimensions can be calculated directly from the probability distribution function. We consider two general examples that serve as models of what might be expected experimentally.

13.3.1 Asymmetric Cantor set

Consider a generalization of the Cantor set (see §11.1) in which the unit interval $0 < X < 1$ is divided into two pieces of length r_1 and r_2 populated with a density P_1 and P_2, respectively, and the process is repeated infinitely many times on ever smaller scales. The *asymmetric Cantor set* satisfies the condition (Schroeder 1991)

$$P_1^q r_1^{(1-q)D_q} + P_2^q r_2^{(1-q)D_q} = 1 \tag{13.19}$$

The generalized dimension D_q can be evaluated for several special cases:

1. Symmetric triadic Cantor set ($r_1 = r_2 = \frac{1}{3}, P_1 = P_2 = \frac{1}{2}$):
   $D_q = \log 2/\log 3 = 0.630929\ldots$ (a monofractal)
2. General symmetric Cantor set ($r_1 = r_2 = r, P_1 = P_2 = \frac{1}{2}$):
   $D_q = -\log 2/\log r$ (a monofractal)
3. Weighted symmetric Cantor set ($r_1 = r_2 = r, P_1 \neq P_2$):
   $D_q = -\frac{\log(P_1^q + P_2^q)}{(1-q)\log r}$ (a multifractal)
4. Unweighted Cantor set with quadratic asymmetry ($r_1 = r_2^2 = r, P_1 = P_2 = \frac{1}{2}$):
   $D_q = \frac{\log[\sqrt{1+2^{q+2}}-1]-\log 2}{(1-q)\log r}$ (a multifractal)
5. Weighted Cantor set with quadratic asymmetry ($r_1 = r_2^2 = r, P_1 \neq P_2$):
   $D_q = \frac{\log[\sqrt{P_1^2+4P_2}-P_1]-\log 2P_2}{(1-q)\log r}$ (a multifractal)

Cases 3 and 4 are shown in Fig. 13.8 for typical choices of r and P. Each case shows an asymptotic D_q for large positive and negative q ($\log P_1/\log r \geq D_q \geq \log P_2/\log r$ for case 3 and $-\log 2/\log r \geq D_q \geq -\log 2/\log r^2$ for case 4) and a monotonic decrease of D_q with increasing q as expected. This behavior is typical of multifractals.

13.3.2 General symmetric map

For the one-dimensional general symmetric map in eqn (13.6), the probability distribution function has a power-law singularity that gives a spectrum of generalized dimensions (Halsey *et al.* 1986)

$$D_q = \frac{1}{q-1}\min(q-1, q/\alpha) \tag{13.20}$$

In particular, the tent map ($\alpha = 1$) is a monofractal with $D_q = 1.0$ for all q (a uniformly dense line segment), and the logistic map ($\alpha = 2$) is a

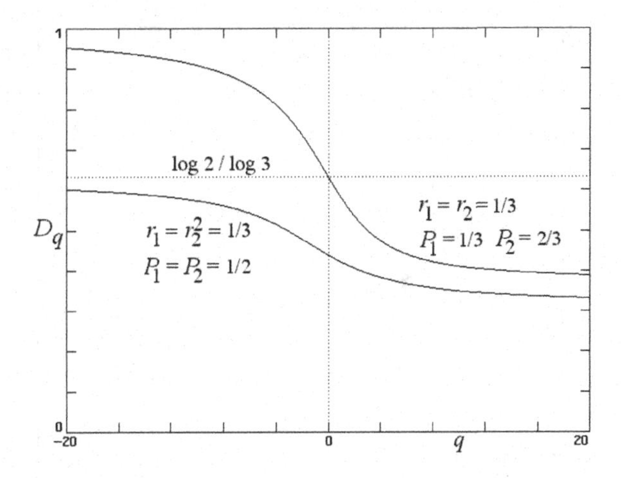

Fig. 13.8 Generalized dimension for two different asymmetric Cantor sets.

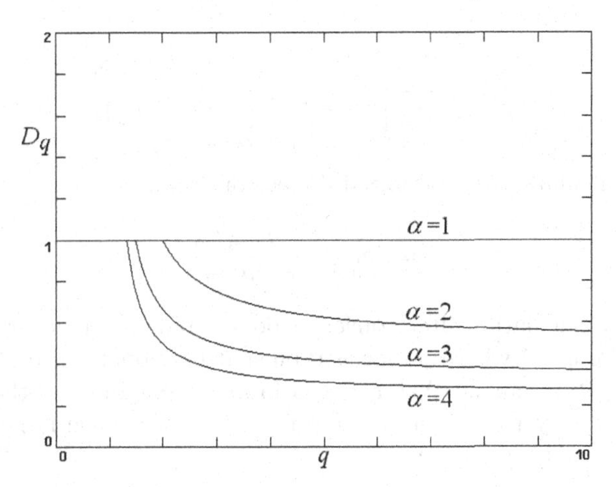

Fig. 13.9 Generalized dimension of a general symmetric map with various α.

multifractal with $D_q = \min[1, q/2(q-1)]$. Other cases are shown in Fig. 13.9. The abrupt change in slope at a critical value of q is analogous to a *phase transition* (Ott *et al.* 1984), such as the melting of a solid or the boiling of a liquid.

13.4 Numerical calculation of generalized dimensions

Calculation of the spectrum of generalized dimensions from eqn (13.14) is hindered by the same limitations as for the capacity dimension (see §12.2). A computationally preferred method is to generalize the correlation sum in eqn (12.3) according to

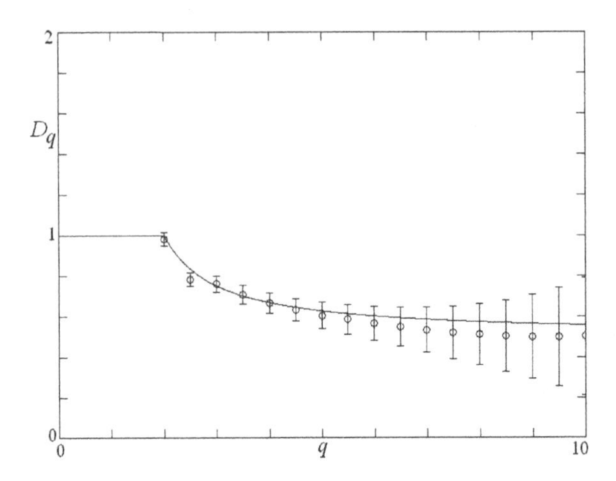

Fig. 13.10 Numerical results for the generalized dimension of the logistic map with $A = 4$.

$$C_q(r) = \frac{1}{N}\sum_{j=1}^{N}\left[\frac{1}{N-1}\sum_{i=1, i\neq j}^{N}\Theta(r - r_{ij})\right]^{q-1} \tag{13.21}$$

and then calculate the generalized dimension from

$$D_q = \lim_{r\to 0}\frac{1}{q-1}\frac{d\log C_q(r)}{d\log r} \tag{13.22}$$

Even so, the calculation often converges poorly and gives inaccurate results, especially for $q \leq 1$ where emphasis is on sparsely populated regions (Badii and Politi 1985, van de Water and Schram 1988). In fact, the quantity raised to the power $q - 1$ can be zero for values of r much larger than the smallest interpoint distance.

13.4.1 Logistic map with $A = 4$

Figure 13.10 shows the result of evaluating eqn (13.22) for 5×10^5 points from the logistic map with $A = 4$ compared with the theoretical prediction from eqn (13.20) with $\alpha = 2$. The logarithmic extrapolation scheme described above was used to correct for the slow convergence for each value of q. Clearly, the calculation is not very accurate for q substantially different from 2.

13.4.2 Logistic map with $A = A_\infty$

The final example is the logistic map at the accumulation point $A = A_\infty = 3.5699456718\ldots$ (see §2.3.4). It can be approximated by an asymmetric Cantor set with $r_1 = 0.408, r_2 = r_1^2$, and $P_1 = P_2 = \frac{1}{2}$, which is the same as case 4 above (Schroeder 1991). The predicted values are $D_{-\infty} \simeq 0.773, D_0 \simeq 0.537, D_1 \simeq 0.515, D_2 \simeq 0.497$, and $D_\infty \simeq 0.387$. The

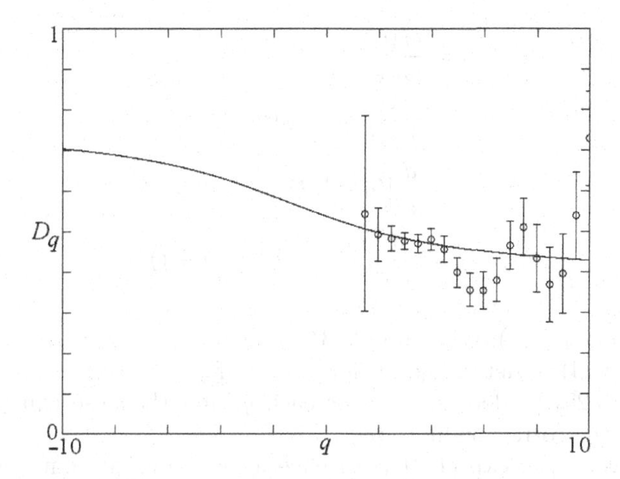

Fig. 13.11 Numerical results for generalized dimension of the logistic map with $A = A_\infty$.

numerical results for $q \geq 1.5$ with $N = 5 \times 10^5$ in Fig. 13.11 show similarly large uncertainties, but roughly consistent with theory. This case is numerically problematic because a small error in A_∞ causes $\nu(r)$ to approach either zero or 1.0 at small r, where the attractor becomes either a finite collection of points (zero-dimensional) or chaotic (one-dimensional), respectively.

13.5 Singularity spectrum

Another way to characterize a multifractal is through its *singularity spectrum* $f(\alpha)$. Suppose we cover the fractal with small boxes of linear size r. The probability measure of box i is assumed to be given by $P_i \propto r^{\alpha_i}$, which defines the *singularity index* α_i, also called the *Lipshitz–Hölder exponent*. Now count the number of boxes for which α_i is between α and $\alpha + \Delta\alpha$. For small r (and hence a large number of boxes) in the continuum limit, the number of boxes with α in the range α to $\alpha + d\alpha$ is proportional to $r^{-f(\alpha)} d\alpha$ (Benzi *et al.* 1984, Frisch and Parisi 1985). The fractal is represented by interwoven sets of singularities, and the quantity $f(\alpha)$ is the fractal dimension of the set whose Lipshitz–Hölder exponent is α (Halsey *et al.* 1986, Bohr and Rand 1987, Collet *et al.* 1987).

Both $f(\alpha)$ and D_q characterize the spectrum of dimensions of a multifractal, and the relationships between them are given by a *Legendre transform*[4]

[4]Adrien-Marie Legendre (1752–1833) was a renowned French mathematical physicist who lost his pension for refusing to vote for the government's candidate in 1824 and died in poverty.

$$q = \frac{df(\alpha)}{d\alpha} \tag{13.23}$$

$$D_q = \frac{1}{q-1}[q\alpha - f(\alpha)] \tag{13.24}$$

$$\alpha = \frac{d}{dq}[(q-1)D_q] \tag{13.25}$$

$$f(\alpha) = q\frac{d}{dq}[(q-1)D_q] - (q-1)D_q \tag{13.26}$$

If you know $f(\alpha)$, use eqn (13.23) to determine q for each α, and then use eqn (13.24) to determine D_q for each q. If you know the D_q spectrum, use eqn (13.25) to determine α for each q, and then use eqn (13.26) to determine the corresponding $f(\alpha)$.

Intuitively, you expect $f(\alpha)$ to have a maximum at some value of α. From eqn (13.23) it follows that this maximum occurs at $q = 0$ where, according to eqn (13.26), $f(\alpha)$ is equal to the capacity dimension (D_0). From eqn (13.25), this maximum occurs at $\alpha = D_0 - \frac{dD_q}{dq}|_{q=0}$. Finally, note that the information dimension is given by $D_1 = \alpha = f(\alpha)$.

There is a useful analogy between the singularity spectrum formulation of multifractals and thermodynamics in which α plays the role of the internal energy per unit volume and $f(\alpha)$ is the corresponding entropy (Ott 1993). In this analogy, discontinuities in $f(\alpha)$ (or D_q) as evident in Figs 13.9 and 13.10 correspond to phase transitions.

A typical example of a singularity spectrum is given by the symmetric weighted Cantor set (case 3 in §13.3.1) with $r_1 = r_2 = \frac{1}{3}, P_1 = \frac{1}{3}$, and $P_2 = \frac{2}{3}$. The generalized dimension is

$$D_q = \frac{\log[(\frac{1}{3})^q + (\frac{2}{3})^q]}{(1-q)\log 3} \tag{13.27}$$

For each q, the corresponding α is

$$\alpha = \frac{(\frac{1}{3})^q \log 3 + (\frac{2}{3})^q \log \frac{3}{2}}{[(\frac{1}{3})^q + (\frac{2}{3})^q]\log 3} \tag{13.28}$$

and the singularity spectrum is

$$f(\alpha) = q\alpha - (q-1)D_q \tag{13.29}$$

A graph of $f(\alpha)$ is in Fig. 13.12. The curve has a maximum at $\alpha_0 = [\log 3 + \log \frac{3}{2}]/2\log 3 = 0.684535\ldots$, where $f(\alpha_0) = D_0 = \log 2/\log 3 = 0.630929\ldots$. The values of α range from $\log\frac{3}{2}/\log 3 = 0.369070\ldots$ (for $q = \infty$) to 1.0 (for $q = -\infty$). This general shape is typical of most singularity spectra.

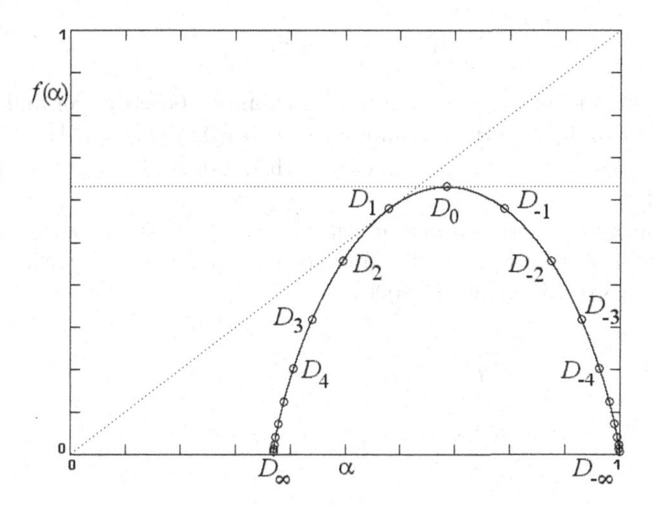

Fig. 13.12 Singularity spectrum for a symmetric weighted Cantor set.

13.6 Generalized entropies

By analogy with the generalized dimension in eqn (13.22), we can generalize the K_2 entropy in eqn (12.8) to arbitrary q

$$K_q = \lim_{m\to\infty} \lim_{r\to 0} \lim_{N\to\infty} \log \frac{C_q(m,r)}{C_q(m+1,r)} \tag{13.30}$$

where $C_q(m,r)$ is the generalized correlation sum in eqn (13.21) for an embedding dimension m. As with the generalized dimension, the *generalized entropy* converges poorly for $q \neq 2$.

The case of $q = 1$ is special because $C_q(m,r)$ involves the division of zero by zero, and hence is undefined. Using l'Hôpital's rule leads to

$$K_1 = \lim_{m\to\infty} \lim_{r\to 0} \lim_{N\to\infty} -\frac{1}{mN} \sum_{i=1}^{N} \log \left[\frac{1}{N-1} \sum_{i=1, i\neq j}^{N} \Theta(r - r_{ij}) \right] \tag{13.31}$$

The quantity K_1 is called the *metric entropy* or *K–S entropy* (after Kolmogorov[5] and Sinai). It measures the rate of loss (or gain) of information in a chaotic system, by analogy with the information dimension D_1. Under some conditions for iterated maps, it can be shown (Bowen 1975) that K_1 is the inverse of the autocorrelation time (see §9.4.5).

The quantity K_0 is called the *topological entropy*. As with the generalized dimension, the generalized entropy has the property that

[5]Andrey Nikolaevich Kolmogorov (1903–1987) was a prolific Russian mathematician with wide interests, who worked on many problems of interest to physicists and devoted much of the last third of his life to the mathematical education of children and the public.

$$K_{q_1} \geq K_{q_2} \quad \text{if } q_2 > q_1 \tag{13.32}$$

with asymptotic values for $q \to \pm\infty$. For example, Grassberger and Procaccia (1984) give for the Hénon map $K_0 = 0.45 \pm 0.02$, $K_1 = 0.4192 \pm 0.0001$, and $K_2 = 0.318 \pm 0.02$. An attractor with uniform measure has the same K_q for all q.

Furthermore, there is a *dynamical spectrum* $g(\beta)$ of entropies analogous to the singularity spectrum for dimensions $f(\alpha)$ in §13.5 with a similar Legendre transform (Hilborn 2000)

$$q = \frac{dg(\beta)}{d\beta} \tag{13.33}$$

$$K_q = \frac{1}{q-1}[q\beta - f(\beta)] \tag{13.34}$$

$$\beta = \frac{d}{dq}[(q-1)K_q] \tag{13.35}$$

$$g(\beta) = q\frac{d}{dq}[(q-1)K_q] - (q-1)K_q \tag{13.36}$$

In a one-dimensional map, $g(\beta)$ is the distribution of local Lyapunov exponents over the attractor.

There are many other ways to define entropy in a dynamical system, but they are all closely related and agree in the appropriate limit. The entropy is generally easy to compute in a one-dimensional system, especially when the system is described symbolically. In higher dimensions, the entropy is more difficult to calculate, which is unfortunate because it is one of the crudest invariants of a dynamical system, suggesting that high-dimensional systems are difficult to classify.

13.7 Unbounded strange attractors

Shaw (1981) has given two examples of chaotic attractors that are unbounded and thus strain the definition of an 'attractor.' The orbits have a high probability of being near the origin but a small probability of being at an arbitrary distance from it, and thus the measure is highly nonuniform. They are attractors in the sense that they attract initial conditions to the vicinity of the origin on average while allowing occasional excursions to arbitrarily large distances.

13.7.1 Spence map

The simplest example is the *Spence map*

$$X_{n+1} = |\ln X_n| \tag{13.37}$$

Values of X near 1.0, where the measure is nearly constant, map into a

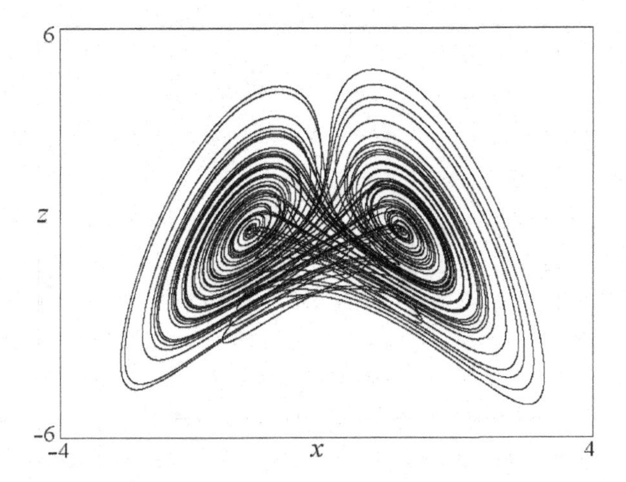

Fig. 13.13 Burke–Shaw attractor projected onto the xz plane.

narrow region near $X = 0$, which then map into regions at arbitrarily large values of X. The Lyapunov exponent given by

$$\lambda = \int_0^\infty P(X) \ln |df/dX| \simeq \Big[P(0)/X\Big]_0^\infty \tag{13.38}$$

is infinite but diverges slowly. The Spence map is one of a large class of such singular maps. For example, the cases in §9.5.3 give approximate Gaussian probability densities with infinite tails.

13.7.2 Burke–Shaw attractor

A similar example in a three-dimensional chaotic flow is the *Burke–Shaw attractor*

$$\frac{dx}{dt} = -Sx - Sy \tag{13.39}$$

$$\frac{dy}{dt} = -Sxz - y \tag{13.40}$$

$$\frac{dz}{dt} = Sxy + V \tag{13.41}$$

with typical values of $S = 10$ and $V = 13$. This system is a variant of the Lorenz attractor and has been studied by Letellier *et al.* (1996). It has no equilibrium points and an unstable solution with constant dz/dt. The probability distribution function is peaked near the origin but asymptotes to zero for large z. The attractor shown in Fig. 13.13 has a highly nonuniform measure, as shown in Fig. 13.14.

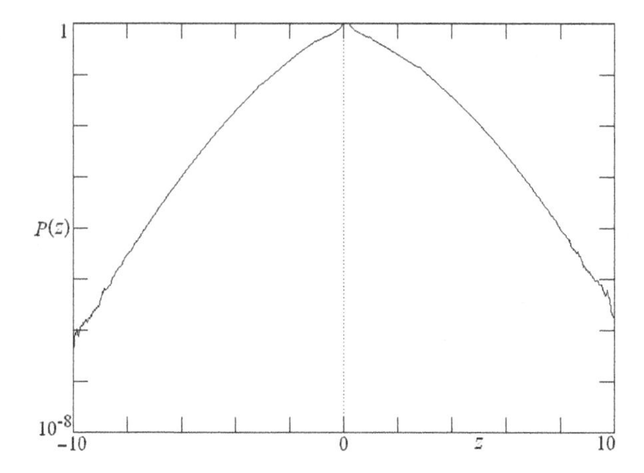

Fig. 13.14 Probability distribution for the z variable of the Burke–Shaw attractor.

13.8 Summary of time-series analysis methods

Since the previous five chapters have dealt primarily with time-series analysis and the identification and quantification of chaos and fractal structure, it is useful to conclude with a summary of methods for analyzing experimental data. Most of these tests are implemented in the *Chaos Data Analyzer* program (Sprott and Rowlands 1995). The good news is that it is possible, in principle, to determine from experimental data whether the underlying mechanism is chaotic or random, but the bad news is that it is hard to do because you need an enormous amount of noise-free data. Time-series analysis is more art than science, and no recipe will guarantee success for every case. Nevertheless, here is a general strategy that might be helpful:

1. Verify the integrity of the data. Examine a printout or graph to identify outrageously silly mistakes, bad or missing points, or formatting errors, and to determine the precision (number of significant digits) of the data.

2. Test for stationarity. Look for obvious trends or low-frequency structure. Compare the first half of the data with the second half. Try detrending if necessary, for example by taking log first differences or fitting to a low-order polynomial or superposition of sine waves.

3. Plot the data in various ways. Plot X_n versus X_{n-1}, return maps (such as each local maximum versus the previous maximum), and Poincaré sections.

4. Determine the correlation time or minimum of the mutual information to see if the sampling rate is adequate but not excessive.

5. If the autocorrelation function oscillates and decays slowly, look for periodicities. Examine the power spectrum for evidence of discrete,

perhaps harmonically-related, frequencies. With a broadband spectrum, see if it is a straight line on a log–log or log–linear plot.

6. Make a space–time separation plot to see if the data record is too short for meaningful analysis.
7. See if there is a low-dimensional embedding using the method of false nearest neighbors or saturation in the correlation dimension.
8. If the embedding is sufficiently low, determine the correlation dimension. Make sure $\log C(r)$ versus $\log r$ has a scaling (linear) region by plotting its derivative $\nu(r)$ versus $\log r$. Make sure the calculated dimension from eqn (13.3) is insensitive to the embedding dimension and time step. Make sure you have sufficiently many data points for the measured dimension, for example using the Tsonis criterion in eqn (12.11).
9. If you have evidence of a low-dimensional attractor, you can then try to calculate at least the largest Lyapunov exponent, entropy, and growth rate of unpredictability. If the dimension is too high, you can try removing some of the noise by integrating the data, using a nonlinear predictor, or using principal component analysis, keeping only a few dominant components.
10. If you think you have found chaos in your data, construct some appropriately randomized surrogate data sets and subject them to the same analysis to verify your conclusion and to test the statistical significance of your estimates of dimension, entropy, predictability, Lyapunov exponents, and so forth.
11. If you find evidence of low-dimensional chaos, you can construct model equations and attempt short-term predictions. However, if the dimension is unmeasurably high, you might have correlated noise for which some predictability is possible, and whose power spectrum and probability distribution allow comparison with theoretical models. In real experiments, the latter case is overwhelmingly more common, apart from specially-constructed laboratory experiments governed by simple physical laws, such as driven pendulums and other nonlinear mechanical or electrical oscillators.

13.9 Exercises

Exercise 13.1 Calculate $\nu(r)$ in eqn (13.3) for a system such as the tent map with a uniform measure $P(X) = 1$ for $0 < X < 1$.

Exercise 13.2 Show that a system such as the tent map with a uniform measure $P(X) = 1$ for $0 < X < 1$ has a correlation dimension of 1.0 according to eqns (13.1)–(13.3).

Exercise 13.3 (difficult) Show that the $\nu(r)$ for $r \ll 1$ in eqn (13.4) follows from eqn (13.1) for the $P(X)$ in eqn (2.11) for the logistic map with $A = 4$. See Sprott and Rowlands (2001).

Exercise 13.4 Show that the probability distribution function near the first iterate of $X = 0.5$ for the logistic map is given by eqn (13.5).

Exercise 13.5 (difficult) Show that the correlation dimension of the general symmetric map in eqn (13.6) is given by eqn (13.7). See Sprott and Rowlands (2001).

Exercise 13.6 Show that the relation for $p(q)$ in eqn (13.8) is consistent with $\nu(r)$ in eqn (13.4).

Exercise 13.7 Verify that the formula in eqn (13.11) is consistent with the definition of the capacity dimension in eqn (12.2).

Exercise 13.8 Verify that the formula in eqn (13.12) is consistent with the definition of the correlation dimension in eqn (12.6).

Exercise 13.9 Verify that the formula in eqn (13.14) for the generalized dimension agrees with the formulas for the capacity and correlation dimension in eqns (13.11) and (13.12), respectively.

Exercise 13.10 Verify that eqn (13.16) for the information dimension follows from eqn (13.14) using l'Hôpital's rule.

Exercise 13.11 Show that $D_{q_1} \leq D_{q_2}$ for $q_1 > q_2$.

Exercise 13.12 Show that eqn (13.19) for the asymmetric Cantor set follows from eqn (13.14) for the case $r_1 = r_2$.

Exercise 13.13 Verify that D_q for each of the special asymmetric Cantor sets in §13.3.1 follows from eqn (13.19).

Exercise 13.14 Show that the asymptotic limits for D_q in the asymmetric Cantor sets in Fig. 13.8 are $1 \geq D_q \geq 0.369070\ldots$ for case 3 and $0.630929\ldots \geq D_q \geq 0.315464\ldots$ for case 4.

Exercise 13.15 Show that the generalized dimension of the logistic map is given by $D_q = \min[1, q/2(q-1)]$.

Exercise 13.16 Show that the generalized dimension in eqn (13.22) is the same as the correlation dimension in eqn (12.6) for $q = 2$.

Exercise 13.17 Show that the generalized dimension in eqn (13.22) is the same as the capacity dimension in eqn (12.2) for $q = 0$ if you use the maximum (supremum) norm.

Exercise 13.18 Show that the generalized dimension in eqn (13.22) is the same as the information dimension in eqn (13.16) for $q = 1$.

Exercise 13.19 Show that the generalized dimension in eqn (13.22) is equivalent to the definition in eqn (13.14) for arbitrary q.

Exercise 13.20 Derive the relations relating D_q and $f(\alpha)$ in eqns (13.23)–(13.26).

Exercise 13.21 Show from eqns (13.23)–(13.26) that the information dimension is given by $D_1 = \alpha = f(\alpha)$.

Exercise 13.22 Derive eqns (13.27)–(13.29) for the symmetric weighted Cantor set with $r_1 = r_2 = \frac{1}{3}$, $P_1 = \frac{1}{3}$, and $P_2 = \frac{2}{3}$.

Exercise 13.23 Show that for the symmetric weighted Cantor set with $r_1 = r_2 = \frac{1}{3}$, and $P_1 = \frac{1}{3}$, the maximum in the singularity spectrum is at $\alpha_0 = [\log 3 + \log \frac{3}{2}]/2\log 3 = 0.684535\ldots$, where $f(\alpha_0) = D_0 = \log 2/\log 3 = 0.630929\ldots$ and that the values of α range from $\log \frac{3}{2}/\log 3 = 0.369070\ldots$ (for $q = \infty$) to 1.0 (for $q = -\infty$).

Exercise 13.24 Verify that eqn (13.31) for the K–S entropy follows from eqn (13.30) using l'Hôpital's rule.

Exercise 13.25 Derive the relations relating K_q and $g(\beta)$ in eqns (13.33)–(13.36).

Exercise 13.26 Show from eqns (13.33)–(13.36) that the K–S entropy is given by $K_1 = \beta = g(\beta)$.

13.10 Computer project: Iterated function systems

The remaining three projects will deal with fractals and related topics. One of the simplest and most powerful techniques for producing fractal patterns (aside from the strange attractors that result from chaotic dynamical systems) is the iterated function system (IFS), described more fully in the next chapter. See also the book by Barnsley (1988) who pioneered IFS techniques.

The procedure is extremely simple. Start with an arbitrary point in the XY plane. Flip an N-sided coin to generate an integer i between 1 and N (on the computer using its random-number generator). Then apply the corresponding affine transformation F_i

$$X_{n+1} = aX_n + bY_n + e \tag{13.42}$$

$$Y_{n+1} = cX_n + dY_n + f \tag{13.43}$$

Repeat the process many thousand times. Discard the first few points, and plot the remaining ones.

1. Use the procedure described above to produce a Sierpinski triangle for which the IFS code is:

	a	b	c	d	e	f
F_1	0.5	0	0	0.5	0	0
F_2	0.5	0	0	0.5	0.5	0
F_3	0.5	0	0	0.5	0.5	0.5

2. Use the procedure described above to produce Barnsley's fern for which the IFS code is:

	a	b	c	d	e	f
F_1	0	0	0	0.16	0	0
F_2	0.849	0.037	−0.037	0.849	0	1.6
F_3	0.197	−0.226	0.226	0.197	0	1.6
F_4	−0.15	0.283	0.26	0.238	0	0.44

3. Produce some interesting patterns of your own, either as variations of the above or by letting the computer choose the IFS code randomly.
4. (optional) Develop a computer code to calculate the capacity (box-counting) dimension of an arbitrary fractal pattern displayed on the computer screen (see §12.8). Test your code using the Sierpinski triangle above, for which this dimension should be $\log 3/\log 2 \simeq 1.585$.

14 Nonchaotic fractal sets

Chaotic dynamical systems and their accompanying strange attractors as described in Chapter 6 are only one way to produce fractal images. Chapter 11 showed a number of fractals produced by other methods. Two of the most important and widely known such methods are *iterated function systems* and *Julia sets* with their relatives such as the *Mandelbrot set*, perhaps the best known and most studied of all fractal sets.

Iterated function systems provide a well-defined method to produce fractals with specific desired characteristics and appearance. They also can be used to encode and compress natural images as well as to test for determinism in a time series.

Julia sets were mentioned in §6.8.4. They are the basin boundary in the space of initial conditions Z_0 for bounded solutions of a complex iterated map for a particular value of the complex parameter C. The Mandelbrot set is a plot of the bounded solutions in the complex C plane for the same iterated map with $Z_0 = 0$. Their main application is for producing computer art, some of which is quite stunning when rendered in full color at high resolution. For reasons of economy, only gray-scale images are included here. If you are interested in such computer art, see the book by Peitgen and Richter (1986) or one of the many Web sites devoted to fractal art.

14.1 The chaos game

The 'chaos game' is an example of an iterated function system (IFS) that was developed by Michael Barnsley (1988). The rules are simple, requiring only a pencil, paper, and ruler, but in practice a simple computer program is used. Put three dots, labeled A, B, and C, at arbitrary places on a sheet of paper. Then start anywhere on the paper and flip a three-sided coin or roll a three-sided die. If you do not have one of those, use a six-sided die with two of the sides labeled A, two labeled B, and two labeled C. If the die comes up A, move half way to dot A and plot a point there, or similarly with B and C. Then roll the die again to determine the next move starting from the current position, and repeat the process many times. The pattern that emerges is the Sierpinski triangle in Fig. 11.9. You may want to discard the first few points or begin with an initial condition on one of the vertices.

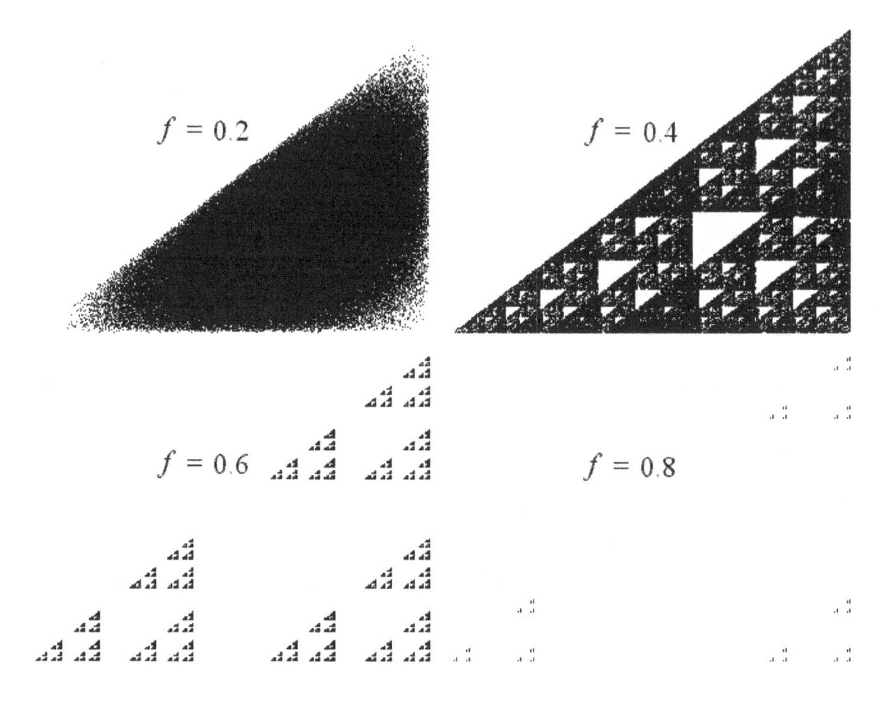

Fig. 14.1 The chaos game with different rules on how far to move with each iteration.

A remarkable property of the game is that the resulting pattern is independent of where you start and the sequence of die rolls. This *random iteration algorithm* produces a deterministic pattern. The pattern is an *attractor* for initial conditions anywhere in the plane. Although the attractor is a fractal, it is not usually called a strange attractor because the orbit is random rather than chaotic. The rule is linear, and thus it cannot produce chaos. If you repeat the process with a slightly different initial condition but with the same sequence of die rolls, then the two orbits will converge, implying a negative Lyapunov exponent. In fact, the contraction is exactly a factor of two for each iteration, and so the Lyapunov exponents are equal to the eigenvalues of the Jacobian matrix, as described in §4.7, giving $\lambda_1 = \lambda_2 = \log 0.5 = -0.693147181\ldots$. If you perform the process with different sequences of random numbers, then the respective orbits will diverge, but the process is not chaotic because the rule is nondeterministic.

This rule is only one example of how you might play the chaos game. Instead of moving half way to each vertex with each iteration, you could move a fraction f anywhere from zero to one. Zero would never take you off your initial condition, and one would make you repeatedly visit the three vertices in random order. Both cases are relatively uninteresting, giving a zero-dimensional object rather than a fractal. Intermediate values give the results in Fig. 14.1 after 10^5 iterations. The fractal dimension decreases

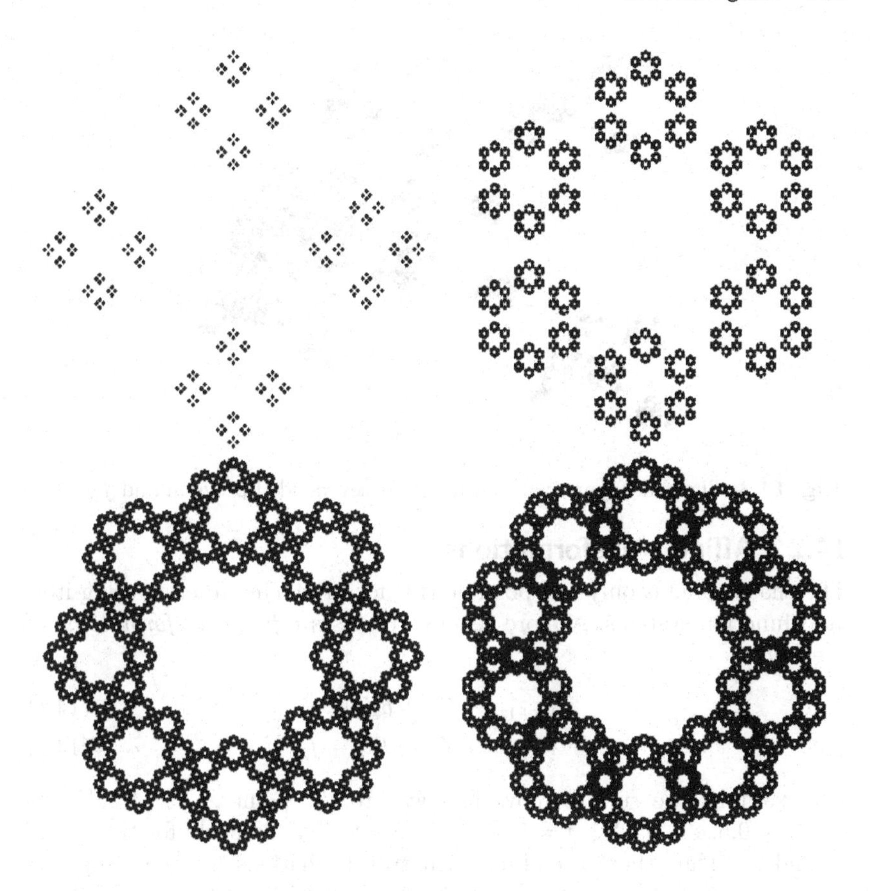

Fig. 14.2 The chaos game with different size regular polygons and $f = 0.3$.

from 2.0 to 0 as f increases.

There is also nothing special about starting with three vertices, which produces a triangular fractal. Figure 14.2 shows examples of regular polygons with 4, 6, 8, and 10 vertices, respectively, for $f = 0.3$. The four-sided polygon is just the Cantor square in Fig. 11.9 rotated 45°. Perhaps it should be called a 'Cantor diamond.' More generally, these shapes are called *Sierpinski polygons*. Starting with *two* points leads to an ordinary Cantor set as in Fig. 11.1. For a given f, the dimension of the fractal increases with the number of vertices, eventually reaching 2.0. Further generalizing to cases where the points are not at the vertices of a regular polygon produces asymmetric patterns, one example of which with seven sides and $f = 0.3$ is in Fig. 14.3.

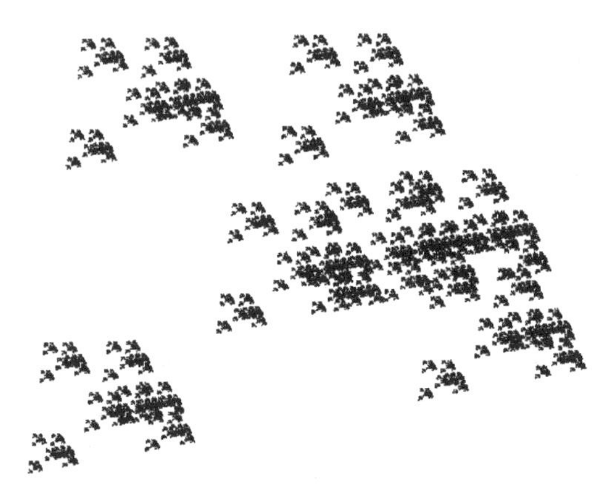

Fig. 14.3 The chaos game with an irregular seven-sided polygon and $f = 0.3$.

14.2 Affine transformations

The chaos game is only one possible rule for producing fractals using iterated function systems. A more general rule uses *affine transformations* of the form

$$X_{n+1} = aX_n + bY_n + e \tag{14.1}$$

$$Y_{n+1} = cX_n + dY_n + f \tag{14.2}$$

For example, the rule to move half way to the point A at (X_A, Y_A) is $a = d = 0.5, e = 0.5X_A, f = 0.5Y_A, b = c = 0$, and similarly for the points B and C. Thus there are three such transformations for the Sierpinski triangle, one of which is chosen randomly at each iteration.

The chaos game involves translation, rotation, and contraction. The general affine map (with $bc \neq 0$) includes *shear* as in Fig. 5.4. With $bc > ad$ there is a mirror reflection, and the three vertices change their orientation from clockwise to counter-clockwise and back with each iteration. If they are ordered ABC going clockwise, they will be ordered ACB going clockwise after the transformation. The affine map in eqns (14.1) and (14.2) is the most general linear mapping in two dimensions. The extension to three and higher dimensions is straightforward but requires additional visualization methods such as color, as described in §6.6.

Because the transformations are linear, the area contraction is constant everywhere in space and given by the determinant of the Jacobian matrix

$$A_{n+1}/A_n = |\det J| = |ad - bc| \tag{14.3}$$

As with strange attractors, the Lyapunov exponents are given by the logarithm of the eigenvalues of the Jacobian matrix, and the sum of the Lyapunov exponents is

$$\lambda_1 + \lambda_2 = \log|\det J| \tag{14.4}$$

If there are N different transformations, each applied with probability P_i, then the net contraction is

$$A_{n+1}/A_n = \sum_{i=1}^{N} P_i |a_i d_i - b_i c_i| \tag{14.5}$$

For the usual case of a bounded system, we expect $A_{n+1}/A_n < 1$. The probabilities can be optimally chosen so that $P_i \propto |a_i d_i - b_i c_i|$. In fact, it is essential to do so if the pattern is to develop in a reasonable number of iterations and with uniform density. Transformations that are strongly contracting do not have to be applied as often as those that are weakly contracting or expanding. Note that area contraction does not guarantee boundedness, since a set can continually contract in one direction and expand in another, approaching a thin filament of zero area and infinite length. Without a nonlinearity, there is no folding, and the orbits extend to infinity, although the probability distribution function may become very small at large distances.

With N different transformations, there are N fixed points, and these fixed points are stable if their eigenvalues are within the unit circle in the complex plane (see §4.7), giving contraction. The solutions can also be saddle points, which are stable in one direction and unstable in the other. However, the limit set is not just N points because the random iterations produce a tug-of-war between the various points. The orbit may approach such a point arbitrarily closely if its transformation is chosen many times in succession, but as soon as another transformation is chosen, it moves abruptly away from that point toward another. The locations of the fixed points are usually not obvious in the pattern because long sequences of identical values are rare. Each point in the resulting pattern can be transformed into another point in the pattern by some sequence of the affine mappings. The collection of all such sequences eventually obtained by the random iteration algorithm is the *iterated function system*. You can choose the initial conditions arbitrarily and discard the first few points or start on one of the fixed points, which is necessarily part of the set.

You can use affine transformations to produce a wide variety of self-similar fractals such as the fractal gaskets in §11.4. The possibilities are limited only by your imagination. Figure 14.4 shows one produced by ap-

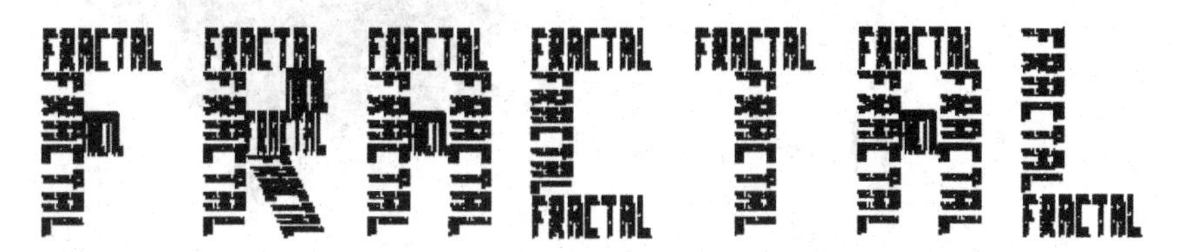

Fig. 14.4 A fractal produced by 23 affine transformations.

plying 23 transformations (count them) with equal probability. You could try making some of your own, perhaps using your name or initials.

14.3 Iterated function systems

Iterated function systems, first studied by Hutchinson (1981), offer a powerful method for producing a wide variety of forms that resemble natural objects and others that are visually interesting. This fact suggests that they can also be used for image compression and perhaps for finding patterns in experimental data.

14.3.1 Examples

Figure 14.5 shows four images produced by a small number of affine transformations applied randomly. The one that resembles a fern is produced by the four transformations

Fig. 14.5 Fractals resembling natural objects produced by a small number of affine transformations.

	a	b	c	d	e	f
F_1	0	0	0	0.16	0	0
F_2	0.849	0.037	−0.037	0.849	0	1.6
F_3	0.197	−0.226	0.226	0.197	0	1.6
F_4	−0.15	0.283	0.26	0.238	0	0.44

The one that resembles a maple leaf is produced by the five transformations

	a	b	c	d	e	f
F_1	0.0036	0	0	0.5783	0.5016	0.0606
F_2	0.3517	0.3554	−0.3554	0.3517	0.3545	0.5
F_3	0.5	0	0	0.5	0.25	0.462
F_4	0.5015	−0.0018	0.0016	0.588	0.2501	0.1054
F_5	0.3534	−0.3537	0.3537	0.3534	0.2879	0.1528

The one that resembles a tree is produced by the four transformations

	a	b	c	d	e	f
F_1	0.195	−0.488	0.344	0.443	0.722	0.536
F_2	0.462	0.414	−0.252	0.361	0.538	1.167
F_3	−0.058	−0.07	0.453	−0.111	1.125	0.185
F_4	−0.045	0.091	−0.469	−0.022	0.863	0.871

and the one that resembles moss is produced by the two transformations

	a	b	c	d	e	f
F_1	−0.4	−0.4	−0.9	0.4	−0.4	−0.2
F_2	−0.7	0	−0.5	0.8	0.6	−0.2

The images in Fig. 14.5 were designed to mimic natural objects. You can also use random numbers, kept the same during the iterations, for the coefficients of the affine maps (Sprott 1994c) as was done with the strange attractors in Fig. 6.1 to produce a variety of interesting shapes. Figure 14.6 shows some images produced this way using only two mappings. Even more interesting and realistic images can be produced using more than two variables to represent depth and color and more than two mappings.

An important application of iterated function systems reverses the method and codes the information required to produce a natural image in a relatively small number of coefficients using the *collage theorem* (Barnsley and

Fig. 14.6 Sample images from iterated function systems using two affine mappings.

Hurd 1993).[1] The method involves identifying non-overlapping portions of the image that resemble the whole, but translated, shrunk, rotated, and sheared, and then finding the corresponding transformations. A 'collage' of the images produced by the transformations is then made. With enough transformations, any image can be produced to arbitrary precision (Jacquin 1992). Such compressed images can be magnified arbitrarily without losing detail, although the detail is artificial and derived from the appearance of the image on a large scale assuming a fractal structure. The compression is lossy (information in the image is destroyed), computationally intensive (slow), and proprietary (a trade secret). Compression factors of ten to a hundred are typical, depending on the image. Decompression is relatively fast and simple. The image produced by the transformations is unique, but a given image can usually be described by more than one set of rules.

These objects are deterministic even though produced by a random process. As with strange attractors, the sequence of numbers determines the orbit, but not the final shape. The rules are non-invertible, since you do not know from which transformation each point in the image came, and thus you cannot iterate backward in time. The patterns are fractal, but they are not generally self-similar because of the nonuniform contraction, rotation, and shear. Instead, they are said to be *self-affine*. They are not chaotic and are not usually called 'strange attractors.' Finding a low-dimensional fractal in experimental data is thus not a proof of underlying chaotic dynamics.

14.3.2 Aesthetics

As with strange attractors, you can characterize IFS patterns by a largest Lyapunov exponent and fractal dimension. The Lyapunov exponent λ_1 is negative because the system is not chaotic, but you can calculate it using the method in §5.5. The similarity dimension can sometimes be calculated from a weighted average of the contractions as described in §12.1, but the calculation is difficult when the mappings overlap. However, you can use the Grassberger–Procaccia algorithm in §12.3 to calculate the correlation dimension D_2 just as with strange attractors.

As with strange attractors (see §6.10), the aesthetic quality correlates with the Lyapunov exponent[2] and correlation dimension (Sprott 1994c) as in Fig. 14.7. There is a clear preference[3] (darker shades) for patterns with correlation dimensions greater than 1.0 and for large negative Lyapunov exponents. For a given dimension, the largest negative Lyapunov expo-

[1]Fractal image compression was patented in 1988 by Michael Barnsley and Alan Sloan who co-founded Iterated Systems, Inc. of Norcross, GA, to commercialize the process. The first major commercial application was *Microsoft Encarta*, a multimedia encyclopedia first published in 1993 on a 600 MB compact disk using .FIF (fractal image format) graphics files.

[2]In this study the Lyapunov exponent was evaluated in base-2.

[3]These evaluations were done by the author in a double-blind experiment in which neither the subject nor the computer knew the parameters until all the evaluations were completed.

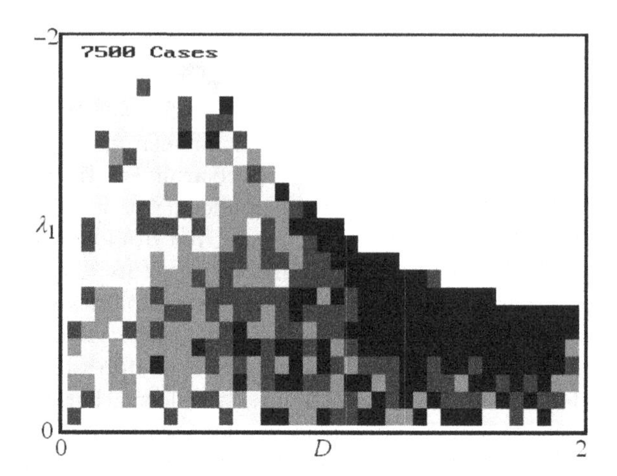

Fig. 14.7 Aesthetic evaluation of 7500 iterated function systems of various dimensions and Lyapunov exponents.

nents occur when the Lyapunov exponents are equal, implying identical contraction in all directions. Thus these patterns preserve their shape and are exactly self-similar. The largest negative Lyapunov exponent λ_1 and fractal dimension D are bounded approximately by the curve

$$D\lambda_1 \simeq -\log N / \log m \tag{14.6}$$

where N is the number of mappings (two in this case) and m is the embedding dimension (two in this case).

The 76 cases rated best have an average correlation dimension of $D = 1.51 \pm 0.43$ and an average Lyapunov exponent of $\lambda_1 = -0.24 \pm 0.15$ bits/iteration, where the errors represent one standard deviation. About 31% of the cases fall within this range. These results suggest that you could program the computer to select cases with a high probability of aesthetic appeal. Another criterion surrounds the best cases with an ellipse

$$[(2 - D)/1.2]^2 + [(2 + \lambda_1/\ln N)/1.6]^2 < 1 \tag{14.7}$$

that excludes about 98% of the two-dimensional cases. The 2% that remain are nearly all visually interesting and mostly different.

14.3.3 IFS clumpiness test

Iterated function systems also suggest a data-analysis method (Peak and Frame 1994). Suppose you play the chaos game with a time series of uncorrelated random numbers $0 < X_n < 1$ on a square with $f = 0.5$. Actually, any parallelogram will work, but let us keep it simple. Label the corners of the square $ABCD$ clockwise from the upper left. Start anywhere in the square, such as the corner A. If the first value in the time series has

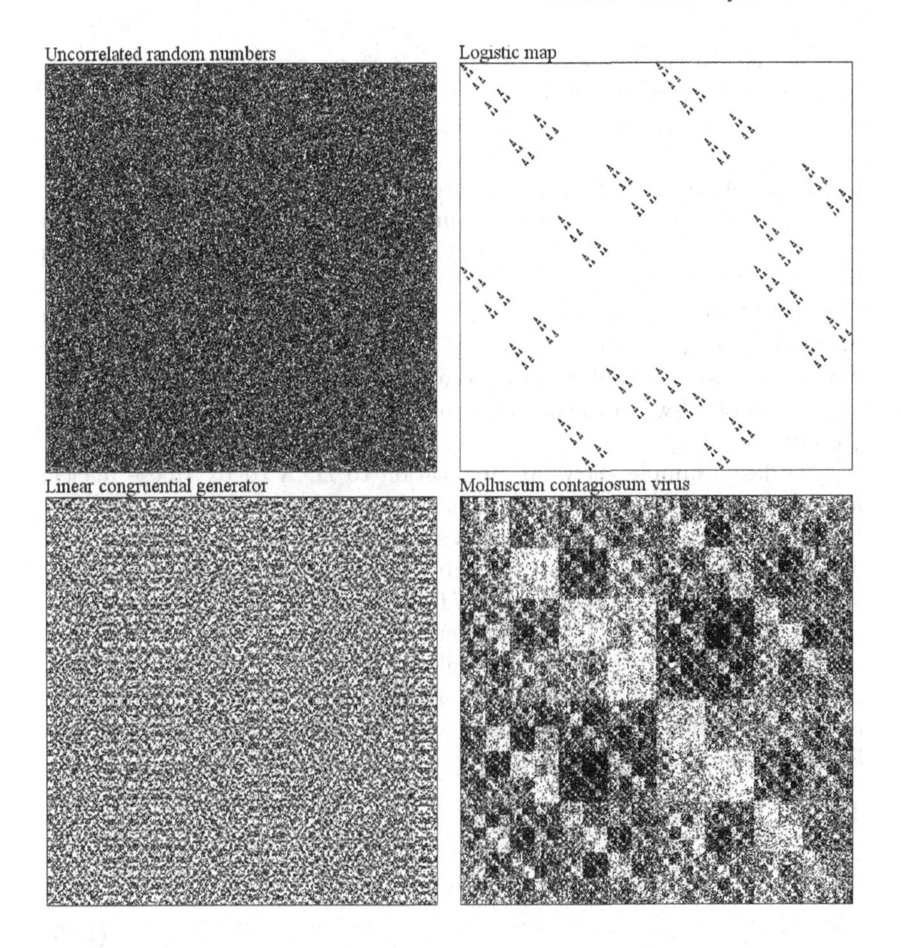

Fig. 14.8 IFS clumpiness test applied to four different data sets.

$0 < X_1 < 0.25$, move half way to A; if $0.25 < X_1 < 0.5$, move half way to B, and so forth. Continue iterating until the square begins to fill in. If the values are uncorrelated, then the points will be uniformly dense as in the upper left of Fig. 14.8 with 2×10^5 points.

Now try the same procedure using the logistic map with $A = 4$, which gives the fractal plot in the upper right of Fig. 14.8 with 2×10^5 points. The *clumpiness* of the plot is an indicator of determinism, whether it be chaos or correlated noise. The method is highly sensitive to determinism in the data. However, it does not very well distinguish chaos from colored (correlated) noise. A nonuniform distribution of points in the plots can result from a different number of values in the four categories. You can eliminate this effect by using *quartiles* (moving the boundaries between the four cases so that each category has the same number of data points). Alternately, you can shuffle the data points, randomizing their order but preserving their distribution, and look for differences in the real and surrogate plots.

The plot in the lower left of Fig. 14.8 shows the result for the linear congruential generator (see §2.6)

$$X_{n+1} = 7141X_n + 54773 \pmod{259200} \tag{14.8}$$

which might be used to produce pseudo-random integers in the range $0 \leq X_n < 259200$. Although the numbers produced this way would pass most tests for randomness, the IFS plot in the lower left of Fig. 14.8 shows horizontal bands not evident in the high-quality random numbers used in the plot above it. Also the pattern looks less dense than the one above it despite having the same number of points, suggesting that more of the iterates overlap. Our ability to discern visual patterns buried in noise presumably evolved when the detection of predators in the wild was essential for survival.

The last example shows an application to DNA (deoxyribose nucleic acid) sequencing (Jeffrey 1992). Since the DNA molecule contains four bases ($ACGT^4$), the chaos game on a square is a natural analysis method.[5] The molluscum contagiosum virus subtype 1 was chosen because it is fully sequenced and has a similar number (190 289) of base pairs, giving the plot in the lower right of Fig. 14.8. Apparently the sequence is not random, but neither is it purely deterministic. Similar plots are produced by other DNA molecules, and the differences could be useful for cataloging and comparison. The method can also be used with speech and music (Meloon and Sprott 1997).

14.4 Julia sets

Julia sets[6] (or *J sets*) were discussed briefly in §6.8.4. They are the basin boundary for bounded solutions of the complex mapping

$$Z_{n+1} = Z_n^2 + C \tag{14.9}$$

in the XY plane with $Z = X + iY$ for a particular value of the complex constant $C = A + iB$. This map, sometimes called the *Mandelbrot map*, can be considered as the generalization of the one-dimensional quadratic map in eqn (2.4) to complex variables, or as the two-dimensional map

$$X_{n+1} = X_n^2 - Y_n^2 + A \tag{14.10}$$

$$Y_{n+1} = 2X_nY_n + B \tag{14.11}$$

with real variables X and Y and real constants A and B. The basin boundary is typically a fractal with intricate structure, one example of which is in

[4]The nitrogen-containing bases of adenine (A), cytosine (C), guanine (G), and thymine (T) form a hydrogen-bonded double helix in which A is paired with C and G is paired with T (Watson 1968).

[5]DNA sequences are publicly available from http://www.ncbi.nlm.nih.gov/Entrez/.

[6]Julia sets were studied by Gaston Maurice Julia (1893–1978), an Algerian-born mathematician and student of Poincaré, who lost his nose during a battle on the French front in the First World War and had to wear a leather strap across his face for the rest of his life.

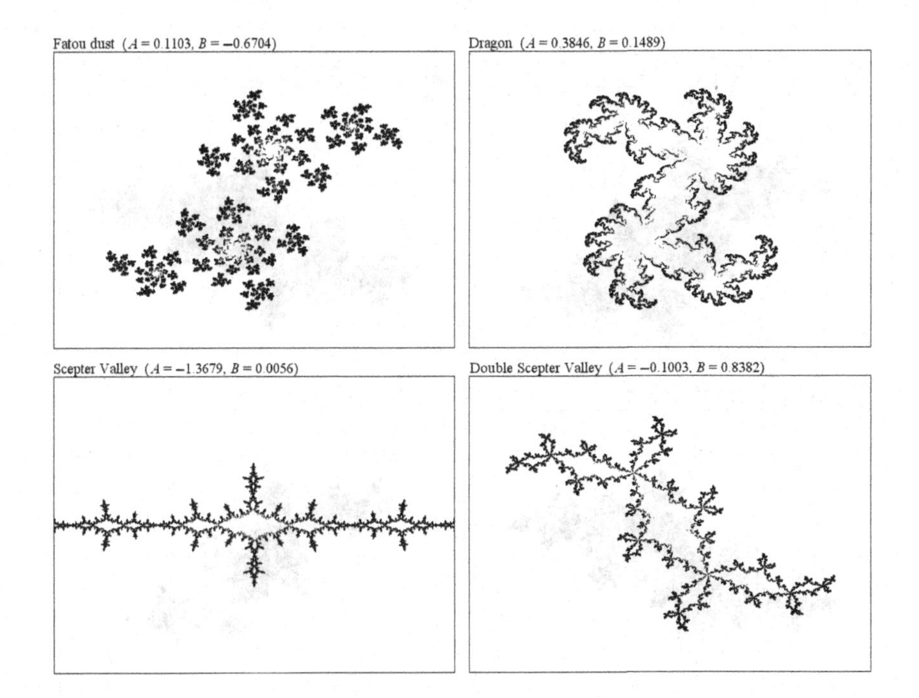

Fig. 14.9 Examples of Julia sets.

Fig. 6.13. Some other examples are in Fig. 14.9, which includes the values of A and B and names that have been given to the various shapes. The first picture of a Julia set was hand-drawn by Cremer (1925), but it was only with the advent of the digital computer in the 1970's that their spectacular appearance became evident and widely known.

These Julia sets typically enclose infinitely many fixed points and periodic orbits of all periods. The fixed points are given by

$$Z^* = \tfrac{1}{2} \pm \tfrac{1}{2}\sqrt{1-4C} \tag{14.12}$$

with one solution attracting and the other repelling. The basin boundary is dense in periodic orbits, but all are unstable (repellors), and they are a set of measure zero in the boundary, which is itself a set of measure zero in the plane. Most initial conditions on the basin boundary have chaotic orbits, and orbits near the basin boundary are transiently chaotic, often wandering near the boundary for thousands of iterations before escaping (if outside) or settling to a periodic cycle (if inside). The boundary is a *nonattracting chaotic set.*

The basin boundary can be found by running time backwards so that the repellors become attractors. You can do this by finding the preimages of the map in eqns (14.10) and (14.11)

$$Z_n = \pm\sqrt{Z_{n+1} - C} \tag{14.13}$$

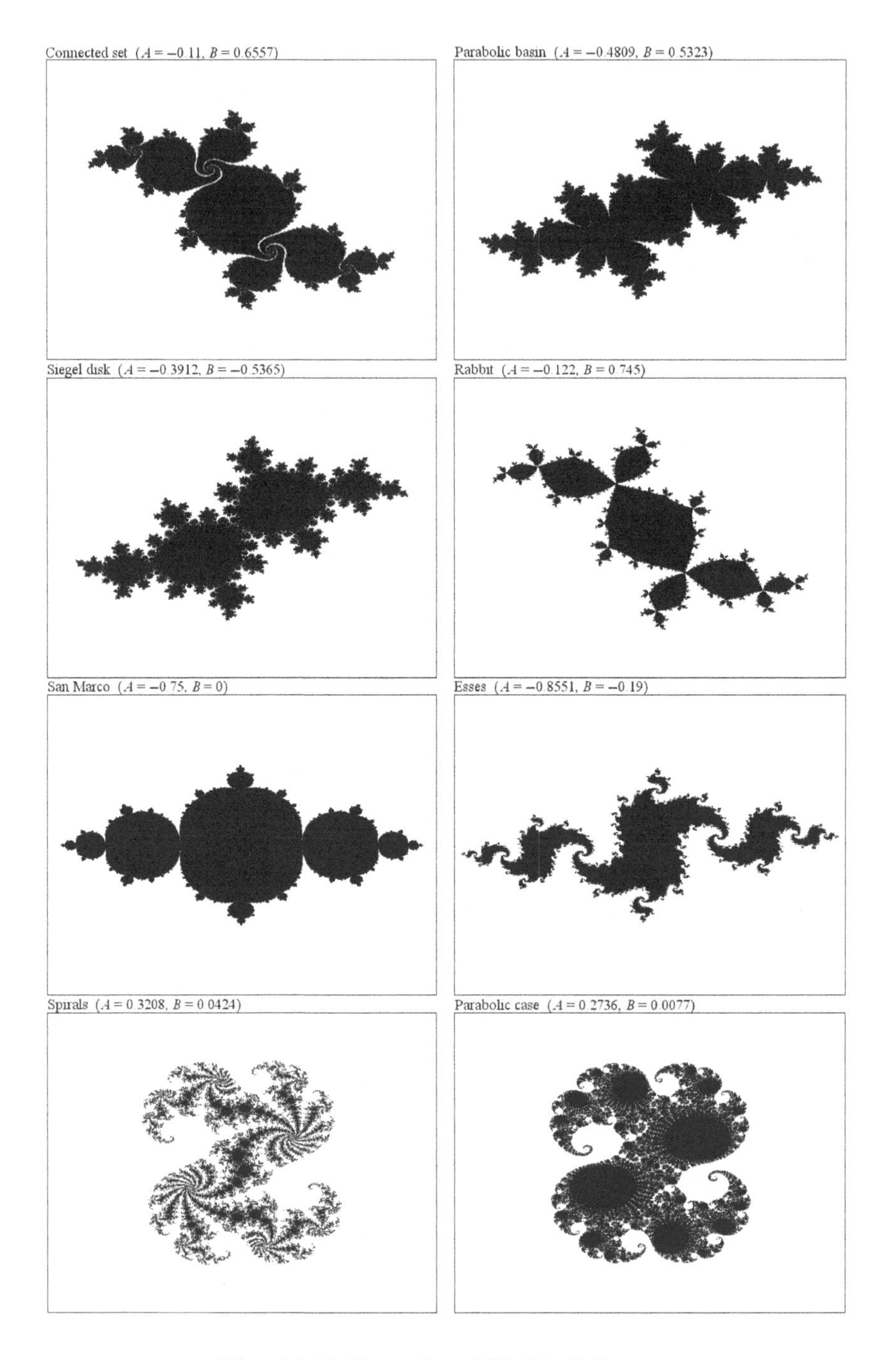

Fig. 14.10 Examples of filled-in Julia sets.

Because each value of Z has two preimages, you can use the random iteration algorithm to find all the points on the boundary by iterating eqn (14.13) in the limit $n \to -\infty$ for an arbitrary initial condition, choosing the plus and minus roots randomly. The stable orbits in the interior of the set become unstable with this procedure and are not observed.

The Julia set is sometimes defined as the whole basin of attraction for bounded solutions. Such sets are more properly called *filled-in Julia sets* or *prisoner sets* since they include the region enclosed by the Julia set. Some examples of filled-in Julia sets with their common names are in Fig. 14.10. Even more confusingly, the Julia set is sometimes defined as the set of *unbounded* solutions, which is the complement of the cases shown here. The complement of the Julia set is called the *Fatou set*[7] (Schroeder 1991). Julia (1918) and Fatou (1919) independently studied these sets, and more detail about them can be found in many modern sources such as Devaney (1992).

You can automate the search for visually interesting Julia sets by seeking values of A and B where the iterates of the initial condition $Z_0 = 0$

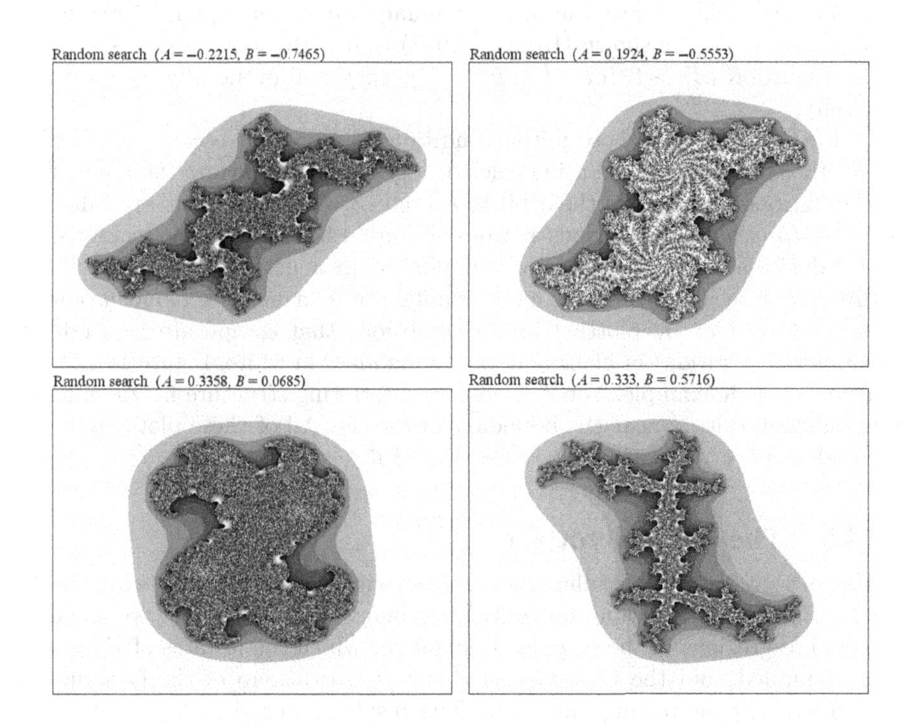

Fig. 14.11 Examples of gray-scale Julia set escape-time plots.

[7]Pierre Joseph Louis Fatou (1878–1929) was a French mathematician and theoretical astronomer who worked simultaneously and independently on some of the same problems as Julia.

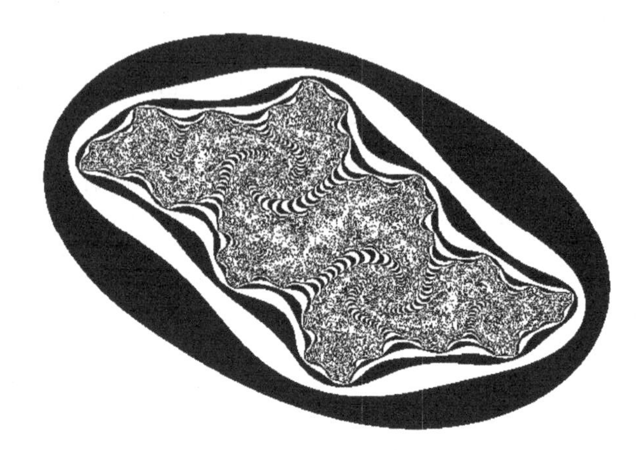

Fig. 14.12 Monochrome escape-time plot of a Julia set with $A = -0.1962$ and $B = 0.6769$.

are unbounded but take many iterations (say a few hundred) to escape, so that the point is near the basin boundary, guaranteeing an interesting structure. It can be shown (Branner 1988) that if the iterates ever satisfy the condition $|Z| > 2$ (or $A^2 + B^2 > 4$), they will eventually escape to infinity.

Rather than plot those initial conditions that do *not* escape, you can plot those that *do* escape using color or a gray scale to denote the number of iterations required for the orbit to exceed $|Z| = 2$. Such plots are called *escape-time plots* for obvious reasons. Figure 14.11 shows four examples of Julia sets obtained this way and plotted in a gray scale that cycles through sixteen levels. Even more simply, you can make a monochrome escape-time plot by plotting initial conditions that escape after an odd number of iterations in black and an even number in white. Figure 14.12 is a typical such example. You can observe interesting structure by zooming in on some feature near the boundary of the set. All of these plots are on a scale $-1.6 < X_0 < 1.6$ and $-1.6 < Y_0 < 1.6$.

14.5 The Mandelbrot set

The visually interesting Julia sets are those with values of C whose iterates with Z_0 near zero escape, but only after many iterations. Thus we could plot those regions in the complex C plane for which the iterates of $Z_0 = 0$ are bounded, and the boundary of this region is a map of the C values with visually interesting Julia sets. The resulting image in Fig. 14.13 is called the *Mandelbrot set* (or *M set*) after Benoit Mandelbrot who studied and popularized it (Mandelbrot 1980). It is probably the most famous fractal. Turned on its side, the set looks like a gingerbread man, which is appropriate since 'mandelbrot' coincidentally means 'almond bread' in German.

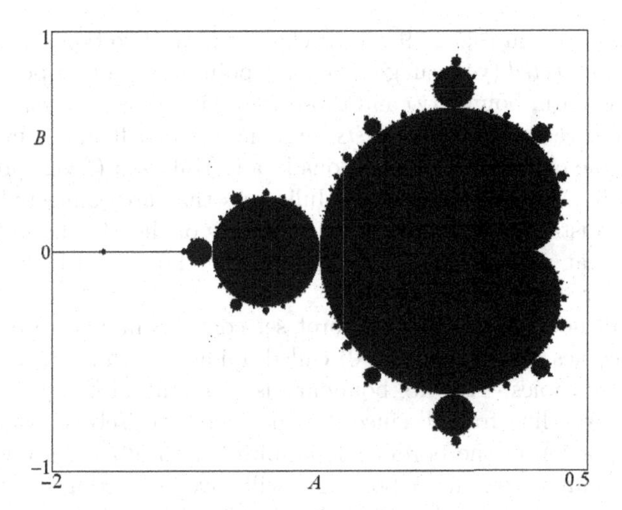

Fig. 14.13 The Mandelbrot set is the set of bounded orbits of $Z^2 + C$ with $Z_0 = 0$ in the complex C plane.

The large cardioid (heart-shaped region) on the right is given by

$$C = \frac{\cos\theta}{2} - \frac{\cos 2\theta}{4} + i\left(\frac{\sin\theta}{2} - \frac{\sin 2\theta}{4}\right) \tag{14.14}$$

and contains orbits that attract to the stable fixed points in eqn (14.12). The cleft in the heart is called 'Elephant Valley' because of the shape of the images that occur there upon magnification. The smaller circular 'bud' to the left of the main cardioid is given by

$$C = -1 + \frac{\cos\theta + i\sin\theta}{4} \tag{14.15}$$

and contains stable period-2 orbits

$$Z^* = -\tfrac{1}{2} \pm \tfrac{1}{2}\sqrt{-3 - 4C} \tag{14.16}$$

that cycle between the positive and negative roots. The notch between it and the main cardioid is called 'Seahorse Valley' because of the shape of the images that occur there upon magnification. The bud to the left of the period-2 bud contains stable period-4 orbits, and so forth. In fact, the real axis ($B = 0$) has a period-doubling route to chaos identical to the logistic map. The buds off the real axis contain stable orbits of other periods. For example, the large buds at the top and bottom are period-3. Located around the set are numerous 'tendrils,' the most prominent of which is 'the spike,' pointing due west and terminating at (-2, 0). Chaotic orbits occur only on the boundary of the set and apparently have zero measure (no area) in the complex C plane. As with the Julia sets, points near the boundary are transiently chaotic.

The Julia sets in Fig. 14.9 can be classified into two types, those whose interior is *connected* (you can get from any point to any other point without crossing the basin boundary) and those whose interior is *disconnected.* The disconnected cases are called *dusts*, and the upper left image in Fig. 14.9 is an example called *Fatou dust.* Douady and Hubbard (1982) proved that points in the Mandelbrot set have Julia sets that are connected, and the others are dusts. Thus points on the boundary of the Mandelbrot set have Julia sets that are just barely connected, and those tend to be the most visually interesting.

You will note that the Mandelbrot set contains miniature but slightly distorted copies of itself (Lei 1990) called 'midgets,' which in turn contain even smaller copies. Thus its boundary is a fractal, although it has been called a 'borderline fractal' since it is nowhere precisely self-similar. Every finite portion of the border has infinite length albeit zero width, and thus a zoom anywhere on the boundary will have interesting structure. The boundary has dimension 2.0 (Shishikura 1998), although numerical experiments suggest it is a set of measure zero (zero area) since chaotic orbits are never found for random choices of C. Thus it is 'infinitely complicated.' The set has an area between 1.50585063 and 1.5613027 with a best estimate of 1.5066, which rules out simple values such as $3/2$ or $\pi/2$ (Hill 1996),[8] but the significance of this result is unknown.

Furthermore, the Mandelbrot set is connected, although only barely so (Douady and Hubbard 1982). However, it is not known whether the Mandelbrot set is *locally connected.* (The interior of a circle from which a single point has been omitted is connected, but not locally connected, because points on opposite sides of the missing point, no matter how close, are only connected through a circular arc.)

If you zoom deeply into some region of the Mandelbrot set, the resulting image will approach that for the deeply zoomed Julia set for that value of C (Lei 1990). Thus the Mandelbrot set in this sense contains all the Julia sets. It has been described as the most complicated mathematical object ever 'seen.' People have zoomed into the set by factors as large at 10^{1600}, and interesting structures continue to appear. Animated zooms are especially engaging and somewhat psychedelic. You can plot escape-time contours for the Mandelbrot set in the same way as for the Julia sets in Figs 14.11 and 14.12. The example in Fig. 14.14 uses a 16-level gray scale. It has also been called the 'ultimate computer virus' since it not only consumes all available computer resources but also countless hours of human time exploring its delights.

The standard Mandelbrot set $Z_{n+1} = Z_n^2 + C$ is just one of an endless variety of sets $Z_{n+1} = f(Z_n)$ with complex Z (Frame and Robertson 1992). Some other interesting examples for you to explore are in Table 14.1

[8]The area of the Mandelbrot set was calculated numerically by a group of over four dozen volunteers around the world using computers connected by the Internet.

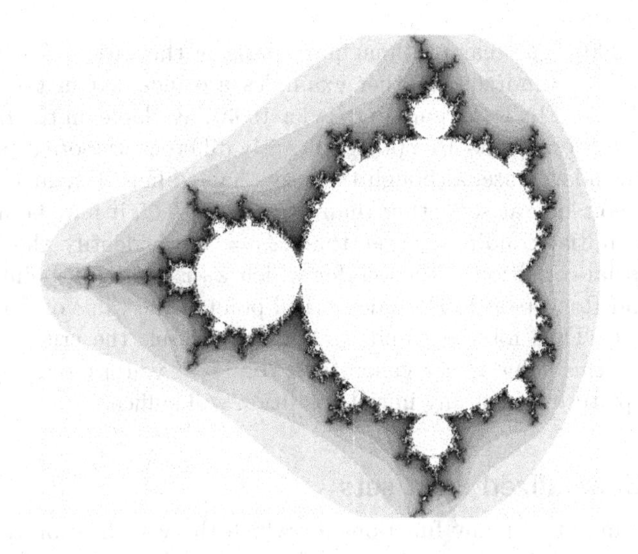

Fig. 14.14 Gray-scale Mandelbrot set escape-time plot.

Table 14.1 Some interesting Mandelbrot/Julia set variants.

Name	$f(Z)$
Standard Mandelbrot	$Z^2 + C$
Arbitrary power	$Z^m + C$
Manzzpwr	$Z^2 + Z^m + C$
Tetrate	C^Z
Dragon curve	$C(1 - Z)^2$
Sine set	$\sin Z + C$
Cosine set	$\cos Z + C$
Exponential set	Ce^Z
Hyperbolic sine set	$\sinh Z + C$
Hyperbolic cosine set	$\cosh Z + C$
Tchebychev T_5 polynomial	$C(16Z^5 - 20Z^3 + 5Z)$
Tchebychev C_6 polynomial	$C(Z^6 - 6Z^4 + 9Z^2 - 2)$
Tchebychev U_5 polynomial	$C(32Z^5 - 32Z^3 + 6Z)$
Legendre P_4 polynomial	$(35Z^4 - 30Z^2 + 3)/8 + C$
Laguerre L_4 polynomial	$(Z^4 - 16Z^3 + 72Z^2 - 96Z + 24)/24 + C$
Hermite H_4 polynomial	$16Z^4 - 48Z^2 + 12 + C$

(Stevens 1990). The real and imaginary parts of the various functions are given in §B.4. Hundreds of other examples are included in the freeware program FRACTINT (Wegner and Tyler 1993), available on the Internet.[9] Generally, these variants are not significantly different or more complicated than the standard case, although their sets have different symmetries.

When you look at sets other than $f(Z) = Z^2 + C$, it may be necessary to use an initial condition other than $Z_0 = 0$ to identify the set. The correct initial condition is the one for which Z_0 is a critical point of $f(Z)$ (Frame and Robertson 1992). The critical point is the value of Z for which $dF/dZ = 0$. Thus for the standard Mandelbrot set, the critical point is $Z = 0$, but this value is not general. Of course, if you are only interested in pretty pictures, most any initial condition will suffice.

14.6 Generalized Julia sets

There are infinitely many functions for which the visualization techniques above can be applied. Even general two-dimensional quadratic maps have twelve parameters

$$X_{n+1} = a_1 + a_2 X_n + a_3 X_n^2 + a_4 X_n Y_n + a_5 Y_n + a_6 Y_n^2 \tag{14.17}$$

$$Y_{n+1} = a_7 + a_8 X_n + a_9 X_n^2 + a_{10} X_n Y_n + a_{11} Y_n + a_{12} Y_n^2 \tag{14.18}$$

You can search for interesting *generalized Julia sets* with these equations by choosing a_1 through a_{12} randomly and iterating until the orbit for initial conditions $X_0 = Y_0 = 0$ grows to some large value such as $X_n^2 + Y_n^2 > 10^6$ (Sprott and Pickover 1995). If this condition occurs for n large but finite (say $100 < n < 1000$), then the resulting Julia set near the origin ($X_0 = Y_0 = 0$) is probably visually interesting.

Figure 14.15 shows examples produced this way. The coefficients are coded into a string with the first character 'E' denoting a two-dimensional quadratic map and the remaining twelve characters representing the respective coefficients, with $A = -1.2$, $B = -1.1$, and so forth. The search program discards about 300 cases for every one that satisfies the escape-time constraint, but it only takes a few seconds of computation to find an interesting case. After finding a candidate, a plot is made for initial conditions in the range $-1 < X_0 < 1, -1 < Y_0 < 1$. The escape value of n is depicted by a 16-level gray scale that continually cycles from pure white to pure black and back again in a manner similar to Figs 14.11 and 14.14.

These examples are not generally fractal since eqns (14.17) and (14.18) do not satisfy the Cauchy–Riemann conditions in eqns (6.17) and (6.18). The most general two-dimensional quadratic map that satisfies these conditions has six parameters and is given by

[9] The FRACTINT home page is http://spanky.triumf.ca/www/fractint/fractint.html.

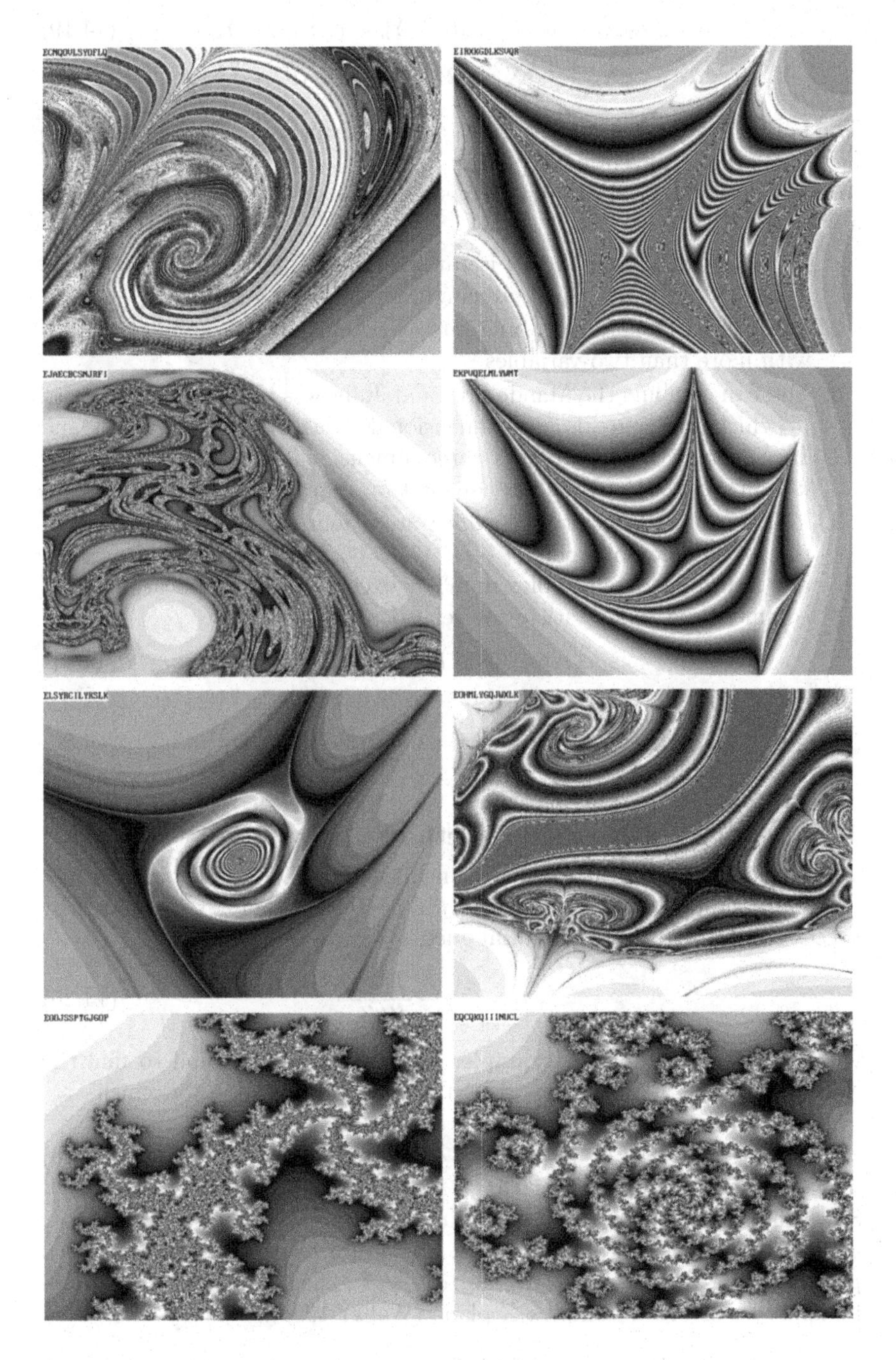

Fig. 14.15 Escape-time plots for generalized Julia sets.

$$X_{n+1} = a_1 + a_2 X_n + a_3 X_n^2 + a_4 X_n Y_n + a_5 Y_n - a_3 Y_n^2 \tag{14.19}$$

$$Y_{n+1} = a_7 - a_5 X_n - \frac{a_4}{2} X_n^2 + 2a_3 X_n Y_n + a_2 Y_n + \frac{a_4}{2} Y_n^2 \tag{14.20}$$

The bottom two images of Fig. 14.15 are constrained in this way and more nearly resemble conventional Julia sets.

Escape-time fractals in dimensions higher than two can be produced in a number of ways:

1. You can plot the number of iterations required for the orbit to escape as the third dimension, producing 'Mandelbrot mountains.'
2. You can use generalized Julia sets as in eqns (14.17) and (14.18) but with more than two variables.
3. You can combine the Mandelbrot and Julia sets to make a 'Juliabrot,' in which the space is four-dimensional, consisting of X_0, Y_0, A, and B. You may want to take a three-dimensional slice of this space by keeping one of the variables constant for ease of visualization.
4. You can use *quaternions*,[10] which are a four-dimensional generalization of complex numbers $A + iB + jC + kD$. Quaternions are added and multiplied like complex numbers except that multiplication is not commutative. They satisfy the relations

$$i^2 = j^2 = k^2 = ijk = -1 \tag{14.21}$$

$$ij = -ji = k \tag{14.22}$$

$$jk = -kj = i \tag{14.23}$$

$$ki = -ik = j \tag{14.24}$$

14.7 Basins of Newton's method

The Newton–Raphson method was mentioned in §4.6.1 as a way to find the solution of an arbitrary equation $f(Z) = 0$ for a complex variable Z by successive iteration from an initial guess Z_0

$$Z_{n+1} = Z_n - \frac{f(Z_n)}{f'(Z_n)} \tag{14.25}$$

where $f'(Z) = df/dZ$. For example, you can use the method to find the square roots of a real constant A by taking $f(Z) = Z^2 - A$ for which

$$Z_{n+1} = \frac{Z_n^2 + A}{2Z_n} \tag{14.26}$$

Since the solution is $Z = \pm\sqrt{A}$, initial conditions with the real part of Z_0 greater than zero converge to $+\sqrt{A}$, and initial conditions with the real part of Z_0 less than zero converge to $-\sqrt{A}$. The basin boundary is the imaginary axis in the complex Z plane (Walter 1994).

[10]Quaternions were invented by William Rowan Hamilton in 1843 while strolling along the Royal Canal in Dublin with his wife, and he spent the rest of his life trying in vain to find a use for them.

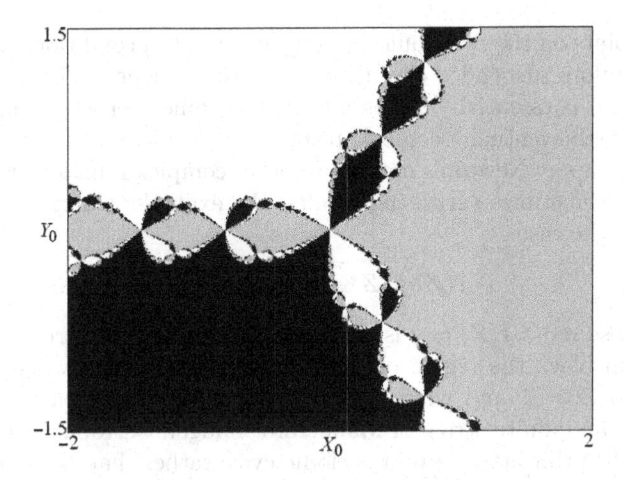

Fig. 14.16 Basins of attraction for the complex values of $\sqrt[3]{1}$ by Newton's method.

The situation is much more interesting (and complicated) for higher roots of a real number[11] (Peitgen *et al.* 1984). For example, the cube root of 1 gives $f(Z) = Z^3 - 1$ for which

$$Z_{n+1} = \frac{2Z_n^3 + 1}{3Z_n^2} \tag{14.27}$$

or in terms of a two-dimensional map for the real variables X and Y with $Z = X + iY$

$$X_{n+1} = \frac{2X_n^5 + 4X_n^3Y_n^2 + 2X_nY_n^4 + X_n^2 - Y_n^2}{3(X_n^2 - Y_n^2)^2 + 12X_n^2Y_n^2} \tag{14.28}$$

$$Y_{n+1} = \frac{2(X_n^4Y_n + 2X_n^2Y_n^3 + Y_n^5 - X_nY_n)}{3(X_n^2 - Y_n^2)^2 + 12X_n^2Y_n^2} \tag{14.29}$$

The three solutions are $Z = 1, (1 \pm \sqrt{3}i)/2$, but the basins of attraction are much more intricate, as shown in Fig. 14.16. They have the curious and seemingly paradoxical property that wherever any two basins meet, the third basin always interjects itself between them. The basin boundary is a multifractal with $D_{-\infty} \simeq 5.419, D_0 \simeq 1.429$, and $D_\infty = 1.146$ (Nauenberg and Schellnhuber 1989). As with the Mandelbrot set, orbits exactly on the basin boundary are chaotic, but they represent a set of measure zero in the space of initial conditions, and they are unstable since the basin boundary is a repellor.

[11]The English mathematician Arthur Cayley (1821–1895), who supported himself as a lawyer for the first fourteen years of his career while publishing about 250 mathematics papers, found the basins for the system $f(Z) = Z^2 - 1$ (Cayley 1879), but he spent the last sixteen years of his life trying unsuccessfully to do the same for $f(Z) = Z^3 - 1$.

The points on the plot could be considered initial conditions for a magnetic pendulum suspended over three attracting magnets. For many initial conditions, it is essentially impossible to determine over which magnet the pendulum will eventually come to rest.

You can apply Newton's method to other complex functions with interesting and sometimes surprising results. For example, Curry *et al.* (1983) considered the case

$$f(Z) = Z^3 + (C-1)Z - C \tag{14.30}$$

one of whose roots $f(Z) = 0$ is $Z = 1$ for any C. If you start with $Z_0 = 0$ and plot in black the region of the complex C plane for which the orbit attracts to $Z = 1$ and white otherwise, you will see a basin boundary containing an infinite string of Mandelbrot midgets surrounded by regions in which the orbit attracts to a periodic cycle rather than to a root of eqn (14.30). Such observations suggest that the Mandelbrot set is perhaps more general than is suggested by eqn (14.9) for which it was first derived.

14.8 Computational considerations

The interest in fractals arose along side the advent of personal computers. Initially, computers were primitive, and rapid calculation of the Mandelbrot and Julia sets was computationally demanding (Rojas 1991). Even now, speed is important when searching for interesting cases, zooming deeply into the sets, and making high resolution images and animations. Here are some ways to speed the computation:

1. Start the first iteration at $Z_1 = C$ rather than $Z_0 = 0$ for the Mandelbrot set.
2. Avoid taking a square root by testing $X^2+Y^2 > 4$ rather than $|Z| > 2$.
3. Optimize the *bail-out condition*, which is the maximum number of iterations before declaring the point to be in the set, to give the best trade-off of speed and precision.
4. Exploit the symmetry of the set to avoid redundant calculations (e.g., Julia sets are symmetric with respect to the transformations $X \to -X$, and $Y \to -Y$, and the Mandelbrot set is symmetric with respect to the transformation $B \to -B$).
5. Perform as many calculations outside the innermost do-loop as possible, and do not repeat calculations such as X^2 and Y^2, which are needed both to iterate X in eqn (14.10) and to test for boundedness.
6. Evaluate $2X$ as $X + X$ and X^2 as $X \times X$ if your compiler does not adequately optimize.
7. Test whether the point has a value known to be in or out of the set such as using eqns (14.14) and (14.15).
8. Test whether the orbit has converged to a fixed point or periodic cycle.
9. Use the connectedness of the set to fill in regions whose boundary is determined to be in the set.

10. Rescale the variables so that all calculations can be done with integer arithmetic. That is the origin of the FRACTINT name.

The Mandelbrot and Julia sets do not seem to have much application besides making pretty pictures and emphasizing the amazing complexity that can result from simple rules. Someone whimsically suggested that a shoe tread in the shape of a Mandelbrot set might have enhanced traction, but there is no good reason that would be true. Such a shoe would certainly make interesting tracks!

14.9 Exercises

Exercise 14.1 Show that the Lyapunov exponents for the standard chaos game described in §14.1 are $\lambda_1 = \lambda_2 = \ln 0.5 = -0.693147181\ldots$.

Exercise 14.2 Calculate the similarity dimension of the Sierpinski triangle in §14.1 as a function of the parameter f.

Exercise 14.3 Calculate the fixed point for the general affine transformation in eqns (14.1) and (14.2).

Exercise 14.4 Calculate the eigenvalues for the general affine transformation in eqns (14.1) and (14.2).

Exercise 14.5 Show that the area contraction of an affine transformation is given by $A_{n+1}/A_n = |ad - bc|$.

Exercise 14.6 Write down the affine transformations for the triadic Cantor set in Fig. 11.1.

Exercise 14.7 Write down the affine transformations that will produce a Cantor set with dimension D for $0 < D < 1$.

Exercise 14.8 Write down the affine transformations for each of the fractal gaskets in Fig. 11.9.

Exercise 14.9 Write down the affine transformations for the Sierpinski tetrahedron in Fig. 11.10.

Exercise 14.10 Write down the affine transformations for the Menger sponge described in §11.5.

Exercise 14.11 Write down the affine transformations for the Sierpinski triangle with arbitrary f as in Fig. 14.1.

Exercise 14.12 Write down the affine transformations for each of the regular polygons in Fig. 14.2.

Exercise 14.13 Write down the affine transformations for the fractal word in Fig. 14.4.

Exercise 14.14 Show that the fractal dimension for an IFS has a maximum value given approximately by eqn (14.6).

Exercise 14.15 Show that the chaos game played on a parallelogram with $f = 0.5$ using the random iteration algorithm with uncorrelated random numbers gives a pattern with uniform density and no structure.

Exercise 14.16 Show that the complex mapping in eqn (14.9) for the Julia set is equivalent to the two-dimensional map in eqns (14.10) and (14.11).

Exercise 14.17 Calculate the location of the fixed points (X^*, Y^*) for the Julia set in eqn (14.12) as a function of A and B.

Exercise 14.18 Prove that one of the fixed points of the Julia set in eqn (14.12) is unstable for all values of C.

Exercise 14.19 Calculate the eigenvalues for the fixed points of the Julia set in eqn (14.12) as a function of A and B, and determine their stability.

Exercise 14.20 Calculate the two preimages (X_n, Y_n) of (X_{n+1}, Y_{n+1}) for the Julia set in eqn (14.9) as a function of A and B.

Exercise 14.21 Show that the iterates of the Julia set in eqn (14.9) escape to infinity if $|Z| > 2$.

Exercise 14.22 Show that the iterates of the Julia set in eqn (14.9) escape to infinity if $|Z| > |C| + 1$.

Exercise 14.23 Show that the Julia set in eqn (14.9) is symmetric with respect to the origin ($Z = 0$) for any value of C and that it is symmetric with respect both the real (X) and imaginary (Y) axes if C is real ($B = 0$).

Exercise 14.24 Show that the Mandelbrot set is symmetric with respect to the $B = 0$ axis.

Exercise 14.25 Show that for $B = 0$ the Mandelbrot set is bounded by $-2 \leq A \leq 0.25$.

Exercise 14.26 Show that the iterates of the Mandelbrot set escape to infinity if $|Z| > 2$.

Exercise 14.27 Prove that in the limit of high magnification, a small region of the Mandelbrot set is identical to the corresponding highly magnified Julia set.

Exercise 14.28 Show that the Mandelbrot set can be written as a one-dimensional time-delay map in the real variable Y as

$$Y_{n+1} = 2Y_n\left[\left(\frac{Y_n - B}{2Y_{n-1}}\right)^2 - Y_{n-1}^2 + A\right] + B \tag{14.31}$$

with initial conditions $Y_0 = B$ and $Y_{-1} = 0$.

Exercise 14.29 Show that the two-dimensional quadratic map in eqns (14.19) and (14.20) satisfy the Cauchy–Riemann equations in eqns (6.17) and (6.18).

Exercise 14.30 Calculate the product of two quaternions $A_1+iB_1+jC_1+kD_1$ and $A_2+iB_2+jC_2+kD_2$, and find its component along each of the four axes.

Exercise 14.31 Derive eqn (14.26) for the Newton–Raphson approximation for $\sqrt{A}$, and calculate the first five iterates with $Z_0 = +1$ and with $Z_0 = -1$ for $A = 2$.

Exercise 14.32 Derive eqn (14.27) for the Newton–Raphson approximation for $\sqrt[3]{1}$, and show that it is equivalent to the two-dimensional map in eqns (14.28) and (14.29).

14.10 Computer project: Mandelbrot and Julia sets

In this project you will explore the Mandelbrot set and its associated Julia sets. The Mandelbrot set is perhaps the most celebrated of all fractal sets, and it provides endless fascination and artistic patterns of unrivaled beauty. It is the simplest example of a two-dimensional escape-time fractal. It has no practical use other than artistic. It is discussed in §14.5, but you may want to refer to almost any book on fractals for more detail and additional sample pictures.

The Mandelbrot set is the set of values of the complex variable C for which the mapping $Z_{n+1} = Z_n^2 + C$ with an initial condition of $Z_0 = 0$ remains bounded as n approaches infinity. The above description is equivalent to the two-dimensional map in eqns (14.10) and (14.11).

1. Plot the Mandelbrot set in the region of the C plane bounded by $-2 < A < 1$ and $-1.5 < B < 1.5$. To do this, you will need to program the computer to laboriously step through each combination of A and B. For each such value of C, start with $Z_0 = 0$ and iterate the above equations until $|Z| > 2$ in which case that value of C is not in the set or until you have iterated 100 times in which case that value of C is presumably in the set. Plot in the C plane those points that are in the set. You should get a solid region of points with a fractal basin boundary as in Fig. 14.13.
2. To make your plot more visually interesting, also plot the points that lie outside the set if the number of iterations required to escape is odd. If you have access to colors, you might color the points that lie outside the set according to some scheme that converts the escape time to a color.
3. Find a region near the boundary of the set that exhibits interesting structure, and zoom in on it by at least a factor of ten. You may want to increase your escape-time limit from 100 to some larger number such as 500, but beware that this will slow down your calculation correspondingly.
4. (optional) Each point in the Mandelbrot set (each C value) has associated with it a filled-in Julia set, which is the set of initial conditions

Z_0 for which the solution is bounded. These sets can be displayed by exactly the same technique you used above. Several such sets are in Fig. 14.10. The interesting cases (choices of C) are those that lie near the boundary of the Mandelbrot set. Make plots of a few such cases.

15

Spatiotemporal chaos and complexity

Most of the previous discussion has concerned low-dimensional chaotic systems, governed by simple iterated maps or ordinary differential equations. By contrast, natural systems such as economic, ecological, social, neural, and meteorological, are inherently high-dimensional, involving many interacting variables. Even relatively simple nonlinear interactions between a limited number of variables can lead to complex behavior. Chaos (in fact, hyperchaos) is ubiquitous in such systems, but new phenomena also occur such as self-organization, evolution, emergence, and adaptation that suggest a degree of organization (Lewin 1992, Waldrop 1992). In some ways complexity is the opposite of chaos, since complex processes can produce simple patterns, sometimes called 'antichaos' (Kauffman 1991). The emergence of order out of the complexity of chaos is called 'simplexity' or 'complicity' by Cohen and Stewart (1994).

Although a completely realistic treatment of complex systems is difficult and computationally intensive, many of the prototypical behaviors occur in simple universal models. These models may not capture *all* the features of a specific system, but they can be surprisingly realistic. If a variety of simple models can reproduce most features of complex dynamics in nature, it may be futile to expect experimental data to shed much light on the underlying mechanisms (Bohr *et al.* 1998).

Some of the simplest models involve a spatial array of cells, each interacting with a small number of nearby cells by simple rules. These local rules can produce global behavior on spatial scales orders of magnitude larger than the range of the interactions. Such models are especially appropriate for biological systems (Murray 1993), and the resulting dynamics often exhibit *spatiotemporal chaos* (Manneville 1990), which means that the spatial structure is aperiodic and unpredictable over long distances, just like the temporal behavior. The complex spatial patterns typically evolve in time (Cross and Hohenberg 1993).

Spatiotemporal systems are but one example of a wide class of complex systems. *Complexity* is a very general term encompassing chaos, fractals, nonlinear science, dynamical systems, self-organization, artificial life, cellular automata, neural networks, and related systems. It is less well defined and developed than the theory of chaotic systems. There is no universally accepted definition and no unifying theory. This chapter will describe a number of simple models of complex systems to give you a taste of the approaches that are used. Think of it as the dessert following the main course of the previous chapters, or perhaps an appetizer for the new frontier in nonlinear dynamics.

15.1 Cellular automata

We begin with a simple class of model called a *cellular automaton* (CA), invented by von Neumann[1] (1966) and classified by Wolfram (1986). A CA has a number of identical cells, each interacting with a few nearby neighbors by simple rules. The rules can be deterministic or probabilistic (random). Each cell is in one of a small number of discrete states. Time advances in distinct steps, and the cell states are updated either *synchronously* (all at once) or *asynchronously* (either randomly or sequentially). Synchronous updating is beneficial for computers with parallel processors, while asynchronous updating conserves memory and simplifies programming. A CA is perhaps the simplest model of a spatiotemporal system, but the dynamics can be very complex and counterintuitive (Wolfram 2002).

15.1.1 One-dimensional example

Imagine a ring of people, holding hands if that makes it easier to visualize. Each person wears a cap with a bill ('peak' in the UK). Initially, all but one of the bills are pointing to the front so as to shade the face, but one nonconformist is wearing his bill to the rear. The rule is simple. At each time step, everyone in the ring observes the neighbors to the left and right. If both your neighbors have their bills forward or both backwards, put (or keep) yours forward, but if only one neighbor has his bill backwards, put (or keep) yours backwards. Figure 15.1 shows the resulting dynamics with 640 cells for 480 time steps.

By convention, CA plots are shown with the spatial dimension horizontal and time increasing downward. The boundary conditions are periodic, so that the right edge is adjacent to the left edge. Thus the perturbation propagates downward from both the right and left edge where it was initially placed. The propagation is at a constant speed of one cell per time step, and thus the pattern resembles a *light cone* in a *spacetime* (or *Minkowski*[2]) *diagram* in which events are only causally related if separated such that the

[1]John von Neumann (1903–1957), known as 'Johnny' by most everyone, was a Hungarian-born mathematician and computer scientist who pioneered the development of computers and was in charge of mathematical calculations during the Manhattan atomic bomb project, later advocating a 'preventive war' against the USSR.

[2]Hermann Minkowski (1864–1909) was a Russian-born mathematical physicist whose theories of space and time influenced his student Einstein in formulating the theory of relativity and who subsequently applied Einstein's ideas to electrodynamics.

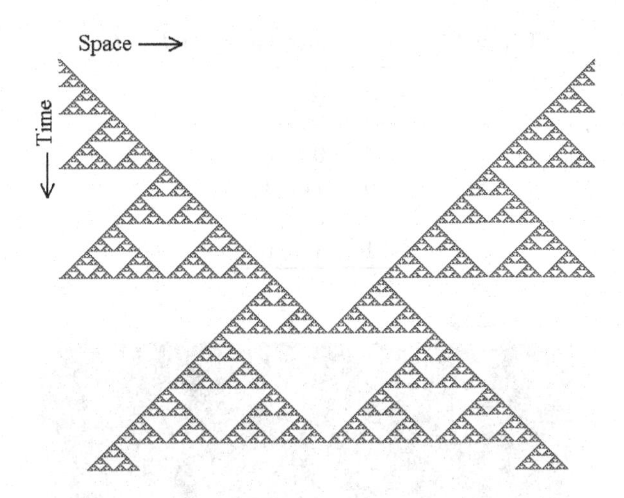

Fig. 15.1 A simple one-dimensional CA that produces Sierpinski triangles.

information propagates no faster than the speed of light, except that time usually advances *upward* in such diagrams. You will recognize Sierpinski triangles (see §11.4.2) in the pattern, at least down to the resolution of an individual cell.

This rule can be expressed in several ways. If $X_n(i)$ is the state of cell i at time step n, with $X = 0$ representing bill forward and $X = 1$ representing bill backwards, then the automaton is binary, and the rule can be written

$$X_{n+1}(i) = X_n(i-1) + X_n(i+1) \pmod 2 \tag{15.1}$$

The rule gives the new value $X_{n+1}(i)$ in terms of the previous values of X_n. Since X is a binary variable, the rule is the *exclusive-or* (xor) *Boolean*[3] operation

$$X_{n+1}(i) = X_n(i-1) \text{ xor } X_n(i+1) \tag{15.2}$$

represented by the truth table in Table 15.1. A digital electronic *xor gate* is also called a *controlled-not gate*. The rule tests whether the sum of the inputs is odd or even and is called a *parity rule*. The patterns in Fig. 15.1 move through each other unperturbed when they collide, and no chaos is evident.

The exclusive-or is just one of $2^4 = 16$ rules you could explore for this simple binary CA with two neighbors, but it gives the most interesting pattern. Different initial conditions also give different results. The 'xor rule' with *random* initial conditions in Fig. 15.2 shows small, barely discernible patterns immersed in noise. Even these patterns would disappear if you only plotted every third time step, and the resulting plot would satisfy

[3]George Boole (1815–1864) was a self-educated English mathematician who created the branch of mathematics known as *symbolic logic*, which is the basis for digital computers.

Table 15.1 Truth table for A xor $B = C$.

A	B	C
0	0	0
0	1	1
1	0	1
1	1	0

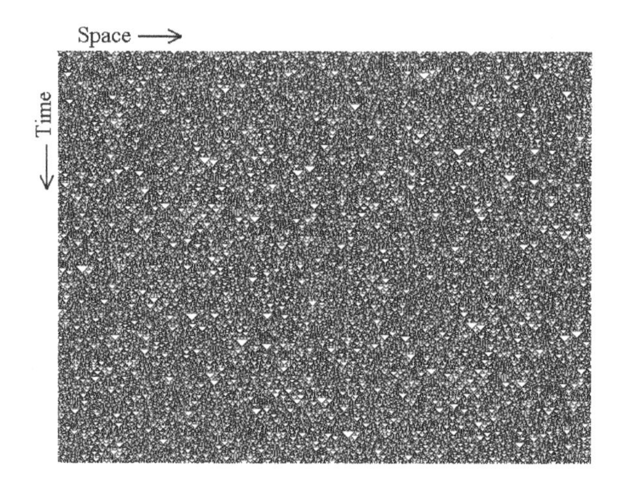

Fig. 15.2 A simple one-dimensional CA with random initial conditions.

most tests of randomness. The pattern is not chaotic but just propagates the random initial condition forward in time. Other rules can cause a random initial condition to become periodic or an ordered initial condition to become disordered after several generations. Carefully chosen simple rules that produce a string of bits satisfying all tests for randomness have been used as computer random number generators.

You can consider larger neighborhoods and include the cell's own state as part of the rule, and the cells can have more than two states and different boundary conditions. For example, a binary CA that depends on the state of the cell and its two nearest neighbors has $2^8 = 256$ possible rules, some of which exhibit quite complex behavior (Wolfram 2002). A CA can be considered as a computer that repetitively performs some operation (the rule) on some data (the initial conditions).

Cellular automata fall into one of four dynamical classes (Wolfram 1984):

1. Class 1 reach a homogeneous state with all cells the same for almost all initial conditions.

2. Class 2 reach a nonuniform state that is either constant or periodic in time, with a pattern that depends on the initial conditions.
3. Class 3 have somewhat random patterns, sensitive dependence on initial conditions, and small-scale structure.
4. Class 4 have relative simple localized structures that propagate and interact in very complicated ways.

The four classes are attractors corresponding roughly to fixed points, periodic cycles, chaos, and quasiperiodicity (or perhaps transient chaos), respectively.

Langton (1986) proposed a quantity λ which is the ratio of the number of neighborhood configurations that produce state 1 divided by the total number of neighborhood configurations. The number that produce state 0 is $1 - \lambda$, and thus it suffices to consider $0 < \lambda < 0.5$ since the results are symmetric with respect to an interchange of 0 and 1. As λ increases from 0 to 0.5 (or decreases from 1.0 to 0.5), most CAs progress from Class 1 to 2 to 4 to 3. The ordering crudely follows the ordering in dynamical systems such as the logistic map (see Chapter 2) when A is increased. The progression is analogous to the transition from a solid at absolute zero (class 1), to a solid at finite temperature (class 2), to the melting of the solid (class 4), to a turbulent fluid (class 3), with λ playing the role of a temperature. In this ordering, class 3 is the most interesting because it lies at the boundary between order and chaos (the 'edge of chaos') and exhibits the most complex behavior.

15.1.2 Game of Life

Cellular automata can also be developed for two or more dimensions. Perhaps the most famous two-dimensional class 4 CA is the *Game of Life* (or simply 'Life'), developed by John Horton Conway (Gardner 1970). It is neither a 'game' nor 'life' in the conventional sense, but it does provide great fascination and emulates some features of biological life. The game is played on a rectangular grid[4] of cells, each of which can have two states, 'dead' (0) or 'alive' (1). Initial conditions are usually taken as random. The neighborhood consists of the eight nearest cells (north, south, east, west, northeast, northwest, southeast, and southwest), called a *Moore neighborhood* (Moore 1962), as in Fig. 15.3 with $r = 1$.

The rule is that a live cell survives to the next generation if it has exactly two or three live neighbors, and a dead cell becomes alive if it has exactly three live neighbors. For all other combinations, the cell dies or remains dead from isolation or overcrowding. Such a rule that involves only the sum of the neighbors of a cell is called an *outer totalistic rule*. The

[4] One-dimensional arrays are called 'chains,' two-dimensional arrays are called 'grids,' and three-dimensional (or higher) arrays are called 'lattices,' although the latter term is often used generically in any dimension.

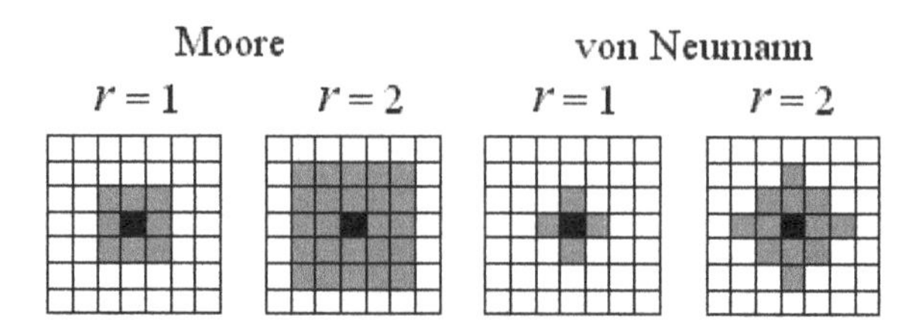

Fig. 15.3 Some neighborhoods used in two-dimensional cellular automata.

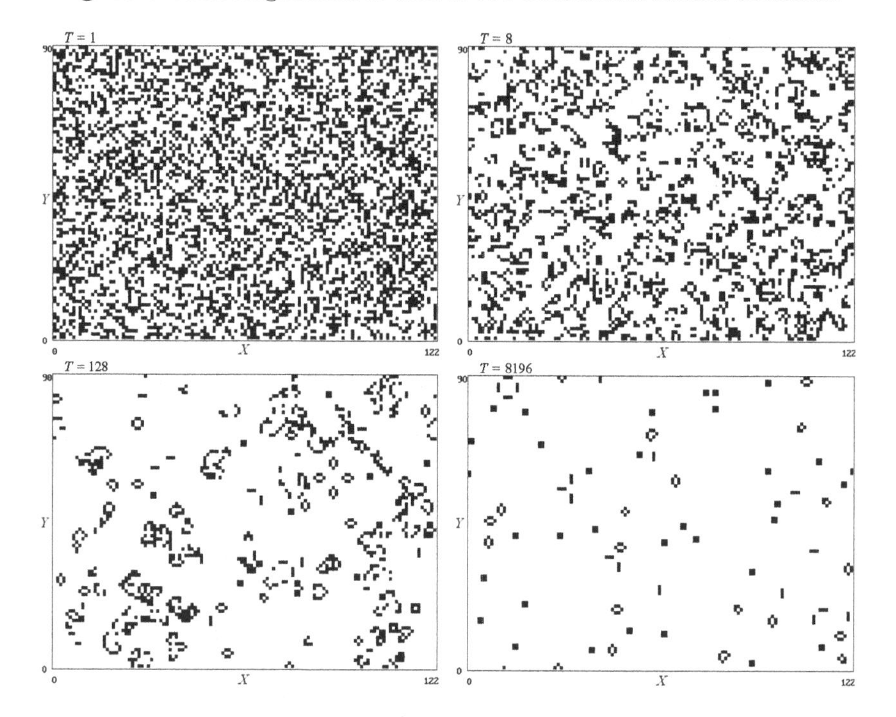

Fig. 15.4 Typical evolution of the Game of Life.

xor rule in eqn (15.1) is also outer totalistic. The cell states are advanced synchronously (simultaneously).

These simple rules cause the initially random pattern to evolve into 'colonies' of various shapes and sizes after a few thousand generations as in Fig. 15.4. This example uses a 122×90 grid with periodic boundary conditions in both X and Y. Most of the colonies are static, but some are periodic. They go by descriptive names such as 'blinkers,' 'gliders,' and 'space ships.' The static figures do not do justice to the time evolution that shows the colonies evolving and interacting in a very life-like manner, at least in the early stages.

A rare but possible configuration is a 'glider gun' that every thirty generations shoots off a 'glider' consisting of five live cells that move diagonally

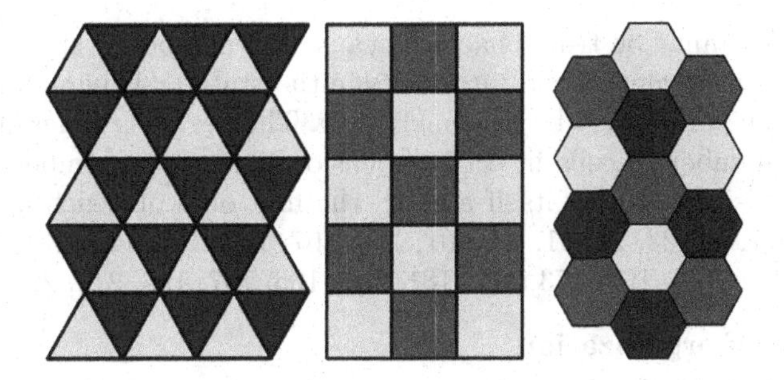

Fig. 15.5 The three possible regular tilings of the plane.

one step every four generations (Berlekamp *et al.* 1982). On an infinite grid, such guns would exist somewhere, and the self-reproducing pattern would thus grow and evolve without limit, mimicking the origin of life.[5]

The game is unpredictable in the sense that the only way to determine its outcome is to play it. This result follows from the *undecidability of the halting theorem*, which states that there cannot exist a computer program that will decide in advance whether a given program run on a given machine will continue forever or eventually stop. The proof follows from the fact that if there were such a program then you could use it to monitor its own progress and to stop only if it did not stop, which is a contradiction.

15.1.3 Extensions

Cellular automata can be extended to dimensions higher than two. In three dimensions, each cell has six nearest neighbors on a cubic lattice. Even in two dimensions, the choice of a square grid is not unique. Sometimes a triangular or hexagonal *tiling*, as in Fig. 15.5, is more convenient. A cell with triangular tiling has three nearest neighbors, and one with hexagonal tiling has six nearest neighbors, whereas the square has four. These are the only possible *regular* tilings of the plane (using regular polygons), although many other irregular tilings are possible and are often used in floor tiles and wallpaper. The cells can also be assigned *vector* values to represent a fluid flow velocity in so-called *lattice gas models* (Doolen 1990). Another extension is to give the CA memory, so that the next state depends not just on the current state but on the past history.

Cellular automata can appear chaotic when a small perturbation leads to a very different final state. However, because of their discrete nature, it is hard to define and calculate a Lyapunov exponent. Perturbations cannot be infinitesimal. Many perturbations have no effect or die after a few generations even in systems for which other perturbations grow. In any case,

[5]When Conway first proposed the Game of Life, he offered $50 to anyone who could find an endlessly growing configuration or prove that none exists. The prize was won by William Gosper and five other MIT students for discovering the 'glider gun,' which enabled Conway to prove the existence of self-replicating patterns in the Game of Life.

a finite CA cannot be truly chaotic because the number of states is finite, and thus it will eventually return to its initial state and thereafter repeat, although the period may be extremely long. The period is generally largest when the number of cells in each dimension is a prime number (a whole number divisible only by itself and 1), the first ones of which are 2, 3, 5, 7, 11, 13, 17, 19, 23, 29, 31, 37, 101, 103, 107, 109, 113, 127, 131, 137, 139, 149, 151, 157, 163, 167, 173, 179, 181, 191, 193, 197, 199, 211, 223, 227, ...

15.1.4 Self-organization

The Game of Life illustrates how simple rules can cause an initial random state to evolve into a highly ordered one, in a process of *self-organization* (Nicolis 1989). Such coherent behavior of an entire system resulting from *cooperative effects* between parts of the system is called 'synergetics' by Haken (1978, 1981, 1983b). Self-organization has long been known and studied by biologists (Solé and Goodwin 2000), but it now appears that nonlinear physical and mathematical systems can also self-organize. It is perhaps not surprising that *deterministic* systems can self-organize, but so can *random* systems.

Consider a simple CA forest model with the trees arranged on a square grid as in the Game of Life with some cells occupied (1) and others empty (0), initialized randomly half and half. At each asynchronous time step, a random cell is chosen and replaced by the contents of one of its randomly chosen four nearest neighbors (north, south, east, and west), called a *von Neumann neighborhood* as in Fig. 15.3 with $r = 1$. This example is also called the 'voter model' (Clifford and Sudbury 1973, Holley and Liggett 1975) because it models an uninformed or highly impressionable electorate in which individuals vote according to the way a randomly chosen neighbor voted in the previous election.

In a more general model, the parameter r could be adjusted to reproduce some desired feature of a landscape (Sprott *et al.* 2002), or the replacements can be chosen from a cell using a probability function (a kernel) that decays with distance such as

$$P(d) \propto \mathrm{e}^{-(d/r)^c} \tag{15.3}$$

The case $c = 1$ is an exponential, and the case $c = 2$ is a Gaussian distribution. Distributions less peaked than Gaussian ($c < 2$) are leptokurtic (see §9.4.1) and are said to have 'fat tails.' An extreme example of a fat tail is a 'small-world' model (Watts 1999) in which most replacements are taken from the near neighborhood, but a few are taken randomly from throughout the array.

You can think of the process as a tree dying ($1 \to 0$), a tree growing in an empty cell from the seed of a nearby tree ($0 \to 1$), one tree being replaced by another ($1 \to 1$), or a cell remaining barren ($0 \to 0$). A cell could also represent a group rather than a single tree. The model does not specify the mechanism for the death and replacement of cells. Trees can die from age, disease, fire, storms, or human activity, and can be replaced from

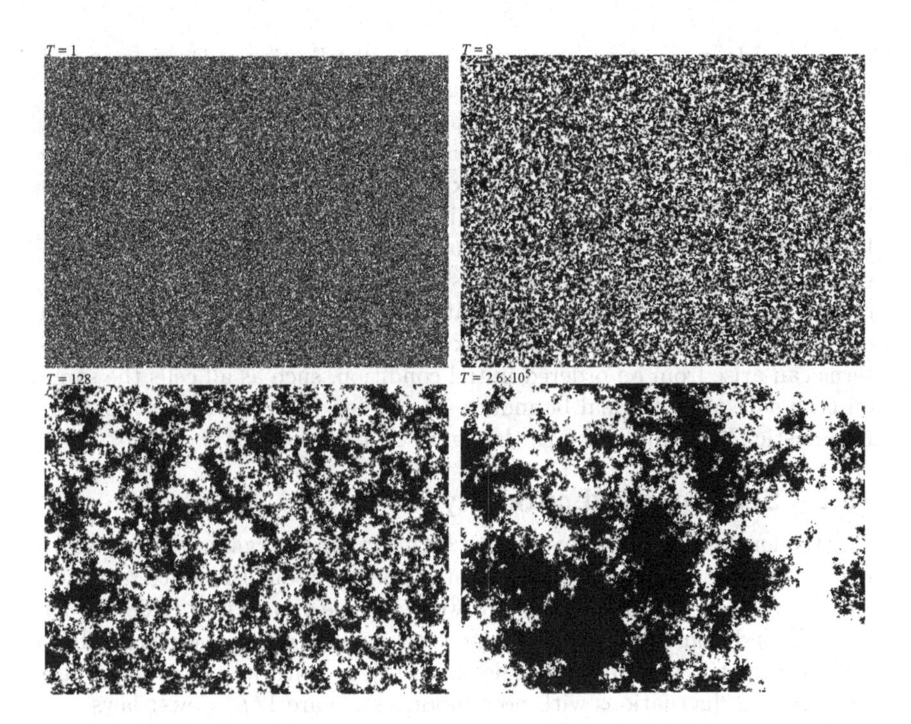

Fig. 15.6 A cellular automaton model of a forest showing self-organization.

seeds dispersed by wind, water, or animals. The rule simulates diffusion of the landscape and does not require detailed knowledge of the relevant biological mechanisms.

A 'generation' is the time for every cell to be updated once on average, corresponding to the average life of a tree, or about 100 years. The rule tends to conserve the proportion of each category, but any particular realization slowly walks away from the initial proportions. After many generations, one state tends to dominate. To prevent that from happening, make replacements with a probability slightly less than 100% that is continually adjusted to keep the number in each state approximately constant.

The evolution of such an artificial forest, as in Fig. 15.6 on a 640×480 grid, shows self-organization to a highly clustered (or 'aggregated') state with an apparent fractal boundary between the regions. Such clusters are called *dissipative structures* by Prigogine (1996). The regions of black and white form a fat fractal with dimension 2.0 but with clusters having a power-law size distribution. A similar final state occurs if the initial condition is highly ordered, such as having all cells north of the equator occupied and all those to the south empty, in which case the evolution disorganizes the pattern. The self-organized state is reminiscent of vegetation patterns on the Earth, but it also looks like clouds. Such a model might represent other systems such as the migration of animals or the spread of disease through a plant or animal population or the distribution of Democrats and Republicans in a United States voting population.

It is remarkable that complex spatial structure arises from a model with no externally imposed heterogeneity and with all cells in the lattice treated equally. You might naively expect spatially homogeneous patterns because of the symmetry of the lattice and the rules governing the evolution. This spontaneous *symmetry-breaking* is an important result, suggesting that complex spatial patterns in natural landscapes may not be a consequence of topographical or climatic variations and are thus not unique or predictable. Furthermore, complex and apparently random spatial patterns can arise from an ordered initial condition, such as all cells the same on each side of a diagonal boundary, even when the rules are entirely deterministic (Wolfram 2002), as the figure on page xx shows.

15.2 Self-organized criticality

The tendency of complex systems to organize (or disorganize if they start highly ordered) leads one to ask what is special about the state to which they are attracted. Bak (1996) and coworkers have argued that under rather general conditions the preferred state is the one for which there is no characteristic spatial or temporal scale size. Objects with no spatial scale are fractals, and fluctuations with no temporal scale are $1/f^{\alpha}$ power laws.

This phenomenon is called *self-organized criticality* (SOC) and is independent of the details of the system. It is ubiquitous and characterizes natural processes as diverse as earthquakes (Gutenberg and Richter 1954, Bak and Tang 1989), quasar luminosities (Press 1978), chemical reactions (Simoyi *et al.* 1982), superconductors (Field *et al.* 1995), plasma transport (Carreras *et al.* 1996), forest fires (Bak *et al.* 1990), rainfall (Peters *et al.* 2002), stock prices (Mandelbrot 1997), traffic jams (Nagel and Herrmann 1993), disease epidemics (Rhodes and Anderson 1996), population dynamics (Miramontes and Rohani 1998), Internet connectivity (Pastor-Satorras *et al.* 2001), DNA base sequences (Voss 1992, Mantegna *et al.* 1994), human cognition (Gilden 2001), wars (Roberts and Turcotte 1998), and biological evolution (Bak and Sneppen 1993). However, some of these examples may not be SOC, since processes other than self-organization can also produce power laws (Jensen 1998, Amaral *et al.* 1998). For example, *Zipf's law*[6] relating the population m of cities to their size rank R obeys a power law $R(m) \propto m^{-\gamma}$ with $\gamma \simeq 1$ and has been attributed to a *least-effort principle of human behavior* (Zipf 1949), although monkeys pressing typewriter keys produce the same distribution (Casti 1995).

[6]George Kingsley Zipf (1902–1950) was a Harvard linguistics professor who found power laws in the frequency and length of words in written text, leading others to examine many natural and human processes that also obey Zipf's law.

15.2.1 Sand piles

The prototypical SOC system is a sand pile. As sand is added, the pile steepens until it reaches the *angle of repose*, whose value depends on the sand (grain size, roughness, wetness, and so forth), whereupon avalanches keep the pile close to this angle. The avalanches have a power-law scaling in both size and duration.

Bak *et al.* (1987, 1988) proposed a simple CA model inspired by the behavior of a sand pile on a square table with sand added slowly, creating a mound with avalanches that occasionally spill over the edge of the table. The model consists of an $N \times N$ square grid of cells, each characterized by a positive integer $Z(i, j)$, which can be considered as the local slope of the pile. Initial conditions are chosen randomly in the range $1 \leq Z \leq 3$, but this choice is arbitrary. At each time step, a randomly chosen cell (i_0, j_0) is incremented asynchronously by one

$$Z_{n+1}(i_0, j_0) = Z_n(i_0, j_0) + 1 \tag{15.4}$$

and any cells with $Z(i, j) > 3$ and their four nearest neighbors are re-adjusted according to

$$Z_{n+1}(i, j) = Z_n(i, j) - 4 \tag{15.5}$$

$$Z_{n+1}(i \pm 1, j) = Z_n(i \pm 1, j) + 1 \tag{15.6}$$

$$Z_{n+1}(i, j \pm 1) = Z_n(i, j \pm 1) + 1 \tag{15.7}$$

The cells outside the boundary $Z(0, j), Z(i, 0), Z(N + 1, j), Z(i, N + 1)$ are kept at zero. Equations (15.5)–(15.7) are iterated until no cells with $Z > 3$ remain. You can think of this rule as rearranging the sand grains to neighboring cells whenever the slope gets too large, although the connection with a real sand pile is rather strained. In fact, real sand piles are not good examples of SOC (Jaeger *et al.* 1989), although rice piles do somewhat better (Frette *et al.* 1996, Malthe-Sørenssen *et al.* 1999).

Figure 15.7 shows on a four-level gray scale $(0 \leq Z \leq 3)$ the value of Z at each cell on a square grid with $N = 256$ after 10^5 generations. It appears random and unremarkable, but nothing could be farther from the truth. It has evolved to a SOC state in which avalanches of all sizes and time-scales occur.

15.2.2 Power-law spectra

Scale invariance can be demonstrated in a number of ways. Define the avalanche size s as the number of cells that topple at each time step, and plot the probability $P(s)$ of avalanches of size s versus s on a log–log scale as in Fig. 15.8. The resulting spectrum is approximately $P(s) \propto 1/s$. Bak *et al.* (1987, 1988) report a value of $1/s^{1.1}$, which is also a good fit. The power-law distribution implies spatial scale invariance for any exponent.

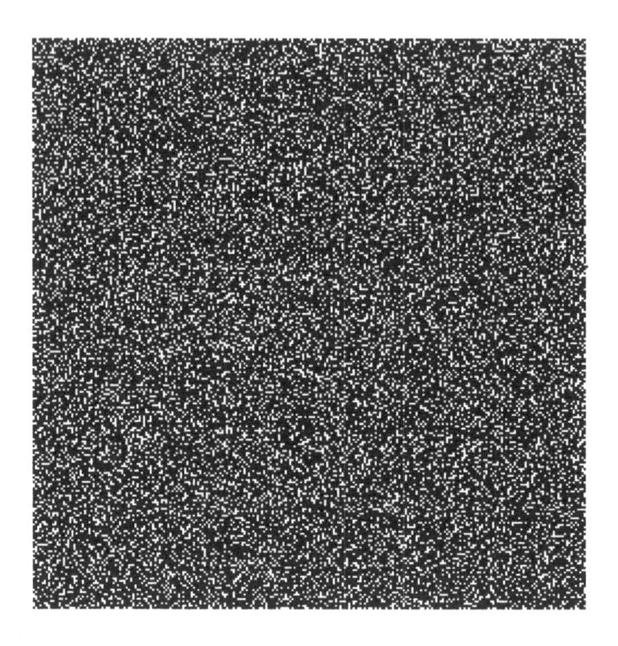

Fig. 15.7 A cellular automaton model of a square sand pile.

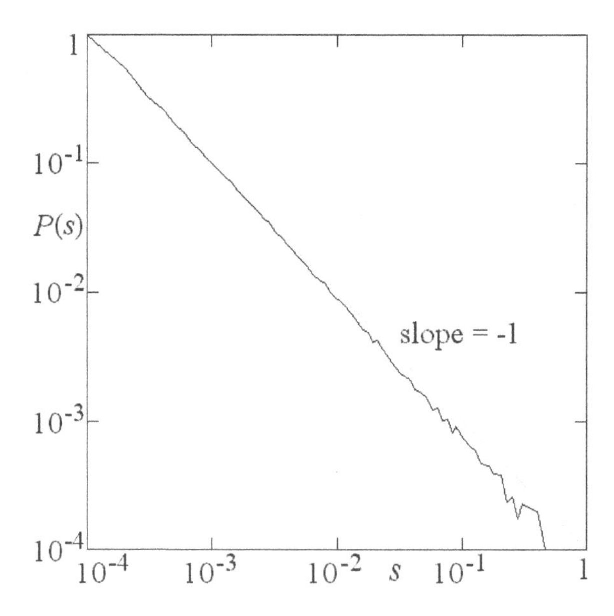

Fig. 15.8 Spectrum of avalanche sizes in a CA sand pile model.

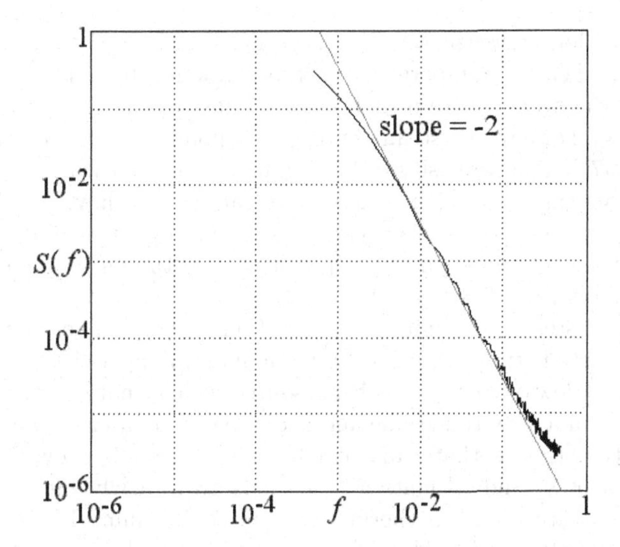

Fig. 15.9 Power spectrum of fluctuating pile size in a CA sand pile model.

To assess temporal scale invariance, record the total size of the pile

$$Z_{\text{tot}} = \sum_{i=1}^{N} \sum_{j=1}^{N} Z(i,j) \tag{15.8}$$

at each time step, and calculate the power spectrum of the resulting time series (see §9.4.4). The result after 3.2×10^4 generations shown in Fig. 15.9 indicates an approximate power law with $S(f) \propto 1/f^2$, in contrast to the $1/f$ reported by Bak *et al.* (1987, 1988). This discrepancy has been noted by others (Jensen *et al.* 1989, Kertész and Kiss 1990). Sometimes the term '$1/f$' is used to mean $1/f^\alpha$ with $0 < \alpha < 2$. The deviation at low frequencies in Fig. 15.9 is caused by the finite size of the array, which also limits the maximum avalanche size in Fig. 15.8. An infinitely large pile, or one with infinitesimal grains, would exhibit power laws over infinitely many decades.

Power laws are very common in nature. For example, both the gravitational and electrical force decrease inversely with the square of the distance from the mass or charge $F \propto 1/r^2$ and thus have no characteristic length scale. You could make a scale model of the Universe, one-tenth the size, and it would work the same. On the other hand, molecular and nuclear forces do not obey power laws, and thus molecules and nuclei do have characteristic sizes. The consequences constitute the science of chemistry.

Criticality is well understood in thermodynamics (Wilson 1979, Binney *et al.* 1992). It occurs at a discontinuous (first-order) *phase transition*, such as the condensation of a gas into a liquid, the freezing of a liquid, or the onset of superconductivity as the temperature is lowered. Sometimes the phase transition is continuous (second-order), such as the onset of ferromagnetism when the temperature of iron is lowered to 1043 K (the *Curie point*[7]) or the point where a high-pressure gas becomes indistinguishable from a liquid. For a water–steam mixture this point occurs at a temperature of about 647 K and a density of 323 kg/m^3. At the critical point, perturbations propagate throughout the system rather than having only a local effect, and particles of all sizes exist with a power-law scaling. However, such criticality requires careful adjustment (or 'tuning') of a parameter such as temperature (see §15.3 below).

In SOC systems, no tuning is required, although you could argue that the driving rate is tuned to zero (Vespignani and Zapperi 1998). The drive is a continual flow of energy or other resource such as matter or information that arrives on a slow time-scale and is dissipated on a fast time-scale. For example, the stresses that cause earthquakes accumulate over years and are relieved in seconds. Plants and animals grow slowly but die quickly. Such systems are far from thermodynamic equilibrium. The equilibrium state of sand is flat and level, and any perturbations decay rapidly. A sand pile maintained by the slow addition of sand from above and dissipated in discrete avalanche events is far from equilibrium and is very sensitive to perturbations. In paleobiology, such processes are called *punctuated equilibria* (Gould and Eldridge 1977). The departure from equilibrium leads to complex spatial structure. Systems can also be SOC when the time-scales are similar, in which case, power-law spectra are not expected (Solé *et al.* 2000). Self-organization seems to contradict the second law of thermodynamics, which states that the entropy of an isolated system cannot decrease, but only because the system is always open to its environment and thus can export excess entropy (Prigogine 1980).

15.3 The Ising model

A simple CA that exhibits a critical phase transition is the *Ising*[8] *model* (Ising 1925), originally developed to explain the onset of ferromagnetism as the temperature is lowered below the Curie point but has been applied to many other phenomena, including the distribution of galaxies in the Universe, the interaction of viruses with their hosts, and the interior of tropical rainforests. The model was originally developed in one dimension

[7]The Curie point is named for the Polish physicist Pierre Curie (1859–1906), husband of Madame Marie Curie (1867–1934) with whom he shared the Nobel Prize (1903) in physics for the discovery of nuclear radioactivity.

[8]The Ising model was invented in 1920 by the German physicist Wilhelm Lenz (1888–1957) (Lenz 1920), but it was named after his student Ernest Ising (1900–1998) whose doctoral dissertation at the University of Hamburg on the one-dimensional model was defended in 1924. Ising never published another research paper but became an accomplished teacher of physics, ending his career in the United States. The two-dimensional case was solved by the Nobel laureate (1968) chemist Lars Onsager (1903–1976) (Onsager 1944), and the three-dimensional case and higher have never been solved.

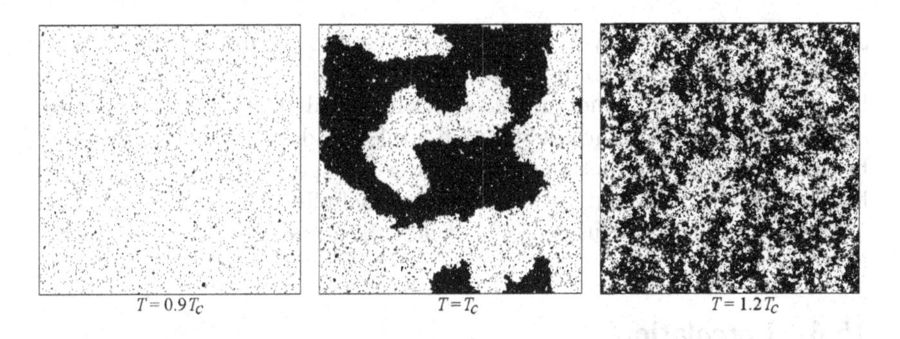

Fig. 15.10 Ising model calculation for three temperatures: below, at, and just above the critical point T_c.

where the critical temperature is at zero degrees, but the two-dimensional case with a nonzero critical point has been widely studied.

The rules are relatively simple. Take a square grid of cells, each representing an atom with its spin oriented in one of two directions, usually taken as ± 1. Assume periodic boundary conditions, corresponding to a torus. Initialize the grid with randomly oriented spins in each cell, and then choose cells randomly to update asynchronously. The energy is a minimum when all the spins are aligned in the same direction, but this can only happen at zero temperature, above which thermal fluctuations cause the spins to flip on occasion.

Assume the total energy is given by the sum of all nearest neighbor pairs $E = -\sum S_i S_j/2$, where the factor of a half prevents double counting. When the spin S_i in cell i flips, the change in total energy is $\Delta E = -S_i \sum S_j$, where the latter sum is over the four nearest neighbors (a von Neumann neighborhood) of cell i. Hence the energy ΔE has integer values in the range $-4 \leq \Delta E \leq 4$. If $\Delta E \leq 0$, accept the change. If $\Delta E > 0$, accept the change with probability $P = \mathrm{e}^{-\Delta E/T}$, where T is the normalized temperature. The easiest way to do this is to pick a random number r in the range $0 \leq r < 1$, and flip the spin only if $r < P$. This procedure is called the *Metropolis algorithm*[9] (Metropolis *et al.* 1953). You can speed the calculation slightly by storing the four allowed values of P corresponding to $1 \leq \Delta E \leq 4$ for the given temperature, since they never change.

The result of the above procedure is shown in Fig. 15.10 on a 480×480 grid for three values of temperature, one slightly below the critical temperature of $T_c = 2/\ln(1+\sqrt{2}) = 2.269185314\ldots$, one at T_c, and one

[9]This dynamical method of generating an arbitrary probability distribution was invented by Nicholas Constantine Metropolis (1915–1999), Edward Teller, and Marshall Nicholas Rosenbluth in 1953 supposedly at a Los Alamos dinner party and is used in the method of simulated annealing (see §10.7.2) to solve the traveling salesman and similar problems. It was cited in *Computing in Science and Engineering* as being among the top ten algorithms having the 'greatest influence on the development and practice of science and engineering in the 20th century.'

slightly above after many iterations. The low-temperature state has most of the atoms aligned, with only a few misaligned. The high-temperature state has a high degree of randomness, with only local structures. The critical state has large-scale structure with a fractal boundary and is characterized by large temporal fluctuations with a power-law spectrum. The Ising model has been applied to many diverse systems including forests (Katori *et al.* 1998), the spread of viruses (Solé *et al.* 1999), and stock market crashes (Vandewalle *et al.* 1998).

15.4 Percolation

To illustrate how a system might self-organize to the critical state without precise tuning, consider the forest model in Fig. 15.6. Suppose the trees grow progressively more dense but are occasionally depleted by fires randomly set by lightning that burn the whole of a cluster. If the clusters are small, fires will affect only local regions. If the clusters are sufficiently large, most of the forest will burn. Thus there is an equilibrium density of trees at which growth is balanced by burning. We expect the critical state to be the one for which the forest is just barely disconnected across the landscape (Gardner et al. 1987). *Connectivity* is an important property of complex systems (Green 1994). Disease and infestation by parasites would spread similarly to fire and in media other than forests, albeit at a slower rate (Kesten 1987). Systems that adapt to changing conditions so as to maintain their complexity are called *complex adaptive systems*. They are characterized by a medium number of agents (hundreds or thousands) and rule-based interactions involving only local information (no central authority).

Such behavior can be observed with a square grid of cells randomly populated with probability P. Assume that fire can spread only to the four nearest neighbors (a von Neumann neighborhood) of a burning cell. Figure 15.11 shows a 128×128 grid with three similar values of P but with very different fractions burned for a fire started along the upper edge of the

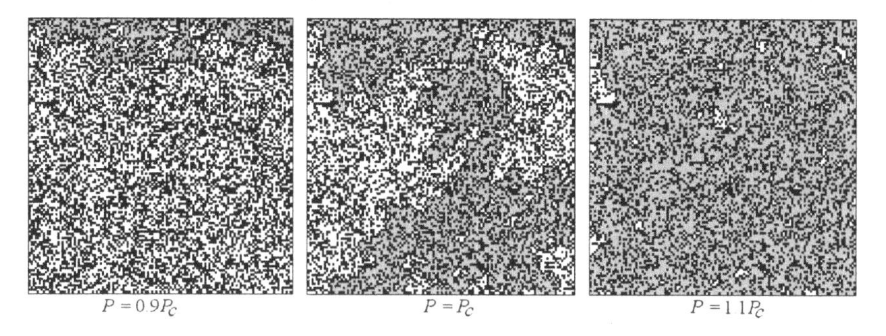

Fig. 15.11 Percolation through a square grid: below, at, and just above the percolation threshold P_c.

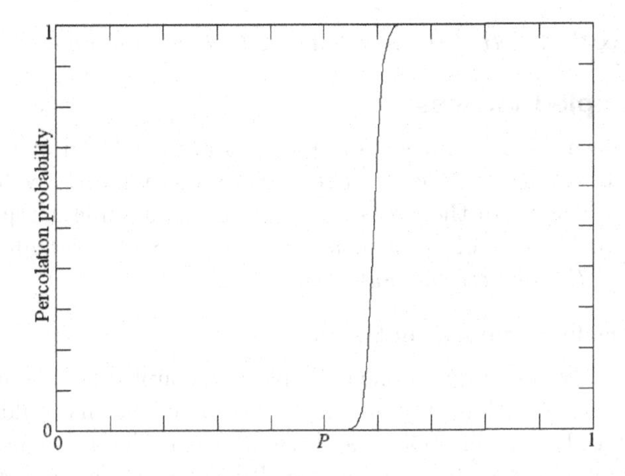

Fig. 15.12 Percolation probability shows a sharp percolation threshold at $P_c \simeq 0.592745$.

plots and allowed to burn to completion. Equivalently, think of this model as the flow of a liquid through a porous medium. Accordingly, it is called *percolation* (as in a coffee percolator, from the Latin, 'to flow through') (Hammersley 1983, Stauffer 1985). The cells that provide the connection are called the *backbone* of the cluster, and the remaining cells are called *dangling bonds*.

How likely it is that a fire started on the top row of cells reaches the bottom row for various P? The result in Fig. 15.12 shows that the likelihood changes abruptly from 0 to 100% at a critical value of P. On an infinite lattice, this value is called the *percolation threshold* P_c. Although no precise theoretical prediction exists, detailed numerical calculations on a large square grid give $P_c \simeq 0.592745$. For an eight-neighbor rule (Moore neighborhood with $r = 1$), the threshold drops to $P_c \simeq 0.407245$, and for a 24-neighbor rule (Moore neighborhood with $r = 2$ as in Fig. 15.3), it is $P_c \simeq 0.168$.

Patterns with $P > P_c$ are dominated by a single large cluster, whereas patterns with $P < P_c$ have numerous, small, fragmented clusters similar to the Ising model. Near the percolation threshold, most quantities have a power-law scaling, and there is no characteristic size or time-scale. The landscape is a fractal with clusters of all sizes from a single cell to the size of the lattice. Temporal fluctuations have a $1/f^{\alpha}$ power spectrum. The position of a fire front advances in time as $r \propto t^{0.87}$ (Albinet *et al.* 1986), which is faster than normal diffusion ($r \propto \sqrt{t}$) but slower than a burning wick ($r \propto t$). The case with $r \propto \sqrt{t}$ is called *Fickian*[10] *diffusion*, and the

[10]Adolph Fick (1829–1901) was a quiet, scholarly German mathematician, physicist, and physiologist who was also interested in philosophy and literature but is best remembered for his law advanced in 1855 that diffusion is proportional to the concentration gradient.

case with $r \propto t^H$ and $H \neq 0.5$ is called *non-Fickian* or *anomalous diffusion*.

15.5 Coupled lattices

The CA is perhaps the simplest spatiotemporal model. An obvious generalization is to let the state of each cell have a continuum of values rather than discrete values. In the case of a forest, the value might represent the number or total mass of trees in a cell. Such models are called *coupled lattices* or *continuous cellular automata*.

15.5.1 Artificial neural networks

We have already encountered one example of a coupled lattice in the feed-forward neural network (see §6.4.2 and §10.7). The neurons could represent different spatial positions (Chua and Yang 1988). These single-layer networks are only one possible architecture. To illustrate the connection to a CA, imagine a circular chain of N neurons, each receiving input from its r nearest neighbors on each side

$$X_{n+1}(i) = \tanh \sum_{j=1}^{r} a(i)[X_n(i-j) + X_n(i+j)] \tag{15.9}$$

For simplicity, the weights $a(i)$ are symmetric and equal on the two sides of a cell but may vary along the chain. Equation (15.9) describes an N-dimensional iterated map.

A typical output for $N = 640, r = 4$, and $a(i)$ uniformly random over the range $-4 < a(i) < 4$ with random initial conditions is in Fig. 15.13. For consistency with cellular automata, position is plotted horizontally, and time increases downward. The value of X is plotted with a 16-level gray scale. Some regions of the lattice are evidently more active than others, but

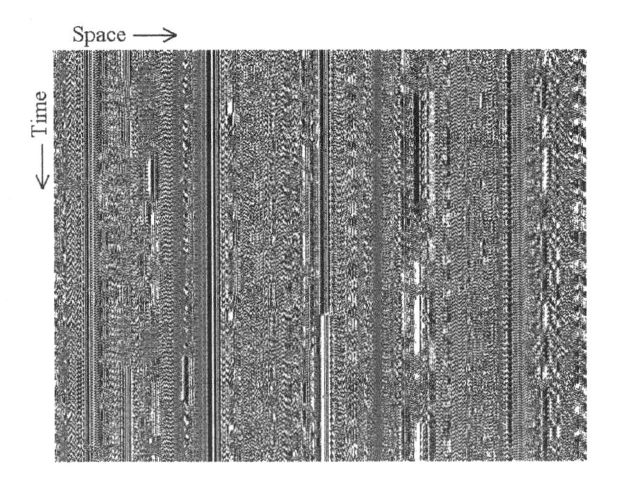

Fig. 15.13 Spatiotemporal chaos in a neural network with $N = 640$ and $r = 4$.

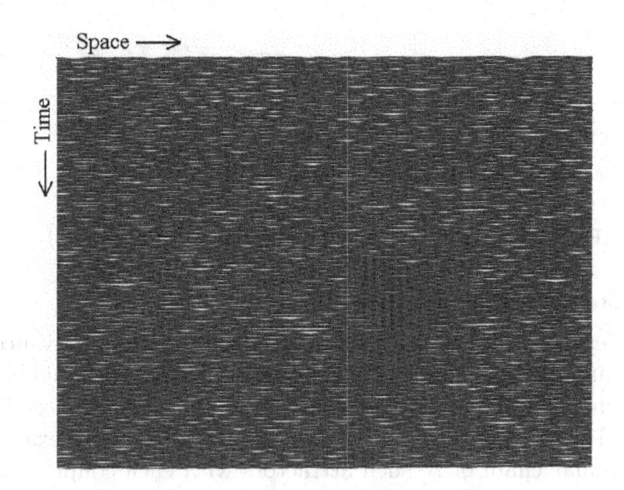

Fig. 15.14 Spatiotemporal chaos in a lattice of 640 coupled logistic maps.

the entire 640-dimensional system is chaotic, as can be verified from the largest Lyapunov exponent (see §5.6). There are also nearly periodic regions and regions that exhibit intermittency. The dynamics have a strange attractor, but it is difficult to plot because its dimension is large.

15.5.2 Coupled map lattices

In the neural network example above, each lattice point responds to its neighbors, but not to itself. Interesting dynamics occur if the lattice points independently exhibit complicated dynamics. For example, each cell in a circular chain could contain a logistic map $f(X) = AX(1-X)$, weakly coupled to r neighbors on each side

$$X_{n+1}(i) = (1-k)f(X_n(i)) + \frac{k}{2r}\sum_{j=1}^{r}[f(X_n(i-j)) + f(X_n(i+j))] \quad (15.10)$$

The constant k determines the strength of the coupling. For $k = 0$, the maps behave independently, and for $k = 1$, each point responds only to its neighbors. Such examples are called *coupled map lattices* (CMLs) (Kaneko 1989), and they are relatively little studied and offer many surprises. For example, adding noise can make them more ordered (Shibata *et al.* 1999), and the shadowing lemma (see §4.8.6) may fail (Lai and Grebogi 1999).

A typical output for $N = 640, r = 4, A = 4$, and $k = 1$ with random initial conditions is in Fig. 15.14. Position is plotted horizontally, and time increases downward, with the values of X plotted with a 16-level gray scale. There are regions in space and time where the system has self-organized, but overall it is chaotic with a high-dimensional strange attractor.

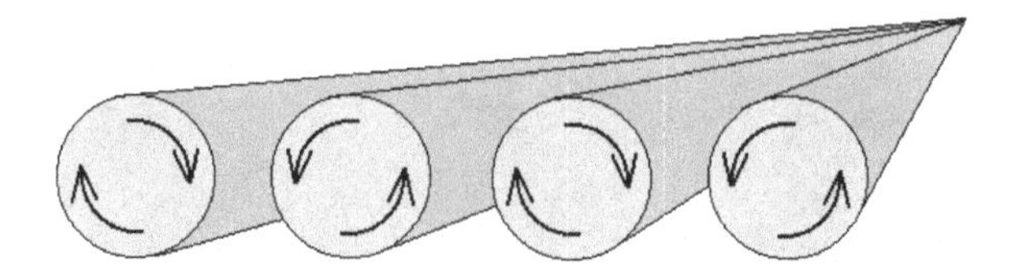

Fig. 15.15 Rayleigh–Bénard thermal convection cells.

15.5.3 Coupled flow lattices

The next generalization relaxes the constraint that time advances in discrete steps and considers lattices of coupled oscillators governed by differential equations. The interesting cases are those for which the oscillators are chaotic, and we take the Lorenz attractor (see §4.8.2) as an example. Construct a circular chain of N such attractors with each coupled *viscously*[11] to its two nearest neighbors (Jackson and Kodogeorgiou 1992)

$$\frac{dx_i}{dt} = \sigma(y_i - x_i) + \mu(x_{i+1} + x_{i-1} + 2x_i) \tag{15.11}$$

$$\frac{dy_i}{dt} = -x_i z_i + r x_i - y_i \tag{15.12}$$

$$\frac{dz_i}{dt} = x_i y_i - b z_i \tag{15.13}$$

The term with coefficient μ is the coupling term, and the strangely appearing signs in that term arise because the adjacent fluids are assumed to counter-rotate to simulate *Rayleigh–Bénard*[12] *thermal convection cells* (Gollub and Benson 1980, Stavans *et al.* 1985) as in Fig. 15.15. Whereas Rayleigh–Bénard convection is driven by a temperature gradient, the same behavior is observed in *Taylor–Couette flow* (Taylor 1923) in which a fluid is confined between two concentric counter-rotating cylinders (see §6.5.3).

A typical result for $N = 640, \sigma = 10, r = 28, b = 8/3$, and $\mu = 1$ with random initial conditions is in Fig. 15.16. The equations were integrated using the Euler method (see §3.9.1) with a step size of $h = 1 \times 10^{-3}$, and 4.8×10^5 time steps are plotted after the initial transient has decayed and the trajectory is on the attractor. Position is plotted horizontally, and time increases downward, with the values of x plotted with a two-level gray scale depending on whether x is positive or negative. Since there is a tendency for adjacent cells in the lattice to have opposite signs of x, black and white are reversed whenever i is odd. Small spatially and temporally coherent structures are evident, but the system is overall chaotic with a high-dimensional strange attractor.

[11] Viscosity is a form of internal friction, especially evident in thick fluids like molasses.

[12] Thermal convection cells were studied experimentally by the French physicist Henri Claude Bénard (1874–1937) in 1900 and explained theoretically by the English physicist and Nobel laureate (1904) Lord Rayleigh (aka John William Strutt, 1842–1919) in 1916.

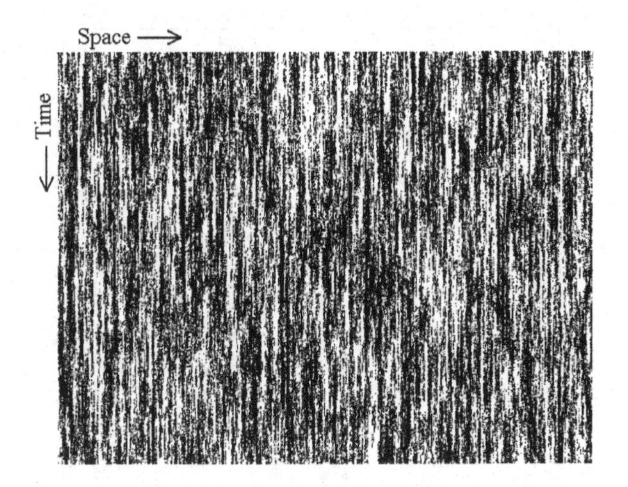

Fig. 15.16 Spatiotemporal chaos in a lattice of 640 coupled Lorenz attractors.

You can generalize these examples of coupled lattices to higher dimensions and use more complicated maps and flows and more complicated coupling rules. Coupling can be *unidirectional* as well as *bidirectional* (or *multi-directional* in higher dimensions). Addison (1995) has studied unidirectional and bidirectional coupling of Duffing's oscillators (see §4.8.1). Umberger *et al.* (1989) examined chains of sinusoidally-driven Duffing's oscillators. Cases in which every lattice cell is coupled to every other are called *globally coupled maps* (GCMs).

Experiments have been constructed with arrays of nonlinear oscillators including electrical circuits (Purwins *et al.* 1988) and coupled masses on a taut string (Moon *et al.* 1991).

15.6 Infinite-dimensional systems

The previous examples use discrete lattices to model spatiotemporal phenomena. With infinitely many lattice points, the discrete models approach the spatially continuous case in the same way maps approach temporally continuous flows with infinitely many infinitesimal time steps. In this sense, continuous spatiotemporal models are *infinite-dimensional*, even when there is only one spatial dimension. You can think of a spatiotemporal system like the atmosphere or ocean as consisting of a nearly infinite number of interacting molecules, each described by a system of ODEs such as Newton's second law. Infinite-dimensional systems can be described mathematically, but when solved numerically, space is usually represented by finitely many discrete variables that advance in small discrete time steps. You can think of the discrete variables as representing small clusters of molecules or small volumes of fluid, although the ideas are not unique to fluids.

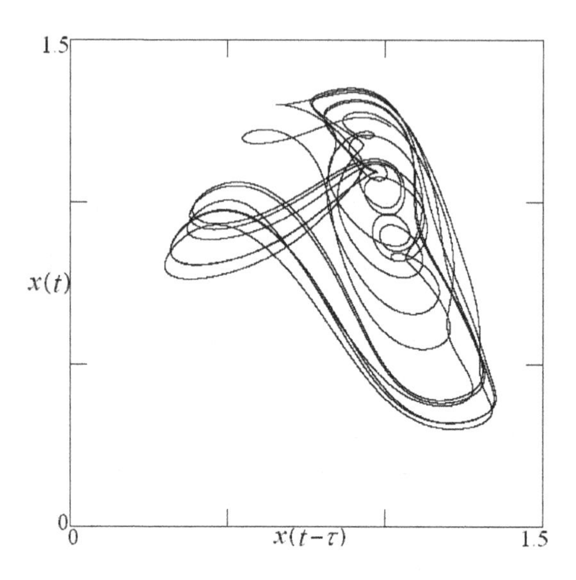

Fig. 15.17 Solution of the Mackey–Glass delay differential equation.

15.6.1 Delay differential equations

An example of an infinite-dimensional system is a *delay differential equation* (DDE) such as the *Mackey–Glass equation* (Glass and Mackey 1988)

$$\frac{dx}{dt} = \frac{ax_\tau}{1 + x_\tau^c} - bx \tag{15.14}$$

where $x_\tau = x(t - \tau)$ is the value of x at time $t - \tau$. This equation was originally proposed to model the production of white blood cells (Mackey and Glass 1977), but it could also apply to the spread of a disease with an incubation period or the growth of an animal population with a gestation period. The solution requires keeping track of a continuum of x values of duration τ. In practice, you would discretize time into N small steps, giving an $(N + 1)$-dimensional map

$$X_{n+1} = X_n + \frac{\tau}{N}\left(\frac{aX_{n-N}}{1 + X_{n-N}^c} - bX_n\right) \tag{15.15}$$

whose solution approaches the continuous case in the limit $N \to \infty$.

The solution of eqn (15.15) with $a = 0.2, b = 0.1$, and $c = 10$ is chaotic for $\tau \gtrsim 16.8$. Figure 15.17 shows a cross-section of the strange attractor for $\tau = 23$ and $N = 1 \times 10^4$. Initial conditions were taken as $X_n = 0.9$ for $-N \le n \le 0$, but they are not critical. The attractor is relatively low-dimensional ($D_2 = 2.49 \pm 0.06$), despite coming from an infinite-dimensional system. Farmer (1982) calculates $D_0 = 2.76 \pm 0.06$ and $D_{\mathrm{KY}} = 2.82 \pm 0.03$. Grassberger and Procaccia (1984) calculate $D_{\mathrm{KY}} = 2.803 \pm 0.01$ and $\lambda_1 = 0.00956 \pm 0.00005$. This system gives attractors with a wide range of dimensions depending on the value of τ.

15.6.2 Partial differential equations

The most general spatiotemporal models involve *partial differential equations* (PDEs). The most elegant formulation of the basic laws of physics involve PDEs. Examples include the *heat equation* (or diffusion equation)

$$k\nabla^2 Q = \frac{\partial Q}{\partial t} \tag{15.16}$$

the *electromagnetic wave equation*

$$c^2\nabla^2 \mathbf{E} = \frac{\partial^2 \mathbf{E}}{\partial t^2} \tag{15.17}$$

and the quantum-mechanical time-dependent *Schrödinger*[13] *wave equation*

$$-\frac{\hbar^2}{2m}\nabla^2\psi + V\psi = i\hbar\frac{\partial \psi}{\partial t} \tag{15.18}$$

Each of these equations involves the *gradient operator* (a three-dimensional derivative)

$$\nabla = \hat{x}\frac{\partial}{\partial x} + \hat{y}\frac{\partial}{\partial y} + \hat{z}\frac{\partial}{\partial z} \tag{15.19}$$

or rather its square (the *Laplacian*[14])

$$\nabla^2 = \nabla \cdot \nabla = \frac{\partial^2}{\partial x^2} + \frac{\partial^2}{\partial y^2} + \frac{\partial^2}{\partial z^2} \tag{15.20}$$

and is linear in its respective variable. Thus they cannot exhibit chaos, spatiotemporal or otherwise, despite being infinite-dimensional.

15.6.3 Navier–Stokes equation

By contrast, the equation of motion for an incompressible fluid of density ρ, pressure p, viscosity μ, and velocity $\mathbf{u}$ is

$$\frac{\partial \mathbf{u}}{\partial t} + (\mathbf{u}\cdot\nabla)\mathbf{u} = \mathbf{f} - \frac{1}{\rho}\nabla p + \frac{\mu}{\rho}\nabla^2\mathbf{u} \tag{15.21}$$

which is the famous *Navier–Stokes*[15] *equation.* The left-hand side is the acceleration in a frame of reference moving with the fluid, and the right-hand

[13]The Austrian theoretical physicist and Nobel laureate (1933) Erwin Schrödinger (1887–1961) devised the fundamental equation of quantum mechanics in 1925 during a Christmas tryst in the Swiss Alps with his Viennese mistress. It is important to realize that the equation is purely *deterministic*, but the quantity determined is the wave function ψ whose square is the *probability* of finding the particle of mass m at a given position and time.

[14]The Laplacian is named after Pierre-Simon Laplace (1749–1827), the renowned French mathematical physicist whose views on determinism and predictability were counter to those advanced by Poincaré a hundred years later.

[15]Claude Louis Marie Henri Navier (1785–1836) was a French engineer, bridge builder, political scientific advisor, and student of Fourier, who in 1821 derived the equation that bears his name despite not fully understanding the underlying physics. George Gabriel Stokes (1819–1903) was a highly religious Irish applied mathematician and Lucasian Professor of Mathematics at Cambridge, who in 1842 independently derived the equation earlier found by Navier.

side consists of the external force $\mathbf{f}$ and the internal forces due to the pressure gradient and viscosity, respectively. The inviscid ($\mu = 0$) limit of eqn (15.21) is called *Euler's equation of motion*,[16] the steady state ($\partial \mathbf{u}/\partial t = 0$) limit of which with $f = 0$ is called *Fourier's law.*

The nonlinearity $(\mathbf{u}\cdot\nabla)\mathbf{u}$ permits chaotic solutions. Chaos in fluid flow is called *turbulence*, and it occurs when the dimensionless *Reynolds number*[17]

$$R = \frac{\rho V D}{\mu} \tag{15.22}$$

where D is a characteristic scale length and V is a characteristic flow velocity, exceeds a critical value, below which the flow is *laminar* (smooth). The Reynolds number is the ratio of inertial to viscous forces (Reynolds 1883). Solution of the Navier–Stokes equation is notoriously difficult, despite some 200 years of intense effort.

15.6.4 Boldrighini–Francheschini equations

The Lorenz equations (see §4.8.2) are one approximation of the Navier–Stokes equation in terms of a truncated system of ordinary differential equations. Another chaotic three-dimensional ODE fluid model (see §A.5.11) was proposed about the same time by Moore and Spiegel (1966). Another such system is given by Boldrighini and Francheschini (1979)

$$\frac{dx_1}{dt} = -2x_1 + 4x_2x_3 + 4x_4x_5 \tag{15.23}$$

$$\frac{dx_2}{dt} = -9x_2 + 3x_1x_3 \tag{15.24}$$

$$\frac{dx_3}{dt} = -5x_3 - 7x_1x_2 + R \tag{15.25}$$

$$\frac{dx_4}{dt} = -5x_4 - x_1x_5 \tag{15.26}$$

$$\frac{dx_5}{dt} = -x_5 - 3x_1x_4 \tag{15.27}$$

This five-dimensional system is strongly damped ($\sum \lambda = -22$) and has a strange attractor shown as a projection on the x_1x_2 plane in Fig. 15.18 for a Reynolds number of $R = 33$.

15.6.5 Burgers' equation

In one spatial dimension, the Navier–Stokes equation does not have chaotic solutions, but there are a number of partial differential equations that

[16] Euler's equation of motion is not to be confused with *Euler's equation*: $e^{i\theta} = \cos\theta + i\sin\theta$.

[17] Osborne Reynolds (1842–1912) was an Irish scientist and engineer who was especially interested in fluid dynamics and formulated the conditions under which fluid flow becomes turbulent.

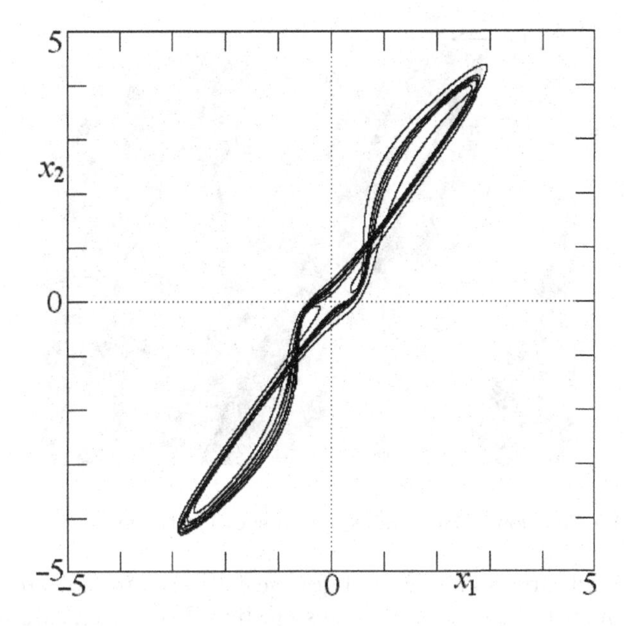

Fig. 15.18 Solution of the Boldrighini–Francheschini equations.

capture many of its features such as *Burgers'*[18] *equation* (Burgers 1974)

$$\frac{\partial u}{\partial t} + u\frac{\partial u}{\partial x} = \frac{1}{R}\frac{\partial^2 u}{\partial x^2} \tag{15.28}$$

where u is the flow velocity. It is the Navier–Stokes equation in one dimension without the pressure and external force.

This equation is integrable and can be reduced to the linear heat equation for Q in eqn (15.16) by the *Hopf–Cole transformation* (Whitham 1974)

$$u = -\frac{2}{Q}\frac{\partial Q}{\partial x} \tag{15.29}$$

Hence it cannot produce chaos despite the nonlinearity, but it exhibits *shocks* (propagating waves with steep wavefronts). It is most easily solved as a system of N difference equations

$$U_{n+1}(i) = U_n(i) + \Delta t\Big\{\frac{1}{R}[U_n(i+1) - 2U_n(i) + U_n(i-1)] - \frac{U_n(i)}{2}[U_n(i+1) - U_n(i-1)]\Big\} \tag{15.30}$$

with the solution in Fig. 15.19 for $N = 640, \Delta t = 1 \times 10^{-4}$, and $R = 10$. As before, x is plotted horizontally with periodic boundary conditions, time

[18]Johannes Martinus Burgers (1895–1981) was a Dutch experimental physicist and hydrodynamicist.

Fig. 15.19 Solution of Burgers' equation showing the formation of shocks.

increases downward, and a 16-level gray scale is used to indicate the value of U in each spatial cell at each time step. Initial conditions are uniformly random with $0 < U_0(i) < 1$. Burgers' equation has also been used to model traffic flow (Nagatani *et al.* 1998).

You can also solve partial differential equations such as Burgers' equation by the *Fourier spectral method* in which the solution is approximated by a sum of sinusoidal spatial modes. With periodic boundary conditions, these modes have wavelengths equal to the system size divided by an integer. Each mode gives an ordinary differential equation, and a good approximation often requires only a few modes. Thus the infinite-dimensional PDE becomes a low-dimensional system of ODEs in which the variables are the mode amplitudes.

15.6.6 Korteweg–de Vries equation

Replacing the $\partial^2 u/\partial x^2$ term in Burgers' equation with $-\partial^3 u/\partial x^3$ gives the *Korteweg–de Vries*[19] (KdV) *equation* (Korteweg and de Vries 1895)

$$\frac{\partial u}{\partial t} + u\frac{\partial u}{\partial x} = -\frac{\partial^3 u}{\partial x^3} \tag{15.31}$$

The added term provides *dispersion* (spatial broadening), but the dispersion is counteracted by the nonlinear term, and there is no damping because the viscous term is absent. Sometimes the KdV equation is written with different coefficients or signs, but the simplest form above will suffice for our purposes.

[19]Gustav de Vries (1866–1934) was a student of the Dutch mathematician Diederik Johannes Korteweg (1848–1941) at the University of Amsterdam where he defended his doctoral thesis on the KdV equation in 1894. His work went largely unnoticed until Zabusky and Kruskal (1965) discovered that two colliding solitary waves emerge unchanged in a numerical solution of the KdV equation.

Like Burgers' equation, the KdV equation is nonlinear but integrable and hence cannot produce chaos. It has an unbounded solution

$$u = 3v\,\mathrm{sech}^2\Big[\frac{\sqrt{v}}{2}(x - vt)\Big] \tag{15.32}$$

where v is the propagation velocity. The amplitude of the disturbance is proportional to its velocity, and the width is inversely proportional to $\sqrt{v}$. The equation also has bounded and spatially periodic solutions. It was the first nonlinear PDE for which an analytic solution was found (Gardner *et al.* 1967).

The importance of the KdV equation is the existence of *solitary waves* (*solitons*) that propagate over long distances without changing shape. Such solitons were first reported by Russel[20] (1845) while observing a boat stop in a narrow channel, but forgotten for a hundred years. A soliton is very robust against perturbations. The bottom of the channel may be uneven, and the soliton passes easily by small objects floating on the water. The solitons behave much like particles, except that one soliton can pass through another without distortion. Linear waves easily pass through one another without distortion, but it was quite a surprise when nonlinear waves with the same property were discovered.

Characteristics of solitons include the following:

1. They are a spatially localized solution of a partial differential equation describing a nonlinear system with infinitely many degrees of freedom.
2. They propagate for a long time or over a large distance without visible change.
3. They are usually, but not necessarily, attributed to integrable systems.
4. They remain unchanged during collisions except for a phase shift.
5. They can be viewed as 'modes' of the system.
6. They can exist in dissipative systems.

In finite difference form, the KdV equation is

$$\begin{aligned} U_{n+1}(i) = U_n(i) - \frac{\Delta t}{2}\{&U_n(i+2) - 2U_n(i+1) + 2U_n(i-1) \\ &- U_n(i-2) + U_n(i)[U_n(i+1) - U_n(i-1)]\} \end{aligned} \tag{15.33}$$

with the solution in Fig. 15.20 for $N = 640$ and $\Delta t = 1 \times 10^{-3}$. As before, x is plotted horizontally with periodic boundary conditions, time increases downward (on a 10× compressed scale), and a 16-level gray scale is used to indicate U. Initial conditions include two solitons, one moving with $v = 0.05$ and the other moving with $v = 0.1$ so as to overtake it.

[20] John Scott Russel (1808–1882) was a Scottish engineer and boat designer who in 1834 followed a soliton on horseback along the bank of the Union Canal near Edinburgh 'rolling on at a rate of some eight or nine miles an hour, preserving its original figure some thirty feet long and a foot to a foot and a half in height' for a distance of 'one or two miles.'

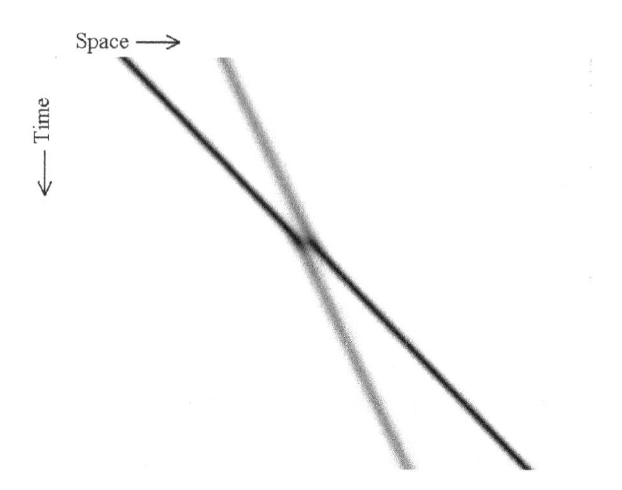

Fig. 15.20 Solution of the KdV equation showing one soliton overtaking and passing through another.

Several other nonlinear PDEs also have soliton-like solutions including two forms of the *modified Korteweg–de Vries* (mKdV) *equation* (Drazin and Johnson 1989)

$$\frac{\partial u}{\partial t} \pm u^2 \frac{\partial u}{\partial x} = -\frac{\partial^3 u}{\partial x^3} \tag{15.34}$$

the *sine–Gordon equation*

$$\frac{\partial^2 u}{\partial t^2} = \frac{\partial^2 u}{\partial x^2} + \sin u \tag{15.35}$$

the *nonlinear Schrödinger equation*

$$i\frac{\partial u}{\partial t} = -\frac{\partial^2 u}{\partial x^2} - |u|^2 u \tag{15.36}$$

and the *Fisher equation*, also called the *KPP equation*

$$\frac{\partial u}{\partial t} = u(1-u) + \frac{\partial^2 u}{\partial x^2} \tag{15.37}$$

The Fisher (1937) or Kolmogoroff–Petrovsky–Piskounoff (1937) equation is an obvious extension of the logistic differential equation (see §3.2) and is the classic and simplest case of a nonlinear *reaction–diffusion equation*, the most general form of which is

$$\frac{\partial u}{\partial t} = f(u) + \nabla^2 u \tag{15.38}$$

where the reaction term $f(u)$ is countered by the diffusion term $\nabla^2 u$. Reaction–diffusion equations have been extensively studied (Fife 1979, Britton 1986) and can generate an astonishing variety of physically and biologically realistic complex spatiotemporal patterns (Meinhardt 1982), called

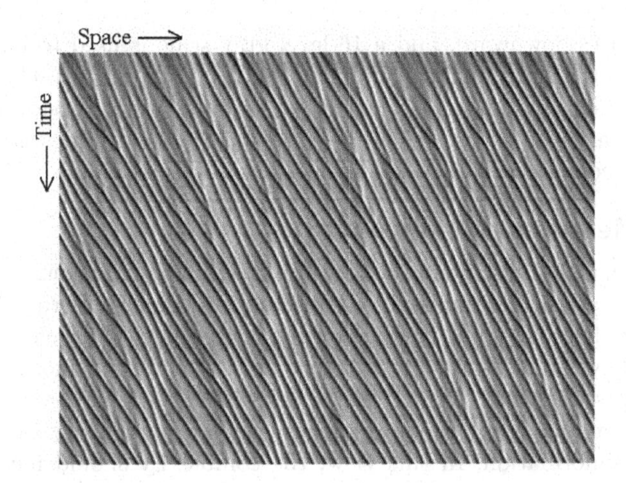

Fig. 15.21 Solution of the Kuramoto–Sivashinsky equation showing turbulence.

Turing structures (Turing 1952). Reaction–diffusion equations also can be generalized to include nonlinear diffusion (Rosenau 2002) such as

$$\frac{\partial u}{\partial t} = u(1-u) + \frac{\partial^2 u^2}{\partial x^2} \tag{15.39}$$

15.6.7 Kuramoto–Sivashinsky equation

A one-dimensional variant of the Navier-Stokes equation that *can* exhibit chaos is the *Kuramoto–Sivashinsky equation*

$$\frac{\partial u}{\partial t} + u\frac{\partial u}{\partial x} = -\frac{1}{R}\frac{\partial^2 u}{\partial x^2} - \frac{\partial^4 u}{\partial x^4} \tag{15.40}$$

The $\partial^2 u/\partial x^2$ term is a negative viscosity leading to the growth of long wavelength modes, and the $\partial^4 u/\partial x^4$ term is a hyperviscosity that damps the short wavelength modes. The nonlinearity $u\partial u/\partial x$ transports energy from the growing modes to the damped modes. This equation has been used to model waves in chemical reactions (Kuramoto and Tsuzuki 1976), flame front modulations (Sivashinsky 1977), and a thin liquid film flowing down an inclined plane (Sivashinsky and Michelson 1980).

In finite difference form, the Kuramoto–Sivashinsky equation is

$$\begin{aligned} U_{n+1}(i) = U_n(i) - \Delta t\Big\{ &\frac{1}{R}[U_n(i+1) - 2U_n(i) + U_n(i-1)] \\ &+ U_n(i+2) - 4U_n(i+1) + 6U_n(i) - 4U_n(i-1) \\ &+ U_n(i-2) + \frac{U_n(i)}{2}[U_n(i+1) - U_n(i-1)]\Big\} \end{aligned} \tag{15.41}$$

with the solution in Fig. 15.21 for $N = 640, \Delta t = 1 \times 10^{-4}$, and $R = 2$. As before, x is plotted horizontally with periodic boundary conditions, time increases downward, and a 16-level gray scale is used to indicate U. Initial conditions are uniformly random with $0 < U_0(i) < 1$. The solution has a dominant spatial period, but it is aperiodic in both space and time, suggesting a turbulent flow. The resemblance to sand dunes is probably more than coincidental.

15.7 Measures of complexity

Whereas chaos is best quantified by the Lyapunov exponent, there is no universal measure of complexity (Badii and Politi 1997). Perhaps the reason is that there is no universal definition of complexity. When an initially disordered system self-organizes, it becomes less complex since its entropy decreases (Dowell and Virgin 1990, Kaneko 1990), although you could also argue that a completely random pattern has low complexity since it conveys little information. In this view, the complexity should increase and then decrease again as a system changes from completely random to highly ordered. One measure that satisfies this condition is the *statistical complexity* (Crutchfield and Young 1989), which is the amount of information about the past required to predict the future. Bandt and Pompe (2002) suggest using the *permutation entropy* as a measure of the complexity of data contaminated by noise.

One measure of the complexity of a system is its *capacity* for self-organization. Since self-organization occurs in a nonequilibrium system with a throughput of energy, a reasonable measure for a physical system is the rate of energy flow per unit mass, such that the human brain is about 10^5 times more complex than the Sun (Chaisson 2001). Complexity seems greatest on human size scales. Atoms and galaxies obey simple laws, whereas life is apparently highly complex. Ruelle (1991) calls an entity complex if it embodies information that is hard to get.

For a chaotic system, the dimension of the attractor is a measure of its complexity since it measures the number of active variables. In a spatiotemporal system with known equations, you can calculate the Kaplan–Yorke dimension (see §5.9). It is more difficult to measure the attractor dimension for spatiotemporal data, and other measures of complexity are preferred.

If you have a succession of patterns, as in Fig. 15.6, one way to quantify the organization is to calculate the probability that an arbitrary point is part of a cluster of identical points. For example, Fig. 15.22 shows the cluster probability for Fig. 15.6 versus time using the four nearest neighbors. For the random pattern at $t = 0$ with half black and half white points, each neighbor of a point has a probability of $1/2$ of being the same as the point. Thus the probability that a point has four identical neighbors is $1/2^4 = 0.0625$. After many generations, this value increases to about 0.75, indicating high organization (or low complexity). The value will depend on the size of the neighborhood and the population of each category in the image.

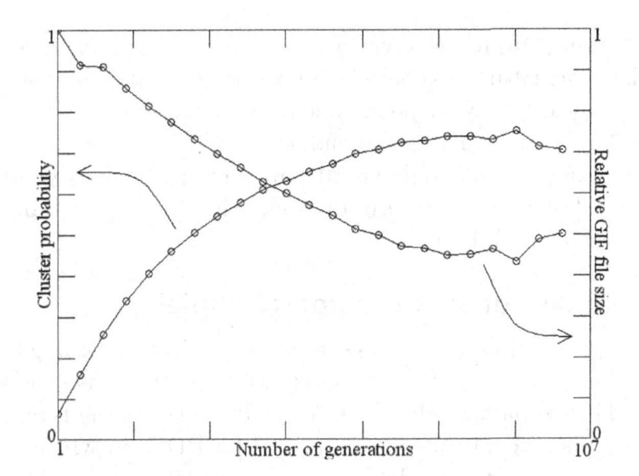

Fig. 15.22 Measures of complexity for the forest model in Fig. 15.6 showing self-organization.

Another measure of complexity is based on the size of the smallest computer program that can produce the pattern (Solomonoff 1964, Kolmogorov 1965, Chaitin 1966), which is a measure of the information required to describe the system.[21] This measure is called the *algorithmic complexity*, *information complexity*, *Kolmogorov complexity*, or *Chaitin complexity*. In this sense, the Mandelbrot set (see §14.5) is not complex since the program is simple, but the patterns it produces are extremely complex. Similarly, the number *pi* (π) is not complex since it can be calculated to arbitrary precision with a simple program even though the digits are apparently random. In general, complexity is not conserved and can change greatly going from cause to effect or from one size or time-scale to another, or more generally, from one context or observer to another.

Algorithms for calculating the algorithmic complexity are available (Lempel and Ziv 1976, Garey and Johnson 1979, Kaspar and Schuster 1987). A simple way to estimate it uses the fact that GIF graphics files are nearly optimally compressed. Thus you only need to determine the size of the GIF file of the image after subtracting the fixed size of the file header. Figure 15.22 shows the result for the data in Fig. 15.6 normalized to the initial random state.

Reductionism, which has served science so well for centuries, assumes that simple laws ultimately govern all natural processes, and the apparent

[21]The idea of using the size of a computer program as a measure of complexity is appealing, but Chaitin (2002) points out that it is impossible to be sure you have identified the smallest such program because of *Gödel's incompleteness theorem.* With this theorem, the Austrian mathematician Kurt Gödel (1906–1978) revolutionized mathematics by showing that there are mathematical propositions that cannot be proven either true or false using the rules and axioms of mathematics (Gödel 1931), but he was also a bit paranoid and starved himself to death, thinking his food was poisoned.

complexity results from the myriad of possibilities. *Emergence* contends that high-level structure is generally unpredictable from the low-level processes, and may not even depend much on their properties. Systems are complex if they exhibit emergent behavior.

These measures are certainly not absolute and may not even relate much to complexity, but they are useful for comparing models with one another and with experimental data.

15.8 Summary of spatiotemporal models

The various spatiotemporal models can be categorized according to whether space, time, and the state of the variables are quantized, as summarized in Table 15.2. The simplest model is a CA in which everything is discrete, and the most complicated models are described by PDEs in which everything is continuous. Some of the models have been described here, but most have been relatively little studied. The choice depends mostly on the nature of the problem, the degree of realism desired, and the available computational resources. When any of the models is solved with a digital computer, it is essentially reduced to a CA, but the step sizes in space, time, and state can be very small and very numerous. The trade-off is the computation time and memory required.

There are also subcategories, depending on the number of spatial dimensions. Some of the dimensions can be discrete, and others continuous, or the continuous spatial information can be approximated by a finite number of Fourier modes. Time can be eliminated from the model, as with the fractal patterns in Chapter 11 not produced by dynamical systems. As computers become more powerful, more realistic models are possible, but realism often comes at the expense of understanding. Sometimes it is better to use a simple model to facilitate exploration and analysis and to identify generic behavior that is independent of the details.

Table 15.2 Summary of spatiotemporal models.

Space	Time	State	Model
Discrete	Discrete	Discrete	Cellular automata
Discrete	Discrete	Continuous	Coupled map lattices
Discrete	Continuous	Discrete	
Discrete	Continuous	Continuous	Coupled flow lattices
Continuous	Discrete	Discrete	
Continuous	Discrete	Continuous	
Continuous	Continuous	Discrete	
Continuous	Continuous	Continuous	Partial differential equations

15.9 Concluding remarks

Nature is complicated, involving a myriad of processes, with countless nonlinearly interacting variables, distributed over a diverse spatial landscape, and often behaving with no apparent rhyme nor reason. It is surprising that such behavior can be described by abstract mathematics (Wigner 1960), and nothing less than miraculous that extremely simple mathematical models often capture the essential features. When the models are chaotic, as often happens, long-term prediction is precluded, but a new opportunity arises for controlling the future with only small changes in the present. Perhaps chaos explains the apparent coexistence of determinism and free will (*Epicurus' dilemma*).[22]

In human terms, we often feel insignificant and powerless in the face of diverse and unexpected natural events. But if we are part of a chaotic system, we resemble the butterfly whose flapping wings can set off tornadoes. Everything we do drastically changes the course of history. Thus it behooves us to behave responsibly. The cumulative effect of many people so acting will make a world that evolves in unexpected and delightful ways. Perhaps Poincaré should have said that life is worth living not just because nature is beautiful, but also because we can each enhance that beauty with only modest effort.

15.10 Exercises

Exercise 15.1 Without using a computer, calculate the first five generations of the one-dimensional CA in Fig. 15.1.

Exercise 15.2 Show that the colonies below in the Game of Life die after one generation.

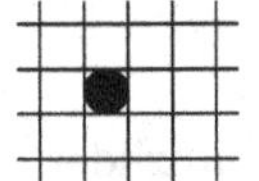 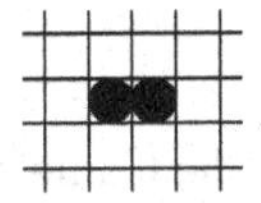 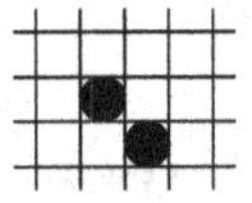

Exercise 15.3 Show that the colonies below in the Game of Life die after two generations.

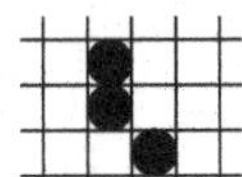 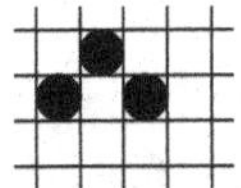 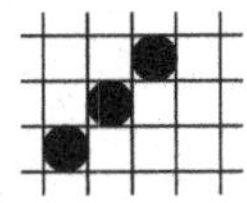

Exercise 15.4 Show that the colony below in the Game of Life dies after three generations.

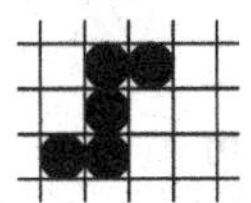

[22]Epicurus (*c.* 341–271 BC) was a Greek philosopher who believed that matter was made up of atoms and who founded Epicureanism, a hedonistic ethic that seeks pleasure through the pursuit of justice, honor, and wisdom rather than the acquisition of goods.

Exercise 15.5 Show that the colonies below in the Game of Life remain stationary. The one in the upper left is called a 'block.'

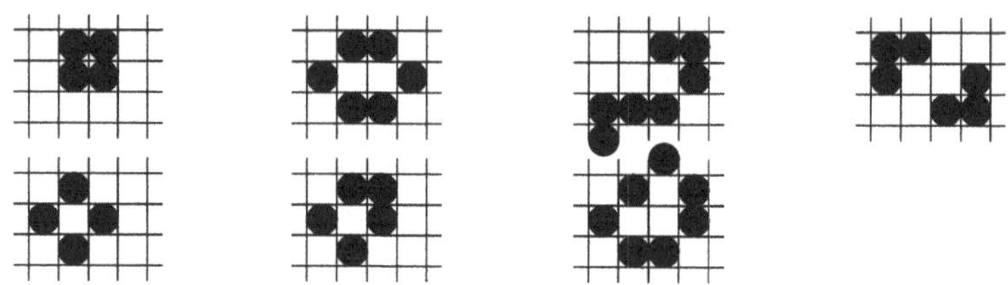

Exercise 15.6 Show that the colony below in the Game of Life is a period-2 'blinker.'

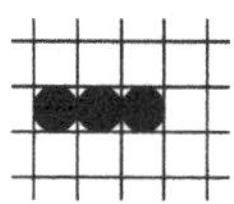

Exercise 15.7 Show that the colony below in the Game of Life evolves into a period-2 'traffic light' after several generations.

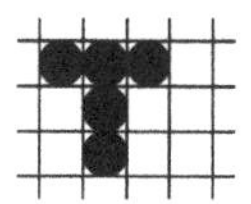

Exercise 15.8 Show that the *r-pentomino* (shaped like an 'r' with five cells) colony below in the Game of Life grows for at least five generations. (In fact, it grows for 1103 generations on an infinite grid before settling down to a pattern that includes three gliders leaving in different directions.)

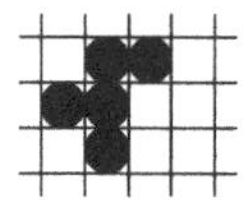

Exercise 15.9 Show that the colony below in the Game of Life is a 'glider' that reproduces itself down and to the right after four generations.

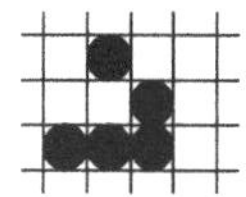

Exercise 15.10 Show that no regular tilings of the plane are possible except those in Fig. 15.5.

Exercise 15.11 Show that a power spectrum of the form $S(f) \propto 1/f^{\alpha}$ has no characteristic time-scale.

Exercise 15.12 Show that the map in eqn (15.15) is equivalent to the Mackey–Glass equation in eqn (15.14) in the limit $N \to \infty$.

Exercise 15.13 Show that the Mackey–Glass equation in eqn (15.14) can be approximated by the map

$$X_{n+1} = \frac{1}{2N + b\tau}\left[(2N - b\tau)X_n + a\tau\left(\frac{X_{n-N}}{1 + X^c_{n-N}} + \frac{X_{n-N+1}}{1 + X^c_{n-N+1}}\right)\right] \quad (15.42)$$

which is more accurate than eqn (15.15) for the same N.

Exercise 15.14 Show that the heat equation in eqn (15.16) has a one-dimensional solution

$$Q(x,t) = \sum_{n=0}^{\infty} A_n e^{n^2\pi^2 k^2 t/L} \sin\frac{n\pi x}{L} \quad (15.43)$$

for $Q(0,t) = Q(L,t) = 0$.

Exercise 15.15 Show that the wave equation in eqn (15.17) has a one-dimensional solution

$$E(x,t) = E_0 \sin(kx - \omega t) \quad (15.44)$$

for a traveling wave of amplitude E_0, frequency ω, and wave number k and that $\omega/k = c$ is the velocity of propagation of the wave.

Exercise 15.16 Show that the Schrödinger wave equation in eqn (15.18) has a one-dimensional solution

$$\psi(x,t) = \psi_0 e^{i(kx-\omega t)} \quad (15.45)$$

for a particle of mass m moving with a constant potential energy V and that the total energy is $\hbar\omega = \hbar^2 k^2/2m + V$.

Exercise 15.17 Show that Burgers' equation in eqn (15.28) can be reduced to the one-dimensional heat equation in eqn (15.16) by use of the Hopf–Cole transformation in eqn (15.30).

Exercise 15.18 Show that eqn (15.32) is a solution of the KdV equation in eqn (15.31).

Exercise 15.19 Show that the *linear* PDE

$$\frac{\partial u}{\partial t} + \frac{\partial u}{\partial x} = -\frac{\partial^3 u}{\partial x^3} \quad (15.46)$$

has a solution $u = Ae^{i(kx-\omega t)}$ with $\omega = k - k^3$ and hence exhibits dispersion.

Exercise 15.20 Show that the sine–Gordon equation in eqn (15.35) has a soliton solution $\partial u/\partial x = 2\sqrt{2/(v^2-1)}\,\mathrm{sech}(x - vt)$.

Exercise 15.21 Show that the nonlinear Schrödinger equation in eqn (15.36) has a solution $u = \sqrt{2\omega}e^{i\omega t}\mathrm{sech}(\sqrt{w}x)$.

15.11 Computer project: Spatiotemporal chaos and complexity

In this project you will produce examples of spatiotemporal chaos in complex systems that exhibit self-organization.

1. Implement the classic 'Game of Life' (a two-dimensional cellular automaton) described in §15.1.2. Use periodic boundary conditions. Start with a random initial condition (about half the cells occupied) and iterate until a steady state is reached. This may take several hundred iterations, depending on the size of your grid. Produce a pattern like Fig. 15.4. [Optional: Start with a slightly different initial condition (one cell different) and see whether the final pattern is sensitively dependent on the initial condition, that is, whether it is chaotic.]
2. Explore the behavior of a one-dimensional coupled-map lattice (CML) using logistic maps as described in §15.5.2. Use random initial conditions in the interval 0 to 1 and periodic boundary conditions. Make a plot of $X(t)$ versus t for one of the lattice points with parameters adjusted to produce chaotic behavior. [Optional: Run the $X(t)$ data through your correlation dimension program and get an estimate of the dimension of the attractor.]
3. Simulate two-dimensional diffusion-limited aggregation (DLA) as described in §11.6.5. Produce a pattern like Fig. 11.16. This is an example of a random fractal. [Optional: Use the box-counting method to estimate the capacity dimension of the pattern you produce.]
4. Plot the trajectory of one of Langton's Ants (Gale 1993), which are governed by the following rules: Begin with a grid of cells (at least 100×100) that are all white. The ant starts on the central cell of the grid. It moves one cell to the east and looks at the color of the cell on which it lands. If it lands on a white cell, it paints it black and turns 90° to the right. If it lands on a black cell, it paints it white and turns 90° to the left. If the ant runs off the edge of the screen, it re-enters at the opposite edge (periodic boundary conditions). The ant follows these same simple rules for ever. Make a plot of the pattern that is produced after 10 000, 100 000, and 1 million steps. [Optional: Sprinkle a few rocks (black squares) around the screen randomly or in some pattern for the ant to bump into, and see how the resulting behavior of the ant is altered.]

A

Common chaotic systems

Below are some common chaotic systems and their parameters, collected here for convenience. In most cases initial conditions are not critical, but values are given in the basin of attraction and near the attractor for the dissipative systems or in the chaotic sea for the conservative systems. Most of the results are original calculations, and all have been independently verified.

The figures show X_{n+1} versus X_n for the one-dimensional maps, Y_n versus X_n for the other maps, and a projection onto the xy plane for the flows. All Lyapunov exponents are base-e and were calculated using the methods in Chapter 5. The Kaplan–Yorke dimension D_{KY} is given by eqn (5.29). The correlation dimension D_2 uses at least 2×10^{12} pairs, corresponding to data sets of over two million points for each case and have been extrapolated to the limit of zero size scale using eqn (13.8).

A.1 Noninvertible maps

A.1.1 Logistic map

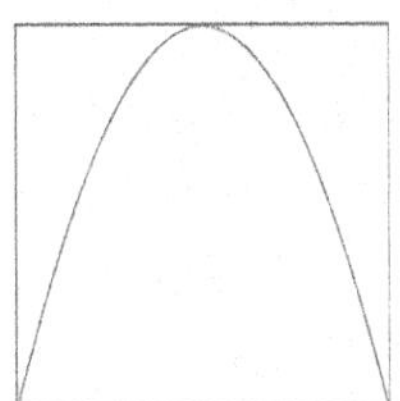

$X_{n+1} = AX_n(1 - X_n)$
Usual parameter: $A = 4$
Initial condition: $X_0 = 0.1$
Lyapunov exponent: $\lambda = \ln 2 = 0.693147181\ldots$
Kaplan–Yorke dimension: $D_{\mathrm{KY}} = 1.0$ (exact value)
Correlation dimension: $D_2 = 1.0$ (exact value, converges slowly)
Ref: May (1976)

A.1.2 Sine map

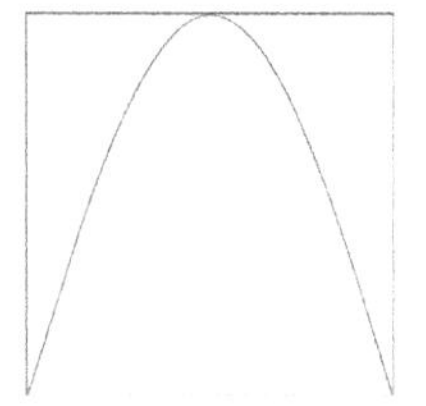

$X_{n+1} = A \sin \pi X_n$
Usual parameter: $A = 1$
Initial condition: $X_0 = 0.1$
Lyapunov exponent: $\lambda \simeq 0.689067$
Kaplan–Yorke dimension: $D_{\mathrm{KY}} = 1.0$ (exact value)
Correlation dimension: $D_2 = 1.0$ (exact value, converges slowly)
Ref: Strogatz (1994)

A.1.3 Tent map

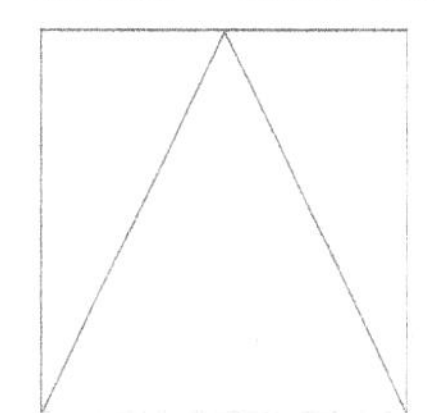

$X_{n+1} = A \min(X_n, 1 - X_n)$
Usual parameter: $A = 2$ (see discussion in §2.5.2)
Initial condition: $X_0 = 1/\sqrt{2}$ (or any irrational $0 < X_0 < 1$)
Lyapunov exponent: $\lambda = \ln |A| = 0.693147181 \ldots$
Kaplan–Yorke dimension: $D_{\mathrm{KY}} = 1.0$ (exact value)
Correlation dimension: $D_2 = 1.0$ (exact value)
Ref: Devaney (1989)

A.1.4 Linear congruential generator

$X_{n+1} = AX_n + B \pmod{C}$
Usual parameters: $A = 7141, B = 54773, C = 259200$
Initial condition: $X_0 = 0$
Lyapunov exponent: $\lambda = \ln |A| = 8.873608101 \ldots$
Kaplan–Yorke dimension: $D_{\mathrm{KY}} = 1.0$ (exact value)
Correlation dimension: $D_2 = 1.0$ (exact value)
Ref: Knuth (1997)

A.1.5 Cubic map

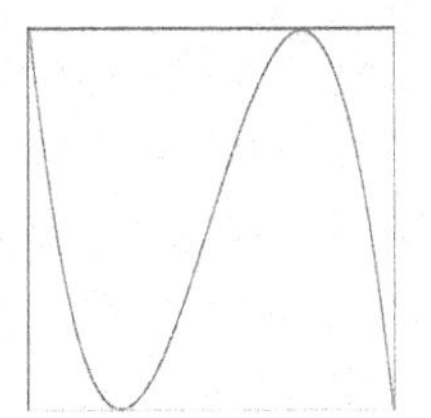

$X_{n+1} = AX_n(1 - X_n^2)$
Usual parameter: $A = 3$
Initial condition: $X_0 = 0.1$
Lyapunov exponent: $\lambda \simeq 1.0986122883$
Kaplan–Yorke dimension: $D_{\mathrm{KY}} = 1.0$ (exact value)
Correlation dimension: $D_2 = 1.0$ (exact value)
Ref: Zeng *et al.* (1985)

A.1.6 Ricker's population model

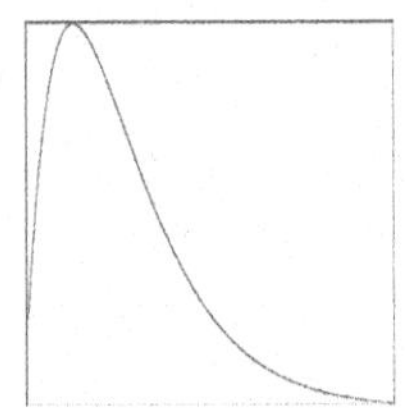

$X_{n+1} = AX_n \mathrm{e}^{-X_n}$
Usual parameter: $A = 20$
Initial condition: $X_0 = 0.1$
Lyapunov exponent: $\lambda \simeq 0.384846$
Kaplan–Yorke dimension: $D_{\mathrm{KY}} = 1.0$ (exact value)
Correlation dimension: $D_2 = 1.0$ (exact value, converges slowly)
Ref: Ricker (1954)

A.1.7 Gauss map

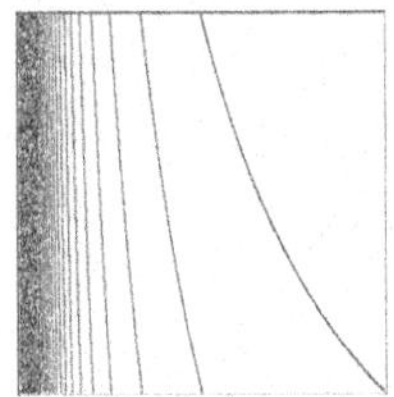

$X_{n+1} = 1/X_n \pmod 1$
Initial condition: $X_0 = 0.1$
Lyapunov exponent: $\lambda \simeq 2.373445$
Kaplan–Yorke dimension: $D_{\mathrm{KY}} = 1.0$ (exact value)
Correlation dimension: $D_2 = 1.0$ (exact value)
Ref: van Wyk and Steeb (1997)

A.1.8 Cusp map

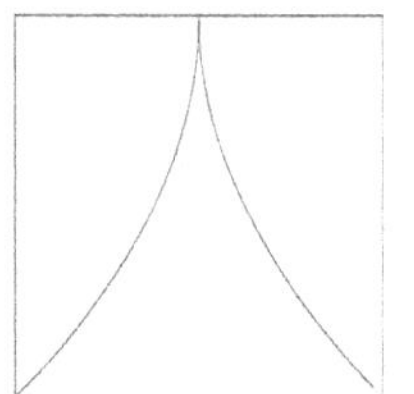

$X_{n+1} = 1 - A\sqrt{|X_n|}$
Usual parameter: $A = 2$
Initial condition: $X_0 = 0.5$
Lyapunov exponent: $\lambda = 0.5$ (exact value)
Kaplan–Yorke dimension: $D_{\mathrm{KY}} = 1.0$ (exact value)
Correlation dimension: $D_2 = 1.0$ (exact value)
Ref: Beck and Schlögl (1995)

A.1.9 Gaussian white chaotic map

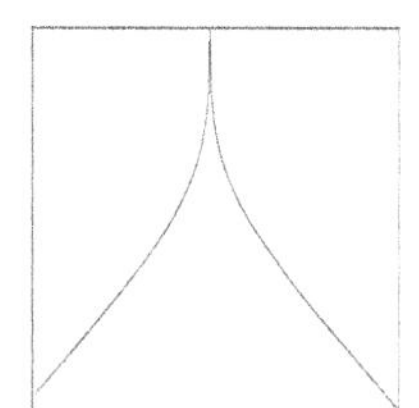

$X_{n+1} = A\ \mathrm{erf}^{-1}[1 - 2\mathrm{erf}(X_n/A)]$
Usual parameter: $A = \sqrt{2}$
Initial condition: $X_0 = 1$
Lyapunov exponent: $\lambda = \ln 2 = 0.693147181\ldots$
Kaplan–Yorke dimension: $D_{\mathrm{KY}} = 1.0$ (exact value)
Correlation dimension: $D_2 = 1.0$ (exact value)
Ref: This text (§9.5.3)

A.1.10 Pinchers map

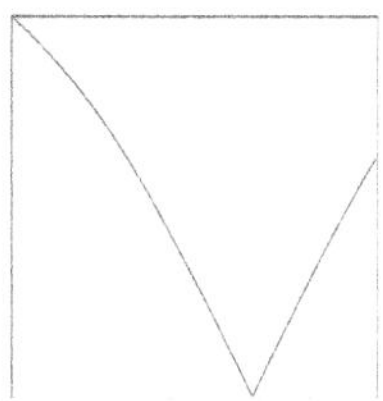

$X_{n+1} = |\tanh s(X_n - c)|$
Usual parameters: $s = 2, c = 0.5$
Initial condition: $X_0 = 0$
Lyapunov exponent: $\lambda \simeq 0.467944$
Kaplan–Yorke dimension: $D_{\mathrm{KY}} = 1.0$ (exact value)
Correlation dimension: $D_2 = 1.0$ (exact value)
Ref: Potapov and Ali (2000)

A.1.11 Spence map

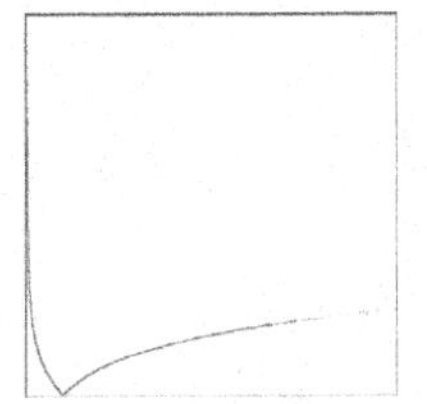

$X_{n+1} = |\ln X_n|$
Initial condition: $X_0 = 0.5$
Lyapunov exponent: $\lambda \to \infty$
Kaplan–Yorke dimension: $D_{\mathrm{KY}} = 1.0$ (exact value)
Correlation dimension: $D_2 = 1.0$ (exact value)
Ref: Shaw (1981)

A.1.12 Sine–circle map

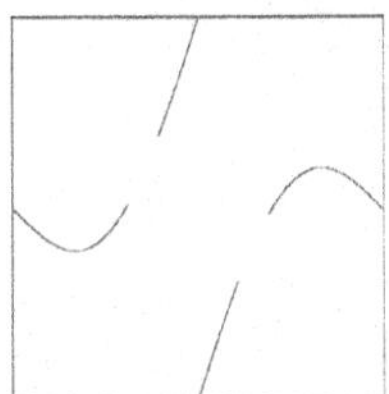

$X_{n+1} = X_n + \Omega - \frac{K}{2\pi} \sin 2\pi X_n \pmod 1$
Usual parameters: $\Omega = 0.5, K = 2$
Initial condition: $X_0 = 0.1$
Lyapunov exponent: $\lambda \simeq 0.353863$
Kaplan–Yorke dimension: $D_{\mathrm{KY}} = 1.0$ (exact value)
Correlation dimension: $D_2 = 1.0$ (exact value, converges slowly)
Ref: Arnold (1965)

A.2 Dissipative maps

A.2.1 Hénon map

$X_{n+1} = 1 - aX_n^2 + bY_n$
$Y_{n+1} = X_n$
Usual parameters: $a = 1.4, b = 0.3$
Initial conditions: $X_0 = 0 : Y_0 = 0.9$
Lyapunov exponents: $\lambda \simeq 0.41922, -1.62319$
Kaplan–Yorke dimension: $D_{\mathrm{KY}} \simeq 1.25827$

Correlation dimension: $D_2 = 1.220 \pm 0.036$
Ref: Hénon (1976)

A.2.2 Lozi map

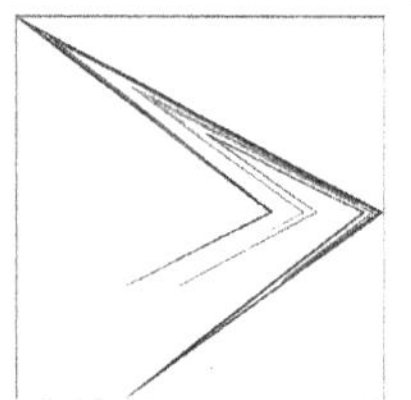

$X_{n+1} = 1 - a|X_n| + bY_n$
$Y_{n+1} = X_n$
Usual parameters: $a = 1.7, b = 0.5$
Initial conditions: $X_0 = -0.1, Y_0 = 0.1$
Lyapunov exponents: $\lambda \simeq 0.47023, -1.16338$
Kaplan–Yorke dimension: $D_{\mathrm{KY}} \simeq 1.40419$
Correlation dimension: $D_2 = 1.384 \pm 0.053$
Ref: Lozi (1978)

A.2.3 Delayed logistic map

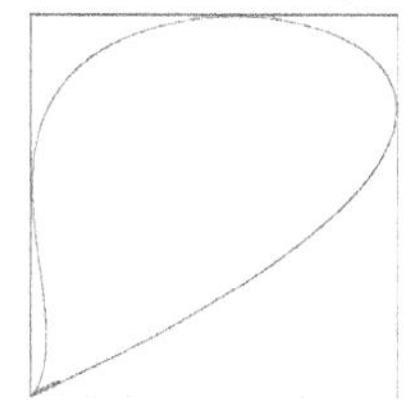

$X_{n+1} = AX_n(1 - Y_n)$
$Y_{n+1} = X_n$
Usual parameter: $A = 2.27$
Initial conditions: $X_0 = 0.001, Y_0 = 0.001$
Lyapunov exponents: $\lambda \simeq 0.18312, -1.24199$
Kaplan–Yorke dimension: $D_{\mathrm{KY}} \simeq 1.14744$
Correlation dimension: $D_2 = 1.144 \pm 0.034$ (converges slowly)
Ref: Aronson *et al.* (1982)

A.2.4 Tinkerbell map

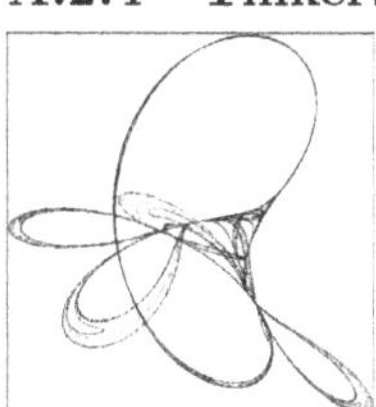

$X_{n+1} = X_n^2 - Y_n^2 + aX_n + bY_n$
$Y_{n+1} = 2X_nY_n + cX_n + dY_n$
Usual parameters: $a = 0.9, b = -0.6, c = 2, d = 0.5$

Initial conditions: $X_0 = 0, Y_0 = 0.5$
Lyapunov exponents: $\lambda \simeq 0.18997, -0.52091$
Kaplan–Yorke dimension: $D_{\mathrm{KY}} \simeq 1.36468$
Correlation dimension: $D_2 = 1.329 \pm 0.036$ (converges slowly)
Ref: Nusse and Yorke (1994)

A.2.5 Burgers' map

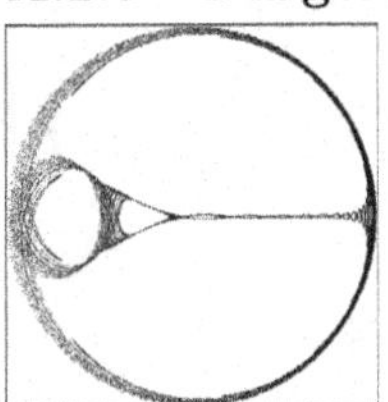

$X_{n+1} = aX_n - Y_n^2$
$Y_{n+1} = bY_n + X_nY_n$
Usual parameters: $a = 0.75, b = 1.75$
Initial conditions: $X_0 = -0.1, Y_0 = 0.1$
Lyapunov exponents: $\lambda \simeq 0.12076, -0.22136$
Kaplan–Yorke dimension: $D_{\mathrm{KY}} \simeq 1.54554$
Correlation dimension: $D_2 = 1.462 \pm 0.054$ (converges slowly)
Ref: Whitehead and Macdonald (1984)

A.2.6 Holmes cubic map

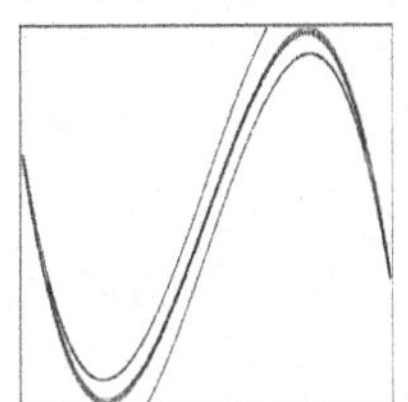

$X_{n+1} = Y_n$
$Y_{n+1} = -bX_n + dY_n - Y_n^3$
Usual parameters: $b = 0.2, d = 2.77$
Initial conditions: $X_0 = 1.6, Y_0 = 0$
Lyapunov exponents: $\lambda \simeq 0.59458, -2.20402$
Kaplan–Yorke dimension: $D_{\mathrm{KY}} \simeq 1.26977$
Correlation dimension: $D_2 = 1.260 \pm 0.039$
Ref: Holmes (1979)

A.2.7 Kaplan–Yorke map

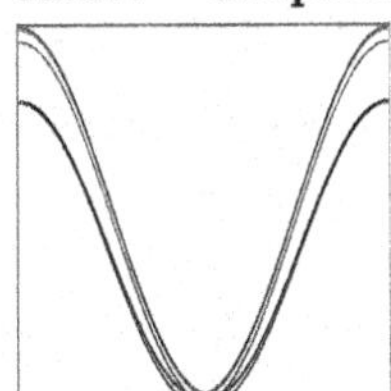

$X_{n+1} = aX_n \pmod 1$
$Y_{n+1} = bY_n + \cos 4\pi X_n \pmod 1$
Usual parameters: $a = 2, b = 0.2$ (see discussion in §2.5.2)
Initial conditions: $X_0 = 1/\sqrt{2}, Y_0 = -0.4$
Lyapunov exponents: $\lambda = \ln a = 0.693147181\ldots, \ln b = -1.609437912\ldots$
Kaplan–Yorke dimension: $D_{\mathrm{KY}} = 1 - \ln a / \ln b = 1.430676558\ldots$
Correlation dimension: $D_2 = 1.432 \pm 0.044$
Ref: Kaplan and Yorke (1979)

A.2.8 Dissipative standard map

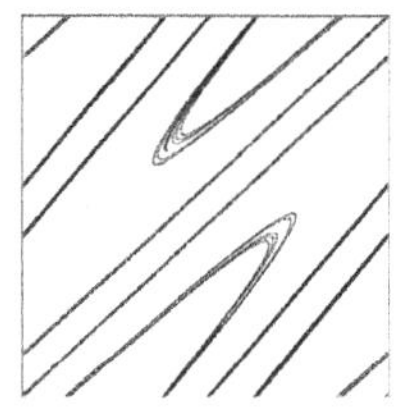

$X_{n+1} = X_n + Y_{n+1} \pmod{2\pi}$
$Y_{n+1} = bY_n + k \sin X_n \pmod{2\pi}$
Usual parameters: $b = 0.1, k = 8.8$
Initial conditions: $X_0 = 0.1, Y_0 = 0.1$
Lyapunov exponents: $\lambda \simeq 1.46995, -3.77254$
Kaplan–Yorke dimension: $D_{\mathrm{KY}} \simeq 1.38965$
Correlation dimension: $D_2 = 1.356 \pm 0.047$
Ref: Schmidt and Wang (1985)

A.2.9 Ikeda map

$X_{n+1} = \gamma + \mu(X_n \cos\phi - Y_n \sin\phi)$
$Y_{n+1} = \mu(X_n \sin\phi + Y_n \cos\phi)$
where $\phi = \beta - \alpha/(1 + X_n^2 + Y_n^2)$
Usual parameters: $\alpha = 6, \beta = 0.4, \gamma = 1, \mu = 0.9$
Initial conditions: $X_0 = 0, Y_0 = 0$
Lyapunov exponents: $\lambda \simeq 0.50760, -0.71832$
Kaplan–Yorke dimension: $D_{\mathrm{KY}} \simeq 1.70665$
Correlation dimension: $D_2 = 1.690 \pm 0.073$
Ref: Ikeda (1979)

A.2.10 Sinai map

$X_{n+1} = X_n + Y_n + \delta \cos 2\pi Y_n \pmod 1$
$Y_{n+1} = X_n + 2Y_n \pmod 1$
Usual parameter: $\delta = 0.1$
Initial conditions: $X_0 = 0.5, Y_0 = 0.5$
Lyapunov exponents: $\lambda \simeq 0.95946, -1.07714$
Kaplan–Yorke dimension: $D_{\text{KY}} \simeq 1.89075$
Correlation dimension: $D_2 = 1.779 \pm 0.063$
Ref: Sinai (1972)

A.2.11 Discrete predator–prey map

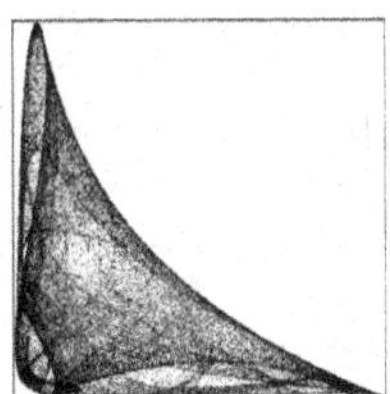

$X_{n+1} = X_n \exp[r(1 - X_n/K) - \alpha Y_n]$ (prey)
$Y_{n+1} = X_n[1 - \exp(-\alpha Y_n)]$ (predator)
Usual parameters: $r = 3, K = 1, \alpha = 5$
Initial conditions: $X_0 = 0.5, Y_0 = 0.5$
Lyapunov exponents: $\lambda \simeq 0.19664, 0.03276$
Kaplan–Yorke dimension: $D_{\text{KY}} = 2.0$ (exact value)
Correlation dimension: $D_2 = 1.903 \pm 0.079$ (converges slowly)
Ref: Beddington *et al.* (1975)

A.3 Conservative maps

A.3.1 Chirikov (standard) map

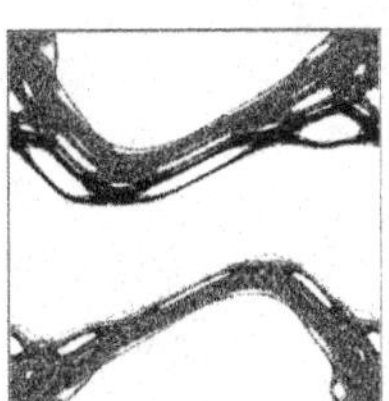

$X_{n+1} = X_n + Y_{n+1} \pmod{2\pi}$
$Y_{n+1} = Y_n + k \sin X_n \pmod{2\pi}$
Usual parameter: $k = 1$

Initial conditions: $X_0 = 0, Y_0 = 6$
Lyapunov exponents: $\lambda \simeq \pm 0.10497$
Kaplan–Yorke dimension: $D_{\mathrm{KY}} = 2.0$ (exact value)
Correlation dimension: $D_2 = 1.954 \pm 0.077$
Ref: Chirikov (1979)

A.3.2 Hénon area-preserving quadratic map

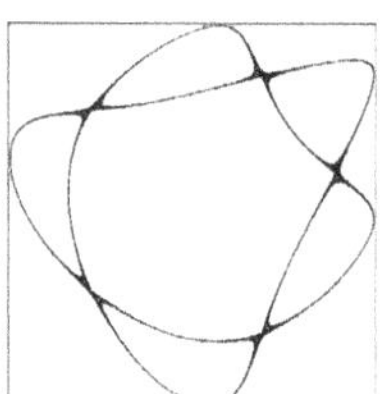

$X_{n+1} = X_n \cos\alpha - (Y_n - X_n^2)\sin\alpha$
$Y_{n+1} = X_n \sin\alpha + (Y_n - X_n^2)\cos\alpha$
Usual parameter: $\cos\alpha = 0.24$
Initial conditions: $X_0 = 0.6, Y_0 = 0.13$
Lyapunov exponents: $\lambda \simeq \pm 0.00643$
Kaplan–Yorke dimension: $D_{\mathrm{KY}} = 2.0$ (exact value)
Correlation dimension: $D_2 = 2.200 \pm 0.063$ (converges slowly)
Ref: Hénon (1969)

A.3.3 Arnold's cat map

$X_{n+1} = X_n + Y_n \pmod 1$
$Y_{n+1} = X_n + kY_n \pmod 1$
Usual parameter: $k = 2$
Initial conditions: $X_0 = 0, Y_0 = 1/\sqrt{2}$
Lyapunov exponents: $\lambda = \pm \ln[\frac{1}{2}(3 + \sqrt{5})] = \pm 0.96242365\ldots$
Kaplan–Yorke dimension: $D_{\mathrm{KY}} = 2.0$ (exact value)
Correlation dimension: $D_2 = 2.0$ (exact value)
Ref: Arnold and Avez (1968)

A.3.4 Gingerbreadman map

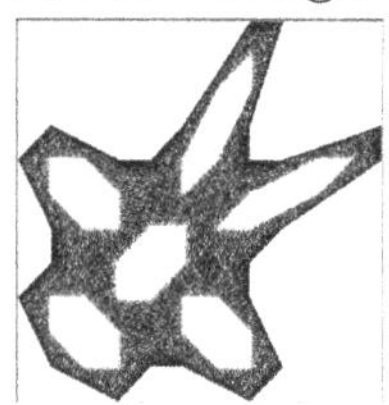

$X_{n+1} = 1 + |X_n| - Y_n$
$Y_{n+1} = X_n$
Initial conditions: $X_0 = 0.5, Y_0 = 3.7$
Lyapunov exponents: $\lambda \simeq \pm 0.07339$
Kaplan–Yorke dimension: $D_{\mathrm{KY}} = 2.0$ (exact value)
Correlation dimension: $D_2 = 2.171 \pm 0.078$ (converges slowly)
Ref: Devaney (1984)

A.3.5 Chaotic web map

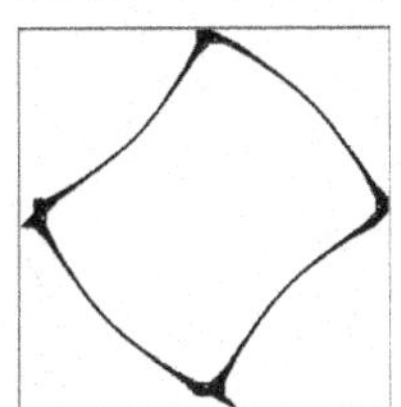

$X_{n+1} = X_n \cos\alpha - (Y_n + k \sin X_n) \sin\alpha$
$Y_{n+1} = X_n \sin\alpha + (Y_n + k \sin X_n) \cos\alpha$
Usual parameters: $\alpha = \pi/2, k = 1$
Initial conditions: $X_0 = 0, Y_0 = 3$
Lyapunov exponents: $\lambda \simeq \pm 0.04847$
Kaplan–Yorke dimension: $D_{\mathrm{KY}} = 2.0$ (exact value)
Correlation dimension: $D_2 = 1.779 \pm 0.059$ (converges slowly)
Ref: Chernikov *et al.* (1988)

A.3.6 Lorenz three-dimensional chaotic map

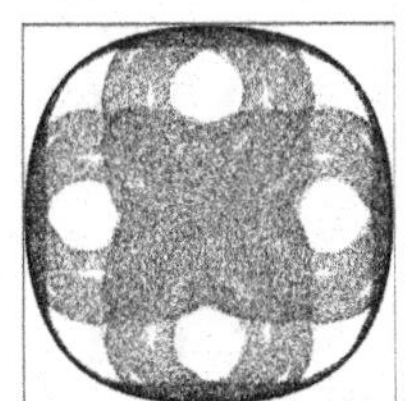

$X_{n+1} = X_n Y_n - Z_n$
$Y_{n+1} = X_n$
$Z_{n+1} = Y_n$
Initial conditions: $X_0 = 0.5, Y_0 = 0.5, Z_0 = -1$
Lyapunov exponents: $\lambda \simeq 0.07456, 0, -0.07456$
Kaplan–Yorke dimension: $D_{\mathrm{KY}} = 2.0$ (exact value)
Correlation dimension: $D_2 = 1.745 \pm 0.057$ (converges slowly)
Ref: Lorenz (1993)

A.4 Driven dissipative flows

A.4.1 Damped driven pendulum

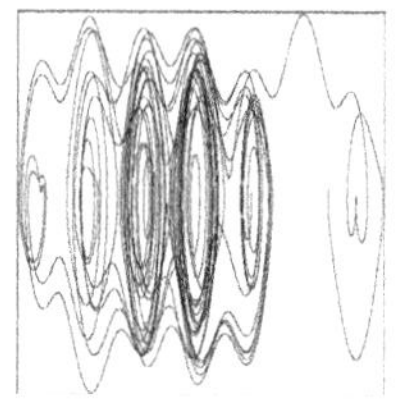

$dx/dt = y$
$dy/dt = -\sin x - by + A \sin \Omega t$
Usual parameters: $b = 0.05, A = 0.6, \Omega = 0.7$
Initial conditions: $x_0 = 0, y_0 = 0, t_0 = 0$
Lyapunov exponents: $\lambda \simeq 0.1414, 0, -0.1914$
Kaplan–Yorke dimension: $D_{\text{KY}} \simeq 2.7387$
Correlation dimension: $D_2 = 2.764 \pm 0.158$
Ref: Baker and Gollub (1996)

A.4.2 Driven van der Pol oscillator

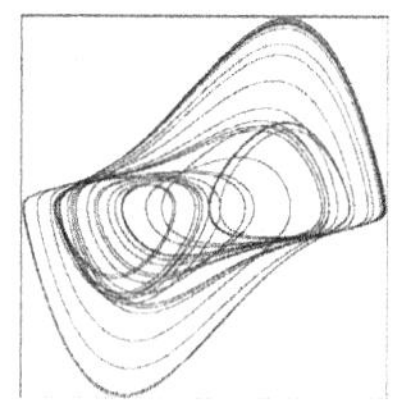

$dx/dt = y$
$dy/dt = -x + b(1 - x^2)y + A \sin \Omega t$
Usual parameters: $b = 3, A = 5, \Omega = 1.788$
Initial conditions: $x_0 = -1.9, y_0 = 0, t_0 = 0$
Lyapunov exponents: $\lambda \simeq 0.1933, 0, -2.0725$
Kaplan–Yorke dimension: $D_{\text{KY}} \simeq 2.0933$
Correlation dimension: $D_2 = 2.190 \pm 0.080$ (converges slowly)
Ref: van der Pol (1926)

A.4.3 Shaw–van der Pol oscillator

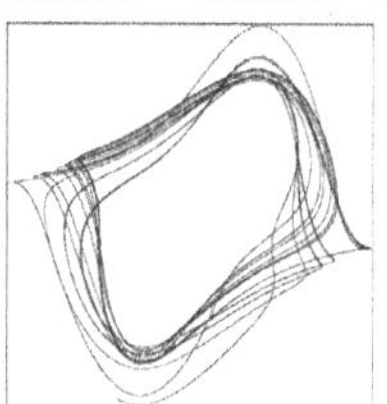

$dx/dt = y + A \sin \Omega t$
$dy/dt = -x + b(1 - x^2)y$
Usual parameters: $b = 1, A = 1, \Omega = 2$
Initial conditions: $x_0 = 1.3, y_0 = 0, t_0 = 0$

Lyapunov exponents: $\lambda \simeq 0.1180, 0, -1.2736$
Kaplan–Yorke dimension: $D_{\text{KY}} \simeq 2.0927$
Correlation dimension: $D_2 = 2.007 \pm 0.091$
Ref: Shaw (1981)

A.4.4 Forced Brusselator

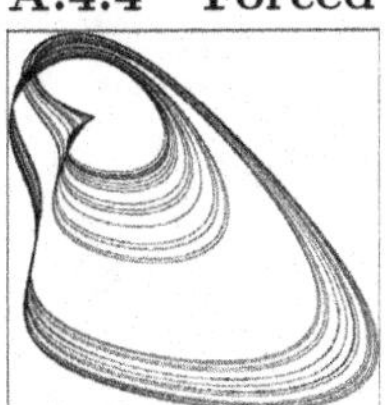

$dx/dt = x^2y - (b+1)x + a + A\sin\Omega t$
$dy/dt = -x^2y + bx$
Usual parameters: $a = 0.4, b = 1.2, A = 0.05, \Omega = 0.8$
Initial conditions: $x_0 = 0.3, y_0 = 2, t_0 = 0$
Lyapunov exponents: $\lambda \simeq 0.0140, 0, -0.2619$
Kaplan–Yorke dimension: $D_{\text{KY}} \simeq 2.0535$
Correlation dimension: $D_2 = 2.224 \pm 0.095$ (converges slowly)
Ref: Tomita and Kai (1978)

A.4.5 Ueda oscillator

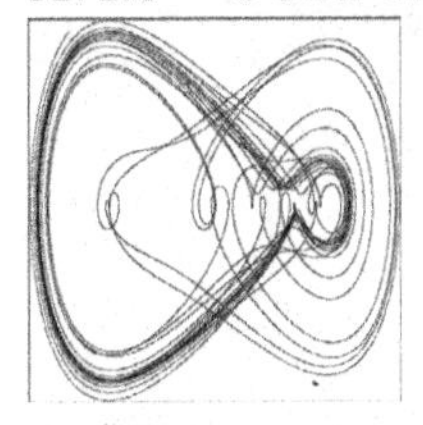

$dx/dt = y$
$dy/dt = -x^3 - by + A\sin\Omega t$
Usual parameters: $b = 0.05, A = 7.5, \Omega = 1$
Initial conditions: $x_0 = 2.5, y_0 = 0, t_0 = 0$
Lyapunov exponents: $\lambda \simeq 0.1034, 0, -0.1534$
Kaplan–Yorke dimension: $D_{\text{KY}} \simeq 2.6741$
Correlation dimension: $D_2 = 2.675 \pm 0.132$ (converges slowly)
Ref: Ueda (1979)

A.4.6 Duffing's two-well oscillator

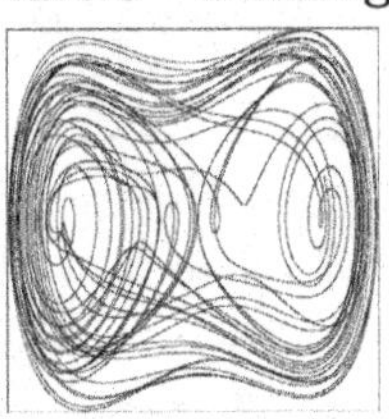

$dx/dt = y$

$dy/dt = -x^3 + x - by + A \sin \Omega t$
Usual parameters: $b = 0.25, A = 0.4, \Omega = 1$
Initial conditions: $x_0 = 0.2, y_0 = 0, t_0 = 0$
Lyapunov exponents: $\lambda \simeq 0.1572, 0, -0.4072$
Kaplan–Yorke dimension: $D_{\mathrm{KY}} \simeq 2.3860$
Correlation dimension: $D_2 = 2.334 \pm 0.114$
Ref: Moon and Holmes (1979)

A.4.7 Duffing–van der Pol oscillator

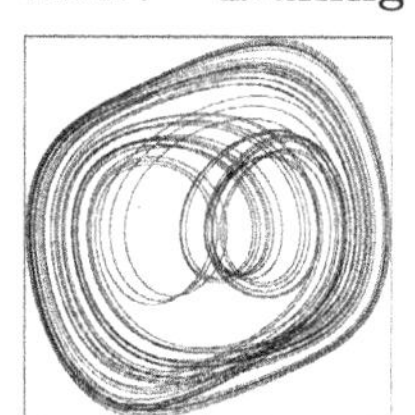

$dx/dt = y$
$dy/dt = \mu(1 - \gamma x^2)y - x^3 + A \sin \Omega t$
Usual parameters: $\mu = 0.2, \gamma = 8, A = 0.35, \Omega = 1.02$
Initial conditions: $x_0 = 0.2, y_0 = -0.2, t_0 = 0$
Lyapunov exponents: $\lambda \simeq 0.0963, 0, -0.2778$
Kaplan–Yorke dimension: $D_{\mathrm{KY}} \simeq 2.3467$
Correlation dimension: $D_2 = 2.333 \pm 0.115$ (converges slowly)
Ref: Ueda (1992)

A.4.8 Rayleigh–Duffing oscillator

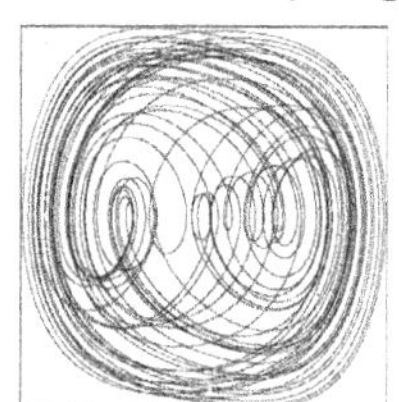

$dx/dt = y$
$dy/dt = \mu(1 - \gamma y^2)y - x^3 + A \sin \Omega t$
Usual parameters: $\mu = 0.2, \gamma = 4, A = 0.3, \Omega = 1.1$
Initial conditions: $x_0 = 0.3, y_0 = 0, t_0 = 0$
Lyapunov exponents: $\lambda \simeq 0.0912, 0, -0.2755$
Kaplan–Yorke dimension: $D_{\mathrm{KY}} \simeq 2.3310$
Correlation dimension: $D_2 = 2.194 \pm 0.120$
Ref: Hayashi *et al.* (1970)

A.5 Autonomous dissipative flows

A.5.1 Lorenz attractor

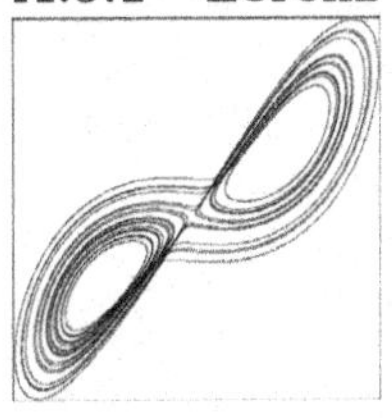

$dx/dt = \sigma(y - x)$
$dy/dt = -xz + rx - y$
$dz/dt = xy - bz$
Usual parameters: $\sigma = 10, r = 28, b = 8/3$
Initial conditions: $x_0 = 0, y_0 = -0.01, z_0 = 9$
Lyapunov exponents: $\lambda \simeq 0.9056, 0, -14.5723$
Kaplan–Yorke dimension: $D_{\text{KY}} \simeq 2.06215$
Correlation dimension: $D_2 = 2.068 \pm 0.086$
Ref: Lorenz (1963)

A.5.2 Rössler attractor

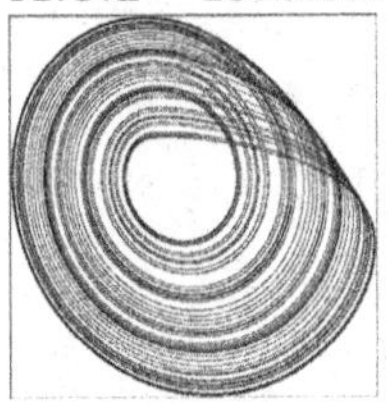

$dx/dt = -y - z$
$dy/dt = x + ay$
$dz/dt = b + z(x - c)$
Usual parameters: $a = b = 0.2, c = 5.7$
Initial conditions: $x_0 = -9, y_0 = 0, z_0 = 0$
Lyapunov exponents: $\lambda \simeq 0.0714, 0, -5.3943$
Kaplan–Yorke dimension: $D_{\text{KY}} \simeq 2.0132$
Correlation dimension: $D_2 = 1.991 \pm 0.065$ (converges slowly)
Ref: Rössler (1976)

A.5.3 Diffusionless Lorenz attractor

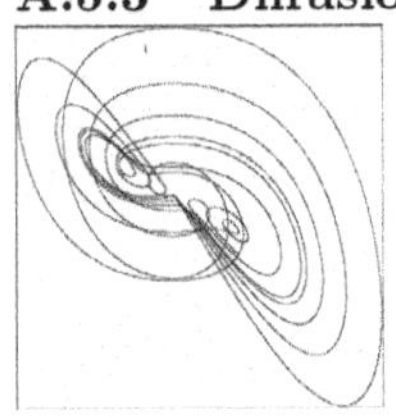

$dx/dt = -y - x$
$dy/dt = -xz$
$dz/dt = xy + R$

Usual parameter: $R = 1$
Initial conditions: $x_0 = 1, y_0 = -1, z_0 = 0.01$
Lyapunov exponents: $\lambda \simeq 0.2101, 0, -1.2101$
Kaplan–Yorke dimension: $D_{\mathrm{KY}} \simeq 2.1736$
Correlation dimension: $D_2 = 2.169 \pm 0.128$
Ref: van der Schrier and Maas (2000)

A.5.4 Complex butterfly

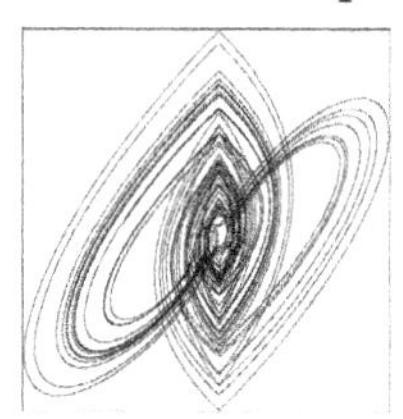

$dx/dt = a(y - z)$
$dy/dt = -z \operatorname{sgn} x$
$dz/dt = |x| - 1$
Usual parameter: $a = 0.55$
Initial conditions: $x_0 = 0.2, y_0 = 0, z_0 = 0$
Lyapunov exponents: $\lambda \simeq 0.1690, 0, -0.7190$
Kaplan–Yorke dimension: $D_{\mathrm{KY}} \simeq 2.2350$
Correlation dimension: $D_2 = 2.491 \pm 0.131$ (converges slowly)
Ref: Elwakil *et al.* (2002)

A.5.5 Chen's system

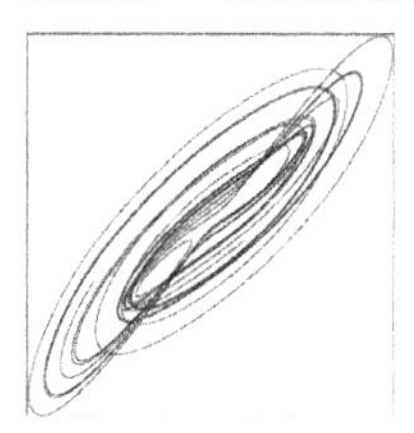

$dx/dt = a(y - x)$
$dy/dt = (c - a)x - xz + cy$
$dz/dt = xy - bz$
Usual parameters: $a = 35, b = 3, c = 28$
Initial conditions: $x_0 = -10, y_0 = 0, z_0 = 37$
Lyapunov exponents: $\lambda \simeq 2.0272, 0, -12.0272$
Kaplan–Yorke dimension: $D_{\mathrm{KY}} \simeq 2.1686$
Correlation dimension: $D_2 = 2.147 \pm 0.117$
Ref: Chen and Ueta (1999)

A.5.6 Hadley circulation

$dx/dt = -y^2 - z^2 - ax + aF$
$dy/dt = xy - bxz - y + G$
$dz/dt = bxy + xz - z$
Usual parameters: $a = 0.25, b = 4, F = 8, G = 1$
Initial conditions: $x_0 = 0, y_0 = 0, z_0 = 1.3$
Lyapunov exponents: $\lambda \simeq 0.1665, 0, -4.4466$
Kaplan–Yorke dimension: $D_{\text{KY}} \simeq 2.0374$
Correlation dimension: $D_2 = 2.162 \pm 0.114$
Ref: Lorenz (1984b)

A.5.7 ACT attractor

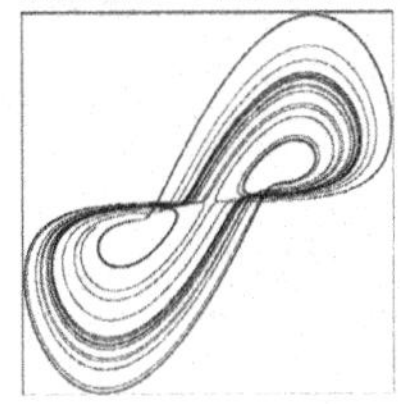

$dx/dt = \alpha(x - y)$
$dy/dt = -4\alpha y + xz + \mu x^3$
$dz/dt = -\delta\alpha z + xy + \beta z^2$
Usual parameters: $\alpha = 1.8, \beta = -0.07, \delta = 1.5, \mu = 0.02$
Initial conditions: $x_0 = 0.5, y_0 = 0, z_0 = 0$
Lyapunov exponents: $\lambda \simeq 0.1634, 0, -9.2060$
Kaplan–Yorke dimension: $D_{\text{KY}} \simeq 2.0177$
Correlation dimension: $D_2 = 2.039 \pm 0.106$
Ref: Arnéodo *et al.* (1981)

A.5.8 Rabinovich–Fabrikant attractor

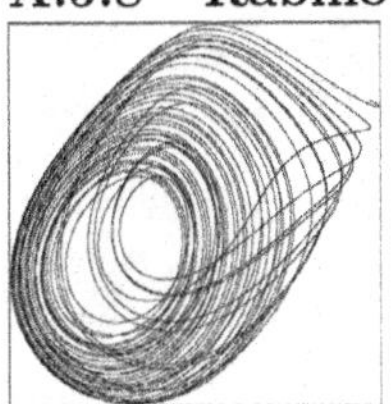

$dx/dt = y(z - 1 + x^2) + \gamma x$
$dy/dt = x(3z + 1 - x^2) + \gamma y$
$dz/dt = -2z(\alpha + xy)$
Usual parameters: $\gamma = 0.87, \alpha = 1.1$

Initial conditions: $x_0 = -1, y_0 = 0, z_0 = 0.5$
Lyapunov exponents: $\lambda \simeq 0.1981, 0, -0.6581$
Kaplan–Yorke dimension: $D_{\text{KY}} \simeq 2.3010$
Correlation dimension: $D_2 = 2.191 \pm 0.113$
Ref: Rabinovich and Fabrikant (1979)

A.5.9 Linear feedback rigid body motion system

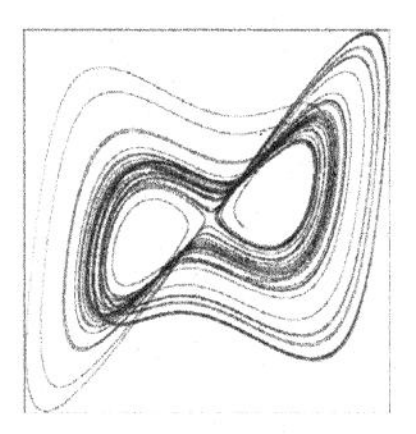

$dx/dt = -0.4x + y + 10yz$
$dy/dt = -x - 0.4y + 5xy$
$dz/dt = \alpha z - 5xy$
Usual parameter: $\alpha = 0.175$
Initial conditions: $x_0 = 0.6, y_0 = 0, z_0 = 0$
Lyapunov exponents: $\lambda \simeq 0.1421, 0, -0.7671$
Kaplan–Yorke dimension: $D_{\text{KY}} \simeq 2.1853$
Correlation dimension: $D_2 = 2.069 \pm 0.121$
Ref: Leipnik and Newton (1981)

A.5.10 Chua's circuit

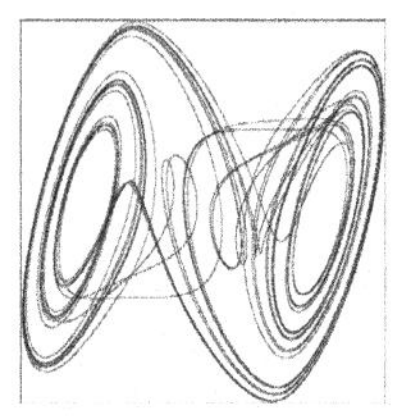

$dx/dt = \alpha[y - x + bx + \frac{1}{2}(a - b)(|x + 1| - |x - 1|)]$
$dy/dt = x - y + z$
$dz/dt = -\beta y$
Usual parameters: $\alpha = 9, \beta = 100/7, a = 8/7, b = 5/7$
Initial conditions: $x_0 = 0, y_0 = 0, z_0 = 0.6$
Lyapunov exponents: $\lambda \simeq 0.3271, 0, -2.5197$
Kaplan–Yorke dimension: $D_{\text{KY}} \simeq 2.1298$
Correlation dimension: $D_2 = 2.125 \pm 0.098$ (converges slowly)
Ref: Matsumoto *et al.* (1985)

A.5.11 Moore–Spiegel oscillator

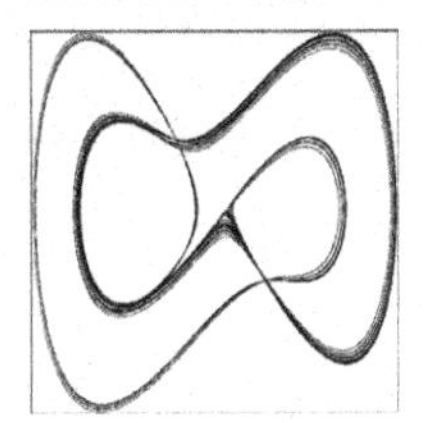

$dx/dt = y$
$dy/dt = z$
$dz/dt = -z - (T - R + Rx^2)y - Tx$
Usual parameters: $T = 6, R = 20$
Initial conditions: $x_0 = 0.1, y_0 = 0, z_0 = 0$
Lyapunov exponents: $\lambda \simeq 0.1119, 0, -1.1119$
Kaplan–Yorke dimension: $D_{\mathrm{KY}} \simeq 2.1006$
Correlation dimension: $D_2 = 2.309 \pm 0.107$ (converges slowly)
Ref: Moore and Spiegel (1966)

A.5.12 Thomas' cyclically symmetric attractor

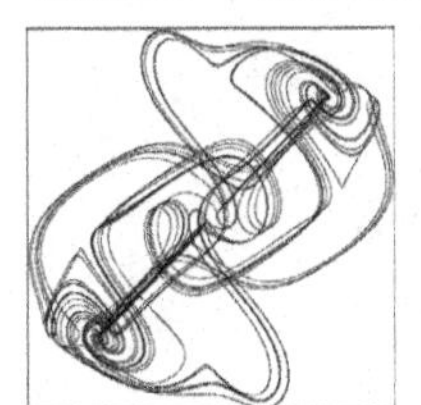

$dx/dt = -bx + \sin y$
$dy/dt = -by + \sin z$
$dz/dt = -bz + \sin x$
Usual parameter: $b = 0.18$
Initial conditions: $x_0 = 0.1, y_0 = 0, z_0 = 0$
Lyapunov exponents: $\lambda \simeq 0.0349, 0, -0.5749$
Kaplan–Yorke dimension: $D_{\mathrm{KY}} \simeq 2.0607$
Correlation dimension: $D_2 = 1.843 \pm 0.075$
Ref: Thomas (1999)

A.5.13 Halvorsen's cyclically symmetric attractor

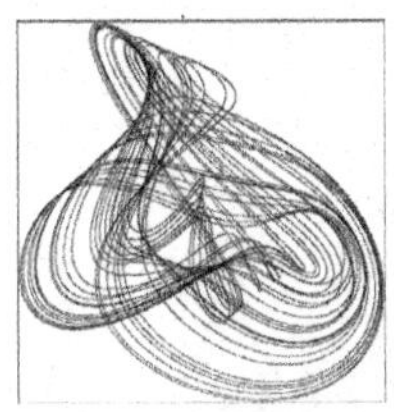

$dx/dt = -ax - 4y - 4z - y^2$
$dy/dt = -ay - 4z - 4x - z^2$
$dz/dt = -az - 4x - 4y - x^2$

Usual parameter: $a = 1.27$
Initial conditions: $x_0 = -5, y_0 = 0, z_0 = 0$
Lyapunov exponents: $\lambda \simeq 0.7899, 0, -4.5999$
Kaplan–Yorke dimension: $D_{\mathrm{KY}} \simeq 2.1717$
Correlation dimension: $D_2 = 2.110 \pm 0.095$
Ref: This text (§8.8.2)

A.5.14 Burke–Shaw attractor

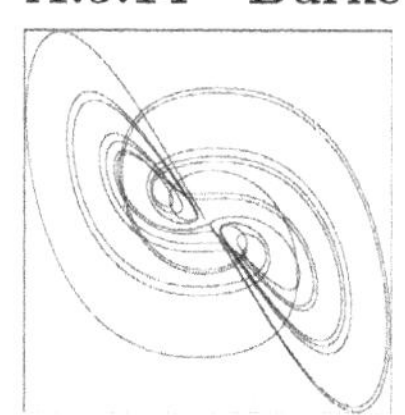

$dx/dt = -Ux - Uy$
$dy/dt = -Uxz - y$
$dz/dt = -Uxy + V$
Usual parameters: $U = 10, V = 13$
Initial conditions: $x_0 = 0.6, y_0 = 0, z_0 = 0$
Lyapunov exponents: $\lambda \simeq 2.2499, 0, -13.2499$
Kaplan–Yorke dimension: $D_{\mathrm{KY}} \simeq 2.1698$
Correlation dimension: $D_2 = 2.211 \pm 0.132$ (converges slowly)
Ref: Shaw (1981)

A.5.15 Rucklidge attractor

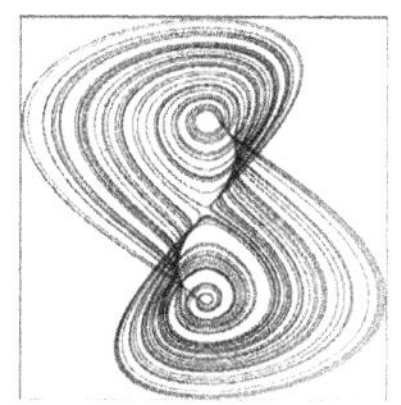

$dx/dt = -\kappa x + \lambda y - yz$
$dy/dt = x$
$dz/dt = -z + y^2$
Usual parameters: $\kappa = 2, \lambda = 6.7$
Initial conditions: $x_0 = 1, y_0 = 0, z_0 = 4.5$
Lyapunov exponents: $\lambda \simeq 0.0643, 0, -3.0643$
Kaplan–Yorke dimension: $D_{\mathrm{KY}} \simeq 2.0210$
Correlation dimension: $D_2 = 2.108 \pm 0.095$ (converges slowly)
Ref: Rucklidge (1992)

A.5.16 WINDMI attractor

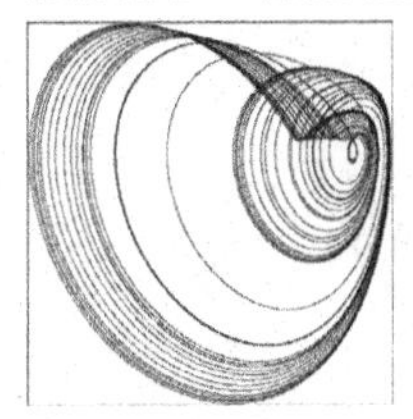

$dx/dt = y$
$dy/dt = z$
$dz/dt = -az - y + b - e^x$
Usual parameters: $a = 0.7, b = 2.5$
Initial conditions: $x_0 = 0, y_0 = 0.8, z_0 = 0$
Lyapunov exponents: $\lambda \simeq 0.0755, 0, -0.7755$
Kaplan–Yorke dimension: $D_{KY} \simeq 2.0974$
Correlation dimension: $D_2 = 2.035 \pm 0.095$ (converges slowly)
Ref: Horton *et al.* (2001)

A.5.17 Simplest quadratic chaotic flow

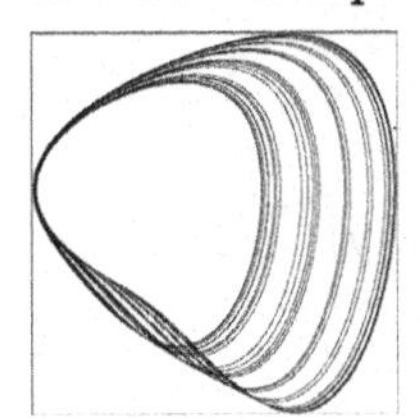

$dx/dt = y$
$dy/dt = z$
$dz/dt = -az + y^2 - x$
Usual parameter: $a = 2.017$
Initial conditions: $x_0 = -0.9, y_0 = 0, z_0 = 0.5$
Lyapunov exponents: $\lambda \simeq 0.0551, 0, -2.0721$
Kaplan–Yorke dimension: $D_{KY} \simeq 2.0266$
Correlation dimension: $D_2 = 2.187 \pm 0.075$ (converges slowly)
Ref: Sprott (1997a)

A.5.18 Simplest cubic chaotic flow

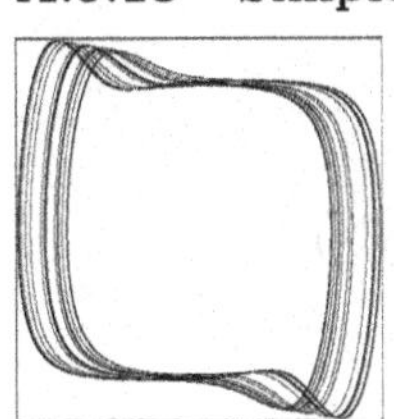

$dx/dt = y$
$dy/dt = z$
$dz/dt = -az + xy^2 - x$

Usual parameter: $a = 2.028$
Initial conditions: $x_0 = 0, y_0 = 0.96, z_0 = 0$
Lyapunov exponents: $\lambda \simeq 0.0837, 0, -2.1117$
Kaplan–Yorke dimension: $D_{\mathrm{KY}} \simeq 2.0396$
Correlation dimension: $D_2 = 2.174 \pm 0.083$ (converges slowly)
Ref: Malasoma (2000)

A.5.19 Simplest piecewise linear chaotic flow

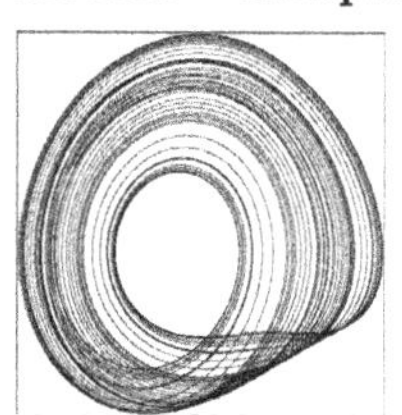

$dx/dt = y$
$dy/dt = z$
$dz/dt = -az - y + |x| - 1$
Usual parameter: $a = 0.6$
Initial conditions: $x_0 = 0, y_0 = -0.7, z_0 = 0$
Lyapunov exponents: $\lambda \simeq 0.0362, 0, -0.6362$
Kaplan–Yorke dimension: $D_{\mathrm{KY}} \simeq 2.0569$
Correlation dimension: $D_2 = 2.131 \pm 0.072$ (converges slowly)
Ref: Linz and Sprott (1999)

A.5.20 Double scroll

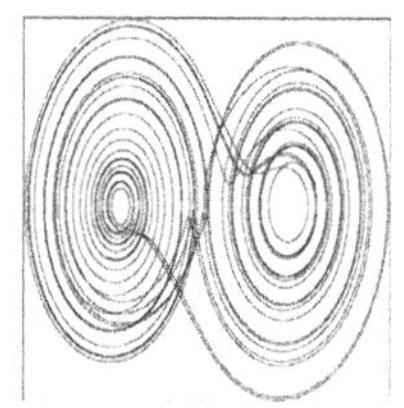

$dx/dt = y$
$dy/dt = z$
$dz/dt = -a[z + y + x - \operatorname{sgn} x]$
Usual parameter: $a = 0.8$
Initial conditions: $x_0 = 0.01, y_0 = 0.01, z_0 = 0$
Lyapunov exponents: $\lambda \simeq 0.0497, 0, -0.8497$
Kaplan–Yorke dimension: $D_{\mathrm{KY}} \simeq 2.0585$
Correlation dimension: $D_2 = 2.184 \pm 0.107$ (converges slowly)
Ref: Elwakil and Kennedy (2001)

A.6 Conservative flows

A.6.1 Driven pendulum

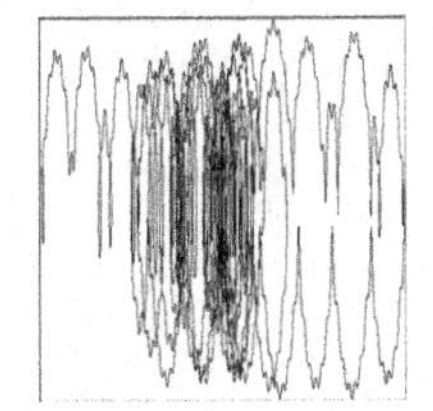

$dx/dt = y$
$dy/dt = -\sin x + A \sin \Omega t$
Usual parameters: $A = 1.0, \Omega = 0.5$
Initial conditions: $x_0 = 0, y_0 = 0, t_0 = 0$
Lyapunov exponents: $\lambda \simeq 0.1633, 0, -0.1633$
Kaplan–Yorke dimension: $D_{\mathrm{KY}} = 3.0$ (exact value)
Correlation dimension: $D_2 = 2.756 \pm 0.149$
Ref: This text (§8.5)

A.6.2 Simplest driven chaotic flow

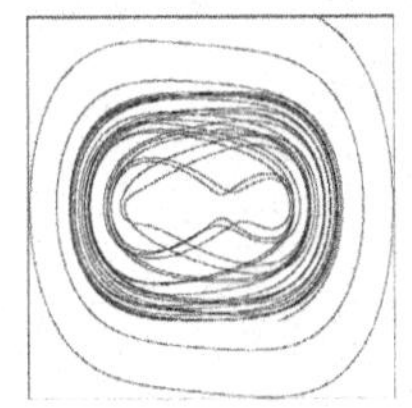

$dx/dt = y$
$dy/dt = -x^3 + \sin \Omega t$
Usual parameter: $\Omega = 1.88$
Initial conditions: $x_0 = 0, y_0 = 0, t_0 = 0$
Lyapunov exponents: $\lambda \simeq 0.0971, 0, -0.0971$
Kaplan–Yorke dimension: $D_{\mathrm{KY}} = 3.0$ (exact value)
Correlation dimension: $D_2 = 2.634 \pm 0.160$
Ref: This text (§8.6)

A.6.3 Nosé–Hoover oscillator

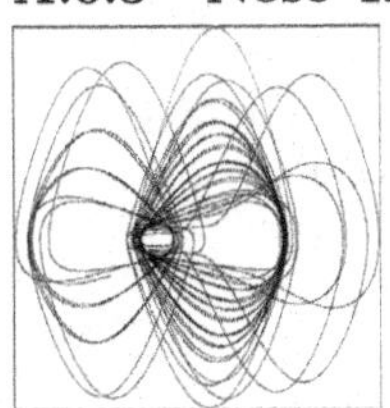

$dx/dt = y$
$dy/dt = -x + yz$
$dz/dt = a - y^2$
Usual parameter: $a = 1$

Initial conditions: $x_0 = 0, y_0 = 5, z_0 = 0$
Lyapunov exponents: $\lambda \simeq 0.0138, 0, -0.0138$
Kaplan–Yorke dimension: $D_{KY} = 3.0$ (exact value)
Correlation dimension: $D_2 = 2.521 \pm 0.146$ (converges slowly)
Refs: Nosé (1991), Hoover (1995)

A.6.4 Labyrinth chaos

$dx/dt = \sin y$
$dy/dt = \sin z$
$dz/dt = \sin x$
Initial conditions: $x_0 = 0.1, y_0 = 0, z_0 = 0$
Lyapunov exponents: $\lambda \simeq 0.1402, 0, -0.1402$
Kaplan–Yorke dimension: $D_{KY} \simeq 3.0$ (exact value)
Correlation dimension: $D_2 = 2.837 \pm 0.173$
Ref: Thomas (1999)

A.6.5 Hénon–Heiles system

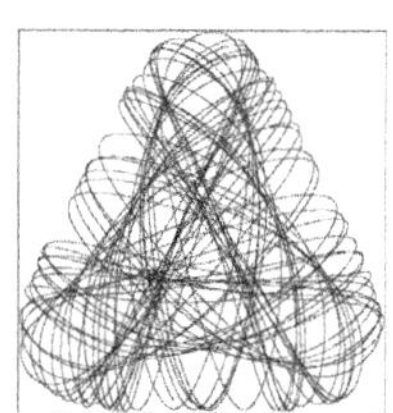

$dx/dt = v$
$dy/dt = w$
$dv/dt = -x - 2xy$
$dw/dt = -y - x^2 + y^2$
Initial conditions: $x_0 = 0.499, y_0 = 0, v_0 = 0, w_0 = 0.03160676\ldots$
Lyapunov exponents: $\lambda \simeq 0.0450, 0, 0, -0.0450$
Kaplan–Yorke dimension: $D_{KY} = 3.0$ (exact value)
Correlation dimension: $D_2 = 2.706 \pm 0.126$ (converges slowly)
Ref: Hénon and Heiles (1964)

B

Useful mathematical formulas

Below are some formulas that may be useful in the analysis of chaotic systems.

B.1 Trigonometric relations

$\sin\theta = -\sin(-\theta) = \cos(\pi/2 - \theta)$
$\cos\theta = \cos(-\theta) = \sin(\pi/2 - \theta)$
$\sin(\theta \pm \phi) = \sin\theta\cos\phi \pm \cos\theta\sin\phi$
$\cos(\theta \pm \phi) = \cos\theta\cos\phi \mp \sin\theta\sin\phi$
$\tan(\theta + \phi) = \frac{\tan\theta + \tan\phi}{1 - \tan\theta\tan\phi}$
$\sin 2\theta = 2\sin\theta\cos\theta = \frac{2\tan\theta}{1+\tan^2\theta}$
$\cos 2\theta = 2\cos^2\theta - 1 = \cos^2\theta - \sin^2\theta$
$\sin\theta\sin\phi = \frac{1}{2}[\cos(\theta - \phi) - \cos(\theta + \phi)]$
$\cos\theta\cos\phi = \frac{1}{2}[\cos(\theta + \phi) + \cos(\theta - \phi)]$
$\sin\theta\cos\phi = \frac{1}{2}[\sin(\theta + \phi) + \sin(\theta - \phi)]$
$\sin^2\theta = \frac{1}{2}(1 - \cos 2\theta)$
$\cos^2\theta = \frac{1}{2}(1 + \cos 2\theta)$
$\tan\theta = \sin\theta / \cos\theta$
$\cot\theta = 1/\tan\theta$
$\sec\theta = 1/\cos\theta$
$\csc\theta = 1/\sin\theta$
$\sin^2\theta + \cos^2\theta = 1$
$A\sin\theta \pm B\cos\theta = \sqrt{A^2 + B^2}\sin(\theta \pm \phi)$, where $\phi = \tan^{-1}(B/A)$
$A\cos\theta \mp B\sin\theta = \sqrt{A^2 + B^2}\cos(\theta \pm \phi)$, where $\phi = \tan^{-1}(B/A)$

B.2 Hyperbolic functions

$\sinh x = \frac{e^x - e^{-x}}{2} = -\sinh(-x) = \sqrt{\cosh^2 x - 1}$
$\cosh x = \frac{e^x + e^{-x}}{2} = \cosh(-x) = \sqrt{\sinh^2 x + 1}$
$\tanh x = \frac{e^x - e^{-x}}{e^x + e^{-x}} = \frac{e^{2x} - 1}{e^{2x} + 1} = 1 - \frac{2}{e^{2x} + 1} = -\tanh(-x) = \frac{\sinh x}{\cosh x}$
$\coth x = \frac{1}{\tanh x} = -\coth(-x) = \sqrt{\mathrm{csch}^2 x + 1}$
$\mathrm{sech}\ x = \frac{1}{\cosh x} = \mathrm{sech}(-x) = \sqrt{1 - \tanh^2 x}$

$\operatorname{csch} x = \frac{1}{\sinh x} = -\operatorname{csch}(-x) = \sqrt{\coth^2 x - 1}$
$\sinh^{-1} x = \ln(x + \sqrt{x^2+1}) = \cosh^{-1} \sqrt{x^2+1}$
$\cosh^{-1} x = \ln(x + \sqrt{x^2-1}) = \sinh^{-1} \sqrt{x^2-1}$
$\tanh^{-1} x = \frac{\ln(1+x)-\ln(1-x)}{2}$
$\coth^{-1} x = \frac{\ln(1+x)-\ln(x-1)}{2}$
$\operatorname{sech}^{-1} x = \ln\left(\frac{1}{x} + \sqrt{\frac{1}{x^2} - 1}\right)$
$\operatorname{csch}^{-1} x = \ln\left(\frac{1}{x} + \sqrt{\frac{1}{x^2} + 1}\right)$

B.3 Logarithms

$\log_a x = \log_b x / \log_b a$
$\log(1/x) = -\log x$
$\log(xy) = \log x + \log y$
$\log(x/y) = \log x - \log y$
$\log x^n = n \log x$
$\log_a a = 1$
$\ln \mathrm{e}^x = x$
$a^{\log_a x} = x$
$\log 0 = -\infty$
$\log 1 = 0$
$\ln e = 1$
$\ln(-1) = i\pi$
$\ln(\pm i) = \pm\frac{1}{2}\pi i$

B.4 Complex numbers

$i = \sqrt{-1},\ i^2 = -1,\ 1/i = -i$
$\mathrm{e}^{i\pi/2} = i,\ \mathrm{e}^{i\pi} = -1,\ \mathrm{e}^{3i\pi/2} = -i,\ \mathrm{e}^{2\pi i} = 1$
$\mathrm{e}^{i\theta} = \cos\theta + i\sin\theta$ (Euler's equation)
$\sin\theta = \frac{1}{2i}(\mathrm{e}^{i\theta} - \mathrm{e}^{-i\theta})$
$\cos\theta = \frac{1}{2}(\mathrm{e}^{i\theta} + \mathrm{e}^{-i\theta})$
$\sin i\theta = i\sinh\theta$
$\cos i\theta = \cosh\theta$
$\sinh i\theta = i\sin\theta$
$\cosh i\theta = \cosh\theta$

These formulas hold for $Z = A + Bi$, $\bar{Z} = A - Bi$, and $\theta = \tan^{-1}(B/A)$:
$|Z| = \sqrt{A^2 + B^2}$ (modulus)
$|Z|^2 = Z\bar{Z} = A^2 + B^2$
$Z = |Z|\mathrm{e}^{i\theta} = |Z|(\cos\theta + i\sin\theta)$
$\frac{1}{Z} = \frac{\bar{Z}}{|Z|^2} = \frac{A-Bi}{A^2+B^2}$
$Z^n = |Z|^n(\cos n\theta + i\sin n\theta)$ (DeMoivre's theorem)
$Z^2 = (A^2 - B^2) + 2iAB$
$Z^3 = (A^3 - 3AB^2) + i(3A^2B - B^3)$

$Z^4 = (A^4 - 6A^2B^2 + B^4) + i(4A^3B - 4AB^3)$
$Z^5 = (A^5 - 10A^3B^2 + 5AB^4) + i(5A^4B - 10A^2B^3 + B^5)$
$e^Z = e^A(\cos B + i\sin B)$
$\ln Z = \ln|Z| + i(\theta + 2\pi N)$, where N is an integer
$\text{Ln } Z = \ln|Z| + i\theta$ (principal value)
$\sin Z = \sin A\cosh B + i\cos A\sinh B$
$\cos Z = \cos A\cosh B - i\sin A\sinh B$
$\sinh Z = \sinh A\cos B + i\cosh A\sin B$
$\cosh Z = \cosh A\cos B + i\sinh A\sin B$

B.5 Derivatives

$d(af) = adf$
$d(f+g) = df + dg$
$d(fg) = fdg + gdf$
$d(f/g) = \frac{gdf - fdg}{g^2}$
$d(af^n) = naf^{n-1}df$
$d(e^{af}) = ae^{af}df$
$d(a^f) = a^f \ln adf$
$d(\ln f) = df/f$
$d\sin f = \cos fdf$
$d\cos f = -\sin fdf$
$d\tan f = \frac{df}{\cos^2 f}$
$d\cot f = -\csc^2 fdf$
$d\sec f = \tan f\sec fdf$
$d\csc f = -\cot f\csc fdf$
$d\sin^{-1} f = (1-f^2)^{-1/2}df$
$d\cos^{-1} f = -(1-f^2)^{-1/2}df$
$d\tan^{-1} f = \frac{df}{1+f^2}$
$d\sinh f = \cosh fdf$
$d\cosh f = \sinh fdf$
$d\tanh f = \text{sech}^2 fdf$
$d\coth f = -\text{csch}^2 fdf$
$d\ \text{sech}\ f = -\text{sech} f\tanh fdf$
$d\ \text{csch}\ f = -\text{csch} f\coth fdf$
$d\sinh^{-1} f = (f^2+1)^{-1/2}df$
$d\cosh^{-1} f = (f^2-1)^{-1/2}df$
$d\tanh^{-1} f = (1-f^2)^{-1}df$

B.6 Integrals

$\int x^n dx = \frac{x^{n+1}}{n+1} + C$
$\int e^{ax}dx = e^{ax}/a + C$
$\int \frac{1}{x}dx = \ln x + C$
$\int \ln xdx = x\ln x - x + C$
$\int \sin xdx = -\cos x + C$

$\int \cos x dx = \sin x + C$
$\int \tan x dx = -\ln(\cos x) + C$
$\int x \sin x dx = \sin x - x \cos x + C$
$\int x \cos x dx = \cos x + x \sin x + C$
$\int \sin x \cos x dx = \frac{1}{2} \sin^2 x + C$
$\int \sin^2 x dx = \frac{1}{2}x - \frac{1}{4} \sin 2x + C$
$\int \cos^2 x dx = \frac{1}{2}x + \frac{1}{4} \sin 2x + C$
$\int \tanh x dx = \ln(\cosh x) + C$

B.7 Approximations

These formulas are valid for $|x| \ll 1$:
$f(a+x) \simeq f(a) + xf'(a)$ (Taylor series)
$(1+x)^n \simeq 1 + nx$
$\sin x \simeq x$
$\cos x \simeq 1 - x^2/2$
$\tan x \simeq x$
$\sin^{-1} x \simeq x$
$\cos^{-1} x \simeq \frac{\pi}{2} - x$
$\tan^{-1} x \simeq x$
$\sinh x \simeq x$
$\cosh x \simeq 1 + x^2/2$
$\tanh x \simeq x$
$\sinh^{-1} x \simeq x$
$\cosh^{-1} x \simeq \ln(x + i)$
$\tanh^{-1} x \simeq x$
$e^x \simeq 1 + x$
$a^x \simeq x \ln a$
$\ln(1+x) \simeq x$

B.8 Matrices and determinants

These formulas hold for two 2×2 matrices $A = \begin{pmatrix} a & b \\ c & d \end{pmatrix}$ and $B = \begin{pmatrix} e & f \\ g & h \end{pmatrix}$:

$A + B = \begin{pmatrix} a+e & b+f \\ c+g & d+h \end{pmatrix}$
$rA = \begin{pmatrix} ra & rb \\ rc & rd \end{pmatrix}$
$AB = \begin{pmatrix} ae+bg & af+bh \\ ce+dg & cf+dh \end{pmatrix} \neq BA$
$\text{trace } A = a + d$
$\det A = ad - bc$
$\det(AB) = (\det A)(\det B)$
$\det(A^{-1}) = 1/\det A$
$A^{-1} = \frac{1}{ad-bc} \begin{pmatrix} d & -b \\ -c & a \end{pmatrix}$ (if $ad \neq bc$)

eigenvalues of A: $\lambda = \frac{1}{2}(a+d) \pm \sqrt{(a-d)^2 + 4bc}$
$\lambda = \frac{1}{2}[\text{trace } A \pm \sqrt{(\text{trace } A)^2 - 4 \det A}]$
$\lambda_1 + \lambda_2 = \text{trace } A$

This formula holds for a 3×3 matrix $A = \begin{pmatrix} a & b & c \\ d & e & f \\ g & h & j \end{pmatrix}$:

$$\det A = a\begin{pmatrix} e & f \\ h & j \end{pmatrix} - d\begin{pmatrix} b & c \\ h & j \end{pmatrix} + g\begin{pmatrix} b & c \\ e & f \end{pmatrix} = aej + bfg + cdh - ceg - bdj - afh$$

This formula holds for a 4×4 matrix $A = \begin{pmatrix} a & b & c & d \\ e & f & g & h \\ j & k & l & m \\ n & o & p & q \end{pmatrix}$:

$$\det A = a\begin{pmatrix} f & g & h \\ k & l & m \\ o & p & q \end{pmatrix} - e\begin{pmatrix} b & c & d \\ k & l & m \\ o & p & q \end{pmatrix} + j\begin{pmatrix} b & c & d \\ f & g & h \\ o & p & q \end{pmatrix} - n\begin{pmatrix} b & c & d \\ f & g & h \\ k & l & m \end{pmatrix}$$

B.9 Roots of polynomials

B.9.1 Linear systems

$f(x) = ax + b = 0$
$x = -b/a$

B.9.2 Quadratic systems

$f(x) = ax^2 + bx + c = 0$
$x = \frac{-b \pm \sqrt{b^2 - 4ac}}{2a}$ (roots are real if $b^2 \geq 4ac$)

B.9.3 Cubic systems

$f(x) = ax^3 + bx^2 + cx + d = 0$
Let $x = y - \frac{b}{3a}$ to get $y^3 + py + q = 0$,
where $p = \frac{1}{3a^2}(3ac - b^2)$, $q = \frac{1}{27a^3}(2b^3 - 9abc + 27a^2 d)$
Let $A = \sqrt[3]{-\frac{q}{2} + \sqrt{\frac{q^2}{4} + \frac{p^3}{27}}}$, $B = \sqrt[3]{-\frac{q}{2} - \sqrt{\frac{q^2}{4} + \frac{p^3}{27}}}$
$y = A + B, -\frac{A+B}{2} \pm \frac{A-B}{2}\sqrt{-3}$

B.9.4 Quartic systems

$f(x) = ax^4 + bx^3 + cx^2 + dx + e = 0$
Let $x = y - \frac{b}{4a}$ to get $y^4 + py^2 + qy + r = 0$,
where $p = \frac{1}{8a^2}(8ac - 3b^2)$, $q = \frac{1}{8a^3}(8a^2 d - 4abc + b^2)$,
$r = \frac{1}{256a^4}(256a^3 e - 64a^2 bd + 16ab^2 c - 3b^4)$
Find one root of $64u^6 + 32pu^4 + 4(p^2 - 4r)u^2 - q = 0$ (cubic in u^2)

$y = -u \pm \sqrt{-u^2 - p/2 \pm q/4u}$

B.9.5 Newton–Raphson method

for $f(x) = ax^k + bx^{k-1} + \cdots = 0$
(Successive approximation of real roots)
$\bar{x} = x - \frac{f(x)}{f'(x)} = x - \frac{ax^k + bx^{k-1} + cx^{k-2} + \cdots}{akx^{k-1} + b(k-1)x^{k-2} + \cdots}$

B.10 Vector calculus

These formulas hold for a scalar function f and a vector function $\mathbf{A}$:

$\nabla = \hat{x}\frac{\partial}{\partial x} + \hat{y}\frac{\partial}{\partial y} + \hat{z}\frac{\partial}{\partial z}$

$\nabla f = \hat{x}\frac{\partial f}{\partial x} + \hat{y}\frac{\partial f}{\partial y} + \hat{z}\frac{\partial f}{\partial z}$

$\nabla^2 = \frac{\partial^2}{\partial x^2} + \frac{\partial^2}{\partial y^2} + \frac{\partial^2}{\partial z^2}$

$\nabla^2 f = \frac{\partial^2 f}{\partial x^2} + \frac{\partial^2 f}{\partial y^2} + \frac{\partial^2 f}{\partial z^2}$

$\nabla \cdot \mathbf{A} = \frac{\partial A_x}{\partial x} + \frac{\partial A_y}{\partial y} + \frac{\partial A_z}{\partial z}$

$\nabla \cdot (f\mathbf{A}) = \mathbf{A} \cdot \nabla f + f \nabla \cdot \mathbf{A}$

$$\nabla \times \mathbf{A} = \begin{pmatrix} \hat{x} & \hat{y} & \hat{z} \\ \frac{\partial}{\partial x} & \frac{\partial}{\partial y} & \frac{\partial}{\partial z} \\ A_x & A_y & A_z \end{pmatrix} = \hat{x}\left(\frac{\partial A_y}{\partial z} - \frac{\partial A_z}{\partial y}\right) + \hat{y}\left(\frac{\partial A_z}{\partial x} - \frac{\partial A_x}{\partial z}\right) + \hat{z}\left(\frac{\partial A_x}{\partial y} - \frac{\partial A_y}{\partial x}\right)$$

$\nabla \times (f\mathbf{A}) = \nabla f \times \mathbf{A} + f \nabla \times \mathbf{A}$

$\nabla \times (\nabla f) = 0$

$\nabla \times (\nabla \times \mathbf{A}) = \nabla(\nabla \cdot \mathbf{A}) - (\nabla \cdot \nabla)\mathbf{A}$

$\nabla \cdot (\nabla \times \mathbf{A}) = 0$

$(\mathbf{A} \cdot \nabla)\mathbf{A} = \nabla(\mathbf{A}^2/2) - \mathbf{A} \times (\nabla \times \mathbf{A})$

C

Journals with chaos and related papers

Chaos and related papers appear in many journals, mostly in physics. Listed below are some of the more common such journals. Links to their Web pages can be found at http://sprott.physics.wisc.edu/chaostsa/journals.htm.

- *Advances in Complex Systems*
- *American Journal of Physics*
- *American Scientist*
- *Annals of the New York Academy of Sciences*
- *Chaos*
- *Chaos, Solitons, and Fractals*
- *Communications in Mathematical Physics*
- *Complex Systems*
- *Complexity*
- *Complexity International*
- *Computers and Graphics*
- *Computers in Physics*
- *Europhysics Letters*
- *Fractals*
- *IEEE Transactions on Circuits and Systems*
- *Interjournal of Complex Systems*
- *International Journal of Bifurcation and Chaos*
- *International Journal of Chaos Theory and Applications*
- *Journal of Mathematical Physics*
- *Journal of Nonlinear Science*
- *Journal of the Franklin Institute*
- *Journal of Statistical Physics*
- *Nature*
- *Nonlinear Dynamics and Systems Theory*
- *Nonlinear Dynamics, Psychology, & Life Sciences*
- *Nonlinear Science Today*

- *Nonlinearity*
- *Physica D*
- *Physical Review E* (formerly *A*)
- *Physical Review Letters*
- *Physics Letters A*
- *Physics Today*
- *Progress of Theoretical Physics*
- *Regular and Chaotic Dynamics*
- *Reviews of Modern Physics*
- *Science*
- *Scientific American*

Bibliography

Abarbanel, H. D. I. (1996). *Analysis of observed chaotic data.* Springer, New York.

Abraham, R. H. and Shaw, C. D. (1988). *Dynamics: the geometry of behavior. Part 4: bifurcation behavior.* Aerial Press, Santa Cruz, CA.

Abraham, R. and Ueda, Y. (2000). *The chaos avant-garde: memories of the early days of chaos theory.* World Scientific, Singapore.

Abramowitz, M. and Stegun, I. A. (1965). *Handbook of mathematical functions.* Dover, London.

Acton, F. S. (1970). *Numerical methods that work.* Harper and Row, New York.

Addison, P. S. (1995). On the characterization of non-linear oscillator systems in chaotic mode. *Journal of Sound Vibrations* **179**, 385–98.

Aks, D. J. and Sprott, J. C. (1996). Quantifying aesthetic preference for chaotic patterns. *Empirical Studies of the Arts* **14**, 1–16.

Aks, D. J. and Sprott, J. C. (2003). The role of depth and $1/f$ dynamic in perceiving reversible figures. *Nonlinear Dynamics in Psychology and Life Sciences* **7**, 161–80.

Aks, D. J., Zelinsky, G., and Sprott, J. C. (2002). Memory across eye movements: $1/f$ dynamic in visual search. *Nonlinear Dynamics in Psychology and Life Sciences* **6**, 1–25.

Albers, D. J., Sprott, J. C., and Dechert, W. D. (1998). Routes to chaos in neural networks with random weights. *International Journal of Bifurcation and Chaos* **8**, 1463–78.

Albinet, G., Searby, G., and Stauffer, D. (1986). Fire propagation in a 2-D random medium. *Journal de Physique (Paris)* **47**, 1–7.

Alexander, J. C., Yorke, J. A., You, Z., and Kan, I. (1992). Riddled basins. *International Journal of Bifurcation and Chaos* **2**, 795–813.

Amaral, L. A. N., Buldyrev, S. V., Havlin, S., Salinger, M. A., and Stanley, H. E. (1998). Power law scaling for a system of interacting units with complex internal structure. *Physical Review Letters* **80**, 1385–8.

Amaral, L. A. N., Ivanov, P. C., Aoyagi, N., Hidaka, I., Tomono, S., Goldberger, A. L., Stanley, H. E., and Yamamoto, Y. (2001). Behavioral-independent features of complex heartbeat dynamics. *Physical Review Letters* **86**, 6026–9.

Ambravaneswaran, B., Phillips, S. D., and Basaran, O. A. (2000). Theoretical analysis of a dripping faucet. *Physical Review Letters* **85**, 5332–5.

Anosov, D. (1962). Roughness of geodesic flows on compact Riemannian manifolds of negative curvature. *Soviet Mathematics Doklady* **3**, 1068–9.

Anosov, D. (1963). Ergodic properties of geodesic flows on closed Riemannian manifolds of negative curvature. *Soviet Mathematics Doklady* **4**, 1153–6.

Appleby, S. (1996). Multifractal characterization of the distribution pattern of the human population. *Geographical Analysis* **28**, 21–31.

Arfken, G. (1985). *Mathematical methods for physicists* (2nd edn). Academic Press, Orlando, FL.

Arnéodo, A., Coullet, P., and Tresser, C. (1981). A possible new mechanism for the onset of turbulence. *Physics Letters A* **81**, 197–201.

Arnéodo, A., Grasseau, G., and Kostelich, E. J. (1987). Fractal dimensions and f(alpha) spectrum of the Hénon attractor. *Physics Letters A* **124**, 426–32.

Arnold, V. I. (1963). Small denominators and problems of stability of motion in classical and celestial mechanics. *Russian Mathematical Surveys* **18** (6), 85–189.

Arnold, V. I. (1964). Instability of dynamical systems with several degrees of freedom. *Soviet Mathematics Doklady* **5**, 581–5.

Arnold, V. I. (1965). Small denominators, I: mappings of the circumference into itself. *American Mathematical Society Translation Series 2* **46**, 213–84.

Arnold, V. I. (1978). *Mathematical methods of classical mechanics* (2nd edn). Springer, New York.

Arnold, V. I. (1983). *Geometrical methods in the theory of ordinary differential equations.* Springer, New York.

Arnold, V. I. and Avez, A. (1968). *Ergodic problems of classical mechanics.* Benjamin, New York.

Aronson, D. G., Chory, M. A., Hall, G. R., and McGehee, R. P. (1982). Bifurcations from an invariant circle for two-parameter families of maps of the plane: a computer-assisted study. *Communications in Mathematical Physics* **83**, 304–54.

Arthur, D. W. G. (1954). The distribution of lunar craters. *Journal of the British Astronomical Association* **64**, 127–32.

Aubry, S. (1983). The twist map, the extended Frenkel–Kontrova model and the devil's staircase. *Physica D* **7**, 240–58.

Auerbach, D., Cvitanović, P., Eckmann, J. -P., Gunaratne, G. H., and Procaccia, I. (1987). Exploring chaotic motion through periodic orbits. *Physical Review Letters* **58**, 2387–9.

Babloyantz, A. and Destexhe, A. (1986). Low dimensional chaos in an instance of epilepsy. *Proceedings of the National Academy of Sciences* (USA) **83**, 3513–17.

Badii, R. and Politi, A. (1984). Intrinsic oscillations in measuring the fractal dimension. *Physics Letters A* **104**, 303–5.

Badii, R. and Politi, A. (1985). Statistical description of chaotic attractors. *Journal of Statistical Physics* **40**, 725–50.

Badii, R. and Politi, A. (1986). On the fractal dimension of filtered chaotic signals. In *Dimensions and entropies in chaotic systems* (ed. G. Mayer-Kress). Springer, Berlin.

Badii, R. and Politi, A. (1997). *Complexity: hierarchical structures and scaling in physics.* Cambridge University Press.

Badii, R., Broggi, G., Derighetti, B., Ravani, M., Ciliberto, S., Politi, A., and Rubbio, M. A. (1988). Dimension increase in filtered chaotic signals. *Physical Review Letters* **60**, 979–82.

Baillieul, J., Brockett, R. W., and Washburn, R. B. (1980). Chaotic motion in nonlinear feedback systems. *IEEE Transactions on Circuits and Systems* **CS-27**, 990–7.

Bak, P. (1986). The devil's staircase. *Physics Today* **39**, 38–45.

Bak, P. (1996). *How nature works: the science of self-organized criticality.* Corpernicus, New York.

Bak, P. and Chen, K. (2001). Scale dependent dimension of luminous matter in the Universe. *Physical Review Letters* **86**, 4215–18.

Bak, P. and Sneppen, K. (1993). Punctuated equilibrium and criticality in a simple model of evolution. *Physical Review Letters* **24**, 4083–6.

Bak, P. and Tang, C. (1989). Earthquakes as a self-organized critical phenomenon. *Journal of Geophysical Research* **94**, 15635–7.

Bak, P., Tang, C., and Wiesenfeld, K. (1987). Self-organized criticality: an explanation of $1/f$ noise. *Physical Review Letters* **59**, 381–4.

Bak, P., Tang, C., and Wiesenfeld, K. (1988). Self-organized criticality. *Physical Review A* **38**, 364–74.

Bak, P., Chen, K., and Tang, C. (1990). A forest-fire model and some thoughts on turbulence. *Physics Letters A* **147**, 297–300.

Baker, G. L. and Gollub, J. P. (1996). *Chaotic dynamics: an introduction* (2nd edn). Cambridge University Press.

Banerjee, S., Yorke, J. A., and Grebogi, C. (1998). Robust Chaos. *Physical Review Letters* **80**, 3049–52.

Barkai, E. and Klafter, J. (1997). Crossover from dispersive to regular transport in biased maps. *Physical Review Letters* **79**, 2245–8.

Barnsley, M. (1988). *Fractals everywhere.* Academic Press, Boston.

Barnsley, M. F. and Hurd, L. P. (1993). *Fractal image compression.* A. K. Peters, Wellesley, MA.

Barreto, E., Hunt, B. R., Grebogi, C., and Yorke, J. A. (1997). From high dimensional chaos to stable periodic orbits: the structure of parameter space. *Physical Review Letters* **78**, 4561–4.

Barrow-Green, J. (1997). *Poincaré and the three-body problem.* American Mathematical Society, Providence, RI.

Beck, C. and Schlögl, F. (1995). *Thermodynamics of chaotic systems.* Cambridge University Press, New York.

Beddington, J. R., Free, C. A., and Lawton, J. H. (1975). Dynamic complexity in predator–prey models framed in difference equations. *Nature* **255**, 58–60.

Benedicks, M. and Carleson, L. (1991). The dynamics of the Hénon map. *Annals of Mathematics* **133**, 73–169.

Benettin, G., Cercignanni, C., Galgani, L., and Giorgilli, A. (1980a). Universal properties in conservative dynamical systems. *Lettere al Nuovo Cimento* **28**, 1–4.

Benettin, G., Galgani, L., Giorgilli, A., and Strelcyn, J. (1980b). Lyapunov characteristic exponents for smooth dynamical systems and for Hamiltonian systems: a method for computing all of them. *Meccanica* **15**, 9–30.

Benzi, R., Paladin, G., Parisi, G., and Vulpiani, A. (1984). Characterization of intermittency in chaotic systems. *Journal of Physics A* **17**, 3521–31.

Bergé, P., Pomeau, Y., and Vidal, C. (1986). *Order within chaos.* Wiley, New York.

Berlekamp, E., Conway, J., and Guy, R. (1982). *Winning ways (for your mathematical plays).* Academic Press, New York.

Berliner, L. M. (1992). Statistics, probability and chaos. *Statistical Science* **7**, 69–122.

Binney, J. J., Dowrick, N. J., Fischer, A. J., and Newman, M. E. J. (1992). *The theory of critical phenomena.* Oxford University Press.

Birkhoff, G. D. (1927). On the periodic motions of dynamical systems. *Acta Mathematica* **50**, 359–79.

Birkhoff, G. D. (1933). *Aesthetic measure.* Harvard University Press, Cambridge, MA.

Blumenthal, R. M. and Menger, K. (1970). *Studies in geometry.* Freeman, San Francisco.

Bohr, T. and Rand, D. (1987). The entropy function for characteristic exponents. *Physica D* **25**, 387–98.

Bohr, T., Jensen, M., Paladin, G., and Vulpiani, A. (1998). *Dynamical systems approach to turbulence.* Cambridge University Press, New York.

Boldrighini, C. and Francheschini, V. (1979). A five-dimensional truncation of the plane incompressible Navier–Stokes equations. *Communications in Mathematical Physics* **64**, 159–70.

Borovkova, S., Burton, R., and Dehling, H. (1999). Consistency of the Takens estimator for the correlation dimension. *Annals of Applied Probability* **9**, 376–90.

Bowen, R. (1975). *Equilibrium states and the ergodic theory of Anosov diffeomorphisms.* Lecture Notes in Mathematics, Vol. 470. Springer, Berlin.

Box, G. E. P., Jenkins, G. M., and Reinsel, G. C. (1994). *Time series analysis: forecasting and control* (3rd edn). Prentice-Hall, Englewood Cliffs, NJ.

Boyce, W. E. and DiPrima, R. C. (1992). *Elementary differential equations and boundary value problems* (5th edn). McGraw-Hill, New York.

Brandstater, A. and Swinney, H. L. (1987). Strange attractors in weakly turbulent Taylor–Couette flow. *Physical Review A* **35**, 2207–20.

Brandt, C. and Pompe, B. (2002). Permutation entropy: a natural complexity measure for time series. *Physical Review Letters* **88**, 174102-1–4.

Branner, B. (1988). The Mandelbrot set. In *Chaos and fractals: the mathematics behind the computer graphics* (ed. R. Devaney and L. Keen), pp. 75–105. American Mathematical Society, Providence, RI.

Breeden, J. L. and Packard, N. H. (1992). Nonlinear analysis of data sampled nonuniformly in time. *Physica D* **58**, 273–83.

Briggs, J. (1992). *Fractals: the patterns of chaos: discovering a new aesthetic of art, science, and nature.* Thames and Hudson, London.

Briggs, K. M. (1991). A precise calculation of the Feigenbaum constants. *Mathematics of Computation* **57**, 435–9.

Briggs, K., Quispel, G. R. W., and Thompson, C. (1991). Feigenvalues for Mandelsets. *Journal of Physics A* **24**, 3363–8.

Britton, N. F. (1986). *Reaction–diffusion equations and their applications to biology.* Academic Press, New York.

Brock, W. A., Hsieh, D. A., and LeBaron, B. (1991). *Nonlinear dynamics, chaos, and instability: statistical theory and economic evidence.* MIT Press, Cambridge, MA.

Brock, W. A., Dechert, W. D., Scheinkman, J. A., and LeBaron, B. (1996). A test for independence based on the correlation dimension. *Econometric Reviews* **15**, 197–235.

Broomhead, D. S. and King, G. P. (1986). Extracting qualitative dynamics from experimental data. *Physica D* **20**, 217–36.

Broomhead, D. S. and Rowlands, G. (1984). On the use of perturbation theory in the calculation of the fractal dimension of strange attractors. *Physica D* **10**, 340–52.

Broomhead, D. S., Huke, J. P., and Potts, M. A. S. (1996). Cancelling deterministic noise by constructing nonlinear inverses to linear filters. *Physica D* **89**, 439–58.

Brown, R. and Chua, L. O. (1999). Clarifying chaos III. *International Journal of Bifurcation and Chaos*, 785–803.

Brown, R., Bryant, P., and Abarbanel, H. D. I. (1991). Computing the Lyapunov spectrum of a dynamical system from an observed time series. *Physical Review A* **43**, 2787–806.

Bryant, P. (1992). Computation of Lyapunov exponents from experimental data. In *Proceedings of the first experimental chaos conference* (ed. S. Vohra, M. Spano, M. Shlesinger, L. Pecora, and W. Ditto), pp. 11–23. World Scientific, Singapore.

Bryant, P., Brown, R., and Abarbanel, H. D. I. (1990). Lyapunov exponents from observed time series. *Physical Review Letters* **65**, 1523–6.

Buck, J. (1988). Synchronous rhythmic flashing of fireflies. II. *Quarterly Review of Biology* **63**, 265–89.

Burgers, J. M. (1974). *The nonlinear diffusion equation.* Reidel, Dordrecht, Netherlands.

Buzug, T. and Pfister, G. (1992). Comparison of algorithms calculating optimal parameters for delay time coordinates. *Physica D* **58**, 127–37.

Buzug, T., Pawelzik, K., von Stamm, J., and Pfister, G. (1994). Mutual information and global strange attractors in Taylor–Couette flow. *Physica D* **72**, 343–50.

Cantor, G. (1883). Grundlagen einer allgemeinen Mannichfältigkeitslehre. *Mathematische Annalen* **21**, 545–91.

Carreras, B. A., Newman, D. E., Lynch, E., and Diamond, P. H. (1996). A model realization of self-organized criticality for plasma confinement. *Physics of Plasmas* **3**, 2903–11.

Cartwright, M. L. and Littlewood, J. E. (1945). On nonlinear differential equation of the second order. I. The equation $\ddot{y} - k(1 - y^2)\dot{y} + y = b\lambda k \cos(\lambda t + \alpha), k$ large. *Journal of the London Mathematical Society* **20**, 180–9.

Casdagli, M. (1997). Recurrence plots revisited. *Physica D* **108**, 12–44.

Casdagli, M., Eubank, S., Farmer, J. D., and Gibson, J. (1991). State space reconstruction in the presence of noise. *Physica D* **51**, 52–98.

Casti, J. L. (1995). Bell curves and monkey languages. *Complexity* **1**, 12–15.

Casti, J. L. (2000). *Five more golden rules: knots, codes, chaos, and other great theories of 20th-Century mathematics*, pp. 35–99. Wiley, New York.

Cayley, A. (1879). The Newton-Fourier imaginary problem. *American Journal of Mathematics* **2**, 97.

Chaisson, E. J. (2001). *Cosmic evolution: the rise of complexity in nature.* Harvard University Press, Cambridge, MA.

Chaitin, G. J. (1966). On the length of programs for computing finite binary sequences. *Journal of the Association of Computing Machinery* **13**, 547–69.

Chaitin, G. J. (2002). Computers, paradoxes and the foundations of mathematics. *American Scientist* **90**, 164–71.

Chatfield, C. (1988). Apples, oranges and mean squared error. *International Journal of Forecasting* **4**, 515–18.

Chatfield, C. (1996). *The analysis of time series: an introduction* (5th edn). Chapman and Hall, London.

Chatfield, C. (2000). *Time-series forecasting.* Chapman and Hall/CRC, Boca Raton, FL.

Chen, G. and Dong, X. (1993). From chaos to order: perspectives and methodologies in controlling chaotic nonlinear dynamical systems. *International Journal of Bifurcation and Chaos* **3**, 1363–1409.

Chen, G. and Ueta, T. (1999). Yet another chaotic attractor. *International Journal of Bifurcation and Chaos* **9**, 1465–6.

Chen, G., Miola, J. L., and Wang, H. O. (2000). Bifurcation control: theories, methods, and applications. *International Journal of Bifurcation and Chaos* **10**, 511–48.

Chernikov, A. A., Sagdeev, R. Z., and Zaslavsky, G. M. (1988). Chaos: how regular can it be? *Physics Today* **41** (11), 27–35.

Chirikov, B. V. (1979). A universal instability of many-dimensional oscillator systems. *Physics Reports* **52**, 263–379.

Chirikov, B. V. and Vecheslavov, V. V. (1989). Chaotic dynamics of Comet Halley. *Astronomy and Astrophysics* **221**, 146–54.

Christiansen, F. and Rugh, H. H. (1997). Computing Lyapunov spectra with continuous Gram–Schmidt orthonormalization. *Nonlinearity* **10**, 1063–72.

Chua, L. O. and Yang, L. (1988). Cellular neural networks: theory. *IEEE Transactions on Circuits and Systems* **CS-35**, 1257–90.

Clifford, P. and Sudbury, A. (1973). A model for spatial conflict. *Biometrika* **60**, 581–8.

Cohen, A. and Procaccia, I. (1985). Computing the Kolmogorov entropy from time signals of dissipative and conservative dynamical systems. *Physical Review A* **31**, 1872–82.

Cohen, J. and Stewart, I. (1994). *The collapse of chaos.* Penguin, New York.

Collet, P., Lebowitz, J. L., and Porzio (1987). The dimension spectrum of some dynamical systems. *Journal of Statistical Physics* **47**, 609–44.

Cook, A. E. and Roberts, P. H. (1970). The Rikitake two disk dynamo system. *Proceedings of the Cambridge Philosophical Society* **68**, 547–69.

Cooley, J. W. and Tukey, J. W. (1965). An algorithm for the machine calculation of complex Fourier series. *Mathematics of Computation* **19**, 297–301.

Coullet, P., Tresser, C., and Arnéodo, A. (1979). A transition to stochasticity for a class of forced oscillators. *Physics Letters A* **72**, 268–70.

Crawford, J. D. (1991). Introduction to bifurcation theory. *Reviews of Modern Physics* **63**, 991–1037.

Cremer, H. (1925). Über die iteration rationaler funktionen. *Jahresbuch der Deutschen Mathematischen Vereinigung* **33**, 185–210.

Cromer, A. (1981). Stable solutions using the Euler approximation. *American Journal of Physics* **49**, 455–9.

Cross, M. and Hohenberg, P. (1993). Pattern formation outside of equilibrium. *Reviews of Modern Physics* **65**, 851–1112.

Crutchfield, J. P. (1988). Spatio-temporal complexity in nonlinear image processing. *IEEE Transactions on Circuits and Systems* **CS-35**, 770–80.

Crutchfield, J. P. and Young, K. (1989). Inferring statistical complexity. *Physical Review Letters* **63**, 105–8.

Cumming, A. and Linsay, P. S. (1988). Quasiperiodicity and chaos in a system with three competing frequencies. *Physical Review Letters* **60**, 2719–22.

Curry, J., Garnett, L., and Sullivan, D. (1983). On the iteration of rational functions: computer experiments with Newton's method. *Communications in Mathematical Physics* **91**, 267–77.

Cvitanović, P. (1984). *Universality in chaos.* Adam Hilger, Bristol.

Cvitanović, P. (1988). Invariant measures of strange sets in terms of cycles. *Physical Review Letters* **61**, 2729–32.

Darbyshire, A. G. and Broomhead, D. S. (1996). Robust estimation of tangent maps and Liapunov spectra. *Physica D* **89**, 287–305.

Darwin, C. (1859). *The origin of species.* Clowes, London.

Davis, J. C. (1986). *Statistics and data analysis in geology.* (2nd edn). Wiley, New York.

Dawkins, R. (1976). *The selfish gene.* Oxford University Press, New York.

Devaney, R. L. (1984). A piecewise linear model for the zones of instability of an area-preserving map. *Physica D* **10**, 387–93.

Devaney, R. L. (1989). *An introduction to chaotic dynamical systems* (2nd edn). Addison-Wesley, Redwood City, CA.

Devaney, R. L. (1992). *A first course in chaotic dynamical systems: theory and experiment.* Addison-Wesley–Longman, Reading, MA.

Devijver, P. A. and Kittler, J. (1982). *Pattern recognition: a statistical approach.* Prentice-Hall, New York.

Dimitrova, E. S. and Yordanov, O. I. (2001). Statistics of some low-dimensional chaotic flows. *International Journal of Bifurcation and Chaos* **11**, 2675–82.

Ding, M., Grebogi, E., Ott, E., Sauer, T., and Yorke, J. A. (1993). Plateau onset for correlation dimension: when does it occur? *Physical Review Letters* **70**, 3872–5.

Ditto, W. L., Spano, M. L., Savage, H.T., Rauseo, S. N., Heagy, J., and Ott, E. (1990). Experimental observation of a strange nonchaotic attractor. *Physical Review Letters* **65**, 533–6.

Doolen, G. (1990). *Lattice gas methods for partial differential equations* (SFI Studies in the Sciences of Complexity). Addison-Wesley-Longman, Reading, MA.

Douady, A. and Hubbard, J. H. (1982). Itération des polynômes quadratiques complexes. *Compte Rendu Academy of Science Paris I* **249**, 123–6.

Dowell, E. H. and Virgin, L. N. (1990). On spatial chaos, asymptotic modal analysis, and turbulence. *Transactions of the American Society of Mechanical Engineers* **57**, 1094–7.

Doyon, B., Cessac, B., Quoy, M., and Samuelides, M. (1993). Control of the transition to chaos in neural networks with random connectivity. *International Journal of Bifurcation and Chaos* **3**, 279–91.

Drazin, P. G. and Johnson, R. S. (1989). *Solitons: an introduction.* Cambridge University Press.

Duffing, G. (1918). *Erzwungene schwingungen bei veränderlicher eigenfrequenz.* Vieweg, Braunschweig.

Eckmann, J. -P. and Ruelle, D. (1985). Ergodic theory of chaos and strange attractors. *Reviews of Modern Physics* **57**, 617–56.

Eckmann, J. -P., Kamphorst, S. O., and Ruelle, D. (1987). Recurrence plots of dynamical systems. *Europhysics Letters* **4**, 973–7.

Eichhorn, R., Linz, S. J., and Hänggi, P. (1998). Transformations of nonlinear dynamical systems to jerky motion and its application to minimal chaotic flows. *Physical Review E* **58**, 7151–64.

Eichhorn, R., Linz, S. J., and Hänggi, P. (2001). Simple polynomial chaotic jerky dynamics. *Chaos, Solitons, and Fractals* **12**, 1377–83.

Eigen, M. and Schuster, P. (1979). *The hypercycle: a principle of natural self-organization.* Springer, New York.

Elwakil, A. S. and Kennedy, M. P. (2001). Construction of classes of circuit-independent chaotic oscillators using passive-only nonlinear devices. *IEEE Transactions on Circuits and Systems–I: Fundamental Theory and Applications* **CS-48**, 289–307.

Elwakil, A. S., Özoğuz, S., and Kennedy, M. P. (2002). Creation of a complex butterfly attractor using a novel Lorenz-type system. *IEEE Transactions on Circuits and Systems–I: Fundamental Theory and Applications* **CS-49**, 527–30.

Engel, W. (1955). Ein satz über ganze cremona-transformationen der ebene. *Mathematische Annalen* **130**, 11–19.

Essex, C. and Nerenberg, M. A. H. (1990). Fractal dimension: limit capacity or Hausdorff dimension? *American Journal of Physics* **58**, 986–8.

Eubank, S. and Farmer, J. D. (1990). An introduction to chaos and randomness. In *1989 lectures in complex systems* (ed. E. Jen), pp. 75–190. Addison-Wesley, Redwood City, CA.

Falconer, K. (1990). *Fractal geometry: mathematical foundations and applications.* Wiley, Chichester.

Farmer, J. D. (1982). Chaotic attractors of an infinite-dimensional dynamical system. *Physica D* **4**, 366–93.

Farmer, J. D. (1985). Sensitive dependence on parameters in nonlinear dynamics. *Physical Review Letters* **55**, 351–4.

Farmer, J. D. and Sidorowich, J. J. (1987). Predicting chaotic time series. *Physical Review Letters* **59**, 845–8.

Farmer, J. D., Ott, E., and Yorke, J. A. (1983). The dimension of chaotic attractors. *Physica D* **7**, 153–80.

Fatou, M. P. (1919). Sur les équations fonctionelles. *Bulletin de la Societé Mathématique de France* **47**, 161–271.

Feder, J. (1988). *Fractals.* Plenum, New York.

Feigenbaum, M. J. (1978). Quantitative universality for a class of nonlinear transformations. *Journal of Statistical Physics* **19**, 24–52.

Feigenbaum, M. J. (1979). The universal metric properties of nonlinear transformations. *Journal of Statistical Physics* **21**, 669–706.

Feigenbaum, M. J. (1980). Universal behavior in nonlinear systems. *Los Alamos Science* **1**, 4–27.

Feller, W. (1951). The asymptotic distribution of the range of sums of independent random variables. *Annals of Mathematical Statistics* **22**, 427–43.

Feller, W. (1968). *An introduction to probability theory and its applications*, Vol. 1. Wiley, New York.

Field, M. and Golubitsky, M. (1992). *Symmetry in chaos: a search for pattern in mathematics, art, and nature.* Oxford University Press, New York.

Field, S., Witt, J., Nori, F., and Ling, X. (1995). Superconducting vortex avalanches. *Physical Review Letters* **74**, 1206–9.

Fife, P. C. (1979). *Mathematical aspects of reacting and diffusing systems.* Lecture Notes in Biomathematics, Vol. 28. Springer, Berlin.

Fisher, R. A. (1937). The wave of advance of advantageous genes. *Annals of Eugenics* **7**, 353–69.

Ford, J. (1975). The statistical mechanics of classical analytic dynamics. In *Fundamental problems in statistical mechanics* (ed. E. G. D. Cohen), pp. 215–55. North-Holland, Amsterdam.

Ford, J., Stoddard, S. D., and Turner, J. S. (1973). On the integrability of the Toda lattice. *Progress of Theoretical Physics* **50**, 1547–60.

Fournier, d'Albe, E. E. (1907). *Two new worlds: I The infra world; II The supra world.* Longmans Green, London.

Fraedrich, K. (1986). Estimating the dimensions of weather and climate attractors. *Journal of the Atmospheric Sciences* **43**, 419–32.

Frame, M. and Robertson, J. (1992). A generalized Mandelbrot set and the role of critical points. *Computers and Graphics* **16**, 35–40.

Frank, M., Blank, H. R., Heindl, J., Kaltenhäuser, M., Köchner, H., Kreische, W., Müller, N., Poscher, S., Sporer, R., and Wagner, T. (1993). Improvements in K2-entropy calculations by means of dimension scaled distances. *Physica D* **65**, 359–64.

Fraser, A. M. (1989a). Reconstructing attractors from scalar time series: a comparison of singular system and redundancy criteria. *Physica D* **34**, 391–404.

Fraser, A. M. (1989b). Information and entropy in strange attractors. *IEEE Transactions on Information Theory* **35**, 245–52.

Fraser, A. M. and Swinney, H. L. (1986). Independent coordinates for strange attractors from mutual information. *Physical Review A* **33**, 1134–40.

Frautschi, S. C., Olenick, R. P., Apostol, T. M., and Goodstein, D. (1986). *The mechanical universe.* Cambridge University Press.

Frette, V., Christensen, K., Malthe-Sørenssen, A., Feder, J., Jøssang, T., and Meakin, P. (1996). Avalanche dynamics in a pile of rice. *Nature* **379**, 49–52.

Frisch, U. and Parisi, G. (1985). On the singularity structure of fully developed turbulence. In *Turbulence and predictability in geophysical fluid dynamics and climate dynamics* (ed. M. Ghil, R. Benzi, and G. Parisi). North-Holland, New York.

Gale, D. (1993). Mathematical entertainments. *Mathematical Intelligencer* **15**, 54–5.

Gallas, J. A. C. (1993). Structure of the parameter space of the Hénon map. *Physical Review Letters* **70**, 2714–17.

Galton, F. (1894). *Natural inheritance.* Macmillan, New York.

Gardner, C. S., Greene, J. M., Kruskal, M. D., and Miura, R. M. (1967). Method for solving the Korteweg-de Vries equation. *Physical Review Letters* **19**, 1095–7.

Gardner, M. (1970). The fantastic combinations of John Conway's new solitaire game of life. *Scientific American* **223**, 120–3.

Gardner, R. H., Milne, B. T., Turner, M. G., and O'Neill, R. V. (1987). Natural models for the analysis of broad-scale landscape pattern. *Landscape Ecology* **1**, 5–18.

Garey, M. R. and Johnson, D. S. (1979). *Computers and intractability.* Freeman, New York.

Gear, W. C. (1971). *Numerical initial value problems in ordinary differential equations.* Prentice-Hall, Englewood Cliffs, NJ.

Geisel, T. and Nierwetberg, J. (1982). Onset of diffusion and universal scaling in chaotic systems. *Physical Review Letters* **48**, 7–10.

Geist, K., Parlitz, U., and Lauterborn, W. (1990). Comparison of different methods for computing Lyapunov exponents. *Progress of Theoretical Physics* **83**, 875–93.

Gençay, R. and Dechert, W. D. (1992). An algorithm for the n Lyapunov exponents of an n-dimensional unknown dynamical system. *Physica D* **59**, 142–57.

Gençay, R. and Dechert, W. D. (1996). The identification of spurious Lyapunov exponents in Jacobian algorithms. *Studies in Nonlinear Dynamics and Econometrics* **1**, 143–54.

Gilden, D. L. (2001). Cognitive emission of $1/f$ noise. *Psychological Review* **108**, 33–56.

Gillham, N. W. (2001). *A life of Sir Francis Galton: from African exploration to the birth of eugenics.* Oxford University Press.

Gilmore, R. and Lefranc, M. (2002). *The topology of chaos.* Wiley-Interscience, New York.

Gilpin, M. E. (1973). Do hares eat lynx? *American Naturalist* **107**, 727–30.

Glass, L. and Mackey, M. C. (1988). *From clocks to chaos.* Princeton University Press, Princeton, NJ.

Glazier, J. A. and Libchaber, A. (1988). Quasi-periodicity and dynamical systems: an experimentalist's view. *IEEE Transactions on Circuits and Systems* **CS-35**, 790–809.

Gleick, J. (1987). *Chaos: making a new science.* Viking, New York.

Gödel, K. (1931). Über formal unentscheidbare sätze der principia mathematica und verwandter systeme I. *Monatshefte für Mathematik und Physik* **38**, 173–98.

Gollub, J. P., and Benson, S. V. (1980). Many routes to turbulent convection. *Journal of Fluid Mechanics* **100**, 927–30.

Gollub, J. P., and Swinney, H. L. (1975). Onset of turbulence in a rotating fluid. *Physical Review Letters* **35**, 927–30.

Gollub, J. P., Brunner, T. O., and Danly, B. G. (1978). Periodicity and chaos in coupled nonlinear oscillators. *Science* **200**, 48–50.

Golubitsky, M. and Schaeffer, D. G. (1985). *Singularities and groups in bifurcation theory*, Vol. I. Applied Mathematical Sciences, Vol. 51. Springer, New York.

Gottlieb, H. P. W. (1996). What is the simplest jerk function that gives chaos? *American Journal of Physics* **64**, 525.

Gottlieb, H. P. W. and Sprott, J. C. (2001). Simplest driven conservative chaotic oscillator. *Physics Letters A* **291**, 385–8.

Gould, S. J. and Eldridge, N. (1977). Punctuated equilibrium: the tempo and mode of evolution reconsidered. *Paleobiology* **3**, 115–51.

Graczyk, J. and Swiatek, G. (1997). Generic hyperbolicity in the logistic family. *Annals of Mathematics*, 1–52.

Grasman, J. (1986). *Asymptotic methods for relaxation oscillations and applications*. Springer, New York.

Grassberger, P. (1983a). On the fractal dimension of the Hénon attractor. *Physics Letters A* **97**, 224–6.

Grassberger, P. (1983b). Generalized dimensions of strange attractors. *Physics Letters A* **97**, 227–30.

Grassberger, P. (1988). Finite sample corrections to entropy and dimension estimates. *Physics Letters A* **128**, 369–73.

Grassberger, P. and Procaccia, I. (1983a). Characterization of strange attractors. *Physical Review Letters* **50**, 346–9.

Grassberger, P. and Procaccia, I. (1983b). Measuring the strangeness of strange attractors. *Physica D* **9**, 189–208.

Grassberger, P. and Procaccia, I. (1983c). Estimation of the Kolmogorov entropy from a chaotic signal. *Physical Review A* **28**, 2591–3.

Grassberger, P. and Procaccia, I. (1984). Dimensions and entropies of strange attractors from fluctuating dynamics approach. *Physica D* **13**, 34–54.

Grassberger, P., Schreiber, T., and Schaffrath, C. (1991). Non-linear time series analysis. *Internal Journal of Bifurcation and Chaos* **1**, 521–47.

Grebogi, C., Ott, E., and Yorke, J. A. (1982). Chaotic attractors in crisis. *Physical Review Letters* **48**, 1507–10.

Grebogi, C., Ott, E., and Yorke, J. A. (1983). Crises, sudden changes in chaotic attractors and transient chaos. *Physica D* **7**, 181–200.

Grebogi, C., Ott, E., Pelikan, S., and Yorke, J. A. (1984). Strange attractors that are not chaotic. *Physica D* **13**, 261–8.

Grebogi, C., McDonald, S. W., Ott, E., and Yorke, J. A. (1985). Exterior dimension of fat fractals. *Physics Letters A* **110**, 1–4.

Grebogi, C., Ott, E., Romeiras, F., and Yorke, J. A. (1987). Critical exponents for crisis-induced intermittency. *Physical Review A* **36**, 5365–80.

Grebogi, C., Ott, E., and Yorke, J. A. (1988). Unstable periodic orbits and the dimensions of multifractal chaotic attractors. *Physical Review A* **37**, 1711–24.

Grebogi, C., Hammel, S. M., Yorke, J. A., and Sauer, T. (1990). Shadowing of physical trajectories in chaotic dynamics: containment and refinement. *Physical Review Letters* **65**, 1527–30.

Green, D. G. (1994). Connectivity and the evolution of biological systems. *Journal of Biological Systems* **2**, 91–103.

Greene, J. M. (1979). A method for determining a stochastic transition. *Journal of Mathematical Physics* **20**, 1183–201.

Greenside, H. S., Wolf, A., Swift, J., and Pignataro, T. (1982). Impracticality of a box-counting algorithm for calculating the dimensionality of strange attractors. *Physical Review A* **25**, 3453–6.

Grossmann, S. and Thomae, S. (1977). Invariant distributions and stationary correlation functions of one-dimensional discrete processes. *Zeitschrift für Naturforschung A* **32**, 1353–63.

Guastello, S. J. (2001). Nonlinear dynamics in psychology. *Discrete Dynamics in Nature and Society* **6**, 11–29.

Guckenheimer, J. and Buzyna, G. (1983). Dimension measurements for geostrophic turbulence. *Physical Review Letters* **51**, 1438–41.

Guckenheimer, J. and Holmes, P. (1990). *Nonlinear oscillations, dynamical systems, and bifurcations of vector fields* (3rd edn). Springer, New York.

Guevara, M. R., Glass, L., and Schrier, A. (1981). Phase locking, period-doubling bifurcations, and irregular dynamics in periodically stimulated cardiac cells. *Science* **214**, 1350.

Gutenberg, B. and Richter, C. F. (1954). *Seismicity of the Earth and associated phenomena.* Princeton University Press, Princeton, NJ.

Guyon, E. and Stanley H. E. (1991). *Fractal forms.* Elsevier, Amsterdam.

Habib, S. and Ryne, R. D. (1995). Symplectic calculation of Lyapunov exponents. *Physical Review Letters* **74**, 70–3.

Hack, J. T. (1957). Studies of longitudinal streams in Virginia and Maryland. *U. S. Geological Survey Professional Papers* **294B**, 45–97.

Hajek, B. (1988). Cooling schedules for optimal annealing. *Mathematics of Operations Research* **13**, 311–29.

Haken, H. (1978). *Synergetics.* Springer, Berlin.

Haken, H. (1981). *The science of structure: synergetics.* Van Nostrand Reinhold, New York.

Haken, H. (1983a). At least one exponent vanishes if the trajectory of an attractor does not contain a fixed point. *Physics Letters A* **94**, 71–4.

Haken, H. (1983b). *Advanced synergetics.* Springer, Berlin.

Halsey, T., Jensen, M., Kadanoff, L., Procaccia, I., and Shraiman, B. (1986). Fractal measures and their singularities: the characterization of strange sets. *Physical Review A* **33**, 1141–51.

Hamburger, D., Biham, O., and Avnir, D. (1996). Apparent fractality emerging from models of random distributions. *Physical Review E* **33**, 3342–58.

Hammersley, J. M. (1983). Origins of percolation theory. *Annals of the Israel Physical Society* **5**, 47–57.

Harrison, E. (1987). *Darkness at night: a riddle of the Universe.* Harvard University Press, Cambridge, MA.

Harrison, R. G. and Biswas, D. J. (1986). Chaos in light. *Nature* **321**, 394–401.

Hausdorff, F. (1919). Dimension und äusseres mass. *Mathematische Annalen* **79**, 157–79.

Hayashi, C. (1964). *Nonlinear oscillations in physical systems.* McGraw-Hill, New York (reprinted 1984 by Princeton University Press).

Hayashi, C., Ueda, Y., Akamatsu, N., and Itakura, H. (1970). On the behavior of self-oscillatory systems with external force (in Japanese). *Transactions of the Institute of Electronics and Communication Engineers of Japan* **53-A**, 150–8.

Hayes, B. (2001a). Randomness as a resource. *American Scientist* **89**, 300–4.

Hayes, B. (2001b). Third base. *American Scientist* **89**, 490–4.

Hegger, R., Kantz, H., and Schreiber, T. (1999). Practical implementation of nonlinear time series methods: the TISEAN package. *Chaos* **9**, 413–35.

Heidel, J. and Zhang, F. (1999). Nonchaotic behaviour in three-dimensional quadratic systems II. the conservative case. *Nonlinearity* **12**, 617–33.

Hénon, M. (1969). Numerical study of quadratic area-preserving mappings. *Quarterly of Applied Mathematics* **27**, 291–312.

Hénon, M. (1976). A two-dimensional mapping with a strange attractor. *Communication in Mathematical Physics* **50**, 69–77.

Hénon, M. (1982). On the numerical computation of Poincaré maps. *Physica D* **5**, 412–14.

Hénon, M. and Heiles, C. (1964). The applicability of the third integral of motion: some numerical experiments. *Astrophysical Journal* **69**, 73–9.

Hentschel, H. G. E. and Procaccia, I. (1983). The infinite number of generalized dimensions of fractals and strange attractors. *Physica D* **8**, 435–44.

Hentschel, H. G. E. and Procaccia, I. (1984). Relative diffusion in turbulent media: the fractal dimension of clouds. *Physical Review A* **29**, 1461–70.

Hilborn, R. C. (2000). *Chaos and nonlinear dynamics* (2nd edn). Oxford University Press.

Hill, J. R. (1996). Fractals and the Grand Internet Parallel Processing Project. In *Fractal horizons: the future use of fractals* (ed. C. A. Pickover), pp. 299–323. St. Martin's, New York.

Hille, E. and Tamarkin, J. D. (1929). Remarks on a known example of a monotone continuous function. *American Mathematics Monthly* **36**, 255–64.

Hirsch, M. W. and Smale, S. (1974). *Differential equations, dynamical systems and linear algebra.* Academic Press, New York.

Hirst, B. and Mandelbrot, B. (1994). *Fractal landscapes from the real world.* Cornerhouse, Manchester.

Hofstadter, D. R. (1980). *Gödel, Escher, Bach: an eternal golden braid.* Vintage, New York.

Holley, R. and Liggett, T. M. (1975). Ergodic theorems for weakly interacting particle systems and the voter model. *Annals of Probability* **3**, 643–63.

Holmes, P. J. (1979). A nonlinear oscillator with a strange attractor. *Philosophical Transactions of the Royal Society of London Series A* **292**, 419–48.

Holmes, P. J. and Moon, F. C. (1983). Strange attractors and chaos in nonlinear mechanics. *Journal of Applied Mechanics* **50**, 1021–32.

Hoover, W. G. (1995). Remark on 'Some simple chaotic flows'. *Physical Review E* **51**, 759–60.

Hopf, E. (1948). A mathematical example displaying the features of turbulence. *Communications on Pure and Applied Mathematics* **1**, 303–22.

Hornik, K. (1989). Multilayer feedforward networks are universal approximators. *Neural Networks* **2**, 359–66.

Hornik, K., Stinchcombe, M., and White, H. (1990). Universal approximation of unknown mapping and its derivatives using multilayer feedforward networks. *Neural Networks* **3**, 535–49.

Horton, W., Weigel, R. S., and Sprott, J. C. (2001). Chaos and the limits of predictability for the solar-wind-driven magnetosphere–ionosphere system. *Physics of Plasmas* **8**, 2946–52.

Hunt, B. R. and Ott, E. (2001). Fractal properties of robust strange nonchaotic attractors. *Physical Review Letters* **87**, 25401-1–4.

Hurst, H. E., Black, R. P., and Simaika, Y. M. (1965). *Long-term storage: an experimental study.* Constable, London.

Hutchinson, J. (1981). Fractals and self similarity. *Indiana University Mathematics Journal* **30**, 713–47.

Huygens, C. (1673). *Horologium oscillatorium.* Muguet, Paris.

Ikeda, K. (1979). Multiple-valued stationary state and its instability of the transmitted light by a ring cavity system. *Optics Communications* **30**, 257–61.

Ising, E. (1925). Beitrag zur theorie des ferromagnetismus. *Zeitschrift für Physik* **31**, 253–8.

Jackson, E. A. (2001). *Exploring nature's dynamics.* Wiley, New York.

Jackson, E. A. and Kodogeorgiou, A. (1992). A coupled Lorenz-cell model of Rayleigh–Bénard turbulence. *Physics Letters A* **168**, 270–5.

Jacquin, A. E. (1992). Image coding based on a fractal theory of iterated contractive image transformation. *IEEE Transactions on Image Processing* **1**, 18–30.

Jaeger, H. M., Liu, C. -H., and Nagel, S. R. (1989). Relaxation at the angle of repose. *Physical Review Letters* **62**, 40–3.

Jakobson, M. V. (1981). Absolutely continuous invariant measures for one-parameter families of one-dimensional maps. *Communications in Mathematical Physics* **81**, 39–88.

Janaki, T. M., Rangarajan, G., Habib, S., and Ryne, R. D. (1999). Computation of Lyapunov spectrum for continuous time dynamical systems and discrete maps. *Physical Review E* **60**, 6614–26.

Jeffrey, H. J. (1992). Chaos game visualization of sequences. *Computers and Graphics* **16**, 25–33.

Jensen, H. J. (1998). *Self-organized criticality: emergent complex behavior in physical and biological systems.* Cambridge University Press.

Jensen, H. J., Christensen, K., and Fogedby, H. C. (1989). $1/f$ noise, distribution of lifetimes, and a pile of sand. *Physical Review B* **40**, 7425–7.

Jensen, R. V. (1987). Classical chaos. *American Scientist* **75**, 168–81.

Jolliffe, I. T. (1986). *Principal component analysis.* Springer, New York.

Jowett, J. M., Month, M., and Turner, S. (1985). *Nonlinear dynamics aspects of particle accelerators.* Lecture Notes in Physics **247**. Springer, New York.

Judd, K. (1992). An improved estimator of dimension and some comments on providing confidence intervals. *Physica D* **56**, 216–28.

Julia, G. (1918). Memoire sur l'itération des fonctions rationnelles. *Journal de Mathématique Pur et Appliquées* **8**, 47–245.

Kahn, P. and Zarmi, Y. (1997). *Nonlinear dynamics: exploration through normal forms.* Wiley, New York.

Kaneko, K. (1989). Pattern dynamics in spatiotemporal chaos. *Physica D* **34**, 1–41.

Kaneko, K. (1990). Supertransients, spatiotemporal intermittency and stability of fully developed spatiotemporal chaos. *Physics Letters A* **149**, 105–12.

Kantz, H. (1994). A robust method to estimate the maximal Lyapunov exponent of a time series. *Physics Letters A* **185**, 77–87.

Kantz, H. and Schreiber, T. (1997). *Nonlinear time series analysis.* Cambridge University Press.

Kaplan, D. T. and Cohen, R. J. (1990). Is fibrillation chaos? *Circulation Research* **67**, 886–92.

Kaplan, J. and Yorke, J. (1979). Chaotic behavior of multidimensional difference equations. In *Functional differential equations and approximation of fixed points*, Lecture Notes in Mathematics, Vol. 730 (ed. H. -O. Peitgen and H. -O. Walther), pp. 228–37. Springer, Berlin.

Kaspar, F. and Schuster, H. G. (1987). An easily calculable measure for the complexity of spatiotemporal patterns. *Physical Review A* **36**, 842–8.

Katok, A. (1980). Lyapunov exponents, entropy and periodic orbits for diffeomorphisms. *Institut des Hautes Études Scientifiques Publications Mathématiques* **51**, 137–73.

Katori, M., Kizaki, A., Terui, Y., and Kubo, T. (1998). Forest dynamics with canopy gap expansion and stochastic Ising model. *Fractals* **6**, 81–6.

Kauffman, L. H. and Sabelli, H. C. (1998). The process equation. *Cybernetics and Systems: An International Journal* **29**, 345–62.

Kauffman, S. A. (1991). Antichaos and adaptation. *Scientific American* **265** (2), 64–70.

Kennel, M. B., Brown, R., and Abarbanel, H. D. I. (1992). Determining minimum embedding dimension using a geometric construction. *Physical Review A* **45**, 3403–11.

Kertész, J. and Kiss, L. B. (1990). The noise spectrum in the model of self-organized criticality. *Journal of Physics A* **23**, L433–40.

Kesten, H. (1987). *Percolation theory and ergodic theory of infinite particle systems.* Springer, New York.

Kingsland, S. E. (1985). *Modeling nature.* University of Chicago Press, Chicago.

Kirkpatrick, S., Gelatt Jr, C. D., and Vecchi, M. P. (1983). Optimization by simulated annealing. *Science* **220**, 671–80.

Kirkwood, D. (1888). *The asteroids, or minor planets between Mars and Jupiter.* Lippencott, Philadelphia.

Kline, M. (1980). *Mathematics: the loss of certainty.* Oxford University Press.

Knuth, D. E. (1997). *Sorting and searching* (3rd edn), Vol. 3 of *The art of computer programming.* Addison-Wesley–Longman, Reading, MA.

Knuth, D. E. (1998). *Seminumerical algorithms* (2nd edn), Vol. 2 of *The art of computer programming.* Addison-Wesley–Longman, Reading, MA.

Kolb, M., Botet, R., and Jullien, R. (1983). Scaling of kinetically growing clusters. *Physical Review Letters* **51**, 1123–6.

Kolmogoroff, A., Petrovsky, I., and Piscounoff, N. (1937). Étude de l'équation de la diffusion avec croissance de la quantité de matière et son application à un problème biologique. *Moscow University Bulletin of Mathematics* **1**, 1–25.

Kolmogorov, A. N. (1958). A new invariant for transitive dynamical systems. *Doklady Akademii nauk Souiza Sovetskikh Sotsialisticheskikh Respuplik* **119**, 861–4.

Kolmogorov, A. N. (1965). Three approaches to the definition of the concept 'quantity of information'. *Problems in Information Transmission* **1**, 1–7.

Korteweg, D. J. and de Vries, G (1895). On the change of long waves advancing in a rectangular canal, and on a new type of long stationary waves. *Philosophical Magazine* (Series 5) **39**, 422–33.

Kostelich, E. J. and Yorke, J. A. (1988). Measuring filtered chaotic signals. *Physical Review A* **38**, 1649–52.

Kramer, G. (1994). *Auditory display: sonification, audification, and auditory interfaces* (SFI Studies in the Sciences of Complexity). Addison-Wesley-Longman, Reading, MA.

Krischer, K., Lübke, M., Wolf, W., Eiswirth, M., and Ertl, G. (1991). Interior crisis in an electrochemical system. In *Bifurcation and chaos: analysis, algorithms, applications (International series of numerical mathematics)*, Vol. 97 (ed. R. Seydel, F. W. Schneider, T. Kupper, and H.

Troger). Birkhäuser Verlag, Basil.
Kruel, Th. M., Eisworth, M., and Schreider, F. W. (1993). *Introduction to non-linear mechanics*. Princeton University Press, Princeton, NJ.
Kuramoto, Y. and Tsuzuki, T. (1976). Persistent propagation of concentration waves in dissipative media far from thermal equilibrium. *Progress of Theoretical Physics* **55**, 356–69.
Kutta, W. (1901). Beitrang zur naherungsweisen integration totaler differentialgleichunge. *Zeitschrift für Mathematik und Physik* **46**, 435–53.
Kuznetsov, Y. A. (1995). *Elements of applied bifurcation theory* (2nd edn). Springer, New York.
L'Ecuyer, P. (1988). Efficient and portable combined random number generators. *Communications of the ACM* **31**, 742–9, 774.
Lai, Y. C. (1996). Transition from strange nonchaotic to strange chaotic attractors. *Physical Review E* **53**, 57–65.
Lai, Y. C. and Grebogi, C. (1999). Modeling of coupled chaotic oscillators. *Physical Review Letters* **82**, 4803–6.
Landau, L. D. (1944). On the problem of turbulence. *Doklady Akademii nauk Souiza Sovetskikh Sotsialisticheskikh Respuplik* **44**, 339–44.
Landau, L. D. and Lifshitz, E. M. (1959). *Fluid mechanics*. Pergamon, Cambridge.
Landauer, R. (1996). Minimal energy requirements in communication. *Science* **272**, 1914–18.
Langton, C (1986). Studying artificial life with cellular automata. *Physica D* **22**, 120–49.
Langton, C. (1990). Computation at the edge of chaos: phase transitions and emergent computation. *Physica D* **42**, 12–37.
Laskar, J. (1996). Large scale chaos and marginal stability in the solar system. *Celestial Mechanics and Dynamical Astronomy* **64**, 115–62.
Lathrop, D. P. and Kostelich, E. J. (1989). Characterization of an experimental strange attractor by periodic orbits. *Physical Review A* **40**, 4928–31.
Lau, Y. -T. and Finn, J. M. (1992). Dynamics of a three-dimensional incompressible flow with stagnation points. *Physica D* **57**, 283–310.
Ledrappier, F. and Young, L. -S. (1985). The metric entropy of diffeomorphisms. *Annals of Mathematics* **2**, 509–74.
Lehnertz, K. and Elger, C. E. (1998). Can epileptic seizures be predicted? evidence from nonlinear time-series analysis of brain electrical activity. *Physical Review Letters* **80**, 5019–22.
Lei, T. (1990). Similarity between the Mandelbrot and Julia set. *Communications in Mathematical Physics* **134**, 587–617.
Leipnik, R. B. and Newton, T. A. (1981). Double strange attractors in rigid body motion with linear feedback control. *Physics Letters A* **86**, 63–7.
Lempel, A. and Ziv, J. (1976). On the complexity of finite sequences. *IEEE Transactions on Information Theory* **22**, 75–81.

Lenz, W. (1920). Beitrag zum verständnis der magnetischen eigenschaften in festen körpern. *Physikalische Zeitschrift* **21**, 613–15.

Letellier, C., Dutertre, P., Reizner, J., and Gouesbet, G. (1996). Evolution of a multimodal map induced by an equivariant vector field. *Journal of Physics A* **29**, 5359–73.

Lewin, R. (1992). *Complexity: life on the edge of chaos.* Macmillan, New York.

Li, T. -Y. and Yorke, J. A. (1975). Period three implies chaos. *American Mathematical Monthly* **82**, 985–92.

Libchaber, A. (1982). Convection and turbulence in liquid helium I. *Physica B* **109** and **110**, 1583–9.

Lichtenberg, A. J. and Liebermann, M. A. (1992). *Regular and chaotic dynamics* (2nd edn). Springer, New York.

Liebovitch, L. S. and Toth, T. (1989). A fast algorithm to determine fractal dimensions by box counting. *Physics Letters A* **141**, 386–90.

Lindenmayer, A. (1968). Mathematical models for cellular interaction in development I: filaments with one sided inputs. *Journal of Theoretical Biology* **18**, 280–9.

Linsay, P. (1981). Period doubling and chaotic behavior in a driven, anharmonic oscillator. *Physical Review Letters* **47**, 1349–52.

Linz, S. J. (1997). Nonlinear dynamical models and jerky motion. *American Journal of Physics* **65**, 523–6.

Linz, S. J. and Sprott, J. C. (1999). Elementary chaotic flow. *Physics Letters A* **259**, 240–5.

Lorenz, E. N. (1963). Deterministic nonperiodic flow. *Journal of Atmospheric Sciences* **20**, 130–41.

Lorenz, E. N. (1969). Atmospheric predictability as revealed by naturally occurring analogues. *Journal of Atmospheric Sciences* **26**, 636–46.

Lorenz, E. N. (1984a). The local structure of a chaotic attractor in four dimensions. *Physica D* **13**, 90–104.

Lorenz, E. N. (1984b). Irregularity: a fundamental property of the atmosphere. *Tellus* **36A**, 98–110.

Lorenz, E. N. (1993). *The essence of chaos.* University of Washington Press, Seattle, WA.

Lotka, A. J. (1920). Undamped oscillations derived from the law of mass action. *Journal of the American Chemical Society* **42**, 1595–99.

Lotka, A. J. (1925). *Elements of physical biology.* Williams and Wilkins, Baltimore.

Lovejoy, S. (1982). Area-perimeter relation for rain and cloud areas. *Science* **216**, 185–7.

Lozi, R. (1978). Un attracteur étrange? Du type attracteur de Hénon. *Journal de Physique (Paris)* **39** (C5), 9–10.

Lunsford, G. H. and Ford, J. (1972). On the stability of periodic orbits for nonlinear oscillator systems in regions exhibiting stochastic behavior. *Journal of Mathematical Physics* **13**, 700–5.

Mackey, D. and Glass, L. (1977). Oscillation and chaos in physiological control systems. *Science* **197**, 287–9.

Madan, R. N. (1993). *Chua's circuit: a paradigm for chaos.* World Scientific, Singapore.

Maganza, C., Causse, R. and Laloe, F. (1986). Bifurcations, period-doubling and chaos in clarinetlike systems. *Europhysics Letters* **1**, 295–304.

Mainieri, R. (1993). On the equality of Hausdorff and box counting dimensions. *Chaos* **3**, 119–25.

Makridakis, S., Wheelwright, S. C., and McGee, V. E. (1983). *Forecasting methods and applications* (2nd edn). Wiley, New York.

Malasoma, J. -M. (2000). What is the simplest dissipative chaotic jerk equation which is parity invariant? *Physics Letters A* **264**, 383–9.

Malasoma, J. -M. (2002). Countable infinite sequence of attractors. *Chaos, Solitons, and Fractals* **13**, 1835–42.

Malthe-Sørenssen, A., Feder, J., Christensen, K., Frette, V., and Jøssang, T. (1999). Surface fluctuations and correlations in a pile of rice. *Physical Review Letters* **83**, 764–7.

Malthus, T. (1798). *An essay on the principle of population, as it affects the future improvement of society with remarks on the speculations of Mr. Godwin, M. Condorcet, and other writers.* Printed for J. Johnson in St. Paul's Church-Yard, London.

Mandelbrot, B. B. (1975). Stochastic models for the Earth's relief, the shape and fractal dimension of the coastlines, and the number–area rule for islands. *Proceedings of the National Academy of Sciences* (USA) **72**, 3825–8.

Mandelbrot, B. B. (1977). *Fractals: form, chance, and dimension.* Freeman, San Francisco.

Mandelbrot, B. B. (1980). Fractal aspects of the iteration $z \leftarrow \lambda z(1-z)$ for complex λ and z. *Annals of the New York Academy of Sciences* **357**, 249–59.

Mandelbrot, B. B. (1983). *The fractal geometry of nature.* Freeman, New York.

Mandelbrot, B. B. (1997). *Fractals and scaling in finance.* Springer, New York.

Mandelbrot, B. B. and van Ness, J. W. (1968). Fractional Brownian motions, fractional noises and applications. *SIAM Review* **10**, 422–37.

Manneville, P. (1980). Intermittency, self-similarity and $1/f$ spectrum in dissipative dynamical systems. *Journal de Physique (Paris)* **41**, 1235–43.

Manneville, P. (1990). *Dissipative structures and weak turbulence: perspectives in physics.* Academic Press, Boston.

Mantegna, R. N., Buldyrev, S. V., Goldberger, A. L., Havlin, S., Peng, C. K., Simons, M., and Stanley, H. E. (1994). Linguistic features of noncoding DNA sequences. *Physical Review Letters* **73**, 3169–72.

Marsden, J. E. and McCracken, M. (1976). *The Hopf bifurcation and its applications*. Springer, New York.

Matsumoto, T., Chua, L. O., and Tanaka, S. (1984). Simplest chaotic nonautonomous circuit. *Physical Review A* **30**, 1155–7.

Matsumoto, T., Chua, L. O., and Komuro, M. (1985). The double scroll. *IEEE Transactions on Circuits and Systems* **CS-33**, 797–818.

May, R. (1972). Limit cycles in predator–prey communities. *Science* **177**, 900–2.

May, R. (1976). Simple mathematical models with very complicated dynamics. *Nature* **261**, 45–67.

Mayer-Kress, G. and Layne, S. P. (1987). Dimensionality of the human electroencephalogram. In *Perspectives in biological dynamics and theoretical medicine* (ed. S. H. Koslow, A. J. Mandel, and M. F. Shlesinger) *Annals of the New York Academy of Sciences* **504**, 62–87.

McDonald, S. W., Grebogi, C., Ott, E., and Yorke, J. A. (1985). Fractal basin boundaries. *Physica D* **17**, 125–53.

McGuire, R. (1991). *An eye for fractals*. Addison-Wesley, Redwood City, CA.

Meakin, P. (1983a). Diffusion-controlled cluster formation in 2–6 dimensional space. *Physical Review A* **27**, 1495–507.

Meakin, P. (1983b). Formation of fractal clusters and networks by irreversible diffusion-limited aggregation. *Physical Review Letters* **51**, 1119–22.

Mees, A. I. and Sparrow, C. (1987). Some tools for analyzing chaos. *Proceedings of the IEEE* **75**, 1058–70.

Meinhardt, H. (1982). *Models of biological pattern formation*. Academic Press, London.

Melnikov, V. K. (1963). On the stability of the center for time periodic perturbations. *Transactions of the Moscow Mathematical Society* **12**, 1–57.

Meloon, B. and Sprott, J. C. (1997). Quantification of determinism in music using iterated function systems. *Empirical Studies of the Arts* **15**, 3–13.

Metropolis, N., Rosenbluth, A. W., Rosenbluth, M. N., Teller, A. H., and Teller, E. (1953). Equation of state calculations by fast computing machines. *Journal of Chemical Physics* **21**, 1087–92.

Metropolis, N., Stein, M. L., and Stein, P. R. (1973). On finite limit sets for transformations on the unit interval. *Journal of Combinational Theory* **15**, 25–44.

Mettin, R., Parlitz, U., and Lauterborn, W. (1993). Bifurcation structure of the driven van der Pol oscillator. *International Journal of Bifurcation and Chaos* **3**, 1529–55.

Mininni, P. D., Gómez, D. O., and Mindlin, G. B. (2000). Stochastic relaxation oscillator model for the sunspot cycle. *Physical Review Letters* **85**, 5476–9.

Miramontes, O. and Rohani, P. (1998). Intrinsically generated colored noise in laboratory populations. *Proceedings of the Royal Society of London B* **265**, 785–92.

Mirus, K. A. and Sprott, J. C. (1999). Controlling chaos in low- and high-dimensional systems with periodic parametric perturbations. *Physical Review E* **59**, 5313–24.

Misiurewicz, M. (1980). Strange attractors for the Lozi mappings. *Annals of the New York Academy of Sciences* **357**, 348–58.

Mitchison, G. J. and Durbin, R. M. (1989). Bounds on the learning capacity of some multi-layer networks. *Biological Cybernetics* **60**, 345–56.

Mitschke, F., Möller, M., and Lange, W. (1988). Measuring filtered chaotic signals. *Physical Review A* **37**, 4518–21.

Möller, M., Lange, W., Mitschke, F., Abraham, N. B., and Hübner (1989). Errors from digitizing and noise in estimating attractor dimensions. *Physics Letters A* **138**, 176–82.

Moon, F. C. (1992). *Chaotic and fractal dynamics: an introduction for applied scientists and engineers.* Wiley-Interscience, New York.

Moon, F. C. and Holmes, P. J. (1979). A magnetoelastic strange attractor. *Journal of Sound Vibration* **65**, 275–96.

Moon, F. C. and Holmes, W. T. (1985). Double Poincaré sections of a quasiperiodically forced, chaotic attractor. *Physics Letters A* **111**, 157–60.

Moon, F. C., Holmes, W. T., and Khoury, P. (1991). Symbol dynamic maps of spatial–temporal chaotic vibrations in a string of impact oscillators. *Chaos* **1**, 65–8.

Moore, D. W. and Spiegel, E. A. (1966). A thermally excited non-linear oscillator. *Astrophysical Journal* **143**, 871–87.

Moore, E. F. (1962). Machine models of self reproduction. *American Mathematical Society Proceedings of Symposia in Applied Mathematics* **14**, 17-33.

Moran, P. A. P. (1950). Some remarks on animal populations. *Biometrics* **6**, 250–8.

Mori, H. (1980). Fractal dimensions of chaotic flows of autonomous dissipative systems. *Progress of Theoretical Physics* **63**, 1044–7.

Moser, J. (1973). *Stable and random motions in dynamical systems.* Princeton University Press, Princeton, NJ.

Moss, F. and McClintock, P. V. E. (1989). *Noise in nonlinear dynamical systems. Volume 3: experiments and simulations.* Cambridge University Press.

Murao, K., Kohda, T., Noda, K., and Yanase, M. (1992). $1/f$ Noise generator using logarithmic and antilogarithmic amplifiers. *IEEE Transactions on Circuits and Systems–I: Fundamental Theory and Applications* **CS-39**, 851–3.

Murray, J. D. (1993). *Mathematical biology* (2nd edn). Springer, New York.

Nagatani, T., Emmerich, H., and Nakanishi, K. (1998). Burgers equation for kinetic clustering in traffic flow. *Physica A* **255**, 158–62.

Nagel, K. and Herrmann, J. (1993). Deterministic models for traffic jams. *Physica A* **199**, 254–69.

Nauenberg, M. and Schellnhuber, H. J. (1989). Analytic evaluation of the multifractal properties of a Newtonian Julia set. *Physical Review Letters* **62**, 1807–10.

Necker, L. A. (1832). Observations on some remarkable optical phenomena seen in Switzerland, and an optical phenomenon which occurs on viewing a figure of a crystal or geometrical solid. *Philosophical Magazine* (3rd series) **1**, 329–37.

Neimark, J. (1959). On some cases of periodic motions depending on parameters. *Doklady Akademii nauk Souiza Sovetskikh Sotsialisticheskikh Respuplik* **129**, 736–9.

Nemytskii, V. V. and Stepanov, V. V. (1960). *Qualitative theory of differential equations.* Princeton University Press, Princeton, NJ.

Newhouse, S. E., Ruelle, D., and Takens, F. (1978). Occurrence of strange axiom A attractors near quasiperiodic flows on $T^m, m \geq 3$. *Communications in Mathematical Physics* **64**, 35–40.

Newland, D. E. (1993). *An introduction to random vibrations, spectral and wavelet analysis.* Longman, Harlow.

Nicolis, G. (1989). In *The new physics* (ed. P. Davids), pp. 316–47. Cambridge University Press.

Nicolis, J. S., Mayer-Kress, G., and Haubs, G. (1983). Non-uniform chaotic dynamics with implications to information processing. *Zeitschrift für Naturforschung A* **38**, 1157–69.

Nosé, S. (1991). Constant temperature molecular dynamics methods. *Progress of Theoretical Physics Supplement* **103**, 1–46.

Nusse, H. E. and Yorke, J. A. (1994). *Dynamics: numerical explorations.* Springer, New York.

Oldham, K. and Spainer, J. (1974). *Fractional calculus.* Academic Press, New York.

Onsager, L. (1944). Crystal statistics I. A two-dimensional model with an order–disorder transition. *Physical Review* **65**, 117–49.

Osborne, A. R. and Provenzale, A. (1989). Finite correlation dimension for stochastic systems with power-law spectra. *Physica D* **35**, 357–81.

Osborne, A. R., Kirwin, A. D., Provenzale, A., and Bergamasco, L. (1986). A search for chaotic behavior in large mesoscale motions in the Pacific Ocean. *Physica D* **23**, 75–83.

Oseledec, V. I. (1968). A multiplicative ergodic theorem: Lyapunov characteristic numbers for dynamical systems. *Transactions of the Moscow Mathematical Society* **19**, 197–221.

Ott, E. (1993). *Chaos in dynamical systems.* Cambridge University Press.

Ott, E. and Spano, M. L. (1995). Controlling chaos. *Physics Today* **48** (5), 34–40.

Ott, E., Withers, W. D., and Yorke, J. A. (1984). Is the dimension of chaotic attractors invariant under coordinate changes? *Journal of Sta-*

tistical Physics **36**, 687–97.

Ott, E., Grebogi, C., and Yorke, J. A. (1990). Controlling chaos. *Physical Review Letters* **64**, 1196–9.

Ozorio de Almeida, A. M. (1988). *Hamiltonian systems: chaos and quantization.* Cambridge University Press.

Packard, N. H., Crutchfield, J. P., Farmer, J. D., and Shaw, R. S. (1980). Geometry from a time series. *Physical Review Letters* **45**, 712–16.

Paladin, G. and Vulpiani, A. (1987). Anomalous scaling laws in multifractal objects. *Physics Reports* **156**, 147–225.

Paluŝ, M. (1993). Identifying and quantifying chaos by using information-theoretic functionals. In *Time-series prediction: forecasting the future and understanding the past* (ed. A. S. Weigend and N. A. Gershenfeld), pp. 387–413. Addison-Wesley, Reading, MA.

Paparella, F., Provenzale, A., Smith, L. A., Taricco, C., and Vio, R. (1997). Local random analogue prediction of nonlinear processes. *Physics Letters A* **235**, 233–40.

Parlitz, U. (1992). Identification of true and spurious Lyapunov exponents from time series. *International Journal of Bifurcation and Chaos* **2**, 155–65.

Pastor-Satorras, R., Vázquez, A., and Vespignani, A. (2001). Dynamical and correlation properties of the Internet. *Physical Review Letters* **87**, 258701-1–4.

Peak, D. and Frame, M. (1994). *Chaos under control: the art and science of complexity.* Freeman, New York.

Peano, G. (1890). Sur une courbe, qui remplit une aire plane. *Mathematische Annalen* **36**, 157–60 (Translation in Peano 1973).

Peano, G. (1973). *Selected works* (ed. H. C. Kennedy). Toronto University Press.

Peitgen, H. -O. and Richter, P. H. (1986). *The beauty of fractals: images of complex dynamical systems.* Springer, Berlin.

Peitgen, H. -O., Saupe, D., and Haeseler, F. V. (1984). Cayley's problem and Julia sets. *Mathematical Intelligencer* **6**, 11–20.

Peixoto, M. M. (1962). Structural stability of two-dimensional manifolds. *Topology* **1**, 101–20.

Pesin, Ya. B. (1977). Characteristic Lyapunov exponents and smooth ergodic theory. *Russian Mathematical Surveys* **32**, 55–114.

Peters, O., Hertlein, C., and Christensen, K. (2002). A complexity view of rainfall. *Physical Review Letters* **88**, 018701-1–4.

Peterson, I. (1993). *Newton's clock: chaos in the Solar System.* Freeman, New York.

Phatak, S. C. and Rao, S. S. (1995). Logistic map: a possible random number generator. *Physical Review E* **51**, 3670–8.

Pikovsky, A. S. and Feudel, U. (1995). Characterizing strange nonchaotic attractors. *Chaos* **5**, 253–60.

Pineda, F. J. and Sommerer, J. C. (1993). Estimating generalized dimensions and choosing time delays: a fast algorithm. In *Time-series prediction: forecasting the future and understanding the past* (ed. A. S. Weigend and N. A. Gershenfeld), pp. 367–385. Addison-Wesley, Reading, MA.

Pippard, A. B. (1989). *The physics of vibration.* Cambridge University Press.

Poincaré, H. (1890). Sur le problème des trois corps et les équations de la dynamique. *Acta Mathematica* **13**, 1-270.

Poincaré, H. (1914). *Science and method* (trans. Francis Maitland). Nelson, New York.

Poincaré, H. (1921). Analyse des travaux scientifiques de Henri Poincaré faites par lui-même. *Acta Mathematica* **38**, 1–135.

Pomeau, Y. and Manneville, P. (1980). Intermittent transition to turbulence in dissipative dynamical systems. *Communications in Mathematical Physics* **74**, 189–97.

Poon, C. -S. and Merrill, C. K. (1997). Decrease of cardiac chaos in congestive heart failure. *Nature* **389**, 492–5.

Porsch, H. A., Hoover, W. G., and Vesely, F. J. (1986). Canonical dynamics of the Nosé Oscillator: stability, order, and chaos. *Physical Review A* **33**, 4253–65.

Porter, E. and Gleick, J. (1990). *Nature's chaos.* Viking, New York.

Potapov, A. and Ali, M. K. (2000). Robust chaos in neural networks. *Physics Letters A* **277**, 310–22.

Press, W. H. (1978). Flicker noise in astronomy and elsewhere. *Comments on Astrophysics* **7**, 103–19.

Press, W. H., Flannery, B. P., Teukolsky, S. A., and Vetterling, W. T. (1992). *Numerical recipes: the art of scientific computing* (2nd edn). Cambridge University Press.

Pressing, J. (1988). Nonlinear maps as generators of musical design. *Computational Music Journal* **12**, 35–46.

Prichard, D. and Theiler, J. (1995). Generalized redundancies for time series analysis. *Physica D* **84**, 951–93.

Priestley, M. B. (1981). *Spectral analysis and time series.* Academic Press, London.

Priestley, M. B. (1988). *Non-linear and non-stationary time series analysis.* Academic Press, London.

Prigogine, I. (1980). *From being to becoming.* Freeman, San Francisco.

Prigogine, I. (1997). *The end of certainty: time, chaos, and the new laws of nature.* The Free Press, New York.

Provenzale, A., Smith, L. A., Vio, R., and Murante, G. (1992). Distinguishing between low-dimensional dynamics and randomness in measured time series. *Physica D* **58**, 31–49.

Prusinkiewicz, P. and Hanan, J (1989). *Lindenmayer systems, fractals, and plants: lecture notes in biomathematics.* Springer, Berlin.

Purwins, H. G., Radehaus, C., and Berkemeier, J. (1988). Experimental investigation of spatial pattern formation in physical systems of activator inhibitor type. *Zeitschrift für Naturforschung A* **43**, 17–29.

Rabinovich, M. I. and Fabrikant, A. L. (1979). Stochastic self-modulation of waves in nonequilibrium media. *Soviet Physics JETP* **50**, 311–17.

Rangarajan, G., Habib, S. and Ryne, R. D. (1998). Lyapunov exponents without rescaling and reorthogonalization. *Physical Review Letters* **80**, 3747–50.

Rényi, A. (1961). On measures of entropy and information. In *Proceedings of the 4th Berkeley Symposium on Mathematical Statistics and Probability*, Vol. 1, pp. 547–61. University of California Press, Berkeley.

Rényi, A. (1970). *Probability theory*. North-Holland, Amsterdam.

Reynolds, O. (1883). An experimental investigation of the circumstances which determine whether the motion of water shall be direct or sinuous, and of the law of resistance in parallel channels. *Philosophical Transactions of the Royal Society of London Series A* **174**, 935–82.

Rhodes, B. and Anderson, R. M. (1996). Power laws governing epidemics in isolated populations. *Nature* **381**, 600–2.

Richards, R. (1999). The subtle attraction: beauty as a force in awareness, creativity, and survival. In *Affect, creative experience, and psychological adjustment* (ed. S. W. Russ), pp. 195–219. Brunner/Mazel, Philadelphia.

Richardson, L. F. (1920). The supply of energy from and to atmospheric eddies. *Proceedings of the Royal Society of London Series A* **97**, 354–73.

Richardson, L. F. (1961). The problem of contiguity: an appendix of statistics of deadly quarrels. *Yearbook of the Society for General Systems Research* **6**, 139–87.

Ricker, W. (1954). Stock and recruitment. *Journal of the Fisheries Research Board of Canada* **11**, 559–663.

Rinzel, J., Maginu, K., and Vidal C. (1984). *Non-equilibrium dynamics in chemical systems*. Springer, New York.

Robbin, J. W. (1971). A structural stability theorem. *Annals of Mathematics* **4**, 447–93.

Roberts, D. C. and Turcotte, D. L. (1998). Fractality and self-organized criticality of wars. *Fractals* **6**, 351–7.

Rojas, R. (1991). A tutorial on efficient computer representations of the Mandelbrot set. *Computers and Graphics* **15**, 91–100.

Romeiras, F. J. and Ott, E. (1987). Strange nonchaotic attractors of the damped pendulum with quasiperiodic forcing. *Physical Review A* **35**, 4404–13.

Rosen, R. (1970). *Dynamical system theory in biology*. Wiley-Interscience, New York.

Rosenau, P. (2002). Reaction and concentration dependent diffusion model. *Physical Review Letters* **88**, 194501-1–4.

Rosenstein, M. T., Collins, J. J., and De Luca, C. J. (1993). A practical method for calculating largest Lyapunov exponents from small data sets.

Physica D **65**, 117–34.

Rosenstein, M. T., Collins, J. J., and De Luca, C. J. (1994). Reconstruction expansion as a geometry-based framework for choosing proper delay times. *Physica D* **73**, 82–98.

Rössler, O. E. (1976). An equation for continuous chaos. *Physics Letters A* **57**, 397–8.

Rössler, O. E. (1979a). Continuous chaos–four prototype equations. *Annals of the New York Academy of Sciences* **316**, 376–92.

Rössler, O. E. (1979b). An equation for hyperchaos. *Physics Letters A* **71**, 155–7.

Roux, J. C., Rossi, A., Bachelart, S., and Vidal, C. (1980). Representation of a strange attractor from an experimental study of chemical turbulence. *Physics Letters A* **77**, 391–3.

Roux, J. C., Simoyi, R. H., and Swinney, H. L. (1983). Observation of a strange attractor. *Physica D* **8**, 257–66.

Rowlands, G. (1991). A numerical algorithm for Hamiltonian systems. *Journal of Computational Physics* **97**, 235–9.

Rowlands, G. and Sprott, J. C. (1992). Extraction of dynamical equations from chaotic data. *Physica D* **58**, 251–9.

Rucker, R. (1982). *Infinity and the mind.* Birkhauser, Boston.

Rucklidge, A. M. (1992). Chaos in models of double convection. *Journal of Fluid Mechanics* **237**, 209–29.

Ruderman, D. L. and Bialek, W. (1994). Statistics of natural images: scaling in the woods. *Physical Review Letters* **73**, 814–17.

Ruelle, D. (1973). Some comments on chemical oscillations. *Transactions of the New York Academy of Sciences* (Series II) **35**, 66–71.

Ruelle, D. (1976). A measure associated with Axiom A attractors. *American Journal of Mathematics* **98**, 619–54.

Ruelle, D. (1978). *Thermodynamic formalism.* Addison-Wesley-Longman, Reading, MA.

Ruelle, D. (1980). Strange attractors. *The Mathematical Intelligencer* **2**, 126–37.

Ruelle, D. (1982). Characteristic exponents and invariant manifolds in Hilbert space. *Annals of Mathematics* **115**, 243–90.

Ruelle, D. (1989). *Chaotic evolution and strange attractors.* Cambridge University Press.

Ruelle, D. (1990). Deterministic chaos: the science and the fiction. *Proceedings of the Royal Society of London Series A* **427**, 241–8.

Ruelle, D. (1991). *Chance and chaos.* Princeton University Press, Princeton, NJ.

Ruelle, D. and Takens, F. (1971). On the nature of turbulence. *Communications in Mathematical Physics* **20**, 167–92.

Runge, C. (1895). Über die numerische auflösung con differentialgleichungen. *Mathematische Annalen* **46**, 167–78.

Russel, S. (1845). Report on waves. In *Report of the fourteenth meeting of the British Association for the Advancement of Science. York, September 1844*, pp. 311–90. London.

Russell, D. A., Hanson, J. D., and Ott, E. (1980). Dimension of strange attractors. *Physical Review Letters* **45**, 1175–8.

Sacker, R. S. (1965). A new approach to the perturbation theory of invariant surfaces. *Communications on Pure and Applied Mathematics* **18**, 717–32.

Sagan, H. (1994). *Space-filling curves.* Springer, New York.

Saha, P. and Strogatz, S. H. (1994). The birth of period three. *Mathematics Magazine* **68**, 42–7.

Sano, M. and Sawada, Y. (1985). *Physical Review Letters* **55**, 1082–5.

Sanz-Serna, J. M. (1994). *Numerical Hamiltonian problems (Applied mathematics and mathematical computation, No. 7)*. CRC Press, Cleveland, OH.

Sarkovskii, A. N. (1964). Co-existence of cycles of a continuous mapping of a line onto itself. *Ukrainian Mathematical Journal* **16**, 61–71.

Sauer, T. D., and Yorke, J. A. (1993). How many delay coordinates do you need? *International Journal of Bifurcation and Chaos* **3**, 737–44.

Sauer, T. D. and Yorke, J. A. (1999). Reconstructing the Jacobian from data with observational noise. *Physical Review Letters* **83**, 1331–4.

Sauer, T. D., Yorke, J. A., and Casdagli, M. (1991). Embedology. *Journal of Statistical Physics* **65**, 579–616.

Sauer, T. D., Grebogi, C., and Yorke, J. A. (1997). How long do numerical chaotic solutions remain valid? *Physical Review Letters* **79**, 59–62.

Sauer, T. D., Tempkin, J. A., and Yorke, J. A. (1998). Spurious Lyapunov exponents in attractor reconstruction. *Physical Review Letters* **81**, 4341–4.

Schaffer, W. M., Ellner, S., and Kot, M. (1986). Effects of noise on some dynamical models in ecology. *Journal of Mathematical Biology* **24**, 479–523.

Schmelcher, P. and Diakonos, F. K. (1997). Detecting unstable periodic orbits of chaotic dynamical systems. *Physical Review Letters* **78**, 4733–6.

Schmidt, G. and Wang, B. H. (1985). Dissipative standard map. *Physical Review A* **32**, 2994–9.

Schot, S. H. (1978). Jerk: the time rate of change of acceleration. *American Journal of Physics* **65**, 1090–4.

Schreiber, T. (1993). Extremely simple nonlinear noise reduction method. *Physical Review E* **47**, 2401–4.

Schreiber, T. (1995). Efficient neighbor searching in nonlinear time series analysis. *International Journal of Bifurcation and Chaos* **5**, 349–58.

Schreiber, T. (1997). Detecting and analyzing nonstationarity in a time series using nonlinear cross predictions. *Physical Review Letters* **78**, 843–6.

Schreiber, T. (1998). Constrained randomization of time series data. *Physical Review Letters* **80**, 2105–8.

Schreiber, T. and Schmitz, A. (1996). Improved surrogate data for nonlinearity tests. *Physical Review Letters* **77**, 635–8.

Schroeder, M. (1991). *Fractals, chaos, power laws: minutes from an infinite paradise.* Freeman, New York.

Schuster, H. G. (1995). *Deterministic chaos: an introduction* (3rd edn). Wiley, New York.

Seeger, R. J. (1966). *Galileo Galilei, his life and works.* Pergamon, Oxford.

Seimenis, J. (1994). *Hamiltonian mechanics: integrability and chaotic behaviour.* Plenum, New York.

Shannon, C. E. (1948). A mathematical theory of communication. *Bell System Technical Journal* **27**, 379–423.

Shaw, R. (1981). Strange attractors, chaotic behavior, and information flow. *Zeitschrift für Naturforschung A* **36**, 80–112.

Shaw, R. (1984). *The dripping faucet as a model chaotic system.* Aerial Press, Santa Cruz, CA.

Shibata, T., Chawanya, T., and Kaneko, K. (1999). Noiseless collective motion out of noisy chaos. *Physical Review Letters* **82**, 4424–7.

Shimada, I. and Nagashima, T. (1979). A numerical approach to ergodic problem of dissipative dynamical systems. *Progress of Theoretical Physics* **61**, 1605–16.

Shinbrot, T., Grebogi, C., Ott, E., and Yorke, J. A. (1993). Using small perturbations to control chaos. *Nature* **363**, 411–17.

Shishikura, M. (1998). The Hausdorff dimension of the boundary of the Mandelbrot set and Julia sets. *Annals of Mathematics* **147**, 225–67.

Shreider, Y. (1966). *The Monte Carlo method.* Pergamon, Oxford University Press.

Shumway, R. H. and Stoffer, D. S. (2000). *Time-series analysis and applications.* Springer, New York.

Sierpinski, W. (1916). Sur une courbe cantorienne qui contient une image biunivoque et continue de toute courbe donnée. *Comptes Rendus (Paris)* **162**, 629–32.

Sil'nikov, L. P. (1965). A case of the existence of a denumerable set of periodic motions. *Soviet Mathematics Doklady* **6**, 163–6.

Simoyi, R. H., Wolf, A., and Swinney, H. L. (1982). One-dimensional dynamics in a multicomponent chemical reaction. *Physical Review Letters* **49**, 245–8.

Sinai, Ya. G. (1959). On the concept of entropy of a dynamical system. *Doklady Akademii nauk Souiza Sovetskikh Sotsialisticheskikh Respuplik* **124**, 768–71.

Sinai, Ya. G. (1972). Gibbs measures in ergodic theory. *Russian Mathematical Surveys* **27**, 21–69.

Sinclair, R. M., Hosea, J. C., and Sheffield, G. V. (1970). A method for mapping a toroidal magnetic field by storage of phase stabilized elec-

trons. *Review of Scientific Instruments* **41**, 1552–9.
Singer, D. (1978). Stable orbits and bifurcations of maps in the interval. *SIAM Journal of Applied Mathematics* **35**, 260–7.
Sivashinsky, G. I. (1977). Nonlinear analysis of hydrodynamic instability in laminar flames, Part 1. Derivation of basic equations. *Acta Astronautica* **4**, 1177–206.
Sivashinsky, G. I. and Michelson, D. M. (1980). On irregular wavy flow of a liquid film flowing down a vertical plane. *Progress of Theoretical Physics* **63**, 2112–17.
Smale, S. (1967). Differential dynamical systems. *Bulletin of the American Mathematical Society* **73**, 747–817.
Smith, L. A. (1988). Intrinsic limits on dimension calculations. *Physics Letters A* **133**, 283–8.
Smith, L. A. (1992a). Identification and prediction of low dimensional dynamics. *Physica D* **58**, 50–76.
Smith, L. A. (1992b). Comments on the paper of R. Smith, Estimating dimension in noisy chaotic time series. *Journal of the Royal Statistical Society Series B - Methodological* **54**, 329–52.
Smith, L. A. (1994). Local optimal prediction: exploiting strangeness and the variation of sensitivity to initial condition. *Philosophical Transactions of the Royal Society of London Series A* **348**, 371–81.
Smith, L. A. (1997). The maintenance of uncertainty. In *Proceedings of the International School of Physics 'Enrico Fermi', Course CXXXIII*, pp. 177–246. Società Italiana di Fisica, Bologna, Italy.
Smith, L. A. and Spiegel, E. A. (1987). Strange accumulators. *Annals of the New York Academy of Science* **497**, 61–5.
Smith, L. A., Fournier, J. D., and Spiegel, E. A. (1986). Lacunarity and intermittency in fluid turbulence. *Physics Letters A* **114**, 465–8.
So, P., Ott, E., Schiff, S. J., Kaplan, D. T., Sauer, T., and Grebogi, C. (1996). Detecting unstable periodic orbits in chaotic experimental data. *Physical Review Letters* **76**, 4705–8.
Soderblom, L. A. (1980). The Galilean moons of Jupiter. *Scientific American* **242**, 88–100.
Solé, R. and Goodwin, B. (2000). *Signs of life: how complexity pervades biology*. Basic Books, New York.
Solé, R. V. and Manrubia, S. C. (1995). Self-similarity in rain forests: evidence for a critical state. *Physical Review E* **51**, 6250–3.
Solé, R. V., Ferrer, R., Gonzalez-Garcia, I., Quer, J., and Domingo, E. (1999). Red Queen dynamics, competition and critical points in a model of RNA virus quasispecies. *Journal of Theoretical Biology* **198**, 47–59.
Solé, R. V., Alonso, D., and McKane, A. (2000). Scaling in a network model of a multispecies ecosystem. *Physica A* **286**, 337–44.
Solomonoff, R. J. (1964). A formal theory of inductive inference. *Information and Control* **7**, 1–22, 224–54.

Sommerer, J. C. and Ott, E. (1993). A physical system with qualitatively uncertain dynamics. *Nature* **365**, 136–40.

Sparrow, C. (1982). *The Lorenz equations: bifurcations, chaos, and strange attractors.* Springer, New York.

Sprott, J. C. (1992). Simple programs create 3-D images. *Computers in Physics* **6**, 132–8.

Sprott, J. C. (1993a). Automatic generation of strange attractors. *Computers and Graphics* **17**, 325–32.

Sprott, J. C. (1993b). How common is chaos? *Physics Letters A* **173**, 21–4.

Sprott, J. C. (1993c). *Strange attractors: creating patterns in chaos.* M&T Books, New York.

Sprott, J. C. (1994a). Some simple chaotic flows. *Physical Review E* **50**, R647–50.

Sprott, J. C. (1994b). Predicting the dimension of strange attractors. *Physics Letters A* **192**, 355–60.

Sprott, J. C. (1994c). Automatic generation of iterated function systems. *Computers and Graphics* **18**, 417–25.

Sprott, J. C. (1996). Strange attractor symmetric icons. *Computers and Graphics* **20**, 325-32.

Sprott, J. C. (1997a). Simplest dissipative chaotic flow. *Physics Letters A* **228**, 271–4.

Sprott, J. C. (1997b). Some simple chaotic jerk functions. *American Journal of Physics* **65**, 537–43.

Sprott, J. C. (1998). Artificial neural net attractors. *Computers and Graphics* **22**, 143–9.

Sprott, J. C. (2000a). A new class of chaotic circuit. *Physics Letters A* **266**, 19–23.

Sprott, J. C. (2000b). Simple chaotic systems and circuits. *American Journal of Physics* **68**, 758–63.

Sprott, J. C. and Linz, S. (2000). Algebraically simple chaotic flows. *International Journal of Chaos Theory and Applications* **5**, 3–22.

Sprott, J. C. and Pickover, C. A. (1995). Automatic generation of general quadratic map basins. *Computers and Graphics* **19**, 309–13.

Sprott, J. C. and Rowlands, G. (1995). *Chaos data analyzer: the professional version.* Physics Academic Software, Raleigh, NC.

Sprott, J. C. and Rowlands, G. (2001). Improved correlation dimension calculation. *International Journal of Bifurcation and Chaos* **11**, 1861–80.

Sprott, J. C., Bolliger, J., and Mladenoff, D. J. (2002). Self-organized criticality in forest-landscape evolution. *Physics Letters A* **297**, 267–71.

Sproull, R. F. (1991). Refinements to the nearest-neighbor searching in k-dimensional trees. *Algorithmica* **6**, 579–89.

Stauffer, D. (1985). *Introduction to percolation theory.* Taylor and Francis, London.

Stavans, J., Heslot, F., and Libchaber, A. (1985). Fixed winding number and the quasiperiodic route to chaos in a convecting fluid. *Physical Review Letters* **55**, 596–9.

Stetson, H. T. (1937). *Sunspots and their effects.* McGraw-Hill, London.

Stevens, R. T. (1990). *Advanced fractal programming in C.* M&T Books, New York.

Stewart, I. and Golubitsky, M. (1992). *Fearful symmetry: is God a geometer?* Blackwell, Oxford.

Stokes, J. P., Weitz, D. A., Gollub, J. P., Dougherty, A., Robbins, M. O., Chaikin, P. M., and Lindsay, H. M. (1986). Interfacial stability of immiscible displacement in a porous medium. *Physical Review Letters* **57**, 1718–21.

Stooke, P. J. (1996). Topography and geology of Hyperion. *Earth, Moon & Planets* **74**, 61–83.

Stoop, R. and Parisi, J. (1991). Calculation of Lyapunov exponents avoiding spurious elements. *Physica D* **50**, 89–94.

Strogatz, S. H. (1988). Love affairs and differential equations. *Mathematics Magazine* **61**, 35.

Strogatz, S. H. (1994). *Nonlinear dynamics and chaos with applications to physics, biology, chemistry, and engineering.* Addison-Wesley–Longman, Reading, MA.

Strogatz, S. H. and Stewart, I. (1993). Coupled oscillators and biological synchronization. *Scientific American* **269** (6), 1023–9.

Stutzer, M. J. (1980). Chaotic dynamics and bifurcation in a macro model. *Journal of Economic Dynamics and Control* **2**, 353–76.

Sugihara, G. and May, R. M. (1990). Nonlinear forecasting as a way of distinguishing chaos from measurement error in time series. *Nature* **344**, 734–41.

Sussman, G. J. and Wisdom, J. (1992). Chaotic evolution of the Solar System. *Science* **257**, 56–62.

Swinney, H. L. and Gollub, J. P. (1978). The transition to turbulence. *Physics Today* **31** (8), 41–7.

Szu, H. and Hartley, R. (1987). Fast simulated annealing. *Physics Letters A* **122**, 157–62.

Takens, F. (1981). Detecting strange attractors in turbulence. In *Dynamical systems and turbulence.* Warwick 1980 Proceedings (ed. D. A. Rand and L. S. Young). Lecture Notes in Mathematics, No. 898, pp. 366–81. Springer, New York.

Takens, F. (1985). On the numerical determination of the dimension of an attractor. In *Dynamical systems and bifurcations* (ed. B. L. J. Braaksma, H. W. Braer, and F. Takens). Lecture Notes in Mathematics, No. 1125, pp. 99–106. Springer, Berlin.

Tanabe, Y. and Kaneko, K. (1994). Behavior of a falling paper. *Physical Review Letters* **73**, 1372–5.

Taylor, G. I. (1923). Stability of a viscous liquid contained between two rotating cylinders. *Philosophical Transactions of the Royal Society of*

London Series A **223**, 289–343.

Testa, J., Perez, J., and Jeffries, C. (1982). Evidence for universal chaotic behavior of a driven nonlinear oscillator. *Physical Review Letters* **48**, 714–17.

Theiler, J. (1986). Spurious dimension from correlation algorithms applied to limited time series data. *Physical Review A* **34**, 2427–32.

Theiler, J. (1988a). Quantifying chaos: practical estimation of the correlation dimension. *Ph.D. dissertation* (California Institute of Technology, Pasadena, CA).

Theiler, J. (1988b). Lacunarity in a best estimator of fractal dimension. *Physics Letters A* **133**, 195–200.

Theiler, J. (1990). Estimating fractal dimension. *Journal of the Optical Society of America A* **7**, 1055–73.

Theiler, J. (1993). Statistical error in a chord estimator of correlation dimension: the 'rule of five'. *International Journal of Bifurcation and Chaos* **3**, 765–71.

Theiler, J. and Eubank, S. (1993). Don't bleach chaotic data. *Chaos* **3**, 771–82.

Theiler, J. and Smith, L. A. (1995). Anomalous convergence of Lyapunov exponent estimates. *Physical Review E* **51**, 3738–41.

Theiler, J., Eubank, S., Longtin, A., Galdrikian, B., and Farmer, J. D. (1992). Testing for nonlinearity in time series: the method of surrogate data. *Physica D* **58**, 77–94.

Thom, R. (1975). *Structural stability and morphogenesis.* Benjamin, Reading MA.

Thomas, R. (1999). Deterministic chaos seen in terms of feedback circuits: analysis, synthesis, 'labyrinth chaos'. *International Journal of Bifurcation and Chaos* **9**, 1889–905.

Thompson, J. M. T. and Stewart, H. B. (1986). *Nonlinear dynamics and chaos: geometrical methods for engineers and scientists.* Wiley, Chichester.

Tomita, K. and Kai, T. (1978). Chaotic behaviour of deterministic orbits: the problem of turbulent phase. *Progress of Theoretical Physics Supplement* **64**, 280–94.

Tong, H. (1983). *Threshold models in non-linear time series analysis.* Lecture Notes in Statistics, No. 21. Springer, Heidelberg.

Tong, H. (1990). *Non-linear time series: a dynamical system approach.* Oxford University Press.

Trickey, S. T. and Virgin, L. N. (1998). Bottlenecking phenomena near a saddle-node remnant in a Duffing oscillator. *Physics Letters A* **248**, 185–90.

Tsonis, A. A. (1992). *Chaos: from theory to applications.* Plenum, New York.

Tucker, W. (1999). The Lorenz attractor exists. *Comtes Rendus de l'Académie des Sciences, Paris, Série I, Mathématique* **328**, 1197–202.

Turcotte, D. L. (1992). *Fractals and chaos in geology and geophysics.* Cambridge University Press.

Turing, A. M. (1952). The chemical basis of morphogenesis. *Philosophical Transactions of the Royal Society of London Series B* **237**, 37–72.

Ueda, Y. (1979). Randomly transitional phenomena in the system governed by Duffing's equation. *Journal of Statistical Physics* **20**, 181–96.

Ueda, Y. (1980). Explosion of strange attractors exhibited in Duffing's equation. *Annals of the New York Academy of Sciences* **357**, 422–34.

Ueda, Y. (1992). *The road to chaos.* Aerial Press, Santa Cruz, CA.

Ueda, Y. and Akamatsu, N. (1981). Chaotically transitional phenomena in the forced negative-resistance oscillator. *IEEE Transactions on Circuits and Systems* **CS-28** (3), 217–24.

Uhlenbeck, G. E. and Ornstein, L. S. (1930). On the theory of Brownian motion. *Physical Review* **36**, 823–41.

Ulam, S. M. (1983). *Adventures of a mathematician.* Scribner's, New York.

Ulam, S. M. and von Neumann, J. (1947). On combination of stochastic and deterministic processes. *Bulletin of the American Mathematical Society.* **53**, 1120.

Umberger, D. K. and Farmer, J. D. (1985). Fat fractals on the energy surface. *Physical Review Letters* **55**, 661–4.

Umberger, D. K., Grebogi, C., Ott, E., and Afeyan, B. (1989). Spatiotemporal dynamics in a dispersively coupled chain of nonlinear oscillators. *Physical Review A* **39**, 4835–42.

van de Water, W. and Schram, P. (1988). Generalized dimensions from near-neighbor information. *Physical Review A* **37**, 3118–25.

van den Dool, H. M. (1994). Searching for analogues, how long must we wait? *Tellus A* **46**, 314–24.

van der Pol, B. (1926). On relaxation oscillations. *Philosophical Magazine* **2**, 978–92.

van der Pol, B. and van der Mark, J. (1927). Frequency demultiplication. *Nature* **120**, 363–4.

van der Pol, B. and van der Mark, J. (1928). The heartbeat considered as a relaxation oscillation and an electrical model of the heart. *Philosophical Magazine* **6**, 763–75.

van der Schrier, G. and Maas, L. R. M. (2000). The diffusionless Lorenz equations; Shil'nikov bifurcations and reduction to an explicit map. *Physica D* **141**, 19–36.

van Wyk, M. A. and Steeb, W.-H. (1997). *Chaos in electronics.* Kluwer, Dordrecht.

Vandewalle, N., Boveroux, Ph., Minguet, A., and Ausloos, M. (1998). The crash of October 1987 seen as a phase transition: amplitude and universality. *Physica A* **255**, 201–10.

Vautard, R. and Ghil, M. (1989). Singular spectrum analysis in nonlinear dynamics, with applications to paleoclimatic time series. *Physica D* **35**, 395–424.

Verhulst, P. F. (1845). Récherches mathématiques sur la loi d'accrossement de la population. *Noveaux Memoires de l'Academie Royale des Sciences et Belles-Lettres de Bruxelles* **18**, 1–45.

Vespignani, A. and Zapperi, S. (1998). How self-organized criticality works: a unified mean-field picture. *Physical Review E* **57**, 6345–62.

Vicsek, T. (1983). Fractal models for diffusion controlled aggregation. *Journal of Physics A* **16**, L647–52.

Virgin, L. N. (1987). The nonlinear rolling response of a vessel including chaotic motions leading to capsize in regular seas. *Applied Ocean Research* **9**, 89–95.

Virgin, L. N. (2000). *Introduction to experimental nonlinear dynamics: a case study in mechanical vibration.* Cambridge University Press.

Volterra, V. (1926). Variazioni e fluttuazioni del numero d'individui in specie animali conviventi. *Memorie dell'Accademia dei Lincei* **2**, 31–113.

Volterra, V. (1931). *Lecònse sur la théorie mathematique de la lutte pour la via.* Gauthier-Villars, Paris.

Volterra, V. (1959). *Theory of functionals and of integral and integro-differential equations.* Dover, New York.

von Bremen, H. F., Udwadia, F. E., and Porskurowski, W. (1997). An efficient QR based method for the computation of Lyapunov exponents. *Physica D* **101**, 1–16.

von Koch, H. (1904). Sur une courbe continue sans tangente, obtenue par une construction géométrique élémentaire. *Arkiv för Matematik, Astronomi och Fysik* **1**, 681–704.

von Neumann, J. (1966). *Theory of self-reproducing automata* (ed. A. W. Burks). University of Illinois Press, Urbana, IL.

Voss, R. F. (1985). Random fractal forgeries. In *Fundamental algorithms in computer graphics* (ed. R. A. Earnshaw), pp. 805–35. Springer, Berlin.

Voss, R. F. (1992). Evolution of long-range fractal correlations and $1/f$ noise in DNA base sequences. *Physical Review Letters* **68**, 3805–8.

Voss, R. F. and Clarke, J. (1978). "$1/f$ noise" in music: music from $1/f$ noise. *The Journal of the Acoustical Society of America* **63**, 258–63.

Waldrop, M. (1992). *Complexity: the emerging science at the edge of order and chaos.* Simon and Schuster, New York.

Wales, D. J. (1991). Calculating the rate of loss of information from chaotic time series by forecasting. *Nature* **350**, 485–8.

Walter, D. (1994). Computer art from Newton's, secant, and Richardson's method. *Computers and Graphics* **18**, 127–33.

Watkins, C. D. and Sharp, L. (1992). *Programming in 3 dimensions.* M&T Books, New York.

Watson, J. (1968). *The double helix.* Sigmet, New York.

Watts, D. J. (1999). *Small worlds: the dynamics of networks between order and randomness.* Princeton University Press, Princeton, NJ.

Wegner, T. and Tyler, B. (1993). *Fractal creations* (2nd edn). Waite Group Press, Corte Madera, CA.

Weibel, E. (1979). *Stereological methods.* Academic Press, London.

Weierstrass, K. (1895). *Mathematische werke.* Mayer and Müler, Berlin.

Weigend, A. S., Huberman, B. A., and Rumelhart, D. E. (1990). Predicting the future: a connectionist approach. *International Journal of Neural Systems* **1**, 193–209.

Weiss, C. O. and Brock, J. (1986). Evidence for Lorenz-type chaos in a laser. *Physical Review Letters* **57**, 2804–6.

Westfall, R. (1980). *Never at rest.* Cambridge University Press.

Weyl, H. (1952). *Symmetry.* Princeton University Press, Princeton, NJ.

Whitehead, R. R. and Macdonald, N. (1984). A chaotic mapping that displays its own homoclinic structure. *Physica D* **13**, 401–7.

Whitham, G. B. (1974). *Linear and nonlinear waves.* Wiley, New York.

Whitney, H. (1936). Differentiable manifolds. *Annals of Mathematics* **37**, 648–80.

Wiggins, S. (1988). *Global bifurcations and chaos, analytical methods.* Springer, New York.

Wiggins, S. (1990). *Introduction to applied nonlinear dynamical systems and chaos.* Springer, New York.

Wigner, E. (1960). The unreasonable effectiveness of mathematics in the natural sciences. *Communications on Pure and Applied Mathematics* **13**, 1–14.

Williams, G. P. (1997). *Chaos theory tamed.* Joseph Henry Press, Washington.

Wilson, K. G. (1979). Problems in physics with many scales of length. *Scientific American* **241**, 158–79.

Winch, D. (1987). *Malthus.* Oxford University Press.

Winfree, A. T. (1980). *The geometry of biological time.* Springer, New York.

Wisdom, J. (1987). Urey prize lecture: chaotic dynamics in the Solar System. *Icarus* **72**, 241–75.

Wisdom, J., Peale, S. J., and Mignard, F. (1984). The chaotic motion of Hyperion. *Icarus* **58**, 137–52.

Witten, T. A. and Sander, L. M. (1983). Diffusion-limited aggregation. *Physical Review B* **27**, 5686–97.

Wolf, A., Swift, J. B., Swinney, H. L., and Vastano, J. A. (1985). Determining Lyapunov exponents from a time series. *Physica D* **16**, 285–317.

Wolfram, S. (1984). Universality and complexity in cellular automata. *Physica D* **10**, 1–35.

Wolfram, S. (1986). *Theory and applications of cellular automata.* World Scientific, Singapore.

Wolfram, S. (2002). *A new kind of science.* Wolfram Media, Champaign, IL.

Wu, K. K. S., Lahav, O., and Rees, M. J. (1999). The large-scale smoothness of the Universe. *Nature* **397**, 225–30.

Young, L. -S. (1984). Dimension, entropy, and Lyapunov exponents in differentiable dynamical systems. *Physica A* **124**, 639–46.

Yule, G. U. (1927). On a method of investigating periodicities in distributed series, with special reference to Wolfer's sunspot numbers. *Philosophical Transactions of the Royal Society of London Series A* **226**, 267–98.

Zabusky, N. J. and Kruskal, M. D. (1965). Integration of 'solitons' in a collisionless plasma and recurrence of initial states. *Physical Review Letters* **15**, 240–3.

Zeng, W., Ding, M., and Li, J. (1985). Symbolic description of periodic windows in the antisymmetric cubic map. *Chinese Physics Letters* **2**, 293–6.

Zhabotinsky, A. M. (1964). Periodical oxidation of malonic acid in solution (a study of the Belousov reaction kinetics). *Biofizika* **9**, 306–11.

Zhang, F. and Heidel, J. (1997). Non-chaotic behaviour in three-dimensional quadratic systems. *Nonlinearity* **10**, 1289–303.

Zhu, L., Raghu, A., and Lai, Y. -C. (2001). Experimental observation of superpersistent chaotic transients. *Physical Review Letters* **86**, 4017–20.

Zipf, G. K. (1949). *Human behavior and the principle of least effort: an introduction to human ecology.* Addison-Wesley, Cambridge, MA.

Index

Printed in the United States
By Bookmasters